Understanding and utilising soil microbiomes for a more sustainable agriculture

It is widely recognised that agriculture is a significant contributor to global warming and climate change. Agriculture needs to reduce its environmental impact and adapt to current climate change whilst still feeding a growing population, i.e. become more 'climate-smart'. Burleigh Dodds Science Publishing is playing its part in achieving this by bringing together key research on making the production of the world's most important crops and livestock products more sustainable.

Based on extensive research, our publications specifically target the challenge of climate-smart agriculture. In this way we are using 'smart publishing' to help achieve climate-smart agriculture.

Burleigh Dodds Science Publishing is an independent and innovative publisher delivering high quality customer-focused agricultural science content in both print and online formats for the academic and research communities. Our aim is to build a foundation of knowledge on which researchers can build to meet the challenge of climate-smart agriculture.

For more information about Burleigh Dodds Science Publishing simply call us on +44 (0) 1223 839365, email info@bdspublishing.com or alternatively please visit our website at www.bdspublishing.com.

Related titles:

Advances in measuring soil health
Print (ISBN 978-1-78676-426-3); Online (ISBN 978-1-78676-428-7, 978-1-78676-429-4)

Improving soil health
Print (ISBN 978-1-78676-670-0); Online (ISBN 978-1-78676-672-4, 978-1-78676-673-1)

Managing soil health for sustainable agriculture Volume 1
Print (ISBN 978-1-78676-188-0); Online (ISBN 978-1-78676-190-3, 978-1-78676-191-0)

Managing soil health for sustainable agriculture Volume 2
Print (ISBN 978-1-78676-192-7); Online (ISBN 978-1-78676-194-1, 978-1-78676-195-8)

Understanding and fostering soil carbon sequestration
Print (ISBN 978-1-78676-969-5); Online (ISBN 978-1-78676-971-8, 978-1-78676-972-5)

Understanding and preventing soil erosion
Print (ISBN 978-1-80146-379-9); Online (ISBN 978-1-80146-380-5, 978-1-80146-381-2)

Chapters are available individually from our online bookshop:
https://shop.bdspublishing.com

climate•SMART•publishing
IN AGRICULTURAL SCIENCE

BURLEIGH DODDS SERIES IN AGRICULTURAL SCIENCE
NUMBER 151

Understanding and utilising soil microbiomes for a more sustainable agriculture

Edited by: Professor Kari E. Dunfield, University of Guelph, Canada

burleigh dodds
SCIENCE PUBLISHING

Published by Burleigh Dodds Science Publishing Limited
82 High Street, Sawston, Cambridge CB22 3HJ, UK
www.bdspublishing.com

Burleigh Dodds Science Publishing, 1518 Walnut Street, Suite 900, Philadelphia, PA 19102-3406, USA

First published 2025 by Burleigh Dodds Science Publishing Limited

Library of Congress Control Number: 2024940514

British Library Cataloguing in Publication Data
A catalogue record for this book is available from the British Library

ISBN 978-1-80146-474-1 (print)
ISBN 978-1-80146-475-8 (ePub)
ISBN 978-1-80146-476-5 (PDF)
ISSN 2059-6936 (print)
ISSN 2059-6944 (online)

DOI: 10.19103/AS.2024.0136

Typeset by Deanta Global Publishing Services, Dublin, Ireland

Contents

Part 2 Bacteria and fungi in soil

Part 5 Promoting soil microbiomes and their contribution to soil ecosystem services

Series list

Title	Series number
Achieving sustainable cultivation of maize - Vol 1 From improved varieties to local applications Edited by: Dr Dave Watson, CGIAR Maize Research Program Manager, CIMMYT, Mexico	001
Achieving sustainable cultivation of maize - Vol 2 Cultivation techniques, pest and disease control Edited by: Dr Dave Watson, CGIAR Maize Research Program Manager, CIMMYT, Mexico	002
Achieving sustainable cultivation of rice - Vol 1 Breeding for higher yield and quality Edited by: Prof. Takuji Sasaki, Tokyo University of Agriculture, Japan	003
Achieving sustainable cultivation of rice - Vol 2 Cultivation, pest and disease management Edited by: Prof. Takuji Sasaki, Tokyo University of Agriculture, Japan	004
Achieving sustainable cultivation of wheat - Vol 1 Breeding, quality traits, pests and diseases Edited by: Prof. Peter Langridge, The University of Adelaide, Australia	005
Achieving sustainable cultivation of wheat - Vol 2 Cultivation techniques Edited by: Prof. Peter Langridge, The University of Adelaide, Australia	006
Achieving sustainable cultivation of tomatoes Edited by: Dr Autar Mattoo, USDA-ARS, USA and Prof. Avtar Handa, Purdue University, USA	007
Achieving sustainable production of milk - Vol 1 Milk composition, genetics and breeding Edited by: Dr Nico van Belzen, International Dairy Federation (IDF), Belgium	008
Achieving sustainable production of milk - Vol 2 Safety, quality and sustainability Edited by: Dr Nico van Belzen, International Dairy Federation (IDF), Belgium	009
Achieving sustainable production of milk - Vol 3 Dairy herd management and welfare Edited by: Prof. John Webster, University of Bristol, UK	010
Ensuring safety and quality in the production of beef - Vol 1 Safety Edited by: Prof. Gary Acuff, Texas A&M University, USA and Prof. James Dickson, Iowa State University, USA	011
Ensuring safety and quality in the production of beef - Vol 2 Quality Edited by: Prof. Michael Dikeman, Kansas State University, USA	012
Achieving sustainable production of poultry meat - Vol 1 Safety, quality and sustainability Edited by: Prof. Steven C. Ricke, University of Arkansas, USA	013
Achieving sustainable production of poultry meat - Vol 2 Breeding and nutrition Edited by: Prof. Todd Applegate, University of Georgia, USA	014

Introduction

Microbiomes are communities of microorganisms (such as archaea, bacteria, fungi and viruses) living in soil and other habitats. There is a rapidly growing understanding of the fundamental importance of soil microbiomes in soil and crop health and as the cornerstone of the more sustainable regenerative agricultural practices needed to mitigate and adapt to the challenge of climate change.

With many soil microorganisms and their ecological functions still to be fully understood, we could be said to be entering a new golden age of soil microbiological research. New techniques are making it easier to identify individual genotypes, analyse the complex relationships within microbial communities and between these communities and plants, as well as in deciphering their role in delivering soil ecosystem services. This collection summarises the wealth of recent research in this important subject.

Part 1 Advances in analysing soil microbiomes

This first group of chapters addresses advances in analysing soil microorganisms. The first chapter provides a comprehensive overview of advances in soil viromics, highlighting key methodologies such as metagenomics, dsRNA and virus-like particle extraction, and their contributions to understanding soil viral diversity. It explores the intricate interactions between soil viruses and environmental factors, including soil pH, temperature, moisture content, and organic matter, and their influence on microbial dynamics and ecosystem functions.

Chapter 2 discusses the challenges facing protein extraction from soil, purification, determination and assignment to available databases for identification. Proteomics involves determination of the protein expression profile of a particular organism, while metaproteomics relates to the protein profile of a complex biological system such as that found in soil. As the chapter shows, metaproteomics studies can provide important insights into mechanisms underlying interactions among microorganisms as well as with soil fauna and plants.

Part 2 Bacteria and fungi in soil

The next group of chapters focuses on the role of bacteria and fungi in soil. Chapter 3 discusses the latest advances in research on bacteroidetes, with a focus on their unique contribution to phosphorus (P) and carbon (C) metabolism is soil. Chapter 4 provides a comprehensive review of recent findings on soil actinobacteria, reviewing classification, genetics/characteristics, what we know and don't know about their diversity and ecological functions, factors affecting distribution/abundance in soil, interactions within soil microbiomes, as well the implications of this research for optimizing their contribution to soil health. Chapter 5 highlights the varied roles that root-associated soil fungi play in agricultural ecosystems and provides an overview of methods for detection, identification and analysis of communities of root-associated fungi. As the chapter shows, soil mycobiomes vary at multiple spatial scales in response to presence and identity of plant roots and soil abiotic factors.

Part 3 Analysing structure and dynamics of soil microbiomes

Chapters in Part 3 focuses on interactions both within the soil microbial community and with other organisms in soil across trophic chains. Chapter 6 examines factors affecting microbiome assembly and functions, including genetics, types of plant species and cultivars and root morphology. It also discusses rhizosphere microbiome-plant interdependencies for plant health with a special focus on the spatial and temporal impacts of microbial assembly and their role in promoting specific plant functions.

Chapter 7 reviews the key functions of free-living as well as plant parasitic nematodes in the soil and rhizosphere environment, with a focus on arable systems. It addresses their role as central components of food webs and as biological indicators for soil and plant health. It also highlights the importance of root-feeding taxa for plant-microbe interactions in the rhizosphere with potential benefits for plant health. Finally, the chapter discusses the importance of nematodes as vectors for potential foodborne diseases.

Chapter 8 discusses the complex interactions between soil microbiomes and different abiotic stressors related to climatic conditions, including effects on soil microbial diversity. It also highlights the key role of soil microbiomes in ecosystem functioning, particularly the ways microbial interactions with plants can promote growth promotion and higher stress tolerance in plants.

Part 4 Soil microbiomes and ecosystem services

Part 4 highlights the central role of the soil microbiome in nutrient cycling and other key functions in soil. Chapter 9 explores the advances in understanding the role of soil microbiomes in carbon cycling in soil, how they are able to regulate carbon cycling through both catabolic and anabolic processes. It also discusses ways of using soil microbiomes as an indicator of biological soil health. Chapter 10 explains how changes in the soil microbiome can also be driven at the ecosystem scale as a result of interactions between climate, topography and soil abiotic factors. The chapter focuses particularly on the role of the soil microbiome in nitrogen (N) cycling within the rhizosphere, detritusphere and bulk soil.

Chapter 11 explores the role of root exudates in recruiting and enriching beneficial bacteria which, among other things, can combat pathogens and herbivores. Understanding this role also provides opportunities for microbiome engineering. Strategies include plant breeding, amendment of fertilizers/selective substrates, adding beneficial microbes to the soil, use of synthetic microbial consortia as probiotics, and microbiome transplantation. The combination of a consortium of microbes alongside 'microbe-friendly' crops developed from plant breeding/selection is a particularly promising route forward. Chapter 12 assesses what we know about the roles of microorganisms in mitigating abiotic stresses, especially drought stress. Understanding the mechanisms underpinning microbiome diversity and their functions in drought stress mitigation will eventually help improve resilience and increase crop yield.

Part 5 Promoting soil microbiomes and their contribution to soil ecosystem services

The last group of chapters focuses on ways of promoting soil microbiomes and optimizing their role in delivering key ecosystem services. Chapter 13 explores the role of the physical environment of soil in modulating the soil microbiome, and how particular physical properties affect soil microbiota. Chapter 14 explains how the soil microbiome comes into contact with fertilizer in the fertosphere, the nutrient-rich zone <5 cm from a dissolving fertilizer granule. It shows how changes in the fertosphere may disrupt the soil microbiome, the robust survival mechanisms deployed by soil bacteria, archaea and fungi in adapting to stressful conditions, and implications for managing crop nutrition better to promote beneficial soil microbial activity.

Chapter 15, which concludes the book, highlights the urgent need to implement sustainable land management practices, including regenerative agriculture and reforestation, to mitigate land degradation and restore ecosystems. The chapter explore examples of the impact of different land management practices on the composition and functionality of soil microbiomes, with a particular emphasis on diverse biomes within Brazil.

Part 1

Advances in analysing soil microbiomes

Chapter 1

Advances in soil viromics in understanding viral diversity and function

Azza Larafa, Agriculture and Agri-Food Canada, Canada/Université de Montréal, Canada and Mamadou L. Fall[1], Agriculture and Agri-Food Canada, Canada

1 Introduction

In the complex weave of life on our planet, viruses are often overlooked and mistakenly seen only as disease agents (Bergh et al., 1989), but they possess a far more profound and expansive influence. Beyond their reputation, viruses play a crucial role in shaping life on Earth, exerting significant impacts on microbial communities and biogeochemical cycles. The emergence of 'soil viromics' as a relatively recent field holds the potential to revolutionize our understanding of soil diversity and its profound implications for terrestrial ecosystems (Fierer et al., 2007). This exploration into the hidden realm of viruses not only challenges preconceived notions but promises to unveil the complex relationships that contribute to the dynamics of life on our planet.

The discovery of viruses dates back over a century but has gained renewed attention as researchers attempt to unveil their astonishing diversity

1 Corresponding author: Mamadou L. Fall: mamadoulamine.fall@agr.gc.ca

http://dx.doi.org/10.19103/AS.2024.0136.04
Published by Burleigh Dodds Science Publishing Limited, 2025.

and multifaceted roles across various ecosystems (Bergh et al., 1989). Viral ecology, rooted in the observation of virus-like particles (VLPs) and laboratory experiments, has advanced our knowledge through metagenomic approaches (Roux et al., 2017). These approaches have offered a window into the extensive genomic diversity of environmental viruses. Additionally, the exploration of cultivated virus–host pairs has revealed complex dynamics of viral control and reprogramming of host molecular machinery during infection (Feiner et al., 2015).

However, despite these significant advances, the understanding of viruses in soil ecosystems is still in its beginnings. The term 'soil' encompasses diverse terrestrial environments, each hosting distinct microbiomes characterized by differences in hydraulic connectivity, particle sizes, and pore spaces. The inherent complexity of soil microbiomes, marked by high richness and evenness, poses unique challenges when reconstructing microbial and viral genomes from metagenomes (Taş et al., 2021).

Over a decade ago, the field of soil viromics was established to specifically address these challenges (Fierer et al., 2007). However, characterization of soil viral diversity has lagged behind that of other environments, resulting in a significant gap in high-throughput, high-resolution genomic data on soil viruses. Consequently, the soil virosphere has been described as 'neglected' and 'severely underestimated and undersampled' (Fierer et al., 2007) (Box 1). Recent advances in soil viromics are now beginning to bridge this gap, offering unprecedented insights into the global diversity and dynamics of soil viruses (Pratama and van Elsas, 2018). Soil viruses, known to infect a vast array of organisms across all three domains of life, extend their influence beyond human health concerns, playing critical roles in agriculture and global nutrient cycling (Breitbart et al., 2018). The abundance of soil viruses, estimated at millions to billions per gram of soil (Carini et al., 2016), underscores their potential impact on soil carbon cycling and the dynamics of microbial communities in response to environmental changes (Pratama and van Elsas, 2018; Rodriguez et al., 2018).

Box 1

The **virosphere** represents the comprehensive spectrum of viruses that pervade various ecosystems, including soils, waters, and the atmosphere, playing pivotal roles in genetic exchange, microbial regulation, and ecosystem dynamics (Suttle, 2007).

Comparing soil properties with those of marine systems highlights unique challenges for characterization of viral communities due to the varied

Published by Burleigh Dodds Science Publishing Limited, 2025.

physicochemical properties of soil (Trubl et al., 2020) (Fig. 1). The heterogeneity in soil texture, mineral composition, and organic matter content necessitates specialized approaches for the efficient extraction of viral nucleic acid, a task that has long been a bottleneck in soil viral studies (Carini et al., 2016). However, recent developments in soil viromics techniques are making strides in overcoming these challenges, gradually narrowing the knowledge gap between marine and soil viral ecosystems (Rodriguez et al., 2018) (Box 2). The comparison between these two environments – soil and marine – reveals distinct challenges and characteristics essential to each, from the intricacies of biotic interactions within the soil ecosystem to the unique conditions of marine environments, such as salinity gradients and temperature variations (Costello et al., 2018; Williamson et al., 2017) (Fig. 1).

Box 2

Metagenomics is the genomic analysis of microbial DNA that is directly isolated from environmental samples, without the need for culturing the organisms present (Handelsman et al., 1998). It offers a revolutionary lens, enabling the direct genetic analysis of environmental samples to decode the complex microbial and viral communities inhabiting these ecosystems without necessitating cultivation. This approach has unveiled the immense diversity and functional potential of microorganisms, including viruses, highlighting their ecological roles and evolutionary significance (Handelsman, 2004). The **virome**, a term denoting all viruses present within a particular environment or host, offers insights into the viral influence on ecological balance and host health (Virgin, 2014). Extending these concepts, **viromics** employs advanced sequencing technologies to conduct a detailed analysis of viral populations, elucidating their biodiversity, evolutionary trends, and interactions with hosts. This comprehensive approach not only enhances our understanding of viral ecology, it also paves the way for biotechnological applications (Paez-Espino et al., 2016).

2 Evolution and advances in the viromics toolbox in soil ecosystem studies

The field of soil virology has experienced a transformative pathway over time, beginning with its focus on pathogenic viruses, especially those impacting human health. Early research, like that of Dowd et al. (1998), concentrated on understanding the fate and transport of exogenous viruses in soil, which primarily employed basic elution buffers and culture-based methods (Dowd et al., 1998). This phase was crucial for laying the groundwork for future

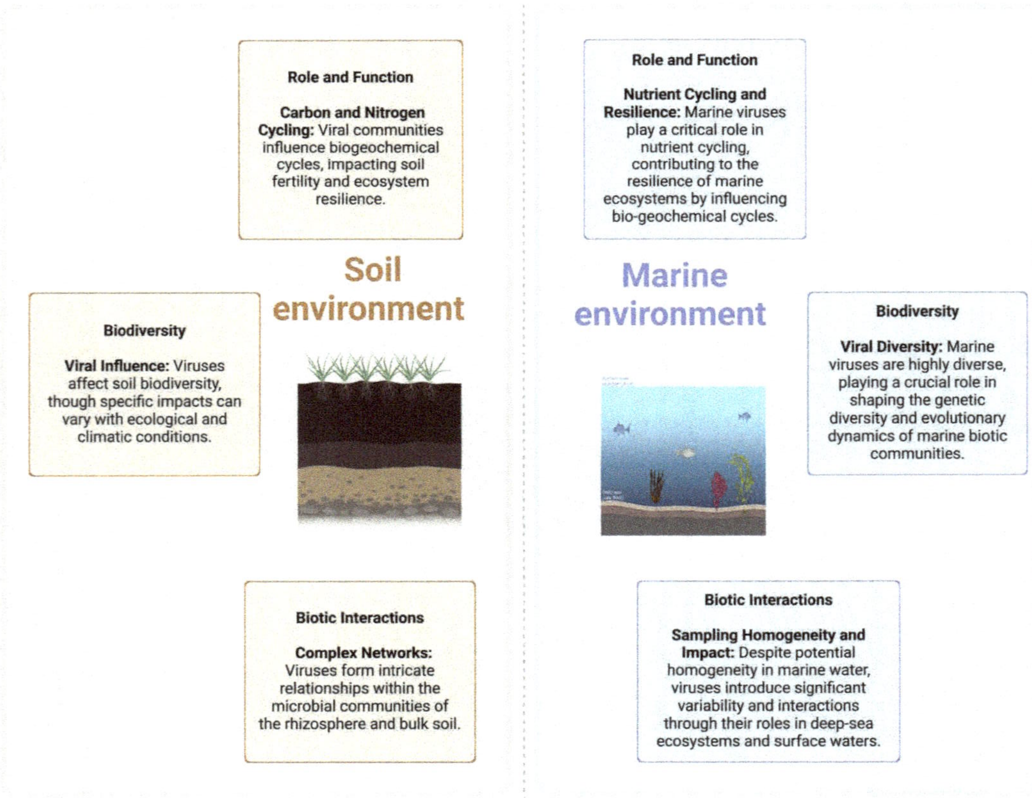

Figure 1 Contrasting virospheric influences on soil and marine ecosystems: a dual analysis of soil and marine virospheres emphasizing biodiversity insights, functional dynamics, and methodological challenges in viral population studies.

advances in the field. A significant turning point in soil virology came with the recognition of the ecological importance of autochthonous soil viruses – those naturally existing in the soil. This shift, inspired by marine studies as noted by Torsvik, Goksøyr, and Daae (1990), led to a transition from culture-dependent methods to culture-independent techniques, which broadened the scope of soil virology to include a wider array of soil viruses (Ashelford et al., 1999; Williamson et al., 2003).

Methodological milestones – visualizing the invisible in soil: The evolution of soil viromics, a field dedicated to studying viral communities in soil environments, has been marked by significant methodological advances and challenges (Box 2). Early studies primarily used techniques like transmission electron microscopy and epifluorescence microscopy to visualize viruses directly from soil samples (Williamson et al., 2003) (Fig. 2). The advent of metagenomics further revolutionized the field, allowing for a more comprehensive analysis of viral communities (Thurber et al., 2009). However, these studies faced significant challenges in the area of sample preparation, particularly in separating viral particles from the complex soil matrix and reducing contamination from non-viral DNA. DNase treatments have been employed to improve the enrichment of viral DNA, enhancing the accuracy of viromic analyses (Sorensen Jackson et al., 2021). Innovations in sample processing and sequencing have also been crucial. Trubl et al. (2020) provided guidance on isolating viruses, detecting them within metagenomes, and applying viromics, thus addressing challenges in soil virus characterization. Concurrently, developments in bioinformatics have been instrumental in deciphering complex soil viromes, aiding in the estimation of viral community composition and diversity (Roux et al., 2017) (Box 2). The integration of metatranscriptomics has enabled the study of RNA viruses in soil, offering insights into their activity and function within microbial communities (Bikel et al., 2015). Addressing the specific challenges of soil environments, such as the efficient extraction of viral nucleic acids, has led to the development of methods like size-specific filtration and desorption processes, which have significantly enhanced the recovery and analysis of soil viromes (Langenfeld et al., 2021; Martinez Martinez et al., 2014) (Fig. 2). These collective advances underscore the dynamic nature of soil viromics and its critical role in understanding viral ecology and their impact on soil ecosystems.

Current methodologies in soil viral ecology – a comparative view: Lately, to study soil viral ecology, researchers have been predominantly relying on two primary methodologies: metagenomic and VLP extraction methods, each of which offers distinct insights into soil virome composition and dynamics. Metagenomic extraction provides a comprehensive approach by analyzing the collective genetic material extracted from soil samples (Roux et al., 2017) (Fig. 3) (Box 2). Despite its broad coverage, this method faces challenges due to the absence of a universal viral marker gene, which often results in the

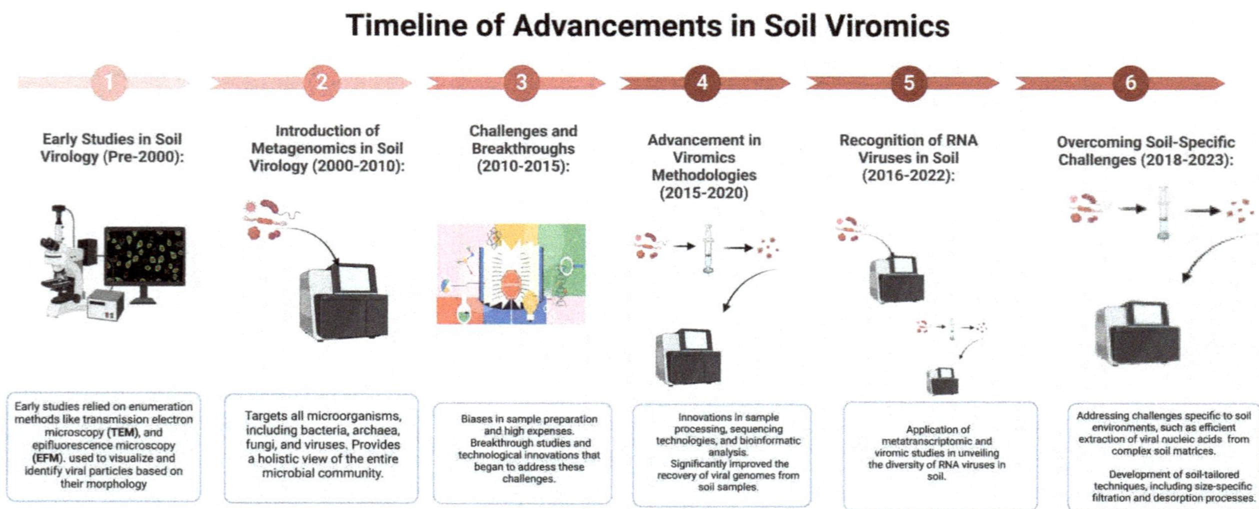

Figure 2 Evolutionary timeline of soil viromics technology: illustration of detailed timeline capturing the evolution of technology and methods in the field of soil viromics. Source: *Created with BioRender.com*.

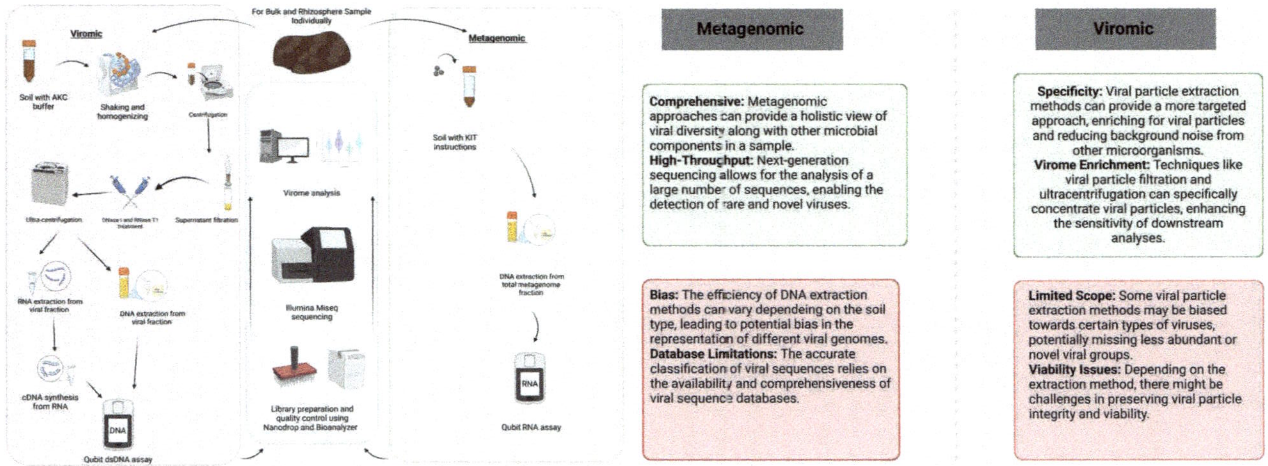

Figure 3 Overview of sil viromic techniques: metagenomics versus VLP (virus-like particle) extraction and comparative study in soil research – a two-part comparison. Section 2 contrasts metagenomic analysis with VLP extraction techniques in soil virology, focusing on procedural differences and strengths. Section 3 demonstrates the practical applications of both methods in soil research. Source: *Created with BioRender.co m, inspired from Poursalavati et al. (2023)*

co-extraction of a mix of genetic sequences, including bacterial and eukaryotic genetic materials (Santos-Medellin et al., 2022) (Fig. 3). Consequently, while effective in identifying the most prevalent viruses, metagenomic extraction may underrepresent the diversity of less abundant viral species (Santos-Medellin et al., 2022).

In contrast, VLP extraction focuses on a more targeted approach (Breitbart et al., 2018). It involves the physical separation of virions from larger soil particles and microbes, primarily through techniques such as sonication and chemical treatments, notably employing solutions like potassium citrate (Williamson, Radosevich, and Wommack, 2005) (Fig. 3). This process selectively enriches viral sequences, thereby facilitating a more detailed and focused investigation of viral diversity within the soil ecosystem (Trubl et al., 2020). The VLP method is particularly adept at uncovering a broader spectrum of viral diversity, including the detection of rarer viral species that might be overlooked in metagenomic datasets (Santos-Medellin et al., 2022) (Fig. 3). However, this technique is not without its challenges. For instance, VLP purification can lead to contamination with bacterial, archaeal, and fungal genetic material, posing a significant hurdle in isolating pure viral sequences and potentially leading to misleading conclusions about soil viral diversity (Zolfo et al., 2019). Moreover, the efficiency of VLP extraction can vary significantly, depending on soil properties such as moisture content, pH, and organic matter, which may limit the ability to capture a comprehensive representation of the soil virome (Narr et al., 2017) (Box 2). Additionally, while VLP methods are effective in identifying a wide range of viruses, they may still miss certain viral species, especially those in low abundance or with unique characteristics that are less amenable to standard extraction protocols (Mugnai et al., 2021). Advances in both methodologies have further refined the accuracy and efficiency of soil virome studies. For instance, improvements in metagenomic sequencing technologies have enhanced the depth and breadth of viral detection, while methodological refinements in VLP extraction, such as optimized buffer compositions, have improved the recovery and purity of viral particles from diverse soil types (Roux and Emerson, 2022).

The integration of these two approaches is thus critical in soil viral ecology research. Metagenomic extraction provides a broad overview of the soil virome, capturing the dominant viral populations and many viral genomes that are not detected using other viromic methods (Kosmopoulos et al., 2023). However, this method may not fully represent the diversity of the viral community, particularly its less abundant members. In contrast, VLP extraction offers a more detailed insight into these less abundant viral constituents, complementing the macroscopic view provided by metagenomic extraction (Santos-Medellin et al., 2022). Employing both methodologies together enables a more comprehensive understanding of soil virome dynamics. This includes understanding of their ecological roles, impacts on soil health, influences on microbial community structure, and contributions to nutrient cycling processes. Given the unique challenges presented by the heterogeneous nature of soil – which varies in texture, mineral composition, and organic matter content – the development of specialized extraction and analysis methods has been crucial (Trubl et al., 2020; Williamson et al., 2013). These advances are instrumental in enhancing our capacity to isolate and study both DNA and RNA viruses present in soil environments.

Transitioning from DNA to RNA viruses – a renewed focus in soil virology: Historically, research in soil virology has been dominated by studies on DNA viruses, primarily due to the relative ease of DNA extraction and sequencing technologies that were initially more suited for DNA than RNA. DNA viruses generally exhibit greater genetic stability compared with RNA viruses, simplifying the analysis and interpretation of their genomic data (Kim et al., 2008; Pekosz et al., 1999). Moreover, the broader array of DNA viruses, their longer history of discovery, and their clearer links to disease in animals and humans have made them a more immediate focus of virological research (Paez-Espino et al., 2016; Porter et al., 2019).

However, recent advances in metatranscriptomic techniques have triggered a renewed interest in soil RNA viruses. These advances have overcome previous technical limitations, allowing for in-depth exploration of the diversity and functions of RNA viruses in soil ecosystems (Hillary et al., 2022; Starr et al., 2019). RNA viruses in soil are now recognized for their dynamic nature, rapid mutation rates, and significant roles in regulating microbial communities and biogeochemical cycles. Studies like Hillary et al. (2022) have shown that RNA viruses are not only more diverse than previously thought but also play crucial roles in ecological processes and microbial dynamics within soils (Hillary et al., 2022). This shift in focus is revealing new insights into the complexity of soil ecosystems and the intricate web of interactions within the soil virome. Recent research developments have introduced a novel method in soil virology, focusing on double-stranded RNA (dsRNA) extraction. This innovative approach marks a significant advance in the toolbox available to virologists. By concentrating on dsRNA, this technique targets a key aspect of many RNA viruses, enabling more precise identification and characterization within soil ecosystems (Poursalavati et al., 2023; Starr et al., 2019). This focus is particularly critical given the elusive nature of RNA viruses and the existing challenges in their detection and analysis (Zhang et al., 2020).

The emphasis on dsRNA offers several benefits. Primarily, it facilitates the effective differentiation and detailed characterization of RNA viruses, thereby deepening our understanding of their diversity and evolutionary patterns. More importantly, this method paves the way for a new understanding of the role of RNA viruses in biogeochemical processes. By improving the identification of RNA viruses and their interactions within soil microbiomes, researchers can gain a better understanding of how these viruses influence nutrient cycling, the structure of microbial communities, and overall soil health. Such a comprehensive approach is essential for a full understanding of the ecological impact of soil viruses and for formulating strategies to manage and safeguard soil ecosystems. Consequently, this dsRNA-focused method, as also explored in the environmental fate study of RNA interference biopesticides, significantly enhances our knowledge of soil virology and its interplay with broader ecological functions (Parker et al., 2019).

In conclusion, the evolution and integration of metagenomics and viromics have dramatically reshaped our methodological approach to studying viruses in soil ecosystems. As soil virology continues to advance, it promises to reveal more about the complex interactions and functions within soil, offering a profound understanding of the hidden world of viruses beneath our feet. The practical implications of advances in metagenomics and viromics are vast. In agriculture, they can inform sustainable farming practices and soil health management. In public health, insights into soil-borne diseases can aid in developing prevention and control strategies. In environmental science, this knowledge can enhance our understanding of biodiversity, ecosystem functions, and resilience.

3 Diversity of soil viruses

Below the Earth's surface, there is an unsuspected viral diversity, often underestimated yet crucial to terrestrial ecosystems (Edwards and Rohwer, 2005). As mentioned, using advanced methodologies, viromic studies have played and continue to play a crucial role in unravelling the mysteries of soil viral diversity, and in exploring the genetic complexities and taxonomic richness of viruses in soil ecosystems (Callanan et al., 2020; Edwards and Rohwer, 2005; Neri et al., 2022; Roux and Emerson, 2022).

For more than a decade, research predominantly focused on pathogenic viruses, especially those with potential risks to human health (Dowd et al., 1998). The emphasis was on understanding the fate, transport, and detection of exogenous viruses, particularly those infecting enteric bacteria, so-called phages (Lance and Gerba, 1984; Williamson et al., 2005; Williamson et al., 2003), offering a central understanding of factors influencing viral adsorption to soil surfaces.

A pivotal shift in soil virology occurred with the recognition of autochthonous soil viruses – those naturally existing in soils – inspired by studies in marine environments (Torsvik et al., 1990; Williamson et al., 2005). This shift highlighted the importance of exploring the ecological relevance of viruses in soils and a transition from culture-dependent methods to culture-independent techniques (Williamson et al., 2003). The adoption of culture-independent methods, particularly metagenomics, marked a significant advance, unveiling a diverse ensemble of viral entities ranging from bacteriophages to RNA viruses and non-encapsidated eukaryotic viruses (Liang et al., 2019; Starr et al., 2019).

A notable contribution to this field was made by Graham et al. (2019). They expanded our knowledge of soil viral diversity through a global distribution analysis of over 24 000 soil viral sequences (Graham et al., 2019). Their research provides insights into the potential hosts of these viruses and their possible roles in biogeochemical cycling, particularly in the decomposition of soil organic carbon (Table 1).

Research has predominantly focused on DNA viruses, potentially creating a misrepresentation of their prevalence and diversity in soil (Roux and Emerson, 2022; Santos-Medellin et al., 2022; Trubl et al., 2018) (Table 1). Metaviromics, which aims to identify viral sequences within metagenomic and metatranscriptomic datasets, expands not only taxonomic knowledge but also demonstrates the functional roles that soil viruses play in shaping microbial communities (Box 3). However, it faces unique challenges due to its reliance on genetic material from organisms with larger genomes, which influences the accurate identification of viruses within this mix (Cobbin et al., 2021; Hillary et al., 2022; Wu et al., 2022) (Table 1).

Box 3

A specialized subset, **metaviromics**, focuses exclusively on the viral components of the metagenome, seeking to understand the diversity, structure, and dynamics of viral populations and their interactions with microbial hosts in environments such as soil. This discipline is crucial for uncovering the ecological impact of viruses, especially in regulating microbial communities and influencing nutrient cycling and disease processes (Rosario and Breitbart, 2011).

Given the extensive potential host diversity in soil, a multitude of RNA viruses is anticipated within this ecosystem (Starr et al., 2019) (Table 1). Studies employing methods such as metatranscriptomics and viromics have explored RNA viral communities in soil, with the recent VLP method revealing a diverse community of soil RNA viruses (Hillary et al., 2022) (Table 1). Over the last 5 years, numerous metaviromic studies have expanded our knowledge of RNA viral diversity in soils. The majority of RNA viruses in soil environments rely on their own RNA-dependent RNA polymerase (RdRp) enzymes for the RNA replication process. This mechanism is pivotal in the replication of RNA viruses and has led to a considerable increase in the identification of diverse

Table 1 Overview of key studies on soil viral diversity

Research study	Authors	Target description
Soil viruses are underexplored players in ecosystem carbon processing	Trubl et al. (2018)	Ecological roles of soil viruses: Explored the impact of soil viruses on ecosystem carbon processing, particularly in thawing permafrost environments. Their findings suggest that soil viruses have a distinct ecology, infect dominant microbial hosts, and may influence ecosystem function in the rapidly changing arctic through top-down and bottom-up controls.
Untapped viral diversity in global soil metagenomes	Graham et al. (2019)	Global soil viral diversity: Expanded our knowledge of soil viral diversity through a global distribution analysis of over 24 000 soil viral sequences. Their study provides insights into the potential hosts of these viruses and their possible roles in biogeochemical cycling, particularly in the decomposition of soil organic carbon.
Metatranscriptomic reconstruction reveals RNA viruses with the potential to shape carbon cycling in soil	Starr et al. (2019)	RNA viruses in soil: Conducted a metatranscriptomic reconstruction to reveal a diverse community of RNA viruses in soil, with potential implications for carbon cycling. Their study tracked RNA viral and host communities in soil environments central to carbon cycling, such as the rhizosphere and detritosphere. This work marks a significant step in understanding the factors driving RNA viral communities in soil, with major implications for ecological processes and carbon cycling.
Viromes outperform total metagenomes in revealing the spatiotemporal patterns of agricultural soil viral communities	Santos-Medellín et al. (2021)	Soil viromics: Characterized the dsDNA viral diversity in agricultural soils, revealing that viromes outperform total metagenomes in uncovering the spatiotemporal patterns of soil viral communities. This study highlights the utility of soil viromics in understanding viral and microbial community dynamics, indicating a partial decoupling of the processes driving their spatial distributions.
RNA-viromics reveals diverse communities of soil RNA viruses with the potential to affect grassland ecosystems across multiple trophic levels	Hillary et al. (2022)	Exploration of soil RNA viral communities: A study exploring the diversity and community patterns of soil RNA viruses, contributing significantly to our understanding of these viruses in complex soil environments. Their research indicates that soil RNA viruses exhibit a high diversity of viral sequences and play an active role in soil ecosystems, which could be crucial for processes like carbon cycling.
Moisture modulates soil reservoirs of active DNA and RNA viruses	Wu et al. (2022)	Soil viruses and environmental change: Used a multi-omics approach to detect active DNA and RNA viruses in soil and determine their responses to variations in soil moisture. The authors found that higher soil moisture increased the transcription of a subset of DNA viruses and identified a high abundance of RNA viruses, such as *Reoviridae* and *Leviviridae*, in the soil. Their study demonstrates the dramatic impacts of environmental changes, like moisture shifts, on the composition, activity, and potential functions of both DNA and RNA soil viruses.

Note: The table presents a summarized overview of significant research studies that have contributed to our understanding of the diversity of soil viruses and their ecological roles. Each entry includes the title of the research study, the authors, and a brief description of the study's focus and findings. The table covers a range of topics, from the ecological impact of soil viruses on carbon cycling and ecosystem functions to the exploration of RNA viruses in soil.

RNA viral sequences across various ecosystems. This discovery underlines the complexity and adaptability of RNA viruses in soil, contributing significantly to our expanding knowledge of viral diversity in terrestrial environments (Wolf et al., 2018).

Soil viral diversity is fundamental in maintaining ecosystem resilience. Viruses contribute to ecological balance by regulating microbial population sizes. They also significantly influence the evolution and adaptation of microbial communities. By acting as vectors of gene transfer, viruses introduce new genetic elements into the microbial gene pool, enabling the emergence of new traits and enhancing microbial adaptability to environmental shifts. This horizontal gene transfer (HGT) can lead to new microbial capabilities, such as antibiotic resistance or metal tolerance, which aid in adaptation to diverse stressors (Balcazar, 2014; Pratama and van Elsas, 2018). For instance, soil phages carrying genes involved in nitrogen cycling impact nitrogen availability, thereby affecting soil fertility and plant growth (Emerson et al., 2018). Soil viruses exhibit a remarkable capacity to adapt to environmental changes, affecting microbial community dynamics. Fluctuations in soil moisture, temperature, and nutrient availability lead to shifts in soil viral diversity. These shifts can significantly influence microbial community structure and function, impacting key ecological processes such as decomposition, pathogen suppression, and nutrient cycling (Williamson et al., 2017).

In summary, the exploration of soil viral diversity through viromic studies has redefined our perception of the hidden soil virome and highlighted the profound impact that soil viral communities have on terrestrial ecosystems. As technology advances, viromic studies hold the promise of uncovering the true soil viral diversity.

4 Comprehensive analysis of environmental factors influencing soil viral communities

4.1 Soil pH and viral dynamics

Soil pH plays a critical role in shaping the microbial structure and dynamic within the soil ecosystem. Its impact on viral populations is profound, influencing both their survival and infective capabilities. For instance, *Achromobacter xylosoxidans* phage phiAxp-3 shows optimal infectivity at neutral pH, but its infectivity significantly diminishes under extremely acidic or alkaline conditions (Ma et al., 2016). Similarly, the interaction between soil viruses and their microbial hosts is also pH-sensitive, as seen in *Bacillus subtilis* phages that prefer a pH range of 7.0–8.0. Soil pH further affects virus–soil particle interactions by altering the electrostatic properties of the soil, thereby influencing virus adhesion to soil colloids (Kapuscinski and Mitchell, 1980; Kimura et al., 2008).

Published by Burleigh Dodds Science Publishing Limited, 2025.

In-depth analysis of viral and prokaryotic populations has further highlighted that soil pH significantly structures these communities with specific viral groups being exclusive to certain pH conditions (Bahram et al., 2018; Lee et al., 2022).

4.2 Temperature: a key regulator of the soil virome

Transitioning from soil pH, temperature emerges as another pivotal factor in soil virology. Temperature variations critically affect the diversity and activity of soil viruses. Higher temperatures often result in faster viral inactivation due to increased microbial activity and enzymatic degradation (Kimura et al., 2008). For example, the survival span of viruses like poliovirus Type 1 is markedly shorter at higher temperatures. Conversely, lower temperatures can prolong the survival of viruses, as evidenced by studies where poliovirus Type 1 remained viable for extended periods at 4°C (Yeager and O'Brien, 1979). Temperature also impacts the relationship between viruses and their bacterial hosts. For instance, specific soil phages infecting *Geobacillus stearothermophilus* exhibit distinct proliferation and inactivation patterns at different temperatures (Reanney and Marsh, 1973). Furthermore, temperature variations can alter viral activity, particularly in their ability to infect hosts. This is observed in the behaviour of *Burkholderia pseudomallei* phages, which display different infection probabilities based on temperature fluctuations (Egilmez et al., 2018). Furthermore, temperature fluctuations are critical in shaping viral communities in varying environments, such as deserts (Makhalanyane et al., 2013; Williamson, 2011).

4.3 Moisture content: influencing soil virus behaviour

Building on temperature's influence, soil moisture content is another significant factor impacting soil virus composition, activity, and survival. Studies have demonstrated that both DNA and RNA viruses respond differently to wet and dry soil conditions, with DNA viruses showing more diversity and activity in moist soils (Wu et al., 2021). Similarly, research in Mediterranean grasslands using viral-size-fraction metagenomes has revealed that dsDNA viruses in soil respond dynamically to precipitation changes, indicating a strong connection between soil moisture and viral community structure (Santos-Medellin et al., 2022). Soil moisture also directly affects virus survival, with varying inactivation rates observed under different moisture levels. For instance, poliovirus Type 1 shows increased inactivation when soil moisture falls below certain levels, highlighting the importance of moisture in maintaining viral longevity (Kimura et al., 2008; Yeager and O'Brien, 1979). Land-use practices that alter soil moisture, including agriculture and forestry, further impact soil viral communities, which highlights the broader environmental implications of soil moisture on soil virology (Bahram et al., 2018; Kimura et al., 2008; Williamson et al., 2005).

4.4 Organic matter: shaping soil viral communities

The concentration of organic matter in soil plays a significant role in determining the behaviour and survival of soil viruses. In environments with limited organic matter, the prevalence of host cells for viruses decreases, leading to reduced viral activity. Nutrient levels in soil can alter the surface characteristics of host cells, which affects their vulnerability to viruses. This relationship is evident in studies showing that nutrient variations, such as phosphate limitation, can change host cell susceptibility to viral attack (Archibald and Coapes, 1976; Delisle and Levin, 1972). The activity of indigenous phages in soil also depends on soil organic content. For example, experiments have shown that dormant *Arthrobacter globiformis* phages become active in nutrient-rich conditions without the need for new host cells (Casida Jr and Liu, 1974). Additionally, viruses influence key microbial functions and the distribution of organic nutrients, shaping microbial diversity and affecting the overall structure of the microbial community in soil (Kimura et al., 2008; Weinbauer, 2004). In marine environments, virus-induced lysis of cyanobacteria significantly affects carbon cycling processes (Kimura et al., 2008). Similarly, in soil ecosystems, viral activities can result in the release of organic compounds from microbial cells, altering the composition of organic matter and influencing the microbial community structure (Weinbauer, 2004; Weinbauer and Rassoulzadegan, 2004). In essence, the concentration of organic matter is a critical factor influencing soil viral community composition. It not only dictates the density and susceptibility of host cells to viral infection but also has broader implications for the microbial ecosystem in soil. Understanding the interplay between organic matter and soil viruses is vital for a comprehensive understanding of soil ecosystems and their ecological functioning.

In summary, the environmental factors, soil pH, temperature, moisture content, and organic matter content collectively exert a significant influence on soil viral community structure and activities. These factors shape the survival, infectivity, and interactions of viruses within the soil ecosystem, affecting everything from individual viral behaviour to the overall structure of microbial and viral communities. A thorough understanding of these factors is essential for advancing our knowledge of soil virology and its ecological implications.

5 Functions of soil viruses

Soil viruses are central to the microbial dynamics in soil ecosystems. These entities profoundly influence both the structure and functionality of microbial communities as evidenced by their interactions with soil bacteria (Pratama and van Elsas, 2018; Zablocki et al., 2014). By interacting with soil bacteria, these viruses exert significant control over microbial population structures and nutrient cycling processes (Allen et al., 2010).

The lytic and lysogenic cycles of these viruses play complementary roles in regulating microbial populations. In the lytic cycle, viruses infect bacterial cells and use their machinery to replicate. This leads to cell lysis and the release of new viruses, a process which not only regulates bacterial population densities but also promotes genetic diversity through HGT (Weinbauer and Rassoulzadegan, 2004). In contrast, the lysogenic cycle involves the integration of the viral genome into the host's genome. Such integration can endow host bacteria with new traits, such as antibiotic resistance, contributing to microbial diversity and evolution (Brüssow and Hendrix, 2002).

These viruses also exhibit a unique dual capacity in microbial regulation. They suppress pathogenic bacteria, like certain *Ralstonia* strains, while promoting beneficial microbial taxa (Wang et al., 2019). This capacity of soil viruses to manipulate microbial populations shows their critical ecological importance. Additionally, soil viruses implement a strategic ecological mechanism known as the 'kill-the-winner' approach (Thingstad, 2000). In this process, they target and lyse rapidly growing bacterial populations, commonly found in environments rich in nutrients (Jover et al., 2014). The lysis of these dominant bacterial groups releases cellular contents, including nutrients and organic matter, into the soil environment. This release is a crucial aspect of nutrient cycling, as it fuels the growth of a diverse array of microbial species and influences the patterns of microbial community succession, a so-called 'Piggyback-the-Winner' mechanism (Liu et al., 2023).

In shaping the assembly and structure of bacterial communities, soil viruses contribute to a blend of random and non-random processes that govern community assembly. High levels of viral activity, for instance, can lead to homogeneous selection within bacterial communities, where viruses selectively target certain bacterial groups for lysis (Zhou et al., 2014). The cellular materials released from these lysed bacteria can then promote the growth of other specific bacterial lineages. This selective pressure exerted by viruses is an integral factor in the assembly of microbial communities in soil ecosystems (Liu et al., 2023).

Moreover, soil viruses, particularly those engaging in lysogenic cycles, are key players in the process of HGT (Touchon et al., 2017). Through HGT, viruses can acquire and pass on essential functional genes crucial for complex carbohydrate degradation, thereby influencing the soil carbon cycle and contributing to the breakdown and turnover of organic materials (Correa et al., 2021; Touchon et al., 2017). The role of viruses in both lysogenic and lytic cycles is fundamental to maintaining the stability and evolution of microbial networks within the soil, highlighting their importance in the ecological functioning of these environments (Liu et al., 2023).

Beyond their direct influence on microbial community dynamics, soil viruses have a significant impact on nutrient cycling, primarily through the mechanism known as the 'viral shunt' (Suttle, 2005; Wilhelm and Suttle, 1999). This process involves the lysis of microbial cells and leads to the release of critical nutrients such as organic carbon, nitrogen, and various micronutrients into the soil. This release revitalizes soil fertility and enhances ecosystem productivity (Fuhrman, 1999; Wilhelm and Suttle, 1999). Viruses in soil environments, particularly those expressing Auxiliary Metabolic Genes (AMGs), play a crucial role in biogeochemical cycling. AMGs encode proteins essential for host metabolic processes and are expressed during infection, altering or directly participating in host metabolic pathways (Kieft et al., 2021). This enhances the metabolic efficiency of host cells in processes like central carbon metabolism, nitrogen metabolism, and phosphorous metabolism. The expression of these AMGs by soil viruses is a key aspect of the biogeochemical cycling of carbon and macronutrients, significantly impacting ecological processes (Ji et al., 2023). Soil viruses significantly contribute to nutrient cycling and soil carbon dynamics. For instance, their role in the nitrogen cycle has been noted, where viruses infect nitrogen-fixing bacteria, impacting their population dynamics and, consequently, the rate of nitrogen fixation (Zehr and Ward, 2002). This affects plant growth and soil fertility, particularly in agricultural ecosystems.

In summary, soil viruses, through their lytic and lysogenic cycles, play multifaceted roles in soil ecosystems. Their regulatory influence on microbial populations, nutrient cycling, and soil carbon dynamics reflects their importance in shaping soil ecosystems. Their activities, including the viral shunt mechanism and the expression of AMGs, significantly influence ecological processes, highlighting their pivotal impact on soil nutrient dynamics and ecosystem productivity.

6 Conclusion

This chapter provides an overview of the current landscape in soil viromics, exploring the established and potential roles of viruses in the functioning and stability of soil ecosystems. It turns out that soil viruses are not only lurking in the shadows; they are actually key players in the ecosystem. These tiny entities, often misunderstood as merely harmful, are in fact crucial in balancing microbial life and cycling nutrients in our soils.

What is really exciting is how we have come to understand their world. Thanks to advances in metagenomics and VLP extraction, we are now peeking into the incredible diversity of soil viruses. We have learned that environmental factors – like the acidity or alkalinity of the soil (pH), its moisture levels, and even temperature – play significant roles in how these viruses behave and interact with their surroundings.

Looking ahead, the horizon is broad and promising. There is a whole unexplored territory in the form of RNA viruses in soil that holds great potential for new discoveries. The implications here stretch far beyond academic curiosity. This growing body of knowledge can revolutionize our approaches in agriculture, help us better conserve our environments, and even inform public policies. By understanding the viral components of the soil ecosystem, we are gaining insights that can help us to make more sustainable choices for our planet.

Ultimately, the study of soil viromics will not only deepen our understanding of the hidden life in soil but also open up numerous avenues for sustainable practices, ecological stewardship, and advances in various scientific fields. As we continue to explore and unravel the complexities of soil viromes, we pave the way for ensuring the health and sustainability of our planet.

7 Where to look for further information

For further information on soil viromics, relevant articles are listed in the references section. These include key studies on viral metagenomics, virus-like particle (VLP) extraction, soil viral ecology, microbial interactions, and biogeochemical cycling. Advances in sequencing technologies and bioinformatics tools for viral identification are also covered in these sources.

8 References

Allen, H. K., Donato, J., Wang, H. H., et al. (2010). Call of the wild: antibiotic resistance genes in natural environments. *Nat. Rev. Microbiol.*, 8(4), 251–259. https://doi.org/10.1038/nrmicro2312

Archibald, A., and Coapes, H. E. (1976). Bacteriophage SP50 as a marker for cell wall growth in Bacillus subtilis. *J. Bacteriol.*, 125(3), 1195–1206.

Ashelford, K. E., Day, M. J., Bailey, M. J., et al. (1999). In situ population dynamics of bacterial viruses in a terrestrial environment. *Appl. Environ. Microbiol.*, 65(1), 169–174. https://doi.org/10.1128/aem.65.1.169-174.1999

Bahram, M., Hildebrand, F., Forslund, S. K., et al. (2018). Structure and function of the global topsoil microbiome. *Nature*, 560(7717), 233–237. https://doi.org/10.1038/s41586-018-0386-6

Balcazar, J. L. (2014). Bacteriophages as vehicles for antibiotic resistance genes in the environment. *PLoS Pathog.*, 10(7), e1004219. https://doi.org/https://doi.org/10.1371/journal.ppat.1004219

Bergh, O., Børsheim, K. Y., Bratbak, G., et al. (1989). High abundance of viruses found in aquatic environments. *Nature*, 340(6233), 467–468. https://doi.org/10.1038/340467a0

Bikel, S., Valdez-Lara, A., Cornejo-Granados, F., et al. (2015). Combining metagenomics, metatranscriptomics and viromics to explore novel microbial interactions: towards a systems-level understanding of human microbiome. *Comput. Struct. Biotechnol. J.*, 13, 390–401. https://doi.org/10.1016/j.csbj.2015.06.001

Breitbart, M., Bonnain, C., Malki, K., et al. (2018). Phage puppet masters of the marine microbial realm. *Nat. Microbiol.*, 3(7), 754–766. https://doi.org/10.1038/s41564-018-0166-y

Brüssow, H., and Hendrix, R. W. (2002). Phage genomics: small is beautiful. *Cell*, 108(1), 13–16. https://doi.org/10.1016/S0092-8674(01)00637-7

Callanan, J., Stockdale, S. R., Shkoporov, A., et al. (2020). Expansion of known ssRNA phage genomes: from tens to over a thousand. *Sci. Adv.*, 6(6), eaay5981. https://doi.org/doi:10.1126/sciadv.aay5981

Carini, P., Marsden, P. J., Leff, J. W., et al. (2016). Relic DNA is abundant in soil and obscures estimates of soil microbial diversity. *Nat. Microbiol.*, 2(3), 16242. https://doi.org/10.1038/nmicrobiol.2016.242

Casida Jr, L., and Liu, K.-C. (1974). Arthrobacter globiformis and its bacteriophage in soil. *J. Appl. Microbiol.*, 28(6), 951–959.

Cobbin, J. C., Charon, J., Harvey, E., et al. (2021). Current challenges to virus discovery by meta-transcriptomics. *Curr. Opin. Virol.*, *51*, 48–55. https://doi.org/10.1016/j.coviro.2021.09.007

Correa, A. M. S., Howard-Varona, C., Coy, S. R., et al. (2021). Revisiting the rules of life for viruses of microorganisms. *Nat. Rev. Microbiol.*, 19(8), 501–513. https://doi.org/10.1038/s41579-021-00530-x

Costello, M. J., Basher, Z., Sayre, R., et al. (2018). Stratifying ocean sampling globally and with depth to account for environmental variability. *Sci. Rep.*, 8(1), 11259. https://doi.org/10.1038/s41598-018-29419-1

Delisle, A., and Levin, R. (1972). Effect of temperature on an obligately psychrophilic phage-host system of Pseudomonas putrefaciens. *Anton. Leeuwe. Int J. G.*, 38(1), 9–15.

Dowd, S. E., Pillai, S. D., Wang, S., et al. (1998). Delineating the specific influence of virus isoelectric point and size on virus adsorption and transport through sandy soils. *Appl. Environ. Microbiol.*, *64*(2), 405–410. https://doi.org/10.1128/aem.64.2.405-410.1998

Edwards, R. A., and Rohwer, F. (2005). Viral metagenomics. *Nat. Rev. Microbiol.*, 3(6), 504–510. https://doi.org/10.1038/nrmicro1163

Egilmez, H. I., Morozov, A. Y., Clokie, M. R. J., et al. (2018). Temperature-dependent virus lifecycle choices may reveal and predict facets of the biology of opportunistic pathogenic bacteria. *Sci. Rep.*, 8(1), 9642. https://doi.org/10.1038/s41598-018-27716-3

Emerson, J. B., Roux, S., Brum, J. R., et al. (2018). Host-linked soil viral ecology along a permafrost thaw gradient. *Nat. Microbiol.*, *3*(8), 870–880. https://doi.org/10.1038/s41564-018-0190-y

Feiner, R., Argov, T., Rabinovich, L., et al. (2015). A new perspective on lysogeny: prophages as active regulatory switches of bacteria. *Nat. Rev. Microbiol.*, 13(10), 641–650. https://doi.org/10.1038/nrmicro3527

Fierer, N., Breitbart, M., Nulton, J., et al. (2007). Metagenomic and small-subunit rRNA analyses reveal the genetic diversity of bacteria, archaea, fungi, and viruses in soil. *Appl. Environ. Microbiol.*, 73(21), 7059–7066. https://doi.org/10.1128/AEM.00358-07

Fuhrman, J. A. (1999). Marine viruses and their biogeochemical and ecological effects. *Nature*, 399(6736), 541–548. https://doi.org/10.1038/21119

Graham, E. B., Paez-Espino, D., Brislawn, C., et al. (2019). Untapped viral diversity in global soil metagenomes. *bioRxiv*, 583997. https://doi.org/10.1101/583997

Handelsman, J. (2004). Metagenomics: application of genomics to uncultured microorganisms. *Microbiol. Mol. Biol. Rev.*, 68(4), 669–685. https://doi.org/10.1128/mmbr.68.4.669-685.2004

Handelsman, J., Rondon, M. R., Brady, S. F., et al. (1998). Molecular biological access to the chemistry of unknown soil microbes: a new frontier for natural products. *Chem. Biol.*, *5*(10), R245–R249. https://doi.org/10.1016/s1074-5521(98)90108-9

Hillary, L. S., Adriaenssens, E. M., Jones, D. L., et al. (2022). RNA-viromics reveals diverse communities of soil RNA viruses with the potential to affect grassland ecosystems across multiple trophic levels. *ISME Commun.*, 2(1), 34. https://doi.org/10.1038/s43705-022-00110-x

Ji, M., Fan, X., Cornell, C. R., et al. (2023). Tundra soil viruses mediate responses of microbial communities to climate warming. *mBio*, 14(2), e0300922. https://doi.org/10.1128/mbio.03009-22

Jover, L. F., Effler, T. C., Buchan, A., et al. (2014). The elemental composition of virus particles: implications for marine biogeochemical cycles. *Nat. Rev. Microbiol.*, 12(7), 519–528. https://doi.org/10.1038/nrmicro3289

Kapuscinski, R. B., and Mitchell, R. (1980). Processes controlling virus inactivation in coastal waters. *Water Res.*, 14(4), 363-371. https://doi.org/https://doi.org/10.1016/0043-1354(80)90084-6

Kieft, K., Zhou, Z., Anderson, R. E., et al. (2021). Ecology of inorganic sulfur auxiliary metabolism in widespread bacteriophages. *Nat. Commun.*, 12(1), 3503. https://doi.org/10.1038/s41467-021-23698-5

Kim, K.-H., Chang, H.-W., Nam, Y.-D., et al. (2008). Amplification of uncultured single-stranded DNA viruses from rice paddy soil. *Appl. Environ. Microbiol.*, 74(19), 5975–5985. https://doi.org/10.1128/AEM.01275-08

Kimura, M., Jia, Z.-J., Nakamaya, N., et al. (2008). Ecology of viruses in soils: past, present and future perspectives. *J. Soil Sci. Plant Nutr.*, 54(1), 1–32. https://doi.org/https://doi.org/10.1111/j.1747-0765.2007.00197.x

Kosmopoulos, J. C., Klier, K. M., Langwig, M. V., et al. (2023). Viromes vs. mixed community metagenomes: choice of method dictates interpretation of viral community ecology. *bioRxiv*. https://doi.org/10.1101/2023.10.15.562385

Lance, J. C., and Gerba, C. P. (1984). Effect of ionic composition of suspending solution on virus adsorption by a soil column. *Appl Environ Microbiol*, 47(3), 484–488. https://doi.org/10.1128/aem.47.3.484-488.1984

Langenfeld, K., Chin, K., Roy, A., et al. (2021). Comparison of ultrafiltration and iron chloride flocculation in the preparation of aquatic viromes from contrasting sample types. *PeerJ*, 9, e11111. https://doi.org/10.7717/peerj.11111

Lee, S., Sorensen, J. W., Walker, R. L., et al. (2022). Soil pH influences the structure of virus communities at local and global scales. *Soil Biol. Biochem.*, *166*, 108569. https://doi.org/https://doi.org/10.1016/j.soilbio.2022.108569

Liang, X., Wagner, R. E., Zhuang, J.,et al. (2019). Viral abundance and diversity vary with depth in a southeastern United States agricultural ultisol. *Soil Biol. Biochem.*, 137, 107546. https://doi.org/https://doi.org/10.1016/j.soilbio.2019.107546

Liu, C., Ni, B., Wang, X., et al. (2023). Effect of forest soil viruses on bacterial community succession and the implication for soil carbon sequestration. *Sci. Total Environ.*, 892, 164800. https://doi.org/10.1016/j.scitotenv.2023.164800

Ma, Y., Li, E., Qi, Z., et al. (2016). Isolation and molecular characterisation of Achromobacter phage phiAxp-3, an N4-like bacteriophage. *Sci. Rep.*, 6, 24776. https://doi.org/10.1038/srep24776

Makhalanyane, T. P., Valverde, A., Lacap, D. C., et al. (2013). Evidence of species recruitment and development of hot desert hypolithic communities. *Environ. Microbiol. Rep.*, 5(2), 219–224. https://doi.org/https://doi.org/10.1111/1758-2229.12003

Martinez Martinez, J., Swan, B. K., and Wilson, W. H. (2014). Marine viruses, a genetic reservoir revealed by targeted viromics. *ISME J*, 8(5), 1079–1088. https://doi.org/10.1038/ismej.2013.214

Mugnai, G., Stuknytė, M., Arioli, S., et al. (2021). Virus-like particles isolated from reactivated biological soil crusts. *Biol. Fertil. Soils*, 57(6), 863–868. https://doi.org/10.1007/s00374-021-01567-z

Narr, A., Nawaz, A., Wick, L. Y., et al. (2017). Soil viral communities vary temporally and along a land use transect as revealed by virus-like particle counting and a modified community fingerprinting approach (fRAPD) [Original Research]. *Front. microbiol.*, 8. https://doi.org/10.3389/fmicb.2017.01975

Neri, U., Wolf, Y. I., Roux, S., et al. (2022). Expansion of the global RNA virome reveals diverse clades of bacteriophages. *Cell*, 185(21), 4023–4037.e4018. https://doi.org/10.1016/j.cell.2022.08.023

Paez-Espino, D., Eloe-Fadrosh, E. A., Pavlopoulos, G. A., et al. (2016). Uncovering earth's virome. *Nature*, 536(7617), 425–430. https://doi.org/10.1038/nature19094

Parker, K. M., Barragán Borrero, V., van Leeuwen, D. M., et al. (2019). Environmental fate of RNA interference pesticides: adsorption and degradation of double-stranded RNA molecules in agricultural soils. *Environ. Sci. Technol.*, 53(6), 3027–3036. https://doi.org/10.1021/acs.est.8b05576

Pekosz, A., He, B., and Lamb, R. A. (1999). Reverse genetics of negative-strand RNA viruses: closing the circle. *Proc. Natl. Acad. Sci. USA*, 96(16), 8804–8806. https://doi.org/10.1073/pnas.96.16.8804

Porter, A. F., Shi, M., Eden, J. S., et al. (2019). Diversity and evolution of novel invertebrate DNA viruses revealed by meta-transcriptomics. *Viruses*, 11(12). https://doi.org/10.3390/v11121092

Poursalavati, A., Larafa, A., and Fall, M. L. (2023). dsRNA-based viromics: a novel tool unveiled hidden soil viral diversity and richness. *bioRxiv*. https://doi.org/10.1101/2023.05.10.540251

Pratama, A. A., and van Elsas, J. D. (2018). The 'neglected' soil virome – potential role and impact. *Trends. Microbiol.*, 26(8), 649–662. https://doi.org/10.1016/j.tim.2017.12.004

Reanney, D., and Marsh, S. (1973). The ecology of viruses attacking Bacillus stearothermophilus in soil. *Soil Biol. Biochem.*, 5(4), 399–408.

Rodriguez, R. L., Gunturu, S., Harvey, W. T., et al. (2018). The Microbial Genomes Atlas (MiGA) webserver: taxonomic and gene diversity analysis of Archaea and Bacteria at the whole genome level. *Nucleic Acids Res.*, 46(W1), W282–W288.https://doi.org/10.1093/nar/gky467

Rosario, K., and Breitbart, M. (2011). Exploring the viral world through metagenomics. *Curr. Opin. Virol.*, 1(4), 289–297. https://doi.org/10.1016/j.coviro.2011.06.004

Roux, S., and Emerson, J. B. (2022). Diversity in the soil virosphere: to infinity and beyond? *Trends Microbiol.*, 30(11), 1025–1035. https://doi.org/10.1016/j.tim.2022.05.003

Roux, S., Emerson, J. B., Eloe-Fadrosh, E. A., et al. (2017). Benchmarking viromics: an in silico evaluation of metagenome-enabled estimates of viral community composition and diversity. *PeerJ*, 5, e3817. https://doi.org/10.7717/peerj.3817

Santos-Medellin, C., Estera-Molina, K., Yuan, M., et al. (2022). Spatial turnover of soil viral populations and genotypes overlain by cohesive responses to moisture in grasslands. *Proc. Natl. Acad. Sci. USA*, 119(45), e2209132119. https://doi.org/10.1073/pnas.2209132119

Sorensen Jackson, W., Zinke Laura, A., ter Horst Anneliek, M., et al. (2021). DNase treatment improves viral enrichment in agricultural soil viromes. *mSystems*, 6(5). https://doi.org/10.1128/msystems.00614-21

Starr, E. P., Nuccio, E. E., Pett-Ridge, J., et al. (2019). Metatranscriptomic reconstruction reveals RNA viruses with the potential to shape carbon cycling in soil. *Proc. Natl. Acad. Sci. USA*, 116(51), 25900–25908. https://doi.org/doi:10.1073/pnas.1908291116

Suttle, C. A. (2005). Viruses in the sea. *Nature*, 437(7057), 356–361. https://doi.org/10.1038/nature04160

Suttle, C. A. (2007). Marine viruses – major players in the global ecosystem. *Nat. Rev. Microbiol.*, 5(10), 801–812. https://doi.org/10.1038/nrmicro1750

Taş, N., de Jong, A. E. E., Li, Y., et al. (2021). Metagenomic tools in microbial ecology research. *Curr. Opin. Biotech.*, 67, 184–191. https://doi.org/https://doi.org/10.1016/j.copbio.2021.01.019

Thingstad, T. F. (2000). Elements of a theory for the mechanisms controlling abundance, diversity, and biogeochemical role of lytic bacterial viruses in aquatic systems. *Limnol. Oceanogr.*, 45(6), 1320–1328. https://doi.org/https://doi.org/10.4319/lo.2000.45.6.1320

Thurber, R. V., Haynes, M., Breitbart, M., et al. (2009). Laboratory procedures to generate viral metagenomes. *Nat. Protoc.*, 4(4), 470–483. https://doi.org/10.1038/nprot.2009.10

Torsvik, V., Goksøyr, J., and Daae, F. L. (1990). High diversity in DNA of soil bacteria. *Appl. Environ. Microbiol.*, 56(3), 782–787. https://doi.org/10.1128/aem.56.3.782-787.1990

Touchon, M., Moura de Sousa, J. A., and Rocha, E. P. (2017). Embracing the enemy: the diversification of microbial gene repertoires by phage-mediated horizontal gene transfer. *Curr. Opin. Microbiol.*, 38, 66–73. https://doi.org/10.1016/j.mib.2017.04.010

Trubl, G., Hyman, P., Roux, S., et al. (2020). Coming-of-age characterization of soil viruses: a user's guide to virus isolation, detection within metagenomes, and viromics. *Soil Syst.* 4(2), 23. https://www.mdpi.com/2571-8789/4/2/23

Trubl, G., Jang, H. B., Roux, S., et al. (2018). Soil viruses are underexplored players in ecosystem carbon processing. *mSystems*, 3(5). https://doi.org/10.1128/mSystems.00076-18

Virgin, H. W. (2014). The virome in mammalian physiology and disease. *Cell*, 157(1), 142–150. https://doi.org/10.1016/j.cell.2014.02.032

Wang, G., Kong, J., Cui, D., et al. (2019). Resistance against Ralstonia solanacearum in tomato depends on the methionine cycle and the γ-aminobutyric acid metabolic pathway. *The Plant J. 97*(6), 1032–1047. https://doi.org/https://doi.org/10.1111/tpj.14175

Weinbauer, M. G. (2004). Ecology of prokaryotic viruses. *FEMS Microbiol. Rev.*, 28(2), 127–181. https://doi.org/10.1016/j.femsre.2003.08.001

Weinbauer, M. G., and Rassoulzadegan, F. (2004). Are viruses driving microbial diversification and diversity? *Environ. Microbiol.*, 6(1), 1–11. https://doi.org/10.1046/j.1462-2920.2003.00539.x

Wilhelm, S. W., and Suttle, C. A. (1999). Viruses and Nutrient Cycles in the Sea: viruses play critical roles in the structure and function of aquatic food webs. *BioScience*, 49(10), 781–788. https://doi.org/10.2307/1313569

Williamson, K. E. (2011). Soil phage ecology: abundance, distribution, and interactions with bacterial hosts. In G. Witzany (Ed.), *Biocommunication in Soil Microorganisms*. Berlin, Germany: Springer Science and Business Media, pp. 113--136.

Williamson, K. E., Corzo, K. A., Drissi, C. L., et al. (2013). Estimates of viral abundance in soils are strongly influenced by extraction and enumeration methods. *Biol. Fertil. Soils*, 49(7), 857–869. https://doi.org/10.1007/s00374-013-0780-z

Williamson, K. E., Fuhrmann, J. J., Wommack, K. E., et al. (2017). Viruses in soil ecosystems: an unknown quantity within an unexplored territory. *Annu. Rev. Virol.*, 4(1), 201–219. https://doi.org/10.1146/annurev-virology-101416-041639

Williamson, K. E., Wommack, K. E., and Radosevich, M. (2003). Sampling natural viral communities from soil for culture-independent analyses. *Appl. Environ. Microbiol.*, 69(11), 6628–6633. https://doi.org/10.1128/AEM.69.11.6628-6633.2003

Williamson, K. E., Radosevich, M., and Wommack, K. E. (2005). Abundance and diversity of viruses in six Delaware soils. *Appl. Environ. Microbiol.*, 71(6), 3119–3125. https://doi.org/10.1128/AEM.71.6.3119-3125.2005

Wolf, Y. I., Kazlauskas, D., Iranzo, J., et al. (2018). Origins and evolution of the global RNA virome. *mBio*, 9(6). https://doi.org/10.1128/mBio.02329-18

Wu, R., Bottos Eric, M., Danna Vincent, G., et al. (2022). RNA viruses linked to eukaryotic hosts in thawed permafrost. *mSystems*, 7(6), e00522–00582. https://doi.org/10.1128/msystems.00582-22

Wu, R., Davison, M. R., Gao, Y., et al. (2021). Moisture modulates soil reservoirs of active DNA and RNA viruses. *Commun. Biol.*, 4(1), 992. https://doi.org/10.1038/s42003-021-02514-2

Yeager, J. G., and O'Brien, R. T. (1979). Enterovirus inactivation in soil. *Appl. Environ. Microbiol.*, 38(4), 694–701. https://doi.org/10.1128/aem.38.4.694-701.1979

Zablocki, O., van Zyl, L., Adriaenssens, E. M. et al. (2014). High-level diversity of tailed phages, eukaryote-associated viruses, and virophage-like elements in the metaviromes of antarctic soils. *Appl. Environ. Microbiol.*, 80(22), 6888–6897. https://doi.org/10.1128/aem.01525-14

Zehr, J. P., and Ward, B. B. (2002). Nitrogen cycling in the ocean: new perspectives on processes and paradigms. *Appl. Environ. Microbiol.*, 68(3), 1015–1024. https://doi.org/10.1128/aem.68.3.1015-1024.2002

Zhang, F., Chase-Topping, M., Guo, C. G., et al. (2020). Global discovery of human-infective RNA viruses: a modelling analysis. *PLoS Pathog.*, 16(11), e1009079. https://doi.org/10.1371/journal.ppat.1009079

Zhou, J., Deng, Y., Zhang, P., et al. (2014). Stochasticity, succession, and environmental perturbations in a fluidic ecosystem. *Proc. Natl. Acad. Sci.*, 111(9), E836–E845.

Zolfo, M., Pinto, F., Asnicar, F., et al. (2019). Detecting contamination in viromes using ViromeQC. *Nat. Biotechnol.*, 37(12), 1408–1412.

Chapter 2

Advances in metaproteomics for analyzing soil microorganisms

Paolo Nannipieri, Environment and Forestry (DAGRI), Italy; Jing Se, Zhejiang University, China; and Laura Giagnoni, Architectural Engineering and Mathematics (DICATAM), Italy

1 Introduction

Thousands of millions of microbial cells inhabit a handful of soil and, thanks to the use of amplicon sequencing and metagenomics, the degree to which we can understand the abundances and diversity of soil microbiomes continues to improve (Dini-Andreote and van Elsas, 2019). DNA-based techniques are more widely used than metatranscriptomics and metaproteomics because they require less investment and are easier to implement, while commercial companies provide relatively cheap and fast sequencing (Fierer, 2017). In contrast, metaproteomics analyses require the use of sensitive and specialist tools and equipment.

However, DNA-based techniques only provide potential indications of *in situ* microbial activities because studies assume that the target microbial species carry out the same activities observed *in vitro* with the detection of a functional gene then proving that the relative function is active in soil (Nannipieri, 2020). However, most microbial species that inhabit soil are inactive (Blagodatskaya

http://dx.doi.org/10.19103/AS.2024.0136.05

and Kuzyakov, 2013). In addition, DNA analyses can give misleading results because they detect 'unknown' genes due to incomplete homology or because they only get the best hits by assigning single reads to databases. Finally, the function of many genes depends on the genetic context in the microbial genome (Nannipieri et al., 2020). Soil DNA does not only exist intracellularly in living cells but also in dead cells, in cell remains and extracellularly – free, immobilized or protected by surface-reactive particles (Nannipieri et al., 2020). Therefore, it is important to determine the final expression of each gene to gain insights into microbial activity *in situ*.

The aim of this chapter is to discuss the challenges facing protein extraction from soil, their purification, determination and assignment to available databases (Fig. 1) and those ascribed to the many protein locations in the soil matrix. It is important to determine the differences between proteomics and metaproteomics in the context of the soil ecosystem. The proteomics approach concerns the determination of protein expression profile of one single organism.

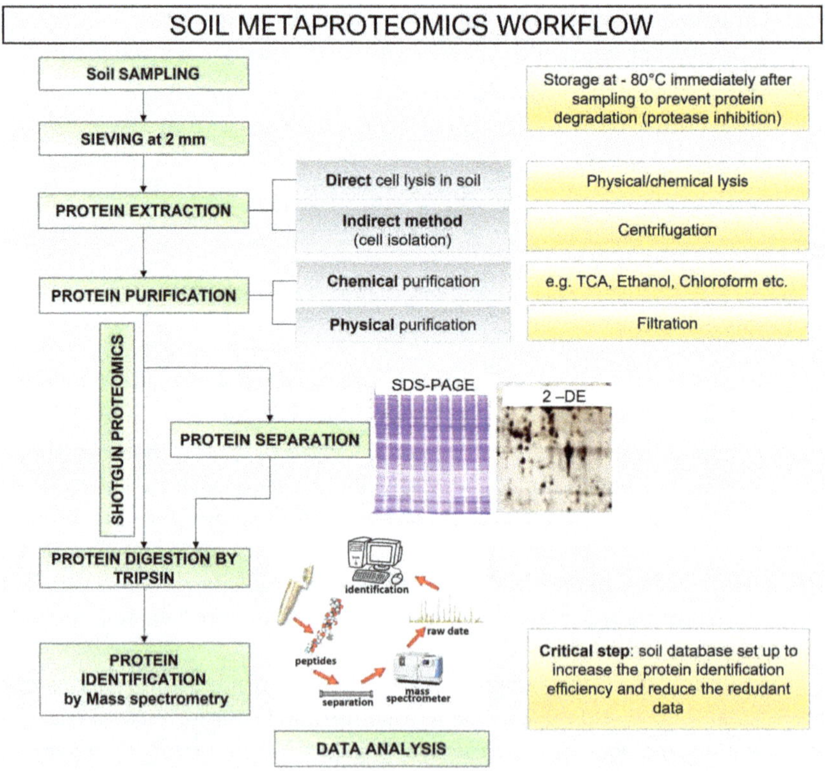

Figure 1 Main steps in soil metaproteomic analysis with examples of used techniques. Source: Modified from Tartaglia et al. (2020).

In contrast, metaproteomics is the protein profile of a complex biological system, as soil, characterized by a wide biodiversity (Nannipieri et al., 2019). The determination of proteins expressed by soil microorganisms (metaproteomics) can provide insights into the effects of pollutants and their degradation products, agricultural practices, deforestation and climate changes on soil functions, which are driven mainly by soil microorganisms (Bastida and Jehmilich, 2016; Nannipieri et al., 2019; Ogunseitan, 2006; Renella et al., 2014, 2015). In addition, metaproteomics studies can provide insights into mechanisms underlying interactions among microorganisms, microorganisms with soil fauna or with plants. Soil proteomics studies can provide insights into mechanisms responsible for the stabilization of extracellular proteins by soil-surface reactive particles, the role of proteins of microbial necromasses in the formation of soil organic nitrogen and the dormancy and resuscitation of single microbial species (Giagnoni et al., 2018).

2 Protein analysis by mass spectrometry (MS) and assignment by databases

The development of highly efficient peptide ionization mass spectrometers has kept pace with advancements in proteomics (Nannipieri et al., 2019). Efficient protein analysis is generally carried out with high mass-accuracy tandem spectrometers (MS/MS), with the first MS determining the relative mass of peptides and the second MS the mass-to-charge ratio; after that each peptide is dissociated by collision with a neutral gas (Keller and Hettich, 2009).

Recently, an ion monitoring technique based on high-resolution and high-precision MS and parallel reaction monitoring (PRM) has quantified proteins and peptides (Bzhozovskiy et al., 2022; Se et al., 2021). The top-down approach involves the loading of intact proteins into MS, while the bottom-up approach requires the trypsin digestion of proteins followed by separation of produced peptides before determination by MS (Nannipieri et al., 2019). The top-down method is preferred to the bottom-up because (Brown et al., 2020):

1 It saves time; and
2 The determination of intact proteins provides insights into the structure of the molecule and post-translational modifications.

In particular, the application of top-down proteomics provides in-depth analysis of the complexity of the protein profile and is suitable for the analysis of spatial and temporal expression (Brown et al., 2020). However, bottom-up proteomics is more commonly used than top-down methods because (Nannipieri et al., 2019):

1 Top-down methods mean it is more complicated to handle proteins than small peptides;

2 The sensitivity of MS is higher with peptides than protein analysis; and
3 The determination of proteins with molecular weights higher than 50 kDa is not possible unless their structure is destroyed.

Mass spectrometry-based proteomic studies have widely used gel electrophoresis for the separation of proteins extracted from cells (Lee et al., 2020). They are usually separated by the molecular mass (one-dimensional gel electrophoresis, 1D) or by both isoelectric point and molecular mass (two-dimensional gel electrophoresis, 2DE), using polyacrylamide gels with sodium dodecyl sulfate (SDS-PAGE) (Brunelle and Green, 2014). In particular, 2DE provides a large-scale proteomic fingerprinting profile with the isolation of a single (or a few) protein as a spot (Giagnoni et al. 2011; Lee et al., 2020).

Following gel electrophoresis, the proteins are visualized as spots by staining, then excised, solubilized and finally subjected to trypsin digestion. However, 2DE has various drawbacks, including:

- Incomplete separation of proteins with similar mass and isoelectric point (Renella et al., 2014); and
- Proteins that are hydrophobic, are of low abundance in the extract, have extreme values of isoelectric point or have not separated molecular weights (Nannipieri et al., 2019; Renella et al., 2014).

2DE, combined with liquid chromatography (LC), provides better separation than 2DE alone. However, 2D-SDS-PAGE gel electrophoresis followed by MALDI-TOF analysis is also used, but a better protein separation and identification is achieved by 1D-SDS-PAGE followed by LC-MS/MS analysis, because it combines LC and tandem MS (Starke et al., 2019).

After MS reading, the last step is the identification of proteins via the use of protein databases (Nannipieri et al., 2019). This is a critical step because:

- Environmental samples contain a multitude of different microbial species and most are unknown, as already mentioned;
- Environmental databases are scarcer than the respective genomic databases (Renella et al., 2014); and
- The selection of databases is subjective and often utilizes specific databases that are weighted in favor of the most researched organisms (Starke et al., 2019).

Insufficient annotation is therefore a critical drawback (Renella et al., 2015), together with scarce information on post-translational modifications (truncation, glycosylation, phosphorylation, etc.) in the available databases. This information is important to improve the identification of proteins extracted from environmental samples and the reconstruction of metabolic pathways (Nannipieri et al., 2019). To overcome these problems, peptides from measured proteins can be matched with those of proteins predicted from fully sequenced genomes, as reported in soil metaproteomics studies by Starke et al. (2016), Huitman et al. (2015) and Lima Parente-Fernandez et al. (2022). The combined use of genomics and proteomics is known as proteogenomics (Bland and Armengaud, 2014). However, in this case, the challenge lies in the presence of unknown species in soil.

3 Soil protein extraction techniques

Soil proteomics studies are influenced by the extraction yield, which depends on the extraction procedure used, which is generally incomplete (Nannipieri et al., 2019; Tartaglia et al., 2020). Studies into soil proteomics have tended to ignore the literature on enzyme extraction from soil. Indeed, buffers, such as citrate or phosphate, combined with sodium dodecyl sulphate (SDS) or SDS-phenol, which are often employed in soil proteomics studies, are less effective than alkaline solutions, such as NaOH or sodium pyrophosphate, in extracting organic compounds, including proteins, from soil (Nannipieri et al., 1996).

As already shown in soil organic matter studies, the increase of the buffer (phosphate) pH increases protein extraction from soil (Murase et al., 2003). However, the increase in protein extraction yield – either by increasing the pH of buffers or by using NaOH or sodium pyrophosphate – is accompanied by an increase in the extraction of humic molecules, followed by strong purification procedures which cause significant protein losses (Schulze et al., 2005). Two different purification procedures (humic coagulation with Al ions and phenol treatment) used to remove extracted humic molecules prior to protein separation have yielded similar results (Mandalakis et al., 2018).

Both Benndorf et al. (2007) and Herruzo-Ruiz et al. (2021) used NaOH to extract proteins from soil, which is the solution generally used to extract humic molecules from soil. Herruzo-Ruiz et al. (2021) removed humic molecules by centrifugation: the supernatant with microbes was centrifuged at high speed and the pellets washed, dissolved in Tris buffer and sonicated to release proteins, which were separated by SDS-PAGE. Following digestion by trypsin and analysis by MS, 2182 proteins were identified by an in-house database constructed from rRNA data. The majority of the proteins extracted were not analyzed due to their removal by humic acids or changes in protein conformation after interacting with humic molecules preventing the trypsin hydrolysis of certain parts of the protein molecule (Arenella et al., 2014).

Proteins are extracted from soil by both direct and indirect methods. The former involves the extraction of proteins following cell breakdown in soil, whereas the latter is based on the extraction of cells from soil, followed by cell breakdown before protein extraction (Renella et al., 2014). The risk of protein losses is minimized by using an indirect extraction method that provides purer protein extracts but lower extraction yields than the direct method, reducing the identification and reconstruction of protein pathway profiles (Renella et al., 2014). In addition, changes in the proteome can occur during the separation of cells from soil particles, as observed in the proteome of *Cupriavidus metallidurans* following its inoculation into a model soil (described later) and immediate separation of cells from soil particles (Giagnoni et al., 2012). The direct method requires a stronger purification procedure than the indirect method.

Both Chourey et al. (2010) and Taylor and Williams (2010) compared the two methods with contrasting results. The extraction of proteins by the indirect method and the successive separation by SDS-PAGE-produced protein bands, which were unresolved in the direct extraction method (Taylor and Williams, 2010). By contrast, the direct method was more effective than the indirect method in detecting proteins extracted from a soil inoculated with *Pseudomonas putida* or *Arthrobacter chlorophenolicus* (Chourey et al., 2010).

However, the detected proteome depends on the extraction procedure, as shown by Keiblinger et al. (2012) who used four different extraction methods with two different soils. The detected proteins ranged from 174 to 541 in the forest soil and from 63 to 326 in the potting soil. The extraction yield ranged from 47% to 49%, as determined by spiking sterilized soils with proteins from *Pectobacterium carotovorium* and *Aspergillus nidulans* (Keiblinger et al., 2012). Therefore, more than 50% of the proteins of the inoculated cells were not detected. Finally, the use of the method of co-extracting DNA, RNA and proteins as proposed by Thorn et al. (2019) to save time and reduce analysis cost is not advisable because it is less effective than methods that extract only nucleic acids or proteins. Indeed, nucleic acid and protein molecules demonstrate different interactions with soil surface-reactive surface particles due to their different physicochemical properties. Proteins released by cell breakdown interact strongly with Na-montmorillonite and only slightly with sand particles (Giagnoni et al., 2012). The presence of humic molecules affects protein separation by gel electrophoresis (Giagnoni et al., 2012), whereas the interactions between proteins and humic molecules affect the protein identification by MS following trypsin digestion (Arenella et al., 2014).

4 Soil studies using metaproteomics and other omics techniques

Earlier studies into soil metaproteomics focused on either specific expressed proteins, such as thermostable cellulase and subsets of proteins in response to pollutants (Ogunseitan, 2006). Later studies dealt with microbial proteins expressed in response to climate change, pollution and management practices, as reviewed by Bastida and Jehmilich (2016), Bastida et al. (2016), Nannipieri et al. (2019), Renella et al. (2014, 2015) and Starke et al. (2019).

A global study of topsoils from four different continents, differing in terms of vegetation cover and climatic and edaphic conditions, demonstrated the dominance of proteins relating to the metabolism of lipid, carbohydrate and one-carbon compounds, tricarboxylic acid cycle, DNA repair, transcription regulation and N fixation (Bastida et al., 2021). The metaproteome analysis provided better taxonomic information than the 16s rRNA sequencing but did not detect specific functional proteins because they were much less abundant than 'housekeeping' proteins, as reported earlier (Starke et al., 2019). The stable isotope probe was used to detect expressed proteins directly by Starke et al. (2016), who observed the dominance of bacterial proteins during the short-term assimilation of tobacco-derived N enriched in ^{15}N, although the number of identified labelled proteins totalled just 26.

Cross-checking of identified proteins with the respective genes can identify and validate new genes, discover post-translational modifications in the protein structure, correct translational starts and provide functional information. Metaproteomics approaches conducted on permafrost soil cores and a young thermokarst bog showed Actinobacteria, Acidobacteria and Proteobacteria as the active microorganisms (Huitman et al., 2015). Proteins identified by the metaproteomics matched those predicted by metagenomics. The multi-omics strategy revealed 7000 proteins, the highest amount among soil metaproteomics studies. The majority of these extracted proteins were unknown, with a predominant presence of those associated with cold adaptation.

The 18S rRNA sequencing and metaproteomics of forest, grassland and shrubland soils showed that fungi in forest soils had richer levels of protein than in shrubland soils. Proteins related to metabolism dominated in the former soils, whereas those related to information processing and storage dominated in the latter (Lima Parente-Fernandez et al., 2022). The fungal richness of the three soil types correlated significantly with the annual temperature and soil pH. The total number of fungal proteins identified in 68 soil samples amounted to 776 with an average richness of 234.

The 16S rRNA sequencing and metaproteomics of 32 topsoils sampled from four different continents, varying in terms of climate conditions and vegetation cover, demonstrated the dominance of both *Nitrosophaerales* and *Thermoplasmata* in the archaea communities (Starke et al., 2021). Of the 7474 identified proteins, just 416 were associated with soil archaea and dominated by those involved in DNA protection during starvation, methyl coenzyme M reductase and tyrosine–tRNA ligase.

5 Other challenges in soil metaproteomics: the huge number of expressed proteins and different protein locations in the soil matrix

All soil metaproteomic studies are biased due to incomplete analysis of total expressed proteins. The highest number of identified proteins in soil metaproteomic studies has totaled 7000 (Huitman et al., 2015) but, considering the huge microbial diversity, the potential number of expressed proteins is certainly much higher. When considering prokaryotes alone, 10 000 000 cells can be found in 1g of soil on average (Raynaud and Nunan, 2014). Assuming that only 10% are active, the total number of expressed proteins would be 2 000 000 000–3 000 000 000, considering that one active microbial cell can express 2000–3000 proteins. It is impossible to extract all these proteins from soil. Even if soil extraction, purification and separation are successful, no MS could cope with this high number.

In addition, studies into soil metaproteomics have generally detected 'housekeeping' proteins, such as those involved in DNA replication, lipid metabolism, Krebs cycle, glycolysis, etc. but rarely proteins with specific functions (Starke et al., 2019) or proteins expressed by rare microbial species which are responsible for the measured soil function, as is the case in alkaline phosphatase activity mediated by rare microbial species bearing the *pho*D gene (Liu et al., 2021; Wei et al., 2019).

Generally, studies of soil proteomics or soil metaproteomics do not consider the unique status of soil N, which is mainly composed of soil organic N (95–99% of the total) with just 4–5% present as microbial biomass N. Therefore, the proteins expressed by the active microbial species represent less than 4–5% of soil N because not all microbial communities inhabiting soil are active (Blagodatskaya and Kuzyakov, 2013).

The use of stable isotope probes can determine the microbial proteins linked to a specific function. Despite C, N, H, O and S having the potential for labeling proteins, labeling N or S gives low sensitivity due to their low abundance in amino acids. This also explains the low number of ^{15}N labeled proteins detected in the study of the microbial assimilation of tobacco-derived N (Starke et al., 2016) discussed earlier. The use of C labeling should be preferred to that of the other abundant atoms, H and O, in amino acids because (Starke et al., 2019)

- The H isotope, deuterium (D), is toxic to bacteria;
- The rapid H and D exchange occurs in cells during sample preparation and compounds with CD bonds behave differently compared to those with CH bonds in the different chromatographic separations of proteins or peptides; and

- The use of O isotope is biased due to the exchange between ^{18}O of the added compounds with ^{16}O of water.

Most soil proteins and peptides are localized extracellularly, accounting for 30–50% of the total soil N (Nannipieri et al., 2019). These proteins are released as enzymes, for example by microbial cells, to hydrolyze high molecular weight polymers in monomers to be taken up by cells. Proteins and peptides are also released into the extracellular soil environment after lysis following the death of the cell. The released proteins or peptides are adsorbed or entrapped by soil particles and protected against proteolysis by the active microorganisms (Nannipieri et al., 2019). Extracellularly stabilized proteins are therefore not related to actual microbial activity but may reflect past microbial events (Nannipieri et al., 2019).

The study of proteins and peptides can provide insights into mechanisms responsible for their interactions with soil particles. Loose adsorption occurs through the formation of weak bonds, such as interactions between hydrophobic moieties of proteins with hydrophobic parts of organic surfaces of soil particles or through hydrogen and electrostatic interactions. Proteins can also be tightly bound to soil particles, through the formation of covalent bonds, or physically entrapped. According to Nannipieri et al. (2019), future research should develop sequential extractions of mild and then strong extracting solutions so as to separate free, loosely and tightly or entrapped proteins.

It is important to emphasize that microbial necromasses are essential in soil organic carbon sequestration and their composition differs from that of living soil microbiomes (Camenzind et al., 2022). Proteomics of soil microbial necromasses is important to monitor the fate of expressed proteins during microbial death and to gain insights into microbial N turnover in soil.

In this context, it is important to note the development of metaexoproteomics with analysis of free extracellular proteins released by active soil microbial cells (Johnson-Rollings et al., 2014). The metaexoproteome extracted from soil by K_2SO_4 displayed chitinolytic activity likely due to *Nocardiopsis chi*A-like chitinase. The combination of metaexoproteomics and metagenomics identified 1885 plant, insect and microbial proteins in both bulk and rhizosphere soil of *Brassica napus* L (oilseed rape) (Lidbury et al., 2022). The study of the rhizosphere metaexoproteome revealed upregulation in both phosphorus and nitrogen mineralization likely to favor plant uptake of both nutrients.

6 Metaproteomics of specific microorganisms inoculated into a model soil

The use of a model soil inoculated with a single microbial species or a simple microbial consortium can provide insights into mechanisms underlying the physiological and metabolic response of inoculants to the presence of pollutants, changes in climate and nutrient concentrations and interactions of soil microorganisms with soil surface-reactive particles. The addition of root exudates to these model systems can provide insights into the effects of plants mediated by root exudates, which can aid in understanding of the rhizosphere behavior of soil microorganisms (Nannipieri, 2020). The use of completely sequenced microorganisms is favored due to the comparison between predicted and detected proteins, which can evidence not only expressed proteins due to the environmental conditions but also show posttranslational modifications (Renella et al., 2014). These mesocosm studies are important but caution is required in terms of generalizing these results to *in situ* conditions.

The model soil used in several proteomic studies was a mixture of sand, goethite, humic acid, kaolinite and montmorillonite. The variable component was the type of clay mineral, differing for cation exchange capacity (CEC), surface area and shrink/swell properties, all playing an important role in the adsorption of organic molecules (Giagnoni et al., 2011; Nannipieri et al., 2019). By increasing the content of montmorillonite in the artificial soil, the amount of extracted proteins by phosphate buffer with SDS from the inoculated *Cupriavidus metallidurans* (a completely sequenced soil-born beta-proteobacterium) cells decreased (Giagnoni et al., 2011). By contrast, the highest protein extraction yield was with the artificial soil comprised of 100% sand, while the extraction yield with that of 100% kaolinite protein was much higher than with montmorillonite, which has a higher surface area and CEC. It also swells and shrinks. Protein patterns of cells of the same bacterium added to sand or kaolinite and incubated for up to 3 days resembled those of bacterium incubated in the pure culture, whereas those extracted from the artificial soil containing humic acids and montmorillonite changed, likely due to protein adsorption by these surface-reactive components. However, it cannot be denied that proteins released following the cell breakdown were extracted but not determined by MS because their interaction with surface-reactive soil particles changed the molecular conformation, preventing the trypsin digestion of some protein moieties, as observed in myoglobin that interacted with humic molecules (Arenella et al., 2014).

Changes in nutrient concentrations can alter the soil proteome profile. An increase in the dissolved organic C increased the abundance of proteins involved in metabolism and replication and nucleic acid integrity (Bastida et al., 2016). Stresses, such as those due to nutrient limitations, can cause the adoption of viable but non-culturable cell (VBNC) status, especially in gram-negative bacteria, whereas gram-positive bacteria respond by producing spores (Atlas and Bartha, 1981). Nutrient limitations to microorganisms are common in soil and some hot spots, such as rhizosphere soil, may not experience nutrient limitations. The soil-borne *C. metallidurans* strain CH 34 cells inoculated into the artificial soil, described earlier, transformed from culturability to the VBNC state under nutrient or water limitation (Giagnoni et al., 2018). The recovery of culturability was obtained by adding water or a C source as gluconate. During transformation to the VBNC state, stress-related proteins were upregulated, whereas proteins related to cell shape and protein synthesis were downregulated. The recovery of the culturability by water showed the upregulation of six proteins, one of which was involved in the expression genes responsible for bacterial resistance to nutrient limitation. By contrast, the recovery by gluconate led to the synthesis of proteins responsible for protein biosynthesis and energy metabolic pathways. Extracted proteins were analyzed by shot-gun proteomics using an advanced proteomic approach (iTRAQ), allowing the identification of a high number of proteins (Giagnoni et al., 2018).

Se et al. (2021) observed that desiccation caused the VBNC state of *Escherichia coli* O157:H7 inoculated in the artificial soil used by Giagnoni et al. (2018); the culturable state was regained after rewetting. The proteomic analysis showed that 2324 proteins were differently expressed when the proteome of the two states was compared. In particular, the proteome of the VBNC state showed the upregulation of proteins involved in nutrient transport, membrane, chemotactic, flagellar, virulence and adhesion, whereas cytoskeletal proteins and enzymes of the Krebs cycle were downregulated and mirrored the morphological changes occurring during the VBNC state. These findings are important to understand the mechanisms that explain the survival of this human pathogen in soil, which can range from several days up to several months (Islam et al., 2014).

Future research should study the effects of plants on the resuscitation of soil-borne bacteria mediated by root exudates. As reported above, caution is required in order not to generalize outcomes from the model soil when testing soil-borne bacteria, such as *Escherichia coli* O157:H7 and *C. metallidurans* strain CH 34 (Giagnoni et al., 2019), whose proteins are widely found in soil proteomes (Bastida et al., 2016).

7 Conclusion and future trends

The amount of expressed microbial proteins in soil can be counted in the order of millions, whereas those extracted are in the thousands. Those identified are even lower for various reasons, such as incomplete databases, unknown proteins and those that go undetected because their interactions with surface-reactive particles, once released after cell breakdown, modify their conformation, preventing trypsin hydrolysis (Arenella et al., 2014). Despite these drawbacks, metaproteomics – determining the expressed proteome by active microbial cells of a soil sample – has great potential to improve our understanding of functions mediated by active soil microorganisms. In addition, metaproteomics can provide not only functional but also taxonomic information. However, future research is required to fill knowledge gaps and resolve the challenges discussed in this review to make soil metaproteomics a powerful tool to study soil functions.

Future soil research will favor a top-down approach due to the MS loading of intact proteins to overcome trypsin hydrolysis of the bottom-up approach, avoiding the bias of the incomplete proteolysis due to changes in protein conformation following interactions of proteins with surface-reactive soil particles (Arenella et al., 2014).

Active microbes are present in soil hot spots, such as rhizosphere soil and detritusphere soil (Blagodatskaya and Kuzyakov, 2013). Future research should focus attention on these hot spots rather than all soil, for example by combining metaproteomics with the use of nanoscale secondary ion mass spectrometry (NanoSIMS), which showed, for example, the transfer of photosynthetically organic C to roots and to mycorrhizal hyphae and then to soil microorganisms (Kaiser et al., 2015). A cascade of transmission electron microscopy (TEM) and NanoSIMS showed the microbial transformation of plant residues via earthworms to surface-reactive particles (Vidal et al., 2019). Focusing on these hot spots reduces the risk of interferences mediated by extracellular proteins not expressed by active soil microorganisms.

Metaexoproteomics is an optimal approach to reduce the number of analyzed proteins, differentiating intracellular from extracellular expressed proteins. It is likely that the use of the K_2SO_4 only extracted the free and loosely bound extracellular proteins. Setting up a sequential extraction by increasing the strength of the solution can differentiate these free and loosely extracellular proteins from those entrapped or strongly bound in the soil.

Experiments with model soils provide insight into the proteomes of microorganisms, which play important roles in the environment and in human health. It is possible to test the effects of different surface-reactive particles, different climatic conditions and different plant exudates. Future research should use artificial soil with a definite composition rather than sterilized soil because it is difficult to kill all soil microorganisms even with gamma irradiation, and the presence of proteins of killed microbial cells affects the analysis of expressed proteins by the inoculated microorganism (Se et al., 2021).

8 References

Arenella, M., Giagnoni, L., Masciandaro, G., Ceccanti, B., Nannipieri, P. and Renella, G. (2014). Interactions between proteins and humic substances affect protein identification by mass spectrometry, *Biol Fertil Soils* 50(3), 447–454.

Atlas, R. M. and Bartha, R. (1981). *Microbial Ecology: Fundamentals and Applications*, Addison-Wesley, Reading, MA.

Bastida, F. and Jehmilich, N. (2016). It's all about functionality: how can metatranscriptomics help us to discuss the attributes of ecological relevance in soil?, *J Proteomics* 144, 159–161.

Bastida, F., Torres, I. F., Moreno, J. L., Baldrian, P., Ondono, S., Ruiz-Navarro, A., Hernandez, T., Richnow, H. H., Starke, R., Garcia, C. and Jehmlich, N. (2016). The active microbial diversity drives ecosystem multifunctionality and is physiologically related to carbon availability in Mediterranean semi-arid soils, *Mol Ecol* 25(18), 4660–4673.

Bastida, F., Jehmlich, N., Starke, R., Schallert, K., Benndorf, D., López-Mondéjar, R., Plaza, C., Freixino, Z., Ramírez-Ortuño, C., Ruiz-Navarro, A., Díaz-López, M., Vera, A., Moreno, J. L., Eldridge, D. J., García, C. and Delgado-Baquerizo, M. (2021). Structure and function of bacterial metaproteomes across biomes, *Soil Biol Biochem* 160, 108331.

Benndorf, D., Balcke, G. U., Harms, H. and von Bergen, M. 2007. Functional metaproteome analysis of proteins extracted from contaminated soil and groundwater, *ISME J* 1(3), 224–234.

Blagodatskaya, E. and Kuzyakov, Y. (2013). Active microorganisms in soil: critical review of estimation criteria and approaches, *Soil Biol Biochem* 67, 192–211.

Bland, C. and Armengaud, J. (2014). Proteogenomics: a new integrative approach for a better description for protein diversity found in soil microflora. In: Nannipieri, P., Pietramellara, G. and Renella, G. (Eds), *Omics in Soil Science*, Custer Academic Press, Norfolk, UK, pp 139–162.

Brown, K. A., Melby, J. A., Roberts, D. S. and Ge, Y. (2020). Top-down proteomics: challenges, innovations, and applications in basic and clinical research, *Expert Rev Proteomics* 17(10), 719–733.

Brunelle, J. L. and Green, R. (2014). One-dimensional SDS-polyacrylamide gel electrophoresis (1D SDS PAGE), *Methods Enzymol* 541, 151–159.

Bzhozovskiy, A., Kononikhim, A., Bugrova, A. E., Kovalev, G. I., Schmit, P. O., Kruppa, G., Nikolaev, E. N. and Borchers, C. H. (2022). The parallel reaction monitoring-parallel accumulation-serial fragmentation (PRM-PASEF). Approach for multiplexed absolute quantification of proteins in human plasma, *Anal Chem* 94(4), 2016–2022.

Camenzind, T., Mason-Jones, K., Mansour, I., Rillig, M. C. and Lehmann, J. (2022). Formation of necromass-derived soil organic determined by microbial death pathways, *Nat Geosci* 16, 115–122.

Chourey, K., Jansson, J. K., Verbekmoes, N., Shah, M., Chavarria, K. L., Tom, L. M., Brodie, E. L. and Hettich, R. L. (2010). Direct lysis/protein extraction for soil metaproteomics, *J Proteome Res* 9(12), 6615–6622.

Dini-Andreote, F. and van Elsas, J. D. (2019). The soil microbiome: an overview. In: van Elsas, J. D., Trevors, J. T., Soares Rosado, A. and Nannipieri, P. (Eds), *Modern Soil Microbiology* (3rd edn.), CRC Press, Boca Raton, FL, pp 37–48.

Fierer, N. (2017). Embracing the unknown: disentangling the complexities of the soil microbiome, *Nat Rev Microbiol* 15(10), 579–590.

Giagnoni, L., Magherini, F., Landi, L., Taghavi, S., Modesti, A., Bini, L., Nannipieri, P., Van der Lelie, D. and Renella, G. (2011). Extraction of microbial proteome from soil: potential and limitations assessed through a model study, *Eur J Soil Sci* 62(1), 74–81.

Giagnoni, L., Magherini, F., Landi, L., Taghavi, S., van der Lelie, D., Puglia, M., Bianchi, L., Bini, L., Nannipieri, P., Renella, G. and Modesti, A. (2012). A soil solid phase effects on the proteomic analysis of *Cupriavidus metallidurans* CH34, *Biol Fertil Soils* 48(4), 425–433.

Giagnoni, L., Arenella, M., Galardi, E., Nannipieri, P. and Renella, G. (2018). Bacterial culturability and the viable but non-culturable (VBNC) state studied by a proteomic approach using an artificial soil, *Soil Boil Biochem* 118, 51–58.

Herruzo-Ruiz, A. M., Fuentes-Almagro, C. A., Jimenez-Pastor, J. M., Perez-Rosa, V. M., Blasco, J., Michan, C. and Alhama, J. (2021). Meta-omic evaluation of bacterial community structure and activity for environment assessment of soils: overcoming protein extraction pitfalls, *Environ Microbiol* 23(8), 4706–4725.

Huitman, J., Waldrop, M. P., Mackelprang, R., David, M. M., McFarland, J., Blazewicz, S. J., Harden, J., Turetsky, M. R., McGuire, A. D., Shah, M. B., VerBerkmoes, N. C., Lee, L. H., Mavrommatis, K. and Jansson, J. K. (2015). Multi-omics of permafrost, active layer and thermokarst bog soil microbiome, *Nature* 521, 208–212.

Islam, M., Doyle, M. P., Phatak, S. C., Millner, P. and Jang, X. P. (2014). Persistence of enterohemorrhagic *Escherichia coli* O157:H7 in soil and on leaf lettuce and parsley grown in fields treated with contaminated manure composts or irrigation water, *J Food Prod* 67, 1365–1370.

Johnson-Rollings, A. S., Wright, E., Masciandaro, G., Macci, C., Doni, S., Calvo-Bado, L. A., Slade, S. E., Vallin Plou, C. and Wellington, E. M. H. (2014). Exploring the functional soil-microbe interface and exoenzymes through soil metaexoproteomics, *ISME J* 8(10), 2148–2150.

Kaiser, C., Kilburn, M. R., Clode, P. L., Fuchslueger, L., Koranda, M., Cliff, J. B., Solamian, Z. M. and Murphy, D. V. (2015). Exploring the transfer of recent photosynthates to soil microbes: mycorrhizal pathway vs direct root exudation, *New Phytol* 205(4), 1537–1551.

Keiblinger, K. M., Wilhartiz, I. C., Schneider, T., Roschitzki, B., Schmid, E., Eberl, L., Riedel, K. and Zechmeister-Boltenstern, S. (2012). Soil metaproteomics – comparative evaluation of protein extraction protocols, *Soil Biol Biochem* 54, 14–24.

Lee, P. Y., Saraygord-Afshari, N. and Low, T. Y. (2020). The evolution of two-dimensional gel electrophoresis – from proteomics to emerging alternative applications, *J Chromatogr A* 1615, 460763.

Lidbury, I. D. E. A., Raguideau, S., Borsetto, C., Murphy, A. R. J., Bottrill, A., Liu, S., Stark, R., Fraser, T., Goodall, A., Jones, A., Bending, G. D., Tibbet, M., Hammond, J. P., Quince, C., Scanlan, D. J., Pandhal, J. and Wellington, E. M. H. (2022). Stimulation of distinct rhizosphere bacteria drivers phosphorus and nitrogen mineralization in oilseed rape under filed conditions, *mSystems* 7, 1–17.

Lima Parente Fernandez, M. L. P., Bastida, F., Jehmlich, N., Martinovic, T., Vetrovisky, T., Baldrian, P., Delgado-Baquerizo, M. and Starke, R. (2022). Functional soil mycobiome across ecosystems, *J Proteomics* 252, 104428.

Liu, S., Zhang, X., Dungait, J. A. J., Quine, T. A. and Razavi, B. S. (2021). Rare microbial taxa rather than *pho*D gene abundance determine hotspots of alkaline phosphatase activity in the karst rhizosphere soil, *Biol Fertil Soils* 57(2), 257–268.

Mandalakis, M., Panikov, N. S., Polymenakou, P. N., Sizova, M. V. and Stamatakis, A. (2018). A simple cleanup method for the removal of humic substances from soil protein extracts, *Environ Sci Poll Res* 25(24), 23845–23856.

Murase, A., Yoneda, M., Ueno, R. and Yonebayashi, K. (2003). Isolation of extracellular protein from greenhouse soil, *Soil Biol Biochem* 35(5), 733–736.

Nannipieri, P., Sequi, P. and Fusi, P. (1996). Humus and enzyme activity. In: Piccolo, A. (Ed.), *Humic Substances in Terrestrial Ecosystems*, Elsevier, Amsterdam, pp 293–328.

Nannipieri, P., Giagnoni, L. and Renella, G. (2019). Metaproteomics of soil microbial communities. In: Van Elsas, J. D., Trevors, J., Soares Rosado, A. and Nannipieri, P. (Eds), *Modern Soil Microbiology* (3rd edn.), CRC Press, Boca Raton, FL, pp 257–268.

Nannipieri, P. (2020). Soil is still an unknown biological system, *Appl Sci* 10(11), 3717, doi:10.3390/app10113717.

Nannipieri, P., Ascher-Jenull, J., Ceccherini, M. T., Pietramellara, G., Renella, G. and Schloter, M. (2020). Beyond microbial diversity for predicting soil functions: a mini review, *Pedosphere* 30(1), 5–17.

Ogunseitan, O. A. (2006). Soil proteomics: extraction and analysis of proteins from soils. In: Nannipieri, P. and Smalla, K. (Eds), *Nucleic Acids and Proteins in Soils*, Springer, Berlin, pp 95–115.

Raynaud, X. and Nunan, N. (2014). Spatial ecology of bacteria at the microscale in soil, *PLoS ONE* 9(1), e87217.

Renella, G., Giagnoni, L., Arenella, M. and Nannipieri, P. (2014). Soil proteomics. In: Nannipieri, P., Pietramellara, G. and Renella, G. (Eds), *Omics in Soil Science*, Custer Academic Press, Norfolk, pp 95–125.

Renella, G., Ogunseitan, O., Giagnoni, L. and Arenella, M. (2015). Environmental proteomics: a long march in the pedosphere, *Soil Biol Biochem* 69, 34–37.

Schulze, W. X., Gleixner, G., Kaiser, K., Guggenberger, G., Mann, M. and Schulze, E. D. (2005). A proteomic fingerprint of dissolved organic carbon and soil particles, *Oecologia* 42, 335–343.

Se, J., Fu, Y., Xie, Y., Xu, F., Shen, C. and Nannipieri, P. (2021). Proteomic changes of viable but nonculturable (VBNC) *Escherichia coli* 0157:H7 induced by low moisture in an artificial soil, *Biol Fertil Soils* 57(2), 219–234.

Starke, R., Kermer, R., Ullmann-Zeunert, L., Baldwin, I. T., Seifert, J., Bastida, F., von Bergen, M. and Jemlich, N. (2016). Bacteria dominate the short-term assimilation of plant-derived N in soil, *Soil Biol Biochem* 96, 30–38.

Starke, R., Jehmlich, N. and Bastida, F. (2019). Using proteins to study how microbes contribute to soil ecosystem services: the current state and future perspectives of soil metaproteomics, *J Proteomics* 198, 50–58.

Starke, R., Siles, J. A., Lima Parente Fernandes, M. L. P., Shallert, K., Benndorf, D., Plaza, C., Jehmelich, N., Delgado-Baquerizo, M. and Bastida, F. (2021). The structure and function of soil archaea across biomes, *J Proteomics* 237, 104147.

Tartaglia, M., Bastida, F., Sciarillo, R. and Guarino, C. (2020). Soil metaproteomics for the study of the relationships between microorganisms and plants: a review of extraction protocols and ecological insights. *J Molec Sci* 21, 8455.

Taylor, E. B. and Williams, M. A. (2010). Microbial protein in soil: influence of extraction method and C amendment on extraction and recovery, *Microb Ecol* 59(2), 390–399.

Thorn, C. E., Bergesch, C., Joyce, A., Sambrano, G., McDonnell, K., Brennan, F., Heyer, R. and Benndorf, D. (2019). A robust, cost-effective method for DNA, RNA and protein co-extraction from soil, other complex microbiome and pure culture, *Mol Ecol Resour* 19, 435–455.

Vidal, A., Watteau, F., Remusat, L., Mueller, C. W., NguyenTu, T. T., Buegger, F., Derenne, S. and Quenea, K. (2019). Earthworm cast formation and development: a shift from plant litter to mineral associated organic matter, *Front Environ Sci* 7:55

Wei, X., Hu, Y., Razavi, B. S., Zhou, J., Shen, J., Nannipieri, P., Wu, J. and Ge, T. (2019). Rare taxa of alkaline phosphomonoesterase-harboring microorganisms mediate soil phosphorus mineralization, *Soil Biol Biochem* 131, 62–70.

Part 2

Bacteria and fungi in soil

Chapter 3

Emerging roles for soil Bacteroidetes in complex carbon and organic phosphorus cycling

Ian D. E. A. Lidbury, Lucy Rogers, Sophie R. M. Groenhof and Andrew Hitchcock, University of Sheffield, UK; and Lauren S. McKee, KTH Royal Institute of Technology, Sweden

1 Introduction

The phylum Bacteroidetes (new nomenclature, Bacteroidota) represents a numerically abundant, yet poorly understood group of Gram-negative bacteria found in soil. The most common belong to the families Sphingobacteraceae, Chitinophagaceae, Flavobacteriaceae and Hymenobacteraceae, with *Flavobacterium* and *Chitinophaga* being the most well-studied genera (Larsbrink and McKee, 2020). Whilst Bacteroidetes are non-flagellated, they can rapidly move across surfaces via 'gliding motility', which allows them to swarm (McBride and Kempf, 1996). In addition, many representatives of this phylum are intrinsically resistant to the antibiotic kanamycin. When grown on a solid medium, *Flavobacterium* and *Chryseobacterium* (family Flavobacteriaceae) form distinct orange-yellow colonies, resulting from the production of flexirubin (Reichenbach et al., 1980; Venil et al., 2014), whilst *Pedobacter* spp. are typically red-pink in colour (Dahal and Kim, 2016).

http://dx.doi.org/10.19103/AS.2024.0136.11

Until recently, research on this phylum has largely focused on the ecology and functional activity of gut (class Bacteroidia) and marine (family Flavobacteraceae) relatives, with soil-dwelling groups receiving less attention (Thomas et al., 2011; Larsbrink and McKee, 2020). This focus on gut and marine representatives has been driven by the role these microbes play in breaking down plant and algal polysaccharides, respectively (Larsbrink et al., 2014a), the potential use of plant cell wall-degrading enzymes in industry and, in the case of the former, the influence on gut health. Perhaps the historic lack of attention on soil representatives results from traditional cultivation approaches frequently selecting against the isolation of Bacteroidetes, with faster-growing soil bacteria from other phyla growing faster on solid media.

However, since the advent of next-generation DNA sequencing, Bacteroidetes have been emerging as abundant and potentially ecologically important members in numerous soil and plant microbiomes (Bulgarelli et al., 2013; Bulgarelli et al., 2012; Lundberg et al., 2012; Bulgarelli et al., 2015; Hilton et al., 2021). In parallel, improvements in technical methodology have enabled their isolation from soils (Nishioka et al., 2016; Lidbury et al., 2021; Bai et al., 2015; Kolton et al., 2012), increasing the number of representatives in culture collections and facilitating genetic and physiological investigations into the mechanisms driving various plant–microbe and microbe–microbe interactions. In this review, we will highlight the latest advances in Bacteroidetes research, with a focus on their unique approaches to phosphorus (P) and carbon (C) metabolism. We also hypothesise how these metabolic strategies link with the emerging importance of these bacteria as plant disease-suppressing agents and discuss their potential for increasing P acquisition in plants.

2 Bacteroidetes: the key polysaccharide degraders in soil?

Polysaccharide degradation requires the possession of specialised gene sets encoding carbohydrate-active enzymes (CAZymes) (Larsbrink and McKee, 2020; McKee et al., 2021). CAZymes initiate and complete degradation and are categorised into broad functional groups, for example glycoside hydrolases (GHs), polysaccharide lyases and carbohydrate esterases (CEs). CAZymes are incredibly diverse, reflecting the enormous variety of naturally occurring carbohydrate structures, particularly polysaccharides. The Bacteroidetes possess an apparently unique adaptation, whereby many of their GHs and other CAZymes are encoded in discrete gene clusters that also contain genes for outer and inner membrane transport systems, as well as glycan-sensing transcriptional response regulators. These gene sets are commonly referred to as polysaccharide utilisation loci (PULs), and their bioinformatic prediction

(Terrapon et al., 2018; Chen et al., 2020) has rapidly outpaced experimental validation of their precise function (McKee et al., 2021).

PULs are not universally employed by all genera, but they are sufficiently widespread throughout the phylum that they can be considered a hallmark of the Bacteroidetes. Through the efficient capture of polysaccharides, PULs provide a competitive advantage for Bacteroidetes in polysaccharide-rich environments, such as the human gut or leaf litter (Thomas et al., 2011). However, other key soil taxa, such as members of the Actinobacteria, Proteobacteria and Firmicutes, harbour varying numbers of CAZymes in their genomes and are capable of degrading various polysaccharides even without the PUL architecture (Attia et al., 2018; Larsbrink et al., 2014b; Gilbert et al., 2008; Yeager et al., 2017; Ochiai et al., 2007).

One clear advantage of possessing PUL is the coordinated production of dedicated CAZymes in response to the detection of low levels of a specific oligosaccharide via efficient starch utilisation system components C and D (SusCD)-like transport complexes (Reintjes et al., 2019; Cuskin et al., 2015; Larsbrink and McKee, 2020). Hence, polysaccharide utilisation in species containing PUL tends to be more efficient than in bacteria that possess the required enzymes but where the cognate genes are not organised in PUL, and which lack high-affinity transport systems. Indeed, members of the Actinobacteria and Firmicutes often contain enzymes that degrade (specific) polysaccharides, but this does not translate into efficient growth. To date, the relative contribution of Bacteroidetes versus non-Bacteroidetes spp. to plant and microbial polysaccharide degradation and utilisation in soils is limited. This is particularly true for complex heteroglycans like hemicelluloses. This knowledge gap relative to their gut-dwelling relatives (Cuskin et al., 2015; McKee et al., 2021) arises from a lack of comprehensive functional validation of their genetic potential, which requires a combination of genomics, transcriptomics, proteomics, biochemistry and bacterial genetics.

3 Ecology in the terrestrial habitats

Bacteroidetes are typically one the most abundant phyla in most terrestrial ecosystems, ranging from deserts to tropical rainforests (Fierer et al., 2007; Fierer et al., 2012). Bacteroidetes are copiotrophic, responding positively to the deposition of labile organic C derived from plant roots and associated microbes, in so-called 'hotspots' of microbial activity (Fierer et al., 2007; Bulgarelli et al., 2013; Kuzyakov and Blagodatskaya, 2015). In addition, Bacteroidetes are typically associated with the decomposition of leaf litter and fungal mycelial necromass (Lladó et al., 2017; Brabcová et al., 2016). Interestingly, distinct genera show contrasting distribution patterns within these various soil environments. In forest soils, where sites of leaf litter deposition or deadwood are common, Chitinophagaceae and Sphingobacteraceae are often enriched (Lladó et al., 2017), whilst in many agricultural cropping systems *Flavobacterium* spp. are highly abundant in the plant microbiome (primarily the rhizosphere and root endosphere) relative to the surrounding bulk soil (Bulgarelli et al., 2015; Bulgarelli et al., 2012; Niu et al., 2017; Tian and Gao, 2014; de Campos et al., 2013; Hilton et al., 2021).

Flavobacterium spp. frequently constitute 5–65% of the microbial community in some crops, including barley, maize, oilseed rape, cucumber and tomato (Kolton et al., 2016; Bulgarelli et al., 2015; Bulgarelli et al., 2013; Hilton et al., 2021; Niu et al., 2017; Tian and Gao, 2014). Likewise, Bulgarelli et al. reported that Flavobacteraceae was the most abundant family in the barley rhizosphere and root endosphere (Bulgarelli et al., 2015). Across agricultural fields in the UK, *Flavobacterium* operational taxonomic units (a proxy for a single species based on 16S rRNA gene sequencing) showed significant enrichment in the rhizosphere and root endosphere of oilseed rape and were associated with 'healthy' or 'good' plant microbiomes, a reduced pathogen load, and higher crop yields (Hilton et al., 2021). Very recently, under agricultural settings, *Flavobacterium* spp. and other Bacteroidetes were shown to be among the most metabolically active members of oilseed rape rhizosphere microbial communities, further confirming specialisation to this niche (Lidbury et al., 2022a).

Remarkably, despite the explosion in plant microbiome research, the mechanisms underpinning the recruitment of *Flavobacterium* spp. and other Bacteroidetes to the rhizosphere and root endosphere remain unknown. Early work suggested that Bacteroidetes were increasingly enriched in the plant microbiome over time, with plants transitioning from secreting organic acids and sugars in earlier stages of growth to nitrogen-rich metabolites, such as amino acids, as well as phenolics in later stages (Chaparro et al., 2014; Chaparro et al., 2013). However, the number of statistically significant correlations between *Arabidopsis* root exudates and Bacteroidetes is significantly lower (24 total) than those correlated to other abundant plant microbiome members, such as Acidobacteria (231), Actinobacteria (149) and Cyanobacteria (373). Manipulation experiments using plant exudate chemicals also failed to identify any compounds that either positively or negatively affect Bacteroidetes abundance (Badri et al., 2013). That study, like many others (Cotton et al., 2019; Hu et al., 2018; Neal et al., 2012; Baran et al., 2015), focussed on low-molecular-weight (LMW) compounds such as recruitment agents and neglected the role of high-molecular-weight (HMW) biopolymers, such as plant-derived complex polysaccharides, in microbiome recruitment and assemblage.

4 Carbon and energy acquisition is focused on complex polymers

Unlike their marine and gut counterparts, knowledge of the role of soil Bacteroidetes in C cycling is relatively limited. In marine systems, Bacteroidetes, especially members of the family Flavobacteriaceae, are the major degraders of algal polysaccharides, and their activity is stimulated during phytoplankton blooms (Kappelmann et al., 2019; Teeling et al., 2012; Zhu et al., 2017; Reintjes et al., 2019; Reintjes et al., 2017; Becker et al., 2020; Mystkowska et al., 2018). Likewise, in gut microbiomes, members of the genus *Bacteroides* are major degraders of dietary polysaccharides, such as heteromannans and xyloglucan, and many other polysaccharides (Cuskin et al., 2015; Larsbrink et al., 2014a; Bolam and van den Berg, 2018). *Bacteroides* spp. also possess the enzymatic potential to degrade dietary plant-derived N-glycans associated with glycoproteins found in the gut (Crouch et al., 2022).

Emerging data reveal that terrestrial Bacteroidetes also specialise in polysaccharide degradation (Larsbrink and McKee, 2020). Pioneering work demonstrated that the genome of *Flavobacterium johnsoniae* harbours numerous genes encoding CAZymes and outer membrane SusCD-like transport systems predicted to degrade complex carbohydrates, which coincided with the ability to grow on various polysaccharides, primarily hemicelluloses and pectins (McBride et al., 2009). In addition, the model soil bacterium *Chitinophaga pinensis* DSM2588 can utilise various fungal- and plant-derived polysaccharides associated with forest soils (McKee et al., 2019; Larsbrink et al., 2017; McKee and Brumer, 2015). In these studies, comparative proteomics and enzymology linked various genes and gene clusters with the metabolism of complex polysaccharides from microbial and plant cell walls. However, the lack of a robust genetic system in this *Chitinophaga pinensis* prevents determining the *in vivo* contribution of these predicted gene products towards polysaccharide breakdown, and so studies have so far focussed on the biochemical characterisation of CAZymes *in vitro*. On the other hand, using the genetically tractable *F. johnsoniae*, the genetic loci required for the degradation of the fungal polymer chitin have been characterised and subsequent analysis of recombinant proteins confirmed the predicted activities, showcasing a remarkable degree of synergistic enzyme activity on a highly recalcitrant crystalline substrate (Larsbrink et al., 2016; Kharade and McBride, 2014). Furthermore, related forest soil-associated Bacteroidetes, such as Sphingobacteraceae spp., produce enzymes required for the breakdown and subsequent utilisation of complex plant polysaccharides (López-Mondéjar et al., 2016).

Whilst these studies primarily focus on Bacteroidetes polysaccharide utilisation in the context of leaf litter decomposition and forest soils (Larsbrink and McKee, 2020), the role this metabolism plays in Bacteroidetes recruitment in the rhizosphere and root endosphere, predominantly in agricultural systems, has received little attention. This is despite the fact that *Flavobacterium* spp. are highly enriched in the plant microbiome, often two to three orders of magnitude greater than in the surrounding bulk soils (Tian and Gao, 2014; Kolton et al., 2016; Bulgarelli et al., 2015). Whilst *Flavobacterium* spp. are not enriched on LMW plant metabolites, comparative genomic analyses revealed that terrestrial *Flavobacterium* spp. harbour significantly more genes encoding GHs predicted to degrade plant polysaccharides compared to their aquatic counterparts (Kolton et al., 2013). Echoing this observation, culture-independent stable-isotope tracer experiments further implement *Flavobacterium* spp. as major *in-situ* utilisers of complex plant-derived organic C in agricultural fields (Kramer et al., 2016). An elegant laboratory study has recently demonstrated that colony expansion of *Flavobacterium* spp. increases in the presence of the plant-associated polysaccharide pectin, suggesting that complex plant-derived C may act as both a nutritional source and an environmental cue (Kraut-Cohen et al., 2021).

Reanalysis of extracellular proteomes sampled from several *Flavobacterium* spp. grown on glucose and contrasting phosphate regimes (Lidbury et al., 2021) revealed that these bacteria constitutively produce the extracellular and periplasmic GHs and SusCD-like outer membrane transport systems required for metabolising plant- and fungal-derived polysaccharides, as well as heavily investing in the synthesis of extracellular proteases (Fig. 1a). This includes production of numerous extracellular lipoproteins that initiate the degradation of polysaccharides, such as ChiA, required for hydrolysis of the *N*-acetylglucosamine-based polysaccharide chitin (Kharade and McBride, 2014) (Fig. 1b). In addition, numerous SusCD-like transport systems and GHs are constitutively synthesised, presumably to rapidly capture and consume the complex plant-derived C commonly associated with the rhizosphere and root endosphere. Furthermore, the deployment of CEs by gut and soil bacteria can also aid in the degradation of recalcitrant plant-based feruloylated and acetylated xylans (Pereira et al., 2021; Kmezik et al., 2020) and may even un-couple lignin from carbohydrates in complex cell walls (Zong et al., 2022; Arnling Bååth et al., 2019; Mazurkewich et al., 2019). Together, these data further confirm a lifestyle focused on the utilisation of complex polymers associated with the plant microbiome, whether they be of plant or microbial origin.

Inspecting the genetic neighbourhoods of a few expressed CAZyme-encoding genes from these field-isolated *Flavobacterium* spp. (Lidbury et al., 2021) reveals that they are colocalised in distinct PULs that presumably enable growth on a range of complex carbohydrates, particularly hemicelluloses (Fig. 2). Several of these PULs encode GH5_2, GH5_4, GH10 and GH26 homologs, suggesting they may target xyloglucan, heteromannans and heteroxylans

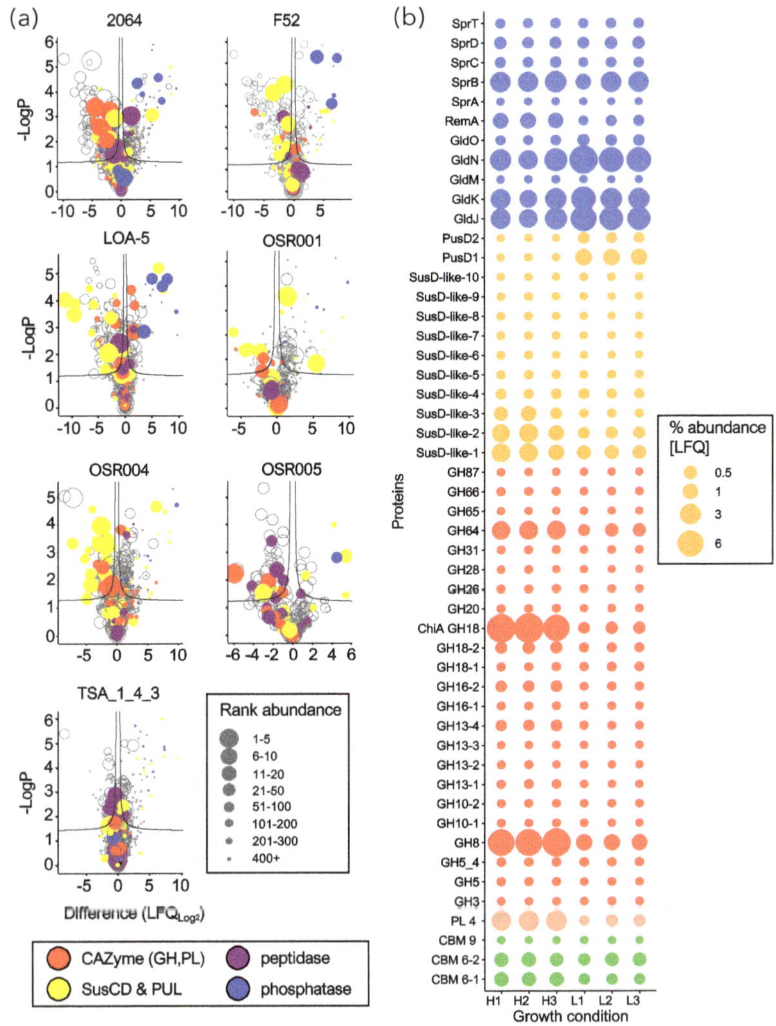

Figure 1 Production of extracellular proteins targeting complex plant and fungal polysaccharides in *Flavobacterium* spp. (a) The exoproteomes (data generated in Lidbury *et al.* (2021)) of seven soil- and plant-associated *Flavobacterium* spp. (2064, F52, LOA-5, OSR001, OSR004, OSR005 and TSA_1_4_3) grown on glucose. The proteome for cells grown under phosphate-deplete (50 ζM) conditions was compared against the proteome of cells grown under phosphate-replete (1.2 mM) conditions. The difference in protein abundance (LFQ$_{Log2}$) between deplete and replete is plotted with a positive number denoting an increase in the former. Note the investment of resources into the production of hydrolytic enzymes and transporters targeting polysaccharides (red and yellow) and protein (purple). (b) The relative abundance of various extracellular CAZymes detected in the exoproteome of *Flavobacterium johnsoniae*. Glycoside hydrolase (GH, red), pectate lyase (PL, salmon) and carbohydrate-binding module protein (CBM, green) families are numbered. Gliding motility proteins (blue) and outer membrane SusCD-like transporters (orange). H1–3 = high phosphate replicates 1–3; L1–3 = low phosphate replicates 1–3.

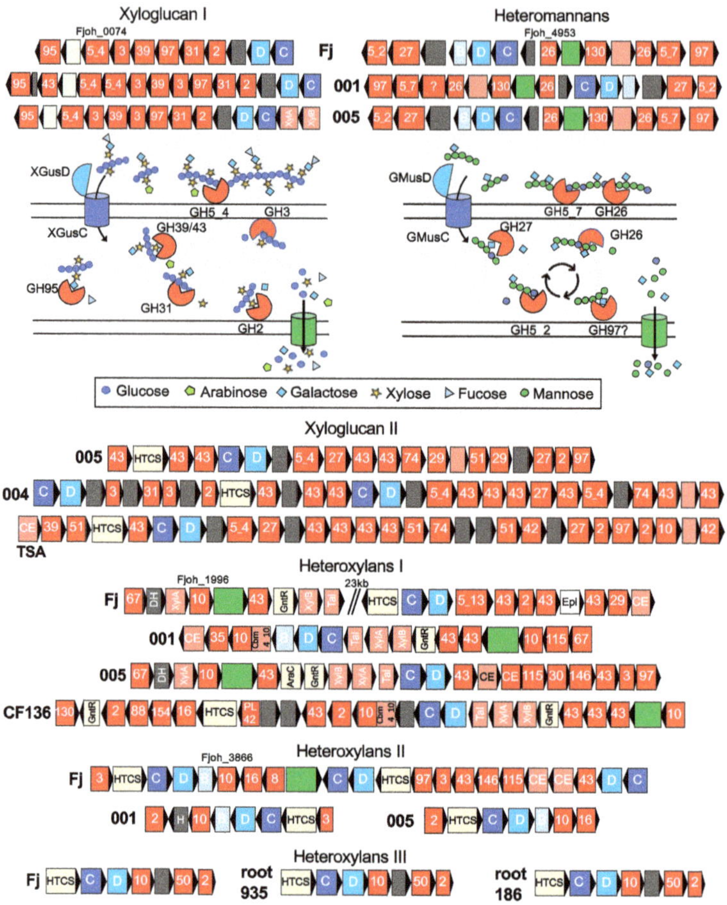

Figure 2 Predicted polysaccharide utilisation loci targeting plant hemicelluloses encoded in the genomes of plant-associated *Flavobacterium* spp. In the hypothetical model of PUL functioning, sub-cellular localisation was determined by the presence of signal peptidase (SP) I or SPII cleavage sites at the N-terminal end of the peptide sequence in combination with data from and based on Larsbrink et al. (2014a). In the schema depicting genetic loci, red = gene encoding glycosyl hydrolases (GH) or pectate lyases (PL), with numbers corresponding to specific families. Pale red = gene encoding cytoplasmic carbohydrate-transforming enzymes. Blue = gene encoding the C, D and B components of the outer membrane transport systems associated with polysaccharide uptake. Green = gene encoding general inner membrane monosaccharide transporters. Pale yellow = gene encoding regulatory proteins. For *F. johnsoniae* PUL, the locus tags for the genes encoding select enzymes are given. These are the extracellular initiator enzymes for each PUL: GH5_5 (endoglucanase), GH26 (endomannanase) and GH10 (endoxylanase). Other genes in the genetic schema are labelled only with their GH family annotation. Fj, *Flavobacterium johnsoniae* DSM2064; 001/5, *Flavobacterium* spp. OSR001/5; TSA, *Flavobacterium* sp. TSA_1_4_3; root 935/186, *Flavobacterium* sp. Root935/186; CF136, *Flavobacterium* sp. CF136; HTCS, hybrid two-component system family; CE, carbohydrate esterase; CBM, carbohydrate-binding modules; AraC/GntR, genetic regulators; DH, dehydrogenase; Tal, transaldolase, Epi, sugar epimerase, XylA, xylose isomerase; XylB, xylulokinase.

(Larsbrink et al., 2014a; Zhang et al., 2014a; Pereira et al., 2021; Larsbrink et al., 2017; Zhang et al., 2014b). Until recently, xyloglucan was typically considered an energy store and component of plant cell walls (Scheller and Ulvskov, 2010), but new studies, driven by technical advances in quantifying these polysaccharides, have revealed this hemicellulose is also a major component of root exudates (Ropitaux et al., 2020; Galloway et al., 2018). Importantly, exudation of xyloglucan appears a common trait among several plant lineages, including crop species, whilst the relative release of heteroxylans and heteromannans varies between plant species. Consequently, plant pathogenic *Xanthomonas* spp. have recently been shown to utilise xyloglucan through a discrete genetic locus (Vieira et al., 2021), which bears similarities to those previously identified in *Cellvibrio japonicus* and gut-dwelling *Bacteroides* spp. (Attia et al., 2018; Larsbrink et al., 2014b). Taken together, soluble xyloglucan and other hemicelluloses may be important sources of nutrients for plant-associated *Flavobacterium* spp. when invading the rhizosphere, providing a distinct metabolic niche and facilitating coexistence with other rhizobacteria that specialise in the acquisition of LMW labile C, such as *Pseudomonas*, *Sinorhizobium*, *Ensifer* and *Burkerholderiales* (Lidbury et al., 2022a; Mauchline et al., 2006; Cotton et al., 2019; Neal et al., 2012).

5 Bacteroidetes possess unique organophosphorus utilisation machinery

As we move towards more sustainable approaches to agriculture, interest in the use of alternative renewable organic fertilisers for replacing non-renewable forms produced from mined rock phosphate is growing (Haygarth et al., 2018; Cordell and White, 2011; Cordell et al., 2009). Thus, identifying hitherto undiscovered and atypical microbial mechanisms for organophosphorus mineralisation presents an exciting opportunity to combat the growing global P crisis (Cordell et al., 2009; Lidbury et al., 2022b; Park et al., 2022).

Environmental bacteria are proficient at scavenging inorganic phosphate from a variety of complexed organic and inorganic sources. Typically, these mechanisms are tightly regulated by the two-component regulatory system, PhoBR, which consists of a transmembrane histidine kinase (PhoR) and a response regulator (PhoB) (Santos-Beneit, 2015). During phosphate-limiting concentrations (typically below 4 ζM), PhoB induces transcription and subsequent production of enzymes and transport systems to mineralise both internal and external organophosphorus complexes. Phosphomonoesters, such as glucose-6-phosphate and inositol-hexaphosphate (phytate), and phosphodiesters, such as DNA and phospholipids, are the predominant forms found in soils, with phosphotriesters and phosphonates also present (Haygarth et al., 2018; Stutter et al., 2012; Turner et al., 2002). Most environmental bacteria synthesise promiscuous extracellular phosphomonoesterases (PhoA, PhoX and PhoD), whilst others specialise in the production of substrate-specific phosphodiesterases and various phosphonatases as well as phosphotriesterases (Sebastian and Ammerman, 2009; Srivastava et al., 2021; Lidbury et al., 2022b; Lidbury et al., 2022a; Lidbury et al., 2021; Murphy et al., 2021; Despotović et al., 2022; Lidbury et al., 2017a; Lidbury et al., 2017b; Lidbury et al., 2016).

In experimental and agricultural cropping systems, Bacteroidetes are associated with phosphatase activity and organophosphorus mineralisation, as well as soil organophosphorus standing stock concentrations (Cuartero et al., 2022; Zhou et al., 2022; Johansen and Binnerup, 2002). A significant advance in the ecological role of soil Bacteroidetes was recently made by investigating the P mobilisation potential of plant-associated strains (Lidbury et al., 2021). In this study, Bacteroidetes isolated from various crop rhizospheres, predominantly oilseed, displayed a unique phenotype whereby individual strains constitutively displayed phosphatase activity, even in the presence of inorganic phosphate. In addition, a comparison with other non-Bacteroidetes rhizobacteria, including *Pseudomonas* spp., demonstrated that plant-associated *Flavobacterium* spp. exhibited superior activity in both the presence and absence of inorganic phosphate at pH concentrations typically found in the rhizosphere. Using the genetically tractable model bacterium *F. johnsoniae* DSM2064, a unique phosphate-insensitive phosphatase PafA was shown to be responsible for this ecologically important phenotype (Lidbury et al., 2022b). PafA continuously remineralises various phosphomonesters, including phosphorylated carbohydrates and phospholipid headgroups, liberating exogenous inorganic phosphate as a by-product (Lidbury et al., 2022b). This mechanism enables *Flavobacterium* spp. to grow on organophosphorus compounds as a sole C and energy source when exogenous levels of inorganic phosphate render classical phosphatases inactive (Lidbury et al., 2021; Lidbury et al., 2016; Monds et al., 2006; Sebastian and Ammerman, 2011).

The phosphomonoesterase PafA was originally identified in 2001 in the opportunistic human pathogen *Elizabethkingia meningoseptica* (Berlutti et al., 2001) (formerly *Chryseobacterium meningoseptica*, class Flavobacteriia), but its occurrence in environmental bacteria had gone unnoticed, perhaps because it is more similar to phosphodiesterases (pfam01663) (Sunden et al., 2016). However, through genomic, metagenomic and metatranscriptomic analyses, PafA was shown to be present in the vast majority of environmental Bacteroidetes, including both marine and soil/plant-associated representatives (Lidbury et al., 2022b). Heterologous production of PafA homologs in a *Pseudomonas putida*-null phosphatase mutant revealed that PafA from *F. johnsoniae* DSM2064 (PafA^{Fj}) is particularly active despite conservation of the key catalytic residues, suggesting that active site architecture and physiochemical properties plus other parts of the protein structure contribute to enhanced catalytic activity (Sunden et al., 2016; Lidbury et al., 2022b). Recombinant production of PafA[Fj] and two PafA variants from the soil bacterium *Chitinophaga pinensis* and subsequent biochemical characterisation have confirmed that the *Flavobacterium* enzyme is more active than its *Chitinophaga* counterparts (Lidbury *et al.*, unpublished data), further suggesting that differences in overall protein folding may dictate enzyme activity. This idea aligns with a recent study showing that mutation of non-catalytic residues in PafA, within or beyond the active site, can alter the enzyme activity (Markin et al., 2021). Whilst this study introduces an excellent tool for rapidly screening mutant libraries, this method does not consider the evolutionary forces refining individual PafA forms. Indeed, several amino acid substitutions adjacent to both structural and catalytic residues can be identified in PafA homologs from different species. Further molecular and structural approaches, as well as a high-throughput screen of environmental PafA variants, would generate precise structure–function relationships that could be subsequently applied to protein engineering.

In addition to the unique PafA phosphatase, transcriptomic and proteomic assessment of *Flavobacterium* spp. responding to phosphate-limiting growth conditions revealed an enhanced repertoire of organophosphorus mineralisation through the possession of both characterised and uncharacterised low-phosphate-inducible proteins (Lidbury et al., 2021). This list included several unique transport systems related to the SusCD-like outer membrane transporters that are typically associated with polysaccharide acquisition (Bolam and van den Berg, 2018; Foley et al., 2016; Martens et al., 2009). Importantly, the genes encoding these components of the newly recognised phosphate utilisation system (PUS), termed the PusCD complex, were expressed during phosphate limitation and cluster with genes encoding phosphodiestereases, phosphomonoesterases, phosphonatases and other hydrolytic genes (Lidbury et al., 2021). Whilst the function of these gene clusters remains unknown, they are enriched among plant-associated Bacteroidetes, including *Flavobacterium* spp., strongly suggesting that they are niche-adapted mechanisms associated with life in the plant microbiome (Lidbury et al., 2021). One example is the PUS1 system identified in all *Flavobacterium* spp., which encodes a putative phosphatidylinositol-specific phospholipase C phosphodiesterase (Fig. 3) capable of degrading this plant-derived structural and regulatory phospholipid (Noack and Jaillais, 2020; Chen et al., 2009; Moroz et al., 2017). This genomic organisation of P-related genes indicates that P metabolism in the phylum may be as well organised as observed for polysaccharide utilisation and may offer similar fitness benefits in terms of colocalising associated proteins and restricting the energetically expensive expression of genes involved in nutrient acquisition. Echoing the CAZyme-rich genomes of certain species in the phylum, *Flavobacterium* spp. also possess high redundancy for inducible phosphomonoesterases, such as PhoX and PhoA (Lidbury et al., 2021).

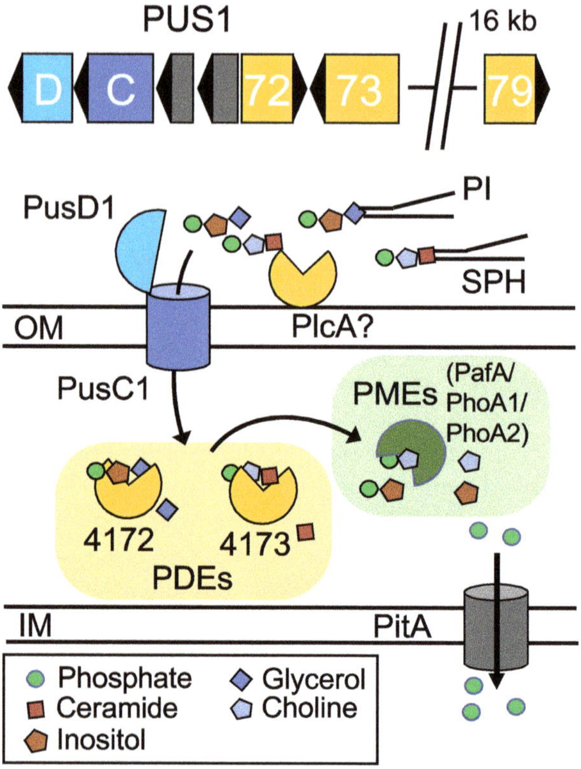

Figure 3 The *Flavobacterium johnsoniae* phosphate utilisation system (PUS)1 predicted to import phospholipids. Blue and yellow boxes represent the genes encoding the outer membrane transporter C and D components or the phosphodiesterases (PDEs), encoded by open reading frames Fjoh_4172 and Fjoh_4173, respectively. The components of phospholipid headgroups are indicated. PitA, low-affinity inner membrane phosphate transporter; PafA, phosphate-insensitive phosphomonesterase (PME); PhoA1/2, two distinct forms of the inducible PMEs from the PhoA family (pfam00249); PI, phosphatidylinositol; SPH, sphingolipids; PlcA, phospholipase family A; OM, outer membrane; IM, inner membrane.

In summary, plant-associated *Flavobacterium* spp. possess niche-adapted and largely unique mechanisms for organophosphorus breakdown, which may render these bacteria key players in soil/rhizosphere P cycling.

6 Emerging role in plant disease suppression

In addition to augmenting plant nutrition, beneficial soil microbes play an integral role in suppressing plant diseases through either direct or indirect mechanisms. Direct mechanisms typically involve antagonism towards a pathogen to prevent infection, whilst indirect approaches can involve 'priming' the plant's immune system and eliciting a systemic response, which reduces pathogen invasion. *Pseudomonas* spp. are the best-studied group of pathogen-suppressing bacteria, and it is now clear that plants can produce several compounds to attract these bacteria during the onset of pathogen infection (Cotton et al., 2019; Cameron et al., 2013; Neal et al., 2012; Mauchline et al., 2015). Furthermore, positive associations of *Pseudomonas* and *Flavobacterium* spp. with disease suppression of bacterial wilt disease caused by *Ralstonia* (Liu et al., 2022) suggest that these bacteria may form mutually beneficial associations and act synergistically to suppress pathogens.

In the past decade, soil Bacteroidetes have emerged as key agents for reducing plant disease in numerous cropping systems (Yin et al., 2013; Carrión et al., 2019; Kwak et al., 2018; Nishioka et al., 2019). Importantly, this suppression is effective on both bacterial and fungal pathogens. For example, in sugar beet seedlings, endophytic Bacteroidetes (Chitinophagaceae and Flavobacteraceae) suppress the pathogen *Rhizoctonia solani* (Carrión et al., 2019), whilst in tomatoes the bacterial pathogen *Ralstonia solanacearum* is suppressed by a rhizosphere-dwelling unclassified member of Flavobacteriaceae (Kwak et al., 2018). Interestingly, in the latter study, several non-specific *Flavobacterium* isolates had no suppressive effect, demonstrating the strain-specificity of bacterial-plant host partnerships that help drive disease suppression. Another study combining cultivation-dependent and cultivation-independent approaches to identify key agents of disease suppression revealed that intercropping cucumber with *Allium* (Welsh onion/onion) reduced the disease severity of the fungal pathogen *Fusarium* by increasing the abundance of suppressive *Flavobacterium* (Nishioka et al., 2019).

The mechanisms behind these antagonistic properties are still being uncovered. However, whilst Bacteroidetes, predominantly Flavobacteriaceae, can be enriched in 'good' or 'disease-suppressive' soil rhizospheres during pathogen invasion, direct antagonism of the pathogen by these beneficial bacteria is not always observed (Yin et al., 2021). Early studies identified that plant-associated isolates related to *Flavobacterium* and *Chryseobacterium* produce 2,4-di-tert-butylphenol and hydrogen cyanide, which can inhibit the growth of the bell pepper oomycete pathogen *Phytophthora capsici* (Marques et al., 2010; Sang and Kim, 2012; Kim et al., 2012). Bacteroidetes account for a large proportion of the biosynthetic gene cluster (BGC) global diversity in soils (Borsetto et al., 2019), which is a good proxy for antimicrobial secondary metabolite production. These BGCs are enriched in sugar beet roots, activated in the presence of pathogen, and play a direct role in the suppression of the fungal pathogen *Rhizoctonia solani* (Carrión et al., 2019). Interestingly, a large BGC is expressed in *F. johnsoniae* during phosphate limitation (Lidbury et al.,

2021), which may imply an intricate link between nutrient status and the disease suppression potential of the plant microbiome.

Another emerging mechanism for direct antagonism by Bacteroidetes is the production of extracellular hydrolytic enzymes that target pathogen cell wall components, such as chitin (Larsbrink et al., 2016), ζ-1,3- and ζ-1,6-linked glucans, ζ-mannans and even cellulose (ζ-1,4-linked glucan, found in oomycete cell walls) (McKee et al., 2019; Mélida et al., 2013). A distinct genomic locus (termed the *Fj*ChiUL, for chitin utilisation locus) induced by chitin and enabling the full breakdown of crystalline chitin has been identified and functionally characterised in *F. johnsoniae* (Larsbrink et al., 2016). This may be an illustration of pathogen-induced pathogen suppression (PIPS) mediated by PULs that are more typically thought of as being primarily involved in nutrient acquisition. Interestingly, the key enzyme in the *Fj*ChiUL responsible for the initial cleavage of chitin in *F. johnsoniae* belongs to the family GH18, and genes encoding similar proteins in Bacteroidetes are activated in sugar beet roots in the presence of the fungal pathogen *Rhizoctonia solani* (Carrión et al., 2019), although it is unknown whether these apparent PIPS phenomena are always mediated by glycan-sensing PUL.

Likewise, the genome of *Chitinophaga pinensis* encodes multiple PULs that are predicted to be induced by, and permit deconstruction of, polysaccharides found in fungal and oomycete cell walls (McKee et al., 2019). The potential for *Chitinophaga pinensis* to be involved in pathogen suppression was bolstered by the discovery and characterisation of an unusual class of lanthionine-containing antibiotics (lantibiotics) produced by this species and termed the 'pinensins', which show striking inhibition of yeasts, filamentous fungi and certain oomycetes (Mohr et al., 2015).

Given that soil Bacteroidetes possess enormous potential for degrading plant polysaccharides (Kolton et al., 2013; Larsbrink et al., 2017; López-Mondéjar et al., 2016), their enrichment in diseased plant root sites of pathogen attack may also be driven by the release of plant cell wall material. Once enriched, one mechanism for suppression may simply be by outcompeting the pathogen for these complex substrates. Indeed, the bacterial root pathogen *Xanthomonas* can utilise the root exudate and cell wall component xyloglucan (Vieira et al., 2021), although, similar to another xyloglucan-degrading proteobacterium *Cellvibrio japonicus* (Larsbrink et al., 2014b; Attia et al., 2018), it possesses an orphan SusC-like/TonB-dependent membrane transporter, lacking the associated binding protein component (SusD like). Thus, it is possible that Bacteroidetes could reduce pathogen attack simply by competitively acquiring xyloglucan and other plant cell wall hemicelluloses in a 'selfish' manner through efficient SusCD-like complexes (Bolam and van den Berg, 2018; Reintjes et al., 2019; Cuskin et al., 2015).

7 Exploring the ecological context of Bacteroidetes carbon and phosphorus cycling

Soils are a large reservoir of C, containing three times more organic C than all land animals and plants combined. Soil C stocks are dependent on the balance of biological production (autotrophic CO_2 fixation and heterotrophic anabolism) versus respiration. Each year, soil respiration releases 10–15 times more C than that emitted from anthropogenic activities (Gougoulias et al., 2014). Therefore, any shift in the balance of production versus respiration in response to global change will have significant ramifications for the global C cycle (Terrer et al., 2021; Naylor et al., 2020).

Central to soil C cycling are the resident soil microbial communities (the soil microbiome) that contribute towards biogeochemical cycling and soil fertility (Naylor et al., 2020). The soil microbiome directly controls the balance of C respired versus C converted into recalcitrant C and the storage of stable soil organic matter (SOM), a process referred to as the microbial C pump (MCP, Fig. 4) (Naylor et al., 2020; Gougoulias et al., 2014; Fierer, 2017). Soil microbes also indirectly affect terrestrial C cycling through the recycling (remineralisation) of essential nutrients, such as nitrogen and P, required to support primary

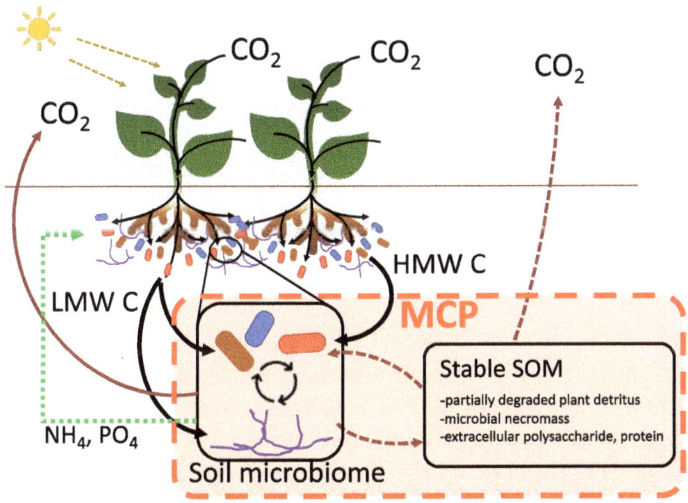

Figure 4 Flow of photosynthetically fixed C through the soil microbiome. The majority of plant-derived organic C is transformed into microbial biomass by the microbial carbon pump (MCP) and ends up either as recalcitrant necromass or as extracellular polymeric substances. These polymers contribute, along with unmetabolised plant polymers, towards stable soil organic matter (SOM). THE MCP controls the amount of SOM remineralised as CO_2 versus that stored as stable SOM. Coloured cells = functional guilds; green arrow = mineralisation; dark red arrow = microbial process. LMW, low molecular weight, HMW, high molecular weight.

production. Importantly, low soil P availability can attenuate the positive response of plant production to elevated atmospheric CO_2 concentrations (Jiang et al., 2020; Terrer et al., 2019; Cunha et al., 2022), highlighting the importance of microbial P cycling in global biogeochemistry and its interaction with C cycling.

Plants provide almost all the 'new' C required to support microbial growth, generating 'hotspots' of activity in regions of C deposition, which are typified by elevated C and P cycling enzyme activity (Lidbury et al., 2022a; Kuzyakov and Blagodatskaya, 2015). Microbial processing of plant-derived C, therefore, represents the entry point for new matter and energy into the MCP (Fig. 4). Plant-derived C is partitioned into two major fractions, LMW C, which can be transformed by microbial enzyme activity within hours, and complex HMW C, including polysaccharides, which can take years to be fully degraded into its monomeric constituents (McKee et al., 2021; Larsbrink and McKee, 2020). HMW plant C can escape microbial attack, initiating the formation of soil aggregates (Larsbrink and McKee, 2020; Galloway et al., 2018) and thus directly contributing to SOM formation. Changes in the belowground deposition of C (rhizodeposition) are predicted to alter the structure and function (enzyme production) of the soil microbiome, with largely unknown consequences for soil fertility and C storage (Naylor et al., 2020; Fierer, 2017).

As discussed in this review, whilst it is known that Bacteroidetes respond to elevated plant C deposition, the underpinning mechanisms are not well understood, although they are likely driven by HMW C deposition. Therefore, two scenarios are possible under future climate conditions. The first is that plants exude a greater proportion of LMW C in response to elevated CO_2, which may select against Bacteroidetes, a phenomenon that has been postulated to be the cause of their decline in several crop microbiomes during the process of domestication (Pérez-Jaramillo et al., 2018). The second possibility is that plant exudation of HMW C also increases under future climate conditions, potentially stimulating Bacteroidetes growth and activity, which in turn could stimulate organophosphorus remineralisation (Lidbury et al., 2022b; Lidbury et al., 2021). In reality, a combination of these two scenarios is most likely, but an important consideration is the type of LMW C exuded by plants. Whilst Bacteroidetes do not respond to organic acid inputs, they do show a positive response to sucrose additions (Fierer et al., 2007) and grow well on glucose (Lidbury et al., 2022b; Lidbury et al., 2021), which are categorised as LMW C sources.

As discussed, Bacteroidetes polysaccharide utilisation is typically PUL associated, and hence more efficient, reducing metabolic cross-feeding (Reintjes et al., 2019; Cuskin et al., 2015; Reintjes et al., 2017). Therefore, any changing biotic and abiotic conditions influencing Bacteroidetes-dominated polysaccharide breakdown relative to other soil bacterial groups may have profound effects on soil C cycling driven through changes in metabolic cross-feeding. Furthermore, through the action of the MCP, most plant polysaccharides ultimately end up as microbial necromass (Fig. 4). Whilst the genes and corresponding enzymes responsible for microbial polymer degradation are poorly characterised, early evidence suggests a role for Bacteroidetes (McKee et al., 2019) and Firmicutes (Schönbichler et al., 2020), and the susceptibility of this soil C pool to microbial attack can be influenced by abiotic conditions (Buckeridge et al., 2020). Hence, global change, be that direct land-use intensity or elevated temperatures and atmospheric CO_2, will likely have important ramifications for soil C cycling and storage (Naylor et al., 2020), with changes in the activity of soil Bacteroidetes likely playing a key role.

8 Conclusion and future trends

An emerging role for soil Bacteroidetes in C and P cycling suggests a distinct metabolic niche for these enigmatic bacteria in soil–plant microbiomes, with implications for nutrient-driven species recruitment as well as inducible contributions to pathogen control and consequences for soil C cycling. Numerous laboratory studies have now firmly revealed Bacteroidetes possess unique enzymes to target complex metabolites associated with plant biomass and exudates. Future research should now focus on determining the contribution of these unique genomic traits towards the success of these microorganisms in various soil and plant niches. Given that *Flavobacterium* and other soil Bacteroidetes hold tremendous potential for recycling organophosphorus and regenerating bioavailable P, developing a predictive framework for correlations between the *in-situ* abundance and activity of Bacteroidetes and organophosphorus remineralisation is now required. Furthermore, determining the fundamental genomic and metabolic drivers that underpin the success of Bacteroidetes in the soil will enable the development of technologies to manipulate ecosystem behaviours such as pathogen suppression, with a better understanding of the mechanisms by which this is induced, being a key target for future research. These advances will build toward the goal of engineering microbiomes through plant breeding, soil management and/or inoculant development to enhance sustainable crop production and buffer against climate change.

9 Where to look for further information

For up-to-date information on the latest news towards sustainability goals surrounding efficient use of phosphorus fertilisers, please visit the European Sustainable Phosphorus Platform (EPSS, https://www.phosphorusplatform.eu/). The EPSS have also launched a nitrogen-recovery working group. The

Irish Nutrient Sustainability Platform (INSP, https://nutrientsustainability.ie/) also provides up-to-date information on reducing our reliance on fertilisers.

In addition, these two seminal papers on the global phosphorus crisis provide excellent introductions to the problems and potential solutions surrounding chemical fertilisers:

- Cordell, D., Drangert, J.-O. and White, S. (2009). The story of phosphorus: Global food security and food for thought, *Global Environmental Change* 19(2), 292–305.
- Cordell, D. and White, S. (2014) Life's bottleneck: Sustaining the world's phosphorus for a food secure future, *Annual Review of Environment and Resources* 39(1), 161-188.

We recommend reading this article 'The Rhizosphere – Roots, Soil and Everything In Between' published in *Nature Education Knowledge* (https://www.nature.com/scitable/knowledge/library/the-rhizosphere-roots-soil-and-67500617/) for background knowledge on the role of the rhizosphere in plant nutrition and health.

Here is a list of top research groups that are pushing forward work on soil Bacteroidetes and sustainability in agriculture and bioenergy:

- The Cytryn Lab (Pi, Eddie Cytryn), Agricultural Research Organization, Volcani Center, Israel (https://cytrynlab.wixsite.com/cytrynlab).
- The McKee Lab (Pi, Lauren McKee), KTH Royal Institute of Technology, Sweden (https://www.kth.se/profile/mckee).
- The Larsbrink Lab (Pi, Johan Larsbrink), Chalmers University of Technology, Sweden (https://research.chalmers.se/en/person/jolarsb).
- The Lidbury Lab (Pi, Ian Lidbury), The University of Sheffield, United Kingdom (https://sites.google.com/sheffield.ac.uk/lidbury-ecophysiology-lab/home).
- The Institute for Sustainable Food, The University of Sheffield, United Kingdom (https://www.sheffield.ac.uk/sustainable-food).
- The Carrion Lab (PI, Victor Carrion), Leiden University, Netherlands, (https://carrionlab.com/).

10 References

Arnling Bååth, J., Mazurkewich, S., Poulsen, J.-C. N., Olsson, L., Lo Leggio, L. and Larsbrink, J. (2019). Structure–function analyses reveal that a glucuronoyl esterase from *Teredinibacter turnerae* interacts with carbohydrates and aromatic compounds, *Journal of Biological Chemistry* 294(16), 6635–6644.

Attia, M. A., Nelson, C. E., Offen, W. A., Jain, N., Davies, G. J., Gardner, J. G. and Brumer, H. (2018). In vitro and in vivo characterization of three *Cellvibrio japonicus* glycoside hydrolase family 5 members reveals potent xyloglucan backbone-cleaving functions, *Biotechnology for Biofuels* 11(1), 45.

Badri, D. V., Chaparro, J. M., Zhang, R., Shen, Q. and Vivanco, J. M. (2013). Application of natural blends of phytochemicals derived from the root exudates of *Arabidopsis* to the soil reveal that phenolic-related compounds predominantly modulate the soil microbiome, *The Journal of Biological Chemistry* 288(7), 4502–4512.

Bai, Y., Müller, D. B., Srinivas, G., Garrido-Oter, R., Potthoff, E., Rott, M., Dombrowski, N., Münch, P. C., Spaepen, S., Remus-Emsermann, M., Hüttel, B., McHardy, A. C., Vorholt, J. A. and Schulze-Lefert, P. (2015). Functional overlap of the *Arabidopsis* leaf and root microbiota, *Nature* 528(7582), 364–369.

Baran, R., Brodie, E. L., Mayberry-Lewis, J., Hummel, E., Da Rocha, U. N., Chakraborty, R., Bowen, B. P., Karaoz, U., Cadillo-Quiroz, H., Garcia-Pichel, F. and Northen, T. R. (2015). Exometabolite niche partitioning among sympatric soil bacteria, *Nature Communications* 6, 8289.

Becker, S., Tebben, J., Coffinet, S., Wiltshire, K., Iversen, M. H., Harder, T., Hinrichs, K. U. and Hehemann, J. H. (2020). Laminarin is a major molecule in the marine carbon cycle, *Proceedings of the National Academy of Sciences of the United States of America* 117(12), 6599–6607.

Berlutti, F., Passariello, C., Selan, L., Thaller, M. C. and Rossolini, G. M. (2001). The *Chryseobacterium meningosepticum* PafA enzyme: Prototype of a new enzyme family of prokaryotic phosphate-irrepressible alkaline phosphatases? The GenBank accession number for pafA reported in this paper is AF157621, *Microbiology* 147(10), 2831–2839.

Bolam, D. N. and van den Berg, B. (2018). TonB-dependent transport by the gut microbiota: Novel aspects of an old problem, *Current Opinion in Structural Biology* 51, 35–43.

Borsetto, C., Amos, G. C. A., da Rocha, U. N., Mitchell, A. L., Finn, R. D., Laidi, R. F., Vallin, C., Pearce, D. A., Newsham, K. K. and Wellington, E. M. H. (2019). Microbial community drivers of PK/NRP gene diversity in selected global soils, *Microbiome* 7(1), 78.

Brabcová, V., Nováková, M., Davidová, A. and Baldrian, P. (2016). Dead fungal mycelium in forest soil represents a decomposition hotspot and a habitat for a specific microbial community, *New Phytologist* 210(4), 1369–1381.

Buckeridge, K. M., Mason, K. E., McNamara, N. P., Ostle, N., Puissant, J., Goodall, T., Griffiths, R. I., Stott, A. W. and Whitaker, J. (2020). Environmental and microbial controls on microbial necromass recycling, an important precursor for soil carbon stabilization, *Communications Earth and Environment* 1(1), 36.

Bulgarelli, D., Garrido-Oter, R., Münch, P. C., Weiman, A., Dröge, J., Pan, Y., McHardy, A. C. and Schulze-Lefert, P. (2015). Structure and function of the bacterial root microbiota in wild and domesticated barley, *Cell Host and Microbe* 17(3), 392–403.

Bulgarelli, D., Rott, M., Schlaeppi, K., Ver Loren van Themaat, E., Ahmadinejad, N., Assenza, F., Rauf, P., Huettel, B., Reinhardt, R., Schmelzer, E., Peplies, J., Gloeckner, F. O., Amann, R., Eickhorst, T. and Schulze-Lefert, P. (2012). Revealing structure and assembly cues for *Arabidopsis* root-inhabiting bacterial microbiota, *Nature* 488(7409), 91–95.

Bulgarelli, D., Schlaeppi, K., Spaepen, S., van Themaat, E. V. L. and Schulze-Lefert, P. (2013). Structure and functions of the bacterial microbiota of plants, *Annual Review of Plant Biology* 64(1), 807–838.

Cameron, D. D., Neal, A. L., van Wees, S. C. M. and Ton, J. (2013). Mycorrhiza-induced resistance: More than the sum of its parts?, *Trends in Plant Science* 18(10), 539–545.

Carrión, V. J., Perez-Jaramillo, J., Cordovez, V., Tracanna, V., de Hollander, M., Ruiz-Buck, D., Mendes, L. W., van Ijcken, W. F. J., Gomez-Exposito, R., Elsayed, S. S., Mohanraju, P., Arifah, A., van der Oost, J., Paulson, J. N., Mendes, R., van Wezel, G. P., Medema, M. H. and Raaijmakers, J. M. (2019). Pathogen-induced activation of disease-suppressive functions in the endophytic root microbiome, *Science* 366(6465), 606–612.

Chaparro, J. M., Badri, D. V., Bakker, M. G., Sugiyama, A., Manter, D. K. and Vivanco, J. M. (2013). Root exudation of phytochemicals in Arabidopsis follows specific patterns that are developmentally programmed and correlate with soil microbial functions, *PLoS ONE* 8(2), e55731.

Chaparro, J. M., Badri, D. V. and Vivanco, J. M. (2014). Rhizosphere microbiome assemblage is affected by plant development, *ISME Journal* 8(4), 790–803.

Chen, I.-M. A., Chu, K., Palaniappan, K., Ratner, A., Huang, J., Huntemann, M., Hajek, P., Ritter, S., Varghese, N., Seshadri, R., Roux, S., Woyke, T., Eloe-Fadrosh, E. A., Ivanova, N. N. and Kyrpides, N. C. (2020). The IMG/M data management and analysis system v.6.0: New tools and advanced capabilities, *Nucleic Acids Research* 49(D1), D751–D763.

Chen, W., Goldfine, H., Ananthanarayanan, B., Cho, W. and Roberts, M. F. (2009). *Listeria monocytogenes* phosphatidylinositol-specific phospholipase C: Kinetic activation and homing in on different interfaces, *Biochemistry* 48(16), 3578–3592.

Cordell, D., Drangert, J.-O. and White, S. (2009). The story of phosphorus: Global food security and food for thought, *Global Environmental Change* 19(2), 292–305.

Cordell, D. and White, S. (2011). Peak phosphorus: Clarifying the key issues of a vigorous debate about long-term phosphorus security, *Sustainability* 3(10), 2027–2049.

Cotton, T. E. A., Pétriacq, P., Cameron, D. D., Meselmani, M. A., Schwarzenbacher, R., Rolfe, S. A. and Ton, J. (2019). Metabolic regulation of the maize rhizobiome by benzoxazinoids, *The ISME Journal* 13(7), 1647–1658.

Crouch, L. I., Urbanowicz, P. A., Baslé, A., Cai, Z. P., Liu, L., Voglmeir, J., Melo Diaz, J. M., Benedict, S. T., Spencer, D. I. R. and Bolam, D. N. (2022). Plant *N*-glycan breakdown by human gut *Bacteroides*, *Proceedings of the National Academy of Sciences of the United States of America* 119(39), e2208168119.

Cuartero, J., Pascual, J. A., Vivo, J.-M., Özbolat, O., Sánchez-Navarro, V., Egea-Cortines, M., Zornoza, R., Mena, M. M., Garcia, E. and Ros, M. (2022). A first-year melon/cowpea intercropping system improves soil nutrients and changes the soil microbial community, *Agriculture, Ecosystems and Environment* 328, 107856.

Cunha, H. F. V., Andersen, K. M., Lugli, L. F., Santana, F. D., Aleixo, I. F., Moraes, A. M., Garcia, S., Di Ponzio, R., Mendoza, E. O., Brum, B., Rosa, J. S., Cordeiro, A. L., Portela, B. T. T., Ribeiro, G., Coelho, S. D., de Souza, S. T., Silva, L. S., Antonieto, F., Pires, M., Salomão, A. C., Miron, A. C., de Assis, R. L., Domingues, T. F., Aragão, L. E. O. C., Meir, P., Camargo, J. L., Manzi, A. O., Nagy, L., Mercado, L. M., Hartley, I. P. and Quesada, C. A. (2022). Direct evidence for phosphorus limitation on Amazon forest productivity, *Nature* 608(7923), 558–562.

Cuskin, F., Lowe, E. C., Temple, M. J., Zhu, Y., Cameron, E. A., Pudlo, N. A., Porter, N. T., Urs, K., Thompson, A. J., Cartmell, A., Rogowski, A., Hamilton, B. S., Chen, R., Tolbert, T. J., Piens, K., Bracke, D., Vervecken, W., Hakki, Z., Speciale, G., Munōz-Munōz, J. L., Day, A., Peña, M. J., McLean, R., Suits, M. D., Boraston, A. B., Atherly, T., Ziemer, C. J., Williams, S. J., Davies, G. J., Abbott, D. W., Martens, E. C. and Gilbert, H. J. (2015).

Human gut Bacteroidetes can utilize yeast mannan through a selfish mechanism, *Nature* 517(7533), 165–169.

Dahal, R. H. and Kim, J. (2016). *Pedobacter humicola* sp. nov., a member of the genus *Pedobacter* isolated from soil, *International Journal of Systematic and Evolutionary Microbiology* 66(6), 2205–2211.

de Campos, S. B., Youn, J. W., Farina, R., Jaenicke, S., Jünemann, S., Szczepanowski, R., Beneduzi, A., Vargas, L. K., Goesmann, A., Wendisch, V. F. and Passaglia, L. M. P. (2013). Changes in root bacterial communities associated to two different development stages of canola (*Brassica napus* L. var *oleifera*) evaluated through next-generation sequencing technology, *Microbial Ecology* 65(3), 593–601.

Despotović, D., Aharon, E., Trofimyuk, O., Dubovetskyi, A., Cherukuri, K. P., Ashani, Y., Eliason, O., Sperfeld, M., Leader, H., Castelli, A., Fumagalli, L., Savidor, A., Levin, Y., Longo, L. M., Segev, E. and Tawfik, D. S. (2022). Utilization of diverse organophosphorus pollutants by marine bacteria, *Proceedings of the National Academy of Sciences of the United States of America* 119(32), e2203604119.

Fierer, N. (2017). Embracing the unknown: Disentangling the complexities of the soil microbiome, *Nature Reviews. Microbiology* 15(10), 579–590.

Fierer, N., Bradford, M. A. and Jackson, R. B. (2007). Toward an ecological classification of soil bacteria, *Ecology* 88(6), 1354–1364.

Fierer, N., Leff, J. W., Adams, B. J., Nielsen, U. N., Bates, S. T., Lauber, C. L., Owens, S., Gilbert, J. A., Wall, D. H. and Caporaso, J. G. (2012). Cross-biome metagenomic analyses of soil microbial communities and their functional attributes, *Proceedings of the National Academy of Sciences of the United States of America* 109(52), 21390–21395.

Foley, M. H., Cockburn, D. W. and Koropatkin, N. M. (2016). The Sus operon: A model system for starch uptake by the human gut Bacteroidetes, *Cellular and Molecular Life Sciences* 73(14), 2603–2617.

Galloway, A. F., Pedersen, M. J., Merry, B., Marcus, S. E., Blacker, J., Benning, L. G., Field, K. J. and Knox, J. P. (2018). Xyloglucan is released by plants and promotes soil particle aggregation, *New Phytologist* 217(3), 1128–1136.

Gilbert, H. J., Stålbrand, H. and Brumer, H. (2008). How the walls come crumbling down: Recent structural biochemistry of plant polysaccharide degradation, *Current Opinion in Plant Biology* 11(3), 338–348.

Gougoulias, C., Clark, J. M. and Shaw, L. J. (2014). The role of soil microbes in the global carbon cycle: Tracking the below-ground microbial processing of plant-derived carbon for manipulating carbon dynamics in agricultural systems, *Journal of the Science of Food and Agriculture* 94(12), 2362–2371.

Haygarth, P. M., Harrison, A. F. and Turner, B. L. (2018). On the history and future of soil organic phosphorus research: A critique across three generations, *European Journal of Soil Science* 69(1), 86–94.

Hilton, S., Picot, E., Schreiter, S., Bass, D., Norman, K., Oliver, A. E., Moore, J. D., Mauchline, T. H., Mills, P. R., Teakle, G. R., Clark, I. M., Hirsch, P. R., van der Gast, C. J. and Bending, G. D. (2021). Identification of microbial signatures linked to oilseed rape yield decline at the landscape scale, *Microbiome* 9(1), 19.

Hu, L., Robert, C. A. M., Cadot, S., Zhang, X., Ye, M., Li, B., Manzo, D., Chervet, N., Steinger, T., van der Heijden, M. G. A., Schlaeppi, K. and Erb, M. (2018). Root exudate metabolites drive plant-soil feedbacks on growth and defense by shaping the rhizosphere microbiota, *Nature Communications* 9(1), 2738.

Jiang, M., Caldararu, S., Zhang, H., Fleischer, K., Crous, K. Y., Yang, J., De Kauwe, M. G., Ellsworth, D. S., Reich, P. B., Tissue, D. T., Zaehle, S. and Medlyn, B. E. (2020). Low phosphorus supply constrains plant responses to elevated CO_2: A meta-analysis, *Global Change Biology* 26(10), 5856–5873.

Johansen, J. E. and Binnerup, S. J. (2002). Contribution of *Cytophaga*-like bacteria to the potential of turnover of carbon, nitrogen, and phosphorus by bacteria in the rhizosphere of barley (*Hordeum vulgare* L.), *Microbial Ecology* 43(3), 298–306.

Kappelmann, L., Krüger, K., Hehemann, J. H., Harder, J., Markert, S., Unfried, F., Becher, D., Shapiro, N., Schweder, T., Amann, R. I. and Teeling, H. (2019). Polysaccharide utilization loci of North Sea *Flavobacteriia* as basis for using SusC/D-protein expression for predicting major phytoplankton glycans, *The ISME Journal* 13(1), 76–91.

Kharade, S. S. and McBride, M. J. (2014). *Flavobacterium johnsoniae* chitinase ChiA is required for chitin utilization and is secreted by the type ix secretion system, *Journal of Bacteriology* 196(5), 961–970.

Kim, H.-S., Sang, M. K., Jung, H. W., Jeun, Y.-C., Myung, I.-S. and Kim, K. D. (2012). Identification and characterization of *Chryseobacterium wanjuense* strain KJ9C8 as a biocontrol agent of *Phytophthora* blight of pepper, *Crop Protection* 32, 129–137.

Kmezik, C., Bonzom, C., Olsson, L., Mazurkewich, S. and Larsbrink, J. (2020). Multimodular fused acetyl–feruloyl esterases from soil and gut Bacteroidetes improve xylanase depolymerization of recalcitrant biomass, *Biotechnology for Biofuels* 13(1), 60.

Kolton, M., Erlacher, A., Berg, G. and Cytryn, E. (2016). The *Flavobacterium* genus in the plant holobiont: Ecological, physiological, and applicative insights. In: Castro-Sowinski, S. (Ed.), *Microbial Models: From Environmental to Industrial Sustainability*. Singapore: Springer, pp. 189–207.

Kolton, M., Green, S. J., Harel, Y. M., Sela, N., Elad, Y. and Cytryn, E. (2012). Draft genome sequence of *Flavobacterium* strain F52, isolated from the rhizosphere of bell pepper *Capsicum annuum* L. cv, *Journal of Bacteriology* 194(19), 5462–5463.

Kolton, M., Sela, N., Elad, Y. and Cytryn, E. (2013). Comparative genomic analysis indicates that niche adaptation of terrestrial Flavobacteria is strongly linked to plant glycan metabolism, *PLoS ONE* 8(9), e76704.

Kramer, S., Dibbern, D., Moll, J., Huenninghaus, M., Koller, R., Krueger, D., Marhan, S., Urich, T., Wubet, T., Bonkowski, M., Buscot, F., Lueders, T. and Kandeler, E. (2016). Resource partitioning between bacteria, fungi, and protists in the detritusphere of an agricultural soil, *Frontiers in Microbiology* 7(1524), 1524.

Kraut-Cohen, J., Shapiro, O. H., Dror, B. and Cytryn, E. (2021). Pectin induced colony expansion of soil-derived *Flavobacterium* strains, *Frontiers in Microbiology* 12, 651891.

Kuzyakov, Y. and Blagodatskaya, E. (2015). Microbial hotspots and hot moments in soil: Concept & review, *Soil Biology and Biochemistry* 83, 184–199.

Kwak, M. J., Kong, H. G., Choi, K., Kwon, S. K., Song, J. Y., Lee, J., Lee, P. A., Choi, S. Y., Seo, M., Lee, H. J., Jung, E. J., Park, H., Roy, N., Kim, H., Lee, M. M., Rubin, E. M., Lee, S. W. and Kim, J. F. (2018). Rhizosphere microbiome structure alters to enable wilt resistance in tomato, *Nature Biotechnology* 36, 1100.

Larsbrink, J. and McKee, L. S. (2020). Bacteroidetes bacteria in the soil: Glycan acquisition, enzyme secretion, and gliding motility. In: Gadd, G. M. and Sariaslani, S. (Eds.), *Advances in Applied Microbiology*. Academic Press 110, 63–98.

Larsbrink, J., Rogers, T. E., Hemsworth, G. R., McKee, L. S., Tauzin, A. S., Spadiut, O., Klinter, S., Pudlo, N. A., Urs, K., Koropatkin, N. M., Creagh, A. L., Haynes, C. A., Kelly, A. G.,

Cederholm, S. N., Davies, G. J., Martens, E. C. and Brumer, H. (2014a). A discrete genetic locus confers xyloglucan metabolism in select human gut Bacteroidetes, *Nature* 506(7489), 498–502.

Larsbrink, J., Thompson, A. J., Lundqvist, M., Gardner, J. G., Davies, G. J. and Brumer, H. (2014b). A complex gene locus enables xyloglucan utilization in the model saprophyte *Cellvibrio japonicus*, *Molecular Microbiology* 94(2), 418–433.

Larsbrink, J., Tuveng, T. R., Pope, P. B., Bulone, V., Eijsink, V. G. H., Brumer, H. and McKee, L. S. (2017). Proteomic insights into mannan degradation and protein secretion by the forest floor bacterium *Chitinophaga pinensis*, *Journal of Proteomics* 156, 63–74.

Larsbrink, J., Zhu, Y., Kharade, S. S., Kwiatkowski, K. J., Eijsink, V. G. H., Koropatkin, N. M., McBride, M. J. and Pope, P. B. (2016). A polysaccharide utilization locus from *Flavobacterium johnsoniae* enables conversion of recalcitrant chitin, *Biotechnology for Biofuels* 9(1), 260.

Lidbury, I. D. E. A., Borsetto, C., Murphy, A. R. J., Bottrill, A., Jones, A. M. E., Bending, G. D., Hammond, J. P., Chen, Y., Wellington, E. M. H. and Scanlan, D. J. (2021). Niche-adaptation in plant-associated Bacteroidetes favours specialisation in organic phosphorus mineralisation, *The ISME Journal* 15(4), 1040–1055.

Lidbury, I. D. E. A., Fraser, T., Murphy, A. R. J., Scanlan, D. J., Bending, G. D., Jones, A. M. E., Moore, J. D., Goodall, A., Tibbett, M., Hammond, J. P. and Wellington, E. M. H. (2017a). The 'known' genetic potential for microbial communities to degrade organic phosphorus is reduced in low-pH soils, *MicrobiologyOpen* 6(4), e00474.

Lidbury, I. D. E. A., Murphy, A. R. J., Fraser, T. D., Bending, G. D., Jones, A. M. E., Moore, J. D., Goodall, A., Tibbett, M., Hammond, J. P., Scanlan, D. J. and Wellington, E. M. H. (2017b). Identification of extracellular glycerophosphodiesterases in *Pseudomonas* and their role in soil organic phosphorus remineralisation, *Scientific Reports* 7(1), 2179–2179.

Lidbury, I. D. E. A., Murphy, A. R. J., Scanlan, D. J., Bending, G. D., Jones, A. M. E., Moore, J. D., Goodall, A., Hammond, J. P. and Wellington, E. M. H. (2016). Comparative genomic, proteomic and exoproteomic analyses of three *Pseudomonas* strains reveals novel insights into the phosphorus scavenging capabilities of soil bacteria, *Environmental Microbiology* 18(10), 3535–3549.

Lidbury, I. D. E. A., Raguideau, S., Borsetto, C., Murphy Andrew, R. J., Bottrill, A., Liu, S., Stark, R., Fraser, T., Goodall, A., Jones, A., Bending Gary, D., Tibbet, M., Hammond John, P., Quince, C., Scanlan David, J., Pandhal, J. and Wellington Elizabeth, M. H. (2022a). Stimulation of distinct rhizosphere bacteria drives phosphorus and nitrogen mineralization in oilseed rape under field conditions, *mSystems* 7(4), e00025-22.

Lidbury, I. D. E. A., Scanlan, D. J., Murphy, A. R. J., Christie-Oleza, J. A., Aguilo-Ferretjans, M. M., Hitchcock, A. and Daniell, T. J. (2022b). A widely distributed phosphate-insensitive phosphatase presents a route for rapid organophosphorus remineralization in the biosphere, *Proceedings of the National Academy of Sciences of the United States of America* 119(5), e2118122119.

Liu, X., Liu, L., Gong, J., Zhang, L., Jiang, Q., Huang, K. and Ding, W. (2022). Soil conditions on bacterial wilt disease affect bacterial and fungal assemblage in the rhizosphere, *AMB Express* 12(1), 110.

Lladó, S., López-Mondéjar, R. and Baldrian, P. (2017). Forest soil bacteria: Diversity, involvement in ecosystem processes, and response to global change, *Microbiology and Molecular Biology Reviews: MMBR* 81(2), e00063-16.

López-Mondéjar, R., Zühlke, D., Becher, D., Riedel, K. and Baldrian, P. (2016). Cellulose and hemicellulose decomposition by forest soil bacteria proceeds by the action of structurally variable enzymatic systems, *Scientific Reports* 6(1), 25279.

Lundberg, D. S., Lebeis, S. L., Paredes, S. H., Yourstone, S., Gehring, J., Malfatti, S., Tremblay, J., Engelbrektson, A., Kunin, V., Glavina del Rio, T. G., Edgar, R. C., Eickhorst, T., Ley, R. E., Hugenholtz, P., Tringe, S. G. and Dangl, J. L. (2012). Defining the core *Arabidopsis thaliana* root microbiome, *Nature* 488(7409), 86–90.

Markin, C. J., Mokhtari, D. A., Sunden, F., Appel, M. J., Akiva, E., Longwell, S. A., Sabatti, C., Herschlag, D. and Fordyce, P. M. (2021). Revealing enzyme functional architecture via high-throughput microfluidic enzyme kinetics, *Science* 373(6553), eabf8761.

Marques, A. P. G. C., Pires, C., Moreira, H., Rangel, A. O. S. S. and Castro, P. M. L. (2010). Assessment of the plant growth promotion abilities of six bacterial isolates using Zea mays as indicator plant, *Soil Biology and Biochemistry* 42(8), 1229–1235.

Martens, E. C., Koropatkin, N. M., Smith, T. J. and Gordon, J. I. (2009). Complex glycan catabolism by the human gut microbiota: The Bacteroidetes Sus-like paradigm, *The Journal of Biological Chemistry* 284(37), 24673–24677.

Mauchline, T. H., Chedom-Fotso, D., Chandra, G., Samuels, T., Greenaway, N., Backhaus, A., McMillan, V., Canning, G., Powers, S. J., Hammond-Kosack, K. E., Hirsch, P. R., Clark, I. M., Mehrabi, Z., Roworth, J., Burnell, J. and Malone, J. G. (2015). An analysis of *Pseudomonas* genomic diversity in take-all infected wheat fields reveals the lasting impact of wheat cultivars on the soil microbiota, *Environmental Microbiology* 17(11), 4764–4778.

Mauchline, T. H., Fowler, J. E., East, A. K., Sartor, A. L., Zaheer, R., Hosie, A. H. F., Poole, P. S. and Finan, T. M. (2006). Mapping the *Sinorhizobium meliloti* 1021 solute-binding protein-dependent transportome, *Proceedings of the National Academy of Sciences of the United States of America* 103(47), 17933–17938.

Mazurkewich, S., Poulsen, J.-C. N., Lo Leggio, L. and Larsbrink, J. (2019). Structural and biochemical studies of the glucuronoyl esterase OtCE15A illuminate its interaction with lignocellulosic components, *Journal of Biological Chemistry* 294(52), 19978–19987.

McBride, M. J. and Kempf, M. J. (1996). Development of techniques for the genetic manipulation of the gliding bacterium *Cytophaga johnsonae*, *Journal of Bacteriology* 178(3), 583–590.

McBride, M. J., Xie, G., Martens, E. C., Lapidus, A., Henrissat, B., Rhodes, R. G., Goltsman, E., Wang, W., Xu, J., Hunnicutt, D. W., Staroscik, A. M., Hoover, T. R., Cheng, Y. Q. and Stein, J. L. (2009). Novel features of the polysaccharide-digesting gliding bacterium *Flavobacterium johnsoniae* as revealed by genome sequence analysis, *Applied and Environmental Microbiology* 75(21), 6864–6875.

McKee, L. S. and Brumer, H. (2015). Growth of Chitinophaga pinensis on plant cell wall glycans and characterisation of a glycoside hydrolase Family 27 β-l-Arabinopyranosidase implicated in arabinogalactan utilisation, *PLoS ONE* 10(10), e0139932.

McKee, L. S., La Rosa, S. L., Westereng, B., Eijsink, V. G., Pope, P. B. and Larsbrink, J. (2021). Polysaccharide degradation by the Bacteroidetes: Mechanisms and nomenclature, *Environmental Microbiology Reports*.

McKee, L. S., Martínez-Abad, A., Ruthes, A. C., Vilaplana, F., Brumer, H. and Vieille, C. (2019). Focused metabolism of β-glucans by the soil Bacteroidetes species *Chitinophaga pinensis*, *Applied and Environmental Microbiology* 85(2), e02231-18.

Mélida, H., Sandoval-Sierra, J. V., Diéguez-Uribeondo, J. and Bulone, V. (2013). Analyses of extracellular carbohydrates in oomycetes unveil the existence of three different cell wall types, *Eukaryotic Cell* 12(2), 194–203.

Mohr, K. I., Volz, C., Jansen, R., Wray, V., Hoffmann, J., Bernecker, S., Wink, J., Gerth, K., Stadler, M. and Müller, R. (2015). Pinensins: The first antifungal lantibiotics, *Angewandte Chemie* 54(38), 11254–11258.

Monds, R. D., Newell, P. D., Schwartzman, J. A. and O'Toole, G. A. (2006). Conservation of the Pho regulon in *Pseudomonas fluorescens* Pf0-1, *Applied and Environmental Microbiology* 72(3), 1910–1924.

Moroz, O. V., Blagova, E., Lebedev, A. A., Norgaard, A., Segura, D. R., Blicher, T. H., Brask, J. and Wilson, K. S. (2017). The structure of a calcium-dependent phosphoinositide-specific phospholipase C from *Pseudomonas* sp. 62186, the first from a Gram-negative bacterium, *Acta Crystallographica. Section D, Structural Biology* 73(1), 32–44.

Murphy, A. R. J., Scanlan, D. J., Chen, Y., Adams, N. B. P., Cadman, W. A., Bottrill, A., Bending, G., Hammond, J. P., Hitchcock, A., Wellington, E. M. H. and Lidbury, I. D. E. A. (2021). Transporter characterisation reveals aminoethylphosphonate mineralisation as a key step in the marine phosphorus redox cycle, *Nature Communications* 12(1), 4554.

Mystkowska, A. A., Robb, C., Vidal-Melgosa, S., Vanni, C., Fernandez-Guerra, A., Höhne, M. and Hehemann, J. H. (2018). Molecular recognition of the beta-glucans laminarin and pustulan by a SusD-like glycan-binding protein of a marine Bacteroidetes, *The FEBS Journal* 285(23), 4465–4481.

Naylor, D., Sadler, N., Bhattacharjee, A., Graham, E. B., Anderton, C. R., McClure, R., Lipton, M., Hofmockel, K. S. and Jansson, J. K. (2020). Soil microbiomes under climate change and implications for carbon cycling, *Annual Review of Environment and Resources* 45(1), 29–59.

Neal, A. L., Ahmad, S., Gordon-Weeks, R. and Ton, J. (2012). Benzoxazinoids in root exudates of maize attract *Pseudomonas putida* to the rhizosphere, *PLoS ONE* 7(4), e35498.

Nishioka, T., Elsharkawy, M. M., Suga, H., Kageyama, K., Hyakumachi, M. and Shimizu, M. (2016). Development of Culture Medium for the isolation of *Flavobacterium* and *Chryseobacterium* from rhizosphere soil, *Microbes and Environments* 31(2), 104–110.

Nishioka, T., Marian, M., Kobayashi, I., Kobayashi, Y., Yamamoto, K., Tamaki, H., Suga, H. and Shimizu, M. (2019). Microbial basis of *Fusarium* wilt suppression by *Allium* cultivation, *Scientific Reports* 9(1), 1715.

Niu, B., Paulson, J. N., Zheng, X. and Kolter, R. (2017). Simplified and representative bacterial community of maize roots, *Proceedings of the National Academy of Sciences of the United States of America* 114(12), E2450–E2459.

Noack, L. C. and Jaillais, Y. (2020). Functions of anionic lipids in plants, *Annual Review of Plant Biology* 71(1), 71–102.

Ochiai, A., Itoh, T., Kawamata, A., Hashimoto, W. and Murata, K. (2007). Plant cell wall degradation by saprophytic Bacillus subtilis strains: Gene clusters responsible for rhamnogalacturonan depolymerization, *Applied and Environmental Microbiology* 73(12), 3803–3813.

Park, Y., Solhtalab, M., Thongsomboon, W. and Aristilde, L. (2022). Strategies of organic phosphorus recycling by soil bacteria: Acquisition, metabolism, and regulation, *Environmental Microbiology Reports* 14(1), 3–24.

Pereira, G. V., Abdel-Hamid, A. M., Dutta, S., D'Alessandro-Gabazza, C. N., Wefers, D., Farris, J. A., Bajaj, S., Wawrzak, Z., Atomi, H., Mackie, R. I., Gabazza, E. C., Shukla, D., Koropatkin, N. M. and Cann, I. (2021). Degradation of complex arabinoxylans by human colonic Bacteroidetes, *Nature Communications* 12(1), 459.

Pérez-Jaramillo, J. E., Carrión, V. J., de Hollander, M. and Raaijmakers, J. M. (2018). The wild side of plant microbiomes, *Microbiome* 6(1), 143–143.

Reichenbach, H., Kohl, W., Böttger-Vetter, A. and Achenbach, H. (1980). Flexirubin-type pigments in *Flavobacterium*, *Archives of Microbiology* 126(3), 291–293.

Reintjes, G., Arnosti, C., Fuchs, B. M. and Amann, R. (2017). An alternative polysaccharide uptake mechanism of marine bacteria, *The ISME Journal* 11(7), 1640–1650.

Reintjes, G., Arnosti, C., Fuchs, B. and Amann, R. (2019). Selfish, sharing and scavenging bacteria in the Atlantic Ocean: A biogeographical study of bacterial substrate utilisation, *The ISME Journal* 13(5), 1119–1132.

Ropitaux, M., Bernard, S., Schapman, D., Follet-Gueye, M. L., Vicré, M., Boulogne, I. and Driouich, A. (2020). Root border cells and mucilage secretions of soybean, *Glycine max* (Merr) L.: Characterization and role in interactions with the oomycete *Phytophthora* parasitica, *Cells* 9(10), 2215.

Sang, M. K. and Kim, K. D. (2012). The volatile-producing *Flavobacterium johnsoniae* strain GSE09 shows biocontrol activity against *Phytophthora capsici* in pepper, *Journal of Applied Microbiology* 113(2), 383–398.

Santos-Beneit, F. (2015). The Pho regulon: A huge regulatory network in bacteria, *Frontiers in Microbiology* 6, 402.

Scheller, H. V. and Ulvskov, P. (2010). Hemicelluloses, *Annual Review of Plant Biology* 61(1), 263–289.

Schönbichler, A., Díaz-Moreno, S. M., Srivastava, V. and McKee, L. S. (2020). Exploring the potential for fungal antagonism and cell wall attack by *Bacillus subtilis natto*, *Frontiers in Microbiology* 11, 521.

Sebastian, M. and Ammerman, J. W. (2009). The alkaline phosphatase PhoX is more widely distributed in marine bacteria than the classical PhoA, *ISME Journal* 3(5), 563–572.

Sebastian, M. and Ammerman, J. W. (2011). Role of the phosphatase PhoX in the phosphorus metabolism of the marine bacterium *Ruegeria pomeroyi* DSS-3, *Environmental Microbiology Reports* 3(5), 535–542.

Srivastava, A., Saavedra, D. E. M., Thomson, B., García, J. A. L., Zhao, Z., Patrick, W. M., Herndl, G. J. and Baltar, F. (2021). Enzyme promiscuity in natural environments: Alkaline phosphatase in the ocean, *The ISME Journal* 15(11), 3375–3383.

Stutter, M. I., Shand, C. A., George, T. S., Blackwell, M. S. A., Bol, R., MacKay, R. L., Richardson, A. E., Condron, L. M., Turner, B. L. and Haygarth, P. M. (2012). Recovering phosphorus from soil: A root solution?, *Environmental Science and Technology* 46(4), 1977–1978.

Sunden, F., AlSadhan, I., Lyubimov, A. Y., Ressl, S., Wiersma-Koch, H., Borland, J., Brown, C. L., Johnson, T. A., Singh, Z. and Herschlag, D. (2016). Mechanistic and evolutionary insights from comparative enzymology of phosphomonoesterases and phosphodiesterases across the alkaline phosphatase superfamily, *Journal of the American Chemical Society* 138(43), 14273–14287.

Teeling, H., Fuchs, B. M., Becher, D., Klockow, C., Gardebrecht, A., Bennke, C. M., Kassabgy, M., Huang, S., Mann, A. J., Waldmann, J., Weber, M., Klindworth, A., Otto, A., Lange, J., Bernhardt, J., Reinsch, C., Hecker, M., Peplies, J., Bockelmann, F. D., Callies, U., Gerdts, G., Wichels, A., Wiltshire, K. H., Glöckner, F. O., Schweder, T. and Amann,

R. (2012). Substrate-controlled succession of marine bacterioplankton populations induced by a phytoplankton bloom, *Science* 336(6081), 608–611.

Terrapon, N., Lombard, V., Drula, É., Lapébie, P., Al-Masaudi, S., Gilbert, H. J. and Henrissat, B. (2018). PULDB: The expanded database of polysaccharide utilization loci, *Nucleic Acids Research* 46(D1), D677–D683.

Terrer, C., Jackson, R. B., Prentice, I. C., Keenan, T. F., Kaiser, C., Vicca, S., Fisher, J. B., Reich, P. B., Stocker, B. D., Hungate, B. A., Peñuelas, J., McCallum, I., Soudzilovskaia, N. A., Cernusak, L. A., Talhelm, A. F., Van Sundert, K., Piao, S., Newton, P. C. D., Hovenden, M. J., Blumenthal, D. M., Liu, Y. Y., Müller, C., Winter, K., Field, C. B., Viechtbauer, W., Van Lissa, C. J., Hoosbeek, M. R., Watanabe, M., Koike, T., Leshyk, V. O., Polley, H. W. and Franklin, O. (2019). Nitrogen and phosphorus constrain the CO_2 fertilization of global plant biomass, *Nature Climate Change* 9(9), 684–689.

Terrer, C., Phillips, R. P., Hungate, B. A., Rosende, J., Pett-Ridge, J., Craig, M. E., van Groenigen, K. J., Keenan, T. F., Sulman, B. N., Stocker, B. D., Reich, P. B., Pellegrini, A. F. A., Pendall, E., Zhang, H., Evans, R. D., Carrillo, Y., Fisher, J. B., Van Sundert, K., Vicca, S. and Jackson, R. B. (2021). A trade-off between plant and soil carbon storage under elevated CO_2, *Nature* 591(7851), 599–603.

Thomas, F., Hehemann, J. H., Rebuffet, E., Czjzek, M. and Michel, G. (2011). Environmental and gut Bacteroidetes: The food connection, *Frontiers in Microbiology* 2, 93.

Tian, Y. and Gao, L. (2014). Bacterial diversity in the rhizosphere of cucumbers grown in soils covering a wide range of cucumber cropping histories and environmental conditions, *Microbial Ecology* 68(4), 794–806.

Turner, B. L., Papházy, M. J., Haygarth, P. M. and McKelvie, I. D. (2002). Inositol phosphates in the environment, *Philosophical Transactions of the Royal Society of London. Series B, Biological Sciences* 357(1420), 449–469.

Venil, C. K., Zakaria, Z. A., Usha, R. and Ahmad, W. A. (2014). Isolation and characterization of flexirubin type pigment from *Chryseobacterium* sp. UTM-3T, *Biocatalysis and Agricultural Biotechnology* 3(4), 103–107.

Vieira, P. S., Bonfim, I. M., Araujo, E. A., Melo, R. R., Lima, A. R., Fessel, M. R., Paixão, D. A. A., Persinoti, G. F., Rocco, S. A., Lima, T. B., Pirolla, R. A. S., Morais, M. A. B., Correa, J. B. L., Zanphorlin, L. M., Diogo, J. A., Lima, E. A., Grandis, A., Buckeridge, M. S., Gozzo, F. C., Benedetti, C. E., Polikarpov, I., Giuseppe, P. O. and Murakami, M. T. (2021). Xyloglucan processing machinery in *Xanthomonas* pathogens and its role in the transcriptional activation of virulence factors, *Nature Communications* 12(1), 4049.

Yeager, C. M., Gallegos-Graves, V., Dunbar, J., Hesse, C. N., Daligault, H. and Kuske, C. R. (2017). Polysaccharide degradation capability of *Actinomycetales* soil isolates from a semiarid grassland of the Colorado plateau, *Applied and Environmental Microbiology* 83(6), e03020-16.

Yin, C., Casa Vargas, J. M., Schlatter, D. C., Hagerty, C. H., Hulbert, S. H. and Paulitz, T. C. (2021). Rhizosphere community selection reveals bacteria associated with reduced root disease, *Microbiome* 9(1), 86.

Yin, C., Hulbert, S. H., Schroeder, K. L., Mavrodi, O., Mavrodi, D., Dhingra, A., Schillinger, W. F. and Paulitz, T. C. (2013). Role of bacterial communities in the natural suppression of *Rhizoctonia solani* bare patch disease of wheat (*Triticum aestivum* L.), *Applied and Environmental Microbiology* 79(23), 7428–7438.

Zhang, M., Chekan, J. R., Dodd, D., Hong, P. Y., Radlinski, L., Revindran, V., Nair, S. K., Mackie, R. I. and Cann, I. (2014a). Xylan utilization in human gut commensal bacteria is orchestrated by unique modular organization of polysaccharide-degrading enzymes, *Proceedings of the National Academy of Sciences of the United States of America* 111(35), E3708–E3717.

Zhang, X., Rogowski, A., Zhao, L., Hahn, M. G., Avci, U., Knox, J. P. and Gilbert, H. J. (2014b). Understanding how the complex molecular architecture of mannan-degrading hydrolases contributes to plant cell wall degradation, *Journal of Biological Chemistry* 289(4), 2002–2012.

Zhou, J., Zhang, L., Feng, G. and George, T. S. (2022). Arbuscular mycorrhizal fungi have a greater role than root hairs of maize for priming the rhizosphere microbial community and enhancing rhizosphere organic P mineralization, *Soil Biology and Biochemistry* 171, 108713.

Zhu, Y., Thomas, F., Larocque, R., Li, N., Duffieux, D., Cladière, L., Souchaud, F., Michel, G. and McBride, M. J. (2017). Genetic analyses unravel the crucial role of a horizontally acquired alginate lyase for brown algal biomass degradation by *Zobellia galactanivorans*, *Environmental Microbiology* 19(6), 2164–2181.

Zong, Z., Mazurkewich, S., Pereira, C. S., Fu, H., Cai, W., Shao, X., Skaf, M. S., Larsbrink, J. and Lo Leggio, L. (2022). Mechanism and biomass association of glucuronoyl esterase: An α/β hydrolase with potential in biomass conversion, *Nature Communications* 13(1), 1449.

Chapter 4

Advances in understanding Actinobacteria in soil[1]

Yuanyuan Bao, Xin Li and Ruirui Chen, Nanjing Forestry University, China; Youzhi Feng[2], Nanjing Forestry University, China and Chinese Academy of Sciences, China

1 Introduction

Actinobacteria are a Gram-positive phylum of bacteria known for their distinctive morphology, growth habits, and ability to produce bioactive compounds, including antibiotics (van Bergeijk et al., 2020). These bacteria are widely distributed in various environments, such as soil, water, and the human body, and play critical roles in a broad range of ecological processes, including plant residue degradation, soil formation, nutrient cycling, and plant growth promotion (Bhatti et al., 2017; Liu et al., 2017; Lacombe-Harvey et al., 2018).

In recent years, the detection and identification of microbial communities have undergone significant advancements (Araujo et al., 2020; Bao et al., 2021). Specifically, there have been two major changes in the methods used

1 Competing interests: The authors declare no conflicts of interest.

2 Corresponding author details: Youzhi Feng, E-mail: fyouzhi@hotmail.com; Tel: +086-025-86881367; Fax: +086-025-86881000.

http://dx.doi.org/10.19103/AS.2024.0136.13

to detect Actinobacteria (Rego et al., 2019): culture-dependent and culture-independent techniques. Traditional methods rely on culturing Actinobacteria in the laboratory, which can be challenging due to their fastidious nature. However, culture-independent techniques that utilize molecular biology and specific primers for DNA or RNA sequences have made it possible to detect Actinobacteria directly from environmental samples, without the need for culturing. These methods include polymerase chain reaction (PCR), amplicon sequencing, and metagenomic sequencing, allowing for high-throughput detection.

The improved detection methods have led to a better understanding of the diversity and ecological functions of Actinobacteria. This chapter aims to provide a comprehensive review of recent findings on soil Actinobacteria, with a focus on their classification, genetics/characteristics, what we know and don't know about the diversity and ecological functions, factors affecting distribution/abundance in soil, interactions within soil microbiomes, implications for optimizing their contribution to soil health, and where next for research.

2 Classification of Actinobacteria

Actinobacteria are a diverse group of bacteria typically shaped like rods, although some species form filaments or branching structures (Hamedi and Poorinmohammad, 2017). They are named after their filamentous growth pattern, which resembles the filaments of fungi. Actinobacteria can be found in a wide range of environments (Dworkin et al., 2006). The phylum Actinobacteria is a major taxon of Bacteria and is gram-positive, aerobic, and typically has high G+C content (65–75%) in its genome DNA (Miao and Davies, 2010). They are also known for their characteristic filamentous growth and branching, which gives them a 'mycelium-like' appearance (Garrett, 2016).

As of September 2021, the National Center for Biotechnology Information (NCBI) Taxonomy database lists over 11 000 formally described and named Actinobacteria species. The classification of Actinobacteria is based on a combination of phylogenetic and phenotypic characteristics (Goodfellow and Williams, 1983). There are six main classes of Actinobacteria that are diverse in ecological and metabolic functions (Gao and Gupta, 2007):

1 Actinobacteria: This class is the largest and most diverse, with over 200 genera. It includes many important soil-dwelling bacteria such as Streptomyces and Mycobacterium, which are known for their ability to produce antibiotics and other bioactive compounds (Barka et al., 2016).
2 Coriobacteriia: This class includes bacteria found in the human gut and involved in the metabolism of dietary compounds, such as bile acids and cholesterol (Auchtung et al., 2018).

3 Acidimicrobiia: This class includes bacteria found in acidic environments, such as peat bogs and mine drainage sites. They play important roles in carbon and nitrogen cycling in these environments (Gittel et al., 2014).
4 Rubrobacteria: This class includes bacteria found in soil and known for their resistance to extreme conditions such as high temperatures, desiccation, and radiation (Vaishampayan et al., 2014).
5 Thermoleophilia: This class includes bacteria found in high-temperature environments such as hot springs and geothermal soils. They are important for the biodegradation of organic matter in these environments (Dong et al., 2014).
6 Nitriliruptoria: This class includes bacteria able to degrade nitriles, a group of toxic organic compounds. They are important for the bioremediation of sites contaminated with nitriles (Sorokin et al., 2009).

It is important to note that the classification of Actinobacteria may change as new information and techniques continue to emerge (Mohammadipanah and Dehhaghi, 2017). For example, recent studies have suggested the existence of new subclasses or orders within Actinobacteria, and some previously classified groups have been reclassified based on new genetic data.

3 Genetics and characteristics of Actinobacteria

Actinobacteria are known for their unique and characteristic morphology, which often involves filamentous or branching structures that can form complex mycelia extending over long distances. Some species also produce spores that aid in their survival under harsh environmental conditions. Actinobacteria have a high G+C content in their genomic DNA, which typically falls in the range of 60–70%. This, combined with their cell wall structure made up of peptidoglycan, arabinogalactan, and mycolic acids, is responsible for their distinct filamentous shape (Schmidt, 2019).

As a group, Actinobacteria are mainly heterotrophic and have a diverse range of metabolic capabilities, thanks to the presence of many genetic features (Barka et al., 2016). A large number of actinobacterial genomes contain gene clusters encoding for the biosynthesis of secondary metabolites, such as antibiotics, antifungals, and anticancer agents (Barka et al., 2016). These gene clusters are often highly conserved, with similar biosynthetic pathways found in different species. This suggests that they may have been acquired through horizontal gene transfer events.

Actinobacteria are equipped with various mobile genetic elements, such as plasmids, transposons, and bacteriophages, which enable gene transfer among different bacterial species (Rückert et al., 2015). These elements play a crucial role in the dissemination of antibiotic-resistance genes, virulence factors, and other adaptive traits. Additionally, many Actinobacteria produce mycolic acids, which are long-chain fatty acids that help them withstand environmental stresses (Hamedi and Poorinmohammad, 2017). These genetic characteristics contribute to their diverse metabolic capabilities and their capacity to adapt to different environmental conditions. Therefore, comprehending the genetic mechanisms that underlie their metabolism and adaptation is critical for the development of novel antibiotics and other bioactive compounds derived from Actinobacteria, ultimately optimizing their contributions to soil ecosystem services.

4 Diversity and ecological functions of Actinobacteria

Actinobacteria are widely distributed in soil and known for their ability to degrade plant residues (Abdulla and El-Shatoury, 2007; Lewin et al., 2016; Bao et al., 2019). Plant residues are composed of cellulose, hemicellulose, and lignin, which require the synergistic action of various enzyme systems encoded by microorganisms, such as cellulases, β-glucosidases, xylanases, and lignin peroxidases, for complete degradation (Abdulla and El-Shatoury, 2007; Bao et al., 2019). Only microorganisms carrying a complete set of residue-degrading functional genes in their genome can independently complete the entire process of residue degradation (Lewin et al., 2016; Wang et al., 2016).

Based on previous research, Actinobacteria is one of the few microorganisms that possess a complete and extensive set of functional genes required for residue degradation (Medie et al., 2012; Wang et al., 2016). Our previous study found that non-dominant Actinobacteria encoded a significantly greater number of functional genes for residue degradation than other groups, such as Proteobacteria and Firmicutes (Bao et al., 2021). Surprisingly, the taxonomic and functional compositions of Actinobacteria remained relatively stable during the entire straw decomposition process at both local (within the decomposition site) and regional (across decomposition sites) scales.

The previously mentioned information provides compelling evidence of the crucial (or stable) ecological and physiological roles that Actinobacteria play in the decomposition of plant residues in soil. Actinobacteria possess a complete set of carbohydrate-active enzymes (CAZymes), which results in a relatively stable taxonomic and functional composition during decomposition, regardless of the environmental conditions, such as soil fertility and climate (Lewin et al., 2016; Bao et al., 2021). Although other phyla, such as Proteobacteria, Firmicutes, and Bacteroidetes, also possess the full set of CAZymes, our previous study revealed that their high functional variability suggests that their contribution to residue decomposition may differ from that of Actinobacteria (Bao et al., 2021). For instance, only certain subsets of taxa within these phyla may possess a subset of classes of CAZymes for plant residue decomposition, which makes them important only at specific time points or specific sites.

It is known that there is a fundamental trade-off between bacterial ecological strategies (speed of growth/populations) and ecological functions (efficiency of substrate utilization) (Pfeiffer et al., 2001; Molenaar et al., 2009; Heinrich and Schuster, 2012). This means that in high-nutrient environments, bacteria tend to grow rapidly but inefficiently concerning substrate utilization, while in low-nutrient environments, bacteria tend to grow slowly but more efficiently (Polz and Cordero, 2016; Roller et al., 2016). In other words, the trade-off between bacterial ecological strategies and ecological functions is regulated by soil nutrient status. Our previous study found that Actinobacteria played a vital role in plant residue decomposition, but their importance varied depending on soil fertility (Bao et al., 2021). This is in line with the trade-off paradigm, as soils with different nutrient statuses can select different microorganisms with contrasting life strategies and functional traits. Although the taxonomic and functional compositions of the Actinobacteria were relatively stable during the whole residue decomposition stages at both local and regional scales, our study suggests that soil fertility is the primary driver of these differences. However, we acknowledge that other environmental factors, such as heavy metals, trace elements, and climate conditions, may also interact with soil fertility and affect the microbial community (Bao et al., 2021).

To investigate the impact of soil fertility on the ecophysiological functions of microorganisms in plant residue decomposition, our recent study utilized two different soils with distinct fertility levels resulting from varying long-term fertilization practices (Bao et al., 2023). We conducted ^{13}C-labeled rice straw decomposition microcosms incubation and discovered that there was a trade-off between the ecological and physiological characteristics of straw-decomposing bacterial communities which were significantly influenced by soil fertility. However, the extent of this trade-off varied depending on the fertility status of the bulk soil (Feng et al., 2018; Bao et al., 2023).

The information presented earlier highlights the crucial ecological and physiological role of Actinobacteria in the degradation of plant residues. Nonetheless, the impact of soil fertility gradients on Actinobacteria-mediated residue degradation remains unclear, and further research is required to conclusively establish the effect of soil fertility on the ecophysiological roles of Actinobacteria in plant residue decomposition.

5 Factors affecting Actinobacterial distribution and abundance in soil

The distribution and abundance of Actinobacteria in soil are influenced by multiple physicochemical factors, including soil pH, organic matter content, moisture, temperature, and texture, as well as plant cover and land use.

1 Soil pH: Soil pH is an important factor that influences the distribution and abundance of Actinobacteria. Actinobacteria are generally more abundant in soils that are neutral to slightly alkaline, with a pH range of 6.5–8.0. However, their growth and activity may be inhibited in acidic soils with a pH below 5.5 (Yadav et al., 2018).
2 Organic matter content: Organic matter provides a source of carbon and energy for these bacteria, and it can also improve soil structure and moisture retention (Dai et al., 2018). It is known that Actinobacteria prefer to thrive in soils with high organic matter content (Barka et al., 2016).
3 Soil moisture: Moisture is a critical factor that affects the survival and growth of Actinobacteria in soil. Drought conditions can reduce the abundance of Actinobacteria in soil, while moist conditions can increase their abundance (Barnard et al., 2013). However, waterlogged or saturated soils can lead to reduced oxygen availability, which may limit actinobacterial growth and activity (Fierer and Schimel, 2002).
4 Temperature: The distribution and abundance of Actinobacteria are greatly influenced by temperature fluctuations (Liu et al., 2017). Actinobacteria are known for their adaptability to a wide range of temperatures, despite being mesophilic bacteria that typically prefer moderate temperatures ranging from 25°C to 30°C (Goodfellow and Williams, 1983). Nevertheless, some species can tolerate extreme temperatures and are found in both hot and cold environments (Yaradoddi and Kontro, 2021).
5 Soil texture: Soil texture plays an important role in the distribution and abundance of Actinobacteria. Actinobacteria are more abundant in soils with a higher proportion of clay and silt particles (Xia et al., 2020). These soil types provide a more stable environment with better water and nutrient retention, which can support their growth and activity.
6 Plant cover: Plant cover can have a significant influence on Actinobacterial growth. Plant roots release organic compounds and other substances into the soil, which can provide a source of nutrients for Actinobacteria (Yadav et al., 2018). This can lead to increased growth and activity of Actinobacteria in the soil. Additionally, plant cover can impact soil moisture, temperature, and pH (Wu et al., 2010), which are all important factors that influence Actinobacterial growth and abundance.
7 Land use: Different land uses such as agricultural or urban areas can affect the abundance and diversity of Actinobacteria in soil (Araujo et al., 2020). Agricultural practices such as tillage, fertilization, and pesticide use can alter soil properties and impact Actinobacterial populations. Urbanization can also result in soil disturbances and changes in soil characteristics that can affect Actinobacterial abundance.

These complex interplay of soil physicochemical properties and environmental conditions highlights the importance of considering multiple factors in understanding Actinobacterial ecology in soil.

6 Interactions within soil microbiomes

Actinobacteria interact with a diverse range of microorganisms within the soil microbiome, including other bacteria, fungi, protists, and viruses, and these interactions can have both positive and negative effects on their abundance and activity. Examples of these interactions include the following:

1. Interactions with other bacteria: Actinobacteria compete with other bacterial species for resources within the soil microbiome, such as nutrients and space (Swarnalakshmi et al., 2016). However, they can also form beneficial relationships with other bacterial species, such as those that produce plant growth-promoting compounds or enzymes that break down complex organic compounds (Steele et al., 2011). In our previous study on the ecophysiological roles of non-dominant Actinobacteria in plant residue decomposition, we found that the nitrogen fixation ability of Actinobacteria increases nitrogen availability for decomposers and thus foster interspecies interactions between Actinobacteria and other soil bacterial taxa with carbohydrate-active functions (Bao et al., 2021). This leads to a positive linkage between them, resulting in a joint promotion of plant residue decomposition (Raes and Bork, 2008).
2. Interactions with fungi: Actinobacteria interact with fungi in the soil microbiome, both as competitors for resources and as mutualistic partners (Kaltenpoth, 2009). Some Actinobacteria produce antibiotics that inhibit the growth of fungi. For example, in one of our recent studies on plant residue decomposition microbiome, we conducted a bacteria–fungi antagonism experiment and found that Actinobacteria significantly inhibited fungal growth (Bao et al., 2024). On the other hand, some Actinobacteria form mutualistic relationships with fungi, such as mycorrhizae, that enhance nutrient uptake by both the fungi and the plant.
3. Interactions with protists: Actinobacteria are subject to predation by protists, including amoebae and ciliates (Matz and Kjelleberg, 2005). This can lead to a decrease in the abundance of Actinobacteria in the soil. Nevertheless, it can also result in the release of nutrients and the promotion of the growth of other microorganisms.
4. Interactions with viruses: Actinobacteria are susceptible to infection by bacteriophages, which are viruses that infect and replicate within

bacterial cells (Garza et al., 2022). These infections can decrease the abundance of Actinobacteria within the soil microbiome. However, bacteriophages can also facilitate genetic diversity by transferring genetic material between bacterial species.

Actinobacteria are crucial members of the soil microbiome and play a critical role in nutrient cycling, soil health, and ecosystem function. Their interactions with other microorganisms within the soil microbiome are diverse and complex. Therefore, understanding these interactions is essential to predict how soil ecosystems will respond to environmental change and to develop strategies that promote the growth and activity of beneficial soil microbes.

7 Implications for optimizing Actinobacterial contribution to soil/crop health and other uses

Actinobacteria, as a diverse group of soil bacteria, play a vital role in maintaining soil and crop health. These bacteria are known for their unique metabolic capabilities, such as the production of secondary metabolites with antimicrobial properties, the decomposition of complex organic matter, and nitrogen fixation (Penn et al., 2009; Bao et al., 2021). These traits make Actinobacteria an essential component of soil microbial communities and underscore their potential for use in sustainable agriculture.

One way to optimize the contribution of Actinobacteria to soil and crop health is through the use of biofertilizers. Actinobacteria-based biofertilizers have been shown to improve plant growth and nutrient uptake, as well as suppress plant diseases caused by pathogenic microorganisms (Kaari et al., 2023). For example, *Streptomyces* sp. has been reported to produce plant growth-promoting substances, such as indole-3-acetic acid (IAA) and gibberellins, which enhance root development and increase the biomass of several crop plants (Ntemafack et al., 2022). In addition, some Actinobacteria, such as *Frankia*, can fix atmospheric nitrogen and provide a source of nitrogen for plants, thereby reducing the need for synthetic fertilizers (Sellstedt and Richau, 2013). These findings suggest that Actinobacteria have the potential to be valuable tools for promoting sustainable agriculture.

Another potential use of Actinobacteria is in the development of novel antibiotics and other bioactive compounds. Actinobacteria are known to produce a vast array of secondary metabolites, many of which have antimicrobial, antifungal, and anticancer properties (Penn et al., 2009). Many of the most widely used antibiotics, such as streptomycin, tetracycline, and erythromycin, were originally isolated from Actinobacteria (Zhao et al., 2018). With the growing problem of antibiotic resistance, the search for new antibiotics has become more urgent than ever. Actinobacteria are an excellent source of new antibiotics and other bioactive compounds, and efforts are currently underway to identify and characterize these compounds for use in human and animal health.

Moreover, Actinobacteria also play critical roles in the remediation of contaminated soils. Actinobacteria are effective in degrading a wide range of environmental pollutants, including polycyclic aromatic hydrocarbons (PAHs), chlorinated hydrocarbons, and heavy metals (Isaac et al., 2015). This is due to the ability of Actinobacteria to produce enzymes that can degrade these pollutants into less harmful compounds (Mishra et al., 2020). The use of Actinobacteria in soil remediation has the potential to reduce the environmental impact of industrial activities and improve the health of both soil and surrounding ecosystems.

To fully realize the potential of Actinobacteria in soil and crop health and other uses, it is important to understand the factors that influence their distribution and abundance in soil. This includes both abiotic factors such as soil pH, moisture, and temperature, as well as biotic factors such as interactions with other microorganisms in the soil. Researchers have also investigated the genetics of Actinobacteria to understand their metabolic capabilities and ability to adapt to different environmental conditions. This knowledge can be used to develop new strategies for enhancing the growth and activity of Actinobacteria in soil.

8 Future trends

Actinobacteria have been extensively studied and have numerous applications in various fields, including agriculture, medicine, and biotechnology. However, there is still much to learn about the diversity, ecology, and biology of Actinobacteria, and many opportunities for future research. Here are some areas of Actinobacteria research that may be of interest:

1. Discovery of novel species: Actinobacteria constitute one of the most diverse bacterial phyla, with an estimated 11 000 species. Nevertheless, many of these species remain uncharacterized and understudied. Future research could aim at identifying and characterizing novel Actinobacteria species, particularly those featuring unique metabolic pathways or biotechnological potential.
2. Genomics and metagenomics: With advances in genomics and metagenomics technologies, our understanding of Actinobacteria diversity, function, and evolution has the potential to be revolutionized. High-throughput sequencing technologies enable the identification and characterization of Actinobacterial communities in diverse environments, offering insights into their ecology and potential biotechnological applications. Comparative genomics provides an avenue for identifying novel genes and pathways that are essential for Actinobacterial

adaptation to specific environments, as well as for identifying novel biosynthetic gene clusters responsible for natural product production.

3 Ecology and biogeography: Actinobacteria are widely distributed in various ecosystems, and their abundance and diversity can be influenced by environmental factors such as pH, temperature, moisture, and nutrient availability. Therefore, further research is needed to understand the biogeography and ecological roles of Actinobacteria in different ecosystems, as well as their interactions with other microorganisms.
4 Understanding the role of Actinobacteria in carbon cycling: Actinobacteria are recognized as key decomposers of organic matter in the soil. However, their contribution to carbon cycling in different ecosystems remains poorly understood. Future research should focus on unraveling the interactions between Actinobacteria and other microorganisms involved in carbon cycling and the effects of environmental changes on these processes.
5 Microbial and plant interactions: Actinobacteria have diverse interactions with other microorganisms within the soil microbiome, making it important to understand these interactions for predicting the response of soil ecosystems to environmental change and for developing strategies to promote the growth and activity of beneficial soil microbes. Further research is needed to elucidate the mechanisms and impacts of these interactions. Additionally, Actinobacteria are known to interact with plants, either as pathogens or as beneficial microbes that promote plant growth and health. Thus, further research is also needed to understand the molecular mechanisms underlying these interactions and how they can be utilized for agricultural purposes.
6 Application in agriculture: Actinobacteria have the potential to enhance plant growth and productivity while also acting as an effective control against plant diseases. In order to harness this potential, future research should focus on the development of Actinobacteria-based biofertilizers and biopesticides, as well as gaining a deeper understanding of the underlying mechanisms behind their beneficial effects on plant growth and disease prevention.
7 Biotechnological applications: Actinobacteria are known to produce a diverse array of bioactive compounds with the potential for use in various fields such as medicine, agriculture, and industry. However, further research is necessary to fully comprehend the mechanisms underlying the biosynthesis of these compounds and to optimize their products for different applications.
8 Climate change and soil health: Climate change is anticipated to exert significant impacts on soil health, as well as the abundance and activity of

soil microorganisms, including Actinobacteria. Investigating the effects of climate change on Actinobacterial communities and their interactions with other soil microorganisms will be critical for comprehending and managing the consequences of climate change on soil ecosystems.

9 Conclusion

Research on Actinobacteria has provided invaluable insights into various fields. The unique characteristics of Actinobacteria, such as its filamentous and branching morphology, have contributed to its widespread presence in diverse environments. The multifaceted nature of Actinobacteria research underscores its significance in addressing global challenges across various domains. As we continue to unravel the complexities of these bacteria, their potential for shaping the future of medicine, agriculture, and industry becomes increasingly apparent, highlighting the need for ongoing exploration and innovation in Actinobacteria-related studies.

10 Where to look for further information

- Dai, Z. M., Su, W. Q., Chen, H. H., Barberán, A., Zhao, H. C., Yu, M. J., Yu, L., Brookes, P. C., Schadt, C. W., Chang, S. X. and Xu, J. M. 2018. Long-term nitrogen fertilization decreases bacterial diversity and favors the growth of Actinobacteria and Proteobacteria in agro-ecosystems across the globe. *Global Change Biology* 24, 3452-3461.
- Doron, L., Sutter, M. and Kerfeld, C. A. 2023. Characterization of a novel aromatic substrate-processing microcompartment in Actinobacteria. *mBio* 14,e01216-01223.
- Jiao, J. Y., Fu, L., Hua, Z. S., Liu, L., Salam, N., Liu, P. F., Lv, A. P., Wu, G., Xian, W. D., Zhu, Q. Y., Zhou, E. M., Fang, B. Z., Oren, A., Hedlund, B. P., Jiang, H. C., Knight, R., Cheng, L. and Li, W. J. 2021. Insight into the function and evolution of the Wood-Ljungdahl pathway in. *ISME Journal* 15, 3005-3018.
- Lee, S. M., Kong, H. G., Song, G. C. and Ryu, C. M. 2021. Disruption of Firmicutes and Actinobacteria abundance in tomato rhizosphere causes the incidence of bacterial wilt disease. *ISME Journal* 15, 330-347.
- Shi, J., Feng, Z. Z., Xu, J. C., Li, F. F., Zhang, Y. Q., Wen, A. J., Wang,F. L., Song, Q., Wang, L., Cui, H., Tong, S. J., Chen, P. Y., Zhu, Y. J., Zhao, G. P., Wang, S., Feng, Y. and Lin, W. 2023. Structural insights into the transcription activation mechanism of the global regulator GlnR from actinobacteria. *Proceedings of the National Academy of Sciences of the United States of America* 120, e2300282120.

11 References

Abdulla, H. M. and El-Shatoury, S. A. 2007. Actinomycetes in rice straw decomposition. *Waste Management* 27(6), 850–853.

Araujo, R., Gupta, V. V. S. R., Reith, F., Bissett, A., Mele, P. and Franco, C. M. M. 2020. Biogeography and emerging significance of Actinobacteria in Australia and northern Antarctica soils. *Soil Biology and Biochemistry* 146, 107805.

Auchtung, T. A., Fofanova, T. Y., Stewart, C. J., Nash, A. K., Wong, M. C., Gesell, J. R., Auchtung, J. M., Ajami, N. J. and Petrosino, J. F. 2018. Investigating colonization of the healthy adult gastrointestinal tract by Fungi. *mSphere* 3(2), e00092–00018.

Bao, Y., Dolfing, J., Chen, R., Li, Z., Lin, X. and Feng, Y. 2023. Trade-off between microbial ecophysiological features regulated by soil fertility governs plant residue decomposition. *Soil and Tillage Research* 229, 105679.

Bao, Y. Y., Dolfing, J., Guo, Z. Y., Chen, R. R., Wu, M., Li, Z. P., Lin, X. G. and Feng, Y. Z. 2021. Important ecophysiological roles of non-dominant *Actinobacteria* in plant residue decomposition, especially in less fertile soils. *Microbiome* 9(1), 84.

Bao, Y. Y., Dolfing, J., Wang, B. Z., Chen, R. R., Huang, M. S., Li, Z. P., Lin, X. G. and Feng, Y. Z. 2019. Bacterial communities involved directly or indirectly in the anaerobic degradation of cellulose. *Biology and Fertility of Soils* 55(3), 201–211.

Barka, E. A., Vatsa, P., Sanchez, L., Gaveau-Vaillant, N., Jacquard, C., Meier-Kolthoff, J. P., Klenk, H. P., Clément, C., Ouhdouch, Y. and van Wezel, G. P. 2016. Taxonomy, physiology, and natural products of Actinobacteria. *Microbiology and Molecular Biology Reviews* 80(1), 1–43.

Barnard, R. L., Osborne, C. A. and Firestone, M. K. 2013. Responses of soil bacterial and fungal communities to extreme desiccation and rewetting. *The ISME Journal* 7(11), 2229–2241.

Bhatti, A. A., Haq, S. and Bhat, R. A. 2017. Actinomycetes benefaction role in soil and plant health. *Microbial Pathogenesis* 111, 458–467.

Dai, Z., Su, W., Chen, H., Barberán, A., Zhao, H., Yu, M., Yu, L., Brookes, P. C., Schadt, C. W., Chang, S. X. and Xu, J. 2018. Long-term nitrogen fertilization decreases bacterial diversity and favors the growth of Actinobacteria and Proteobacteria in agro-ecosystems across the globe. *Global Change Biology* 24(8), 3452–3461.

Dong, Y., Kumar, C. G., Chia, N., Kim, P. J., Miller, P. A., Price, N. D., Cann, I. K. O., Flynn, T. M., Sanford, R. A., Krapac, I. G., Locke Ii, R. A., Hong, P. Y., Tamaki, H., Liu, W. T., Mackie, R. I., Hernandez, A. G., Wright, C. L., Mikel, M. A., Walker, J. L., Sivaguru, M., Fried, G., Yannarell, A. C. and Fouke, B. W. 2014. Halomonas sulfidaeris-dominated microbial community inhabits a 1.8 km-deep subsurface Cambrian Sandstone reservoir. *Environmental Microbiology* 16(6), 1695–1708.

Dworkin, M., Falkow, S., Rosenberg, E., Schleifer, K. H. and Stackebrandt, E. 2006. *The Prokaryotes: Archaea. Bacteria: Firmicutes, Actinomycetes Vol. 3: A Handbook on the Biology of Bacteria*. Springer.

Feng, Y. Z., Chen, R. R., Stegen, J. C., Guo, Z. Y., Zhang, J. W., Li, Z. P. and Lin, X. G. 2018. Two key features influencing community assembly processes at regional scale: Initial state and degree of change in environmental conditions. *Molecular Ecology* 27(24), 5238–5251.

Fierer, N. and Schimel, J. P. 2002. Effects of drying–rewetting frequency on soil carbon and nitrogen transformations. *Soil Biology and Biochemistry* 34(6), 777–787.

Gao, B. and Gupta, R. S. 2007. Phylogenomic analysis of proteins that are distinctive of Archaea and its main subgroups and the origin of methanogenesis. *BMC Genomics* 8, 86.

Garrett, S. D. 1981. *Soil Fungi and Soil Fertility: An Introduction to Soil Mycology*. Pergamon Press. Oxford.

Garza, D. R., Di Blasi, D., Bruns, J. A., Empson, B., Light, I., Ghannam, M., Castillo, S., Quijada, B., Zorawik, M., Garcia-Vedrenne, A. E. and Freise, A. C. 2022. Characterization of genomic diversity in bacteriophages infecting *Rhodococcus*. *bioRxiv*. doi: 10.1101/2022.11.23.517428.

Gittel, A., Bárta, J., Kohoutová, I., Mikutta, R., Owens, S., Gilbert, J., Schnecker, J., Wild, B., Hannisdal, B., Maerz, J., Lashchinskiy, N., Čapek, P., Šantrůčková, H., Gentsch, N., Shibistova, O., Guggenberger, G., Richter, A., Torsvik, V. L., Schleper, C. and Urich, T. 2014. Distinct microbial communities associated with buried soils in the Siberian tundra. *The ISME Journal* 8(4), 841–853.

Goodfellow, M. and Williams, S. T. 1983. Ecology of actinomycetes. *Annual Review of Microbiology* 37, 189–216.

Hamedi, J. and Poorinmohammad, N. 2017. The cellular structure of Actinobacteria. In: Wink, J., Mohammadipanah, F. and Hamedi, J. (Eds.), *Biology and Biotechnology of Actinobacteria*. Cham: Springer International Publishing, pp. 5–28.

Heinrich, R. and Schuster, S. 2012. *The Regulation of Cellular Systems*. Springer. New York.

Isaac, P., Martínez, F. L., Bourguignon, N., Sánchez, L. A. and Ferrero, M. A. 2015. Improved PAHs removal performance by a defined bacterial consortium of indigenous *Pseudomonas* and Actinobacteria from Patagonia, Argentina. *International Biodeterioration and Biodegradation* 101, 23–31.

Kaari, M., Manikkam, R., Annamalai, K. K. and Joseph, J. 2023. Actinobacteria as a source of biofertilizer/biocontrol agents for bio-organic agriculture. *Journal of Applied Microbiology* 134(2), lxac047.

Kaltenpoth, M. 2009. Actinobacteria as mutualists: General healthcare for insects? *Trends in Microbiology* 17(12), 529–535.

Lacombe-Harvey, M. È., Brzezinski, R. and Beaulieu, C. 2018. Chitinolytic functions in Actinobacteria: Ecology, enzymes, and evolution. *Applied Microbiology and Biotechnology* 102(17), 7219–7230.

Lewin, G. R., Carlos, C., Chevrette, M. G., Horn, H. A., McDonald, B. R., Stankey, R. J., Fox, B. G. and Currie, C. R. 2016. Evolution and ecology of *Actinobacteria* and their bioenergy applications. *Annual Review of Microbiology* 70, 235–254.

Liu, X., Cong, J., Lu, H., Xue, Y., Wang, X., Li, D. and Zhang, Y. 2017. Community structure and elevational distribution pattern of soil Actinobacteria in alpine grasslands. *Acta Ecologica Sinica* 37(4), 213–218.

Matz, C. and Kjelleberg, S. 2005. Off the hook–how bacteria survive protozoan grazing. *Trends in Microbiology* 13(7), 302–307.

Medie, F. M., Davies, G. J., Drancourt, M. and Henrissat, B. 2012. Genome analyses highlight the different biological roles of cellulases. *Nature Reviews Microbiology* 10(3), 227–234.

Miao, V. and Davies, J. 2010. Actinobacteria: The good, the bad, and the ugly. *Antonie van Leeuwenhoek* 98(2), 143–150.

Mishra, A., Rathour, R., Singh, R., Kumari, T. and Thakur, I. S. 2020. Degradation and detoxification of phenanthrene by actinobacterium *Zhihengliuella* sp. ISTPL4. *Environmental Science and Pollution Research International* 27(22), 27256–27267.

Mohammadipanah, F. and Dehhaghi, M. 2017. Classification and taxonomy of Actinobacteria. In: Wink, J., Mohammadipanah, F. and Hamedi, J. (Eds.), *Biology and Biotechnology of Actinobacteria*. Cham: Springer International Publishing, pp. 51–77.

Molenaar, D., Van Berlo, R., De Ridder, D. and Teusink, B. 2009. Shifts in growth strategies reflect tradeoffs in cellular economics. *Molecular Systems Biology* 5, 323.

Ntemafack, A., Ahmed, S., Kumar, A., Chouhan, R., Kapoor, N., Bharate, S. B., Hassan, Q. P. and Gandhi, S. G. 2022. Plant growth promoting potential of butyl isobutyl phthalate and *Streptomyces* sp. from *Rumex dentatus* on rice. *Applied Microbiology and Biotechnology* 106(7), 2603–2617.

Penn, K., Jenkins, C., Nett, M., Udwary, D. W., Gontang, E. A., McGlinchey, R. P., Foster, B., Lapidus, A., Podell, S., Allen, E. E., Moore, B. S. and Jensen, P. R. 2009. Genomic islands link secondary metabolism to functional adaptation in marine Actinobacteria. *The ISME Journal* 3(10), 1193–1203.

Pfeiffer, T., Schuster, S. and Bonhoeffer, S. 2001. Cooperation and competition in the evolution of ATP-producing pathways. *Science* 292(5516), 504–507.

Polz, M. F. and Cordero, O. X. 2016. Bacterial evolution: Genomics of metabolic trade-offs. *Nature Microbiology* 1(11), 16181.

Raes, J. and Bork, P. 2008. Molecular eco-systems biology: Towards an understanding of community function. *Nature Reviews Microbiology* 6(9), 693–699.

Rego, A., Raio, F., Martins, T. P., Ribeiro, H., Sousa, A. G. G., Seneca, J., Baptista, M. S., Lee, C. K., Cary, S. C., Ramos, V., Carvalho, M. F., Leao, P. N. and Magalhaes, C. 2019. Actinobacteria and Cyanobacteria diversity in terrestrial Antarctic microenvironments evaluated by culture-dependent and independent methods. *Frontiers in Microbiology* 10, 1018.

Roller, B. R., Stoddard, S. F. and Schmidt, T. M. 2016. Exploiting rRNA operon copy number to investigate bacterial reproductive strategies. *Nature Microbiology* 1(11), 16160.

Rückert, C., Albersmeier, A., Busche, T., Jaenicke, S., Winkler, A., Friðjónsson, Ó. H., Hreggviðsson, G. Ó., Lambert, C., Badcock, D., Bernaerts, K., Anne, J., Economou, A. and Kalinowski, J. 2015. Complete genome sequence of Streptomyces lividans TK24. *Journal of Biotechnology* 199, 21–22.

Schmidt, T. M. 2019. *Encyclopedia of Microbiology*. Academic Press. San Diego, Calif.

Sellstedt, A. and Richau, K. H. 2013. Aspects of nitrogen-fixing Actinobacteria, in particular free-living and symbiotic *Frankia*. *FEMS Microbiology Letters* 342(2), 179–186.

Sorokin, D. Y., van Pelt, S., Tourova, T. P. and Evtushenko, L. I. 2009. *Nitriliruptor alkaliphilus* gen. nov., sp. nov., a deep-lineage haloalkaliphilic actinobacterium from soda lakes capable of growth on aliphatic nitriles, and proposal of *Nitriliruptoraceae* fam. nov. and *Nitriliruptorales* ord. nov. *International Journal of Systematic and Evolutionary Microbiology* 59(2), 248–253.

Steele, J. A., Countway, P. D., Xia, L., Vigil, P. D., Beman, J. M., Kim, D. Y., Chow, C. E. T., Sachdeva, R., Jones, A. C., Schwalbach, M. S., Rose, J. M., Hewson, I., Patel, A., Sun, F. Z., Caron, D. A. and Fuhrman, J. A. 2011. Marine bacterial, archaeal and protistan association networks reveal ecological linkages. *The ISME Journal* 5(9), 1414–1425.

Swarnalakshmi, K., Senthilkumar, M. and Ramakrishnan, B. 2016. Endophytic *Actinobacteria*: Nitrogen fixation, phytohormone production, and antibiosis. In: Subramaniam, G., Arumugam, S. and Rajendran, V. (Eds.), *Plant Growth Promoting Actinobacteria: A Vew Avenue for Enhancing the Productivity and Soil Fertility of Grain Legumes*. Singapore: Springer, pp. 123–145.

Vaishampayan, P., Roberts, A. H., Augustus, A., Pukall, R., Schumann, P., Schwendner, P., Mayilraj, S., Salmassi, T. and Venkateswaran, K. 2014. *Deinococcusphoenicis* sp. nov., an extreme ionizing-radiation-resistant bacterium isolated from the Phoenix Lander assembly facility. *International Journal of Systematic and Evolutionary Microbiology* 64(10), 3441–3446.

van Bergeijk, D. A., Terlouw, B. R., Medema, M. H. and van Wezel, G. P. 2020. Ecology and genomics of Actinobacteria: New concepts for natural product discovery. *Nature Reviews. Microbiology* 18(10), 546–558.

Wang, C., Dong, D., Wang, H. S., Muller, K., Qin, Y., Wang, H. L. and Wu, W. X. 2016. Metagenomic analysis of microbial consortia enriched from compost: New insights into the role of Actinobacteria in lignocellulose decomposition. *Biotechnology for Biofuels* 9, 22.

Wu, X., Yao, Z., Brüggemann, N., Shen, Z. Y., Wolf, B., Dannenmann, M., Zheng, X. and Butterbach-Bahl, K. 2010. Effects of soil moisture and temperature on CO_2 and CH_4 soil–atmosphere exchange of various land use/cover types in a semi-arid grassland in Inner Mongolia, China. *Soil Biology and Biochemistry* 42(5), 773–787.

Xia, Q., Rufty, T. and Shi, W. 2020. Soil microbial diversity and composition: Links to soil texture and associated properties. *Soil Biology and Biochemistry* 149, 107953.

Yadav, A. N., Verma, P., Kumar, S., Kumar, V., Kumar, M., Kumari Sugitha, T. C., Singh, B. P., Saxena, A. K. and Dhaliwal, H. S. 2018. Chapter 2. Actinobacteria from rhizosphere: Molecular diversity, distributions, and potential biotechnological applications. In: Singh, B. P., Gupta, V. K. and Passari, A. K. (Eds.), *New and Future Developments in Microbial Biotechnology and Bioengineering*. Elsevier, pp. 13–41. Amsterdam .

Yaradoddi, J. S. and Kontro, M. H. 2021. Actinobacteria: Basic adaptation to harsh environments. In: Yaradoddi, J. S., Kontro, M. H. and Ganachari, S. V. (Eds.), *Actinobacteria: Ecology, Diversity, Classification and Extensive Applications*. Singapore: Springer Nature Singapore, pp. 69–88.

Zhao, P., Xue, Y., Gao, W., Li, J., Zu, X., Fu, D., Feng, S., Bai, X., Zuo, Y. and Li, P. 2018. Actinobacteria–Derived peptide antibiotics since 2000. *Peptides* 103, 48–59.

Chapter 5

Fungi: advances in understanding the role of root-associated fungi in soil

Nimalka Weerasuriya, University of Western Ontario, Canada and A & L Biologicals, Canada; Noor Saeed Cheema, Marianna Wallace and R. Greg Thorn, University of Western Ontario, Canada; and Saveetha Kandasamy, Soledad Saldias and George Lazarovits, A & L Biologicals, Canada

1 Introduction: root and rhizosphere mycobiomes

The microbiome is a community of microorganisms within a defined habitat, with distinct physicochemical properties and interactions with its environment (Berg et al., 2020). The mycobiome refers to the fungal community present in a habitat and its activities (Berg et al., 2020). The root mycobiome plays a direct role in plant productivity, and the study of community composition within the rhizosphere allows for identification of organisms – both beneficial and pathogenic – that influence plant health (Patel et al., 2015; Rashid et al., 2016). Root-associated mycobiomes are dynamically affected by both the surrounding biotic and abiotic conditions in the soil and the host plant itself (Tkacz and Poole, 2015; Muneer et al., 2021). The rhizosphere consists of the soil that is in direct contact with the roots of a plant and is a hot spot of microbial diversity and dynamic interactions (Hinsinger et al., 2009). Root exudates stimulate the

http://dx.doi.org/10.19103/AS.2024.0136.15

growth of microorganisms in the vicinity of the roots, leading to soil around the roots that is incredibly rich with microbes, including plant-beneficial organisms and plant pathogens (Berg and Smalla, 2009). Through this process of root exudation, the plant feeds the microbial community in the soil, directly influencing its composition and activities. The soil that is not associated with the rhizosphere, bulk soil, is not penetrated by active plant roots and typically has lower microbial activity, reflecting the lower availability of labile carbon sources there (Ai et al., 2012). The microbial community of bulk soil may or may not be less diverse than in the rhizosphere (DeAngelis et al., 2009; Wang et al., 2021), but the members represent a 'candidate pool' awaiting stimulation to activity and may thus have less direct correlation with the health of the plant community.

1.1 Fungal-plant interactions

Fungi are a major component of soil ecosystems and have a long evolutionary history with plants, with fossil records indicating symbiosis between the two groups dating back over 400 million years (Rai and Agarkar, 2016). Plants and associated fungi are often considered as co-evolving units, with microbial diversity and interactions playing a vital role in maintaining plant health and productivity (Vandenkoornhuyse et al., 2015). All plants are thought to interact with the fungi residing within plant tissues (Petrini, 1996; Southworth, 2012). The interactions that occur in the rhizosphere and root can be harmful or beneficial to the plant or fungal associates, or apparently inconsequential to one or the other (Fig. 1). Mutualistic rhizosphere organisms that are beneficial to plant growth may improve plant nutrient acquisition and stress tolerance (Philippot et al., 2013). Numerous studies have shown that crops influence the soil microbial communities by shaping their composition and diversity through root exudates and symbiotic association in a species-specific manner (Staskawicz, 2001; Costa et al., 2006; Garbeva et al., 2008). Thus, diversified crop rotations can improve yield through enrichment of microorganisms selectively nurtured by each crop in the rotation that positively affect growth and productivity (Dias et al., 2015). However, in the rhizosphere are also fungal pathogens that elicit harmful effects on plants, which can result in crop losses (Philippot et al., 2013). While fungal pathogens may be present even in high productivity soils, they are less likely to affect plant growth in a highly diverse rhizosphere, as surrounding microbes, fungi, and plant defenses have developed mechanisms of pathogen resistance (Schnitzer et al., 2011).

1.1.1 Beneficial interactions

There are many beneficial fungal–plant interactions. Beneficial fungi positively affect plant fitness through processes such as aiding in biotic and abiotic stress tolerance, accelerating growth, and controlling reproduction (Petrini, 1996;

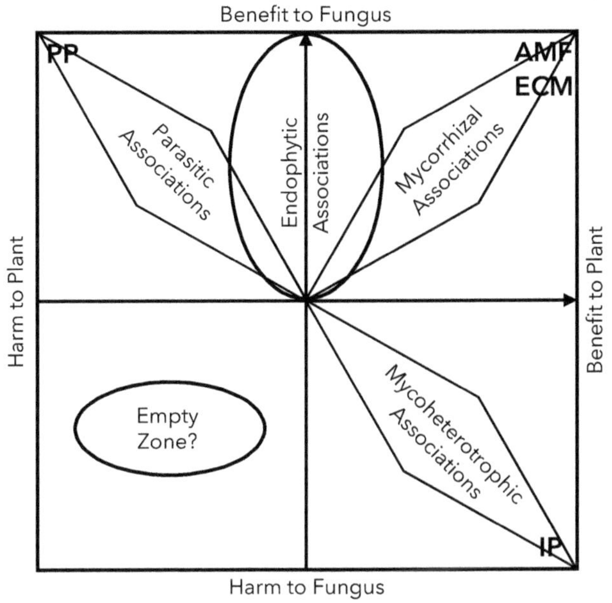

Figure 1 The range of interactions between plant roots and associated fungi (modified from Brundrett, 2004). PP = plant pathogens, AMF = arbuscular mycorrhizal fungi, ECM = ectomycorrhizae, IP = Indian pipe (*Monotropa uniflora*) as an example mycoheterotrophic plant. Note that rhizosphere and endophytic associates may be commensal with respect to the plant (providing neither benefit nor harm, while deriving benefits from the plant such as space or nutrients) but may, alternatively, be beneficial or harmful to the host, with this relationship possibly changing over the lifespan of the plant. When we accept that all mycorrhizal associates are nutritional parasites of their hosts (receiving host carbon in exchange for water, soil nutrients, and possibly other benefits), and that their specific costs and benefits may vary with the fungus and host identity, time or environmental conditions, the distinctions among the different groups illustrated begin to blur, allowing more nuanced analyses of the costs and benefits to each member of the partnership.

Southworth, 2012). Beneficial fungi take the form of mycorrhiza, endophytes, or are free living. Mycorrhizae improve plant productivity and keep plant health robust against various abiotic and biotic stresses (Smith and Read, 2008). These interactions are extremely important for development and sustainability of plant communities in ecosystems where nutrient supply is limited (Buscot et al., 2000). The main categories of mycorrhizal associations are arbuscular mycorrhizae and ectomycorrhizae (Brundrett, 2004; Allen, 2007; Smith and Read, 2008). The mycorrhizal associates of most of our crop plants are members of the Glomeromycota, forming arbuscular mycorrhizae (Brundrett, 2004; Brundrett and Tedersoo, 2020). In arbuscular mycorrhizal associations, hyphae penetrate the plant root cells to form highly branched structures called arbuscules, where nutrient exchange occurs (Smith and Read, 2008). Arbuscular mycorrhizal fungi are particularly important for the delivery of phosphate to plants (Javot et al.,

2007), but also play roles in delivery of water and nitrogen, with mycorrhizal N uptake stimulated by carbon supplied by the host plant (Fellbaum et al., 2012).

Ectomycorrhizal associations, which are much more important in natural and managed forest systems in Canada, are formed by many members of the phyla Ascomycota and Basidiomycota and, unlike arbuscular mycorrhizae, do not penetrate their host's cell walls, instead forming a highly branched network of hyphae around plant root cortical cells called a Hartig net (Brundrett, 2009). Both arbuscular mycorrhizal and ectomycorrhizal interactions can involve multiple fungi interacting with the same plant simultaneously (Allen, 2007). In addition to mycorrhizal fungi, plant roots also interact with endophytic fungi, endophytic bacteria, root pathogens, and fungi and bacteria found in the surrounding soil. Endophytes represent a large component of the microbial community of roots and can become pathogenic or remain mutualistic, helping improve plant growth, health, and stress tolerance (Baron and Rigobelo, 2022). Roots also interact with pathogenic fungi that elicit harmful effects on plants. However, there are relatively fewer parasitic fungi associating with roots than beneficial or neutral fungi as plants have developed diverse mechanisms to protect themselves (Staskawicz, 2001). Recent studies of plant root and rhizosphere mycobiomes have shown us that plant roots may have many 'unexpected' partners – e.g. ectomycorrhizal or ericoid mycorrhizal fungi on or in roots of AMF hosts, or vice versa (e.g. Poelman et al., 2021) – and that these associations may provide ecological benefits to the host plants (Almario et al., 2017).

Growth-promoting fungi help prime the plant's immune system to detect and evade pathogens through the production of volatile organic compounds that act as elicitors, by production of antimicrobial compounds, and by competition with pathogens for space, nutrients or other resources (Enebe and Babalola, 2019). An example of a mutualistic fungal endophyte is *Fusarium oxysporum* strain Fo162, which can act as a biological control against infection by root-parasitic nematodes (Rai and Agarkar, 2016). Some mutualistic fungal endophytes emit beneficial compounds such as phytohormones that stimulate the production of antimicrobial secondary metabolites that result in disease suppression (Meena et al., 2017).

Fungal and bacterial species can help suppress pathogen growth through direct antagonism with pathogens for space and nutrients or by producing antimicrobial metabolites, or indirectly through induction of systemic resistance, increasing resistance against pathogens via upregulation of the host plant's defense genes (White et al., 2019). The interaction of multiple species of microbes may amplify these effects. High numbers in the wheat rhizosphere of a dominant bacterium *Stenotrophomonas rhizophila* helped increase plant growth while reducing the virulence of *Fusarium pseudograminearum* (Liu et al., 2021).

Plant growth-promoting fungi can aid plant tolerance of environmental and biological stressors (Ray et al., 2020). Fungal endophytes help plants become more tolerant to abiotic stressors such as drought, salinity, and temperature (Redman et al., 2002; Márquez et al., 2007; Rodriguez et al., 2008; Southworth, 2012). The ability of fungi to help plants tolerate drought stress was especially important during the initial colonization of plants onto land since symbiotic fungi aided in drought tolerance of many plants; indeed, they formed the water and nutrient uptake structures prior to the development of plant roots (Pirozynski and Malloch, 1975). An extreme example of fungal endophytes co-evolving with plants involves the grass *Dichanthelium lanuginosum*, which thrives in hot geothermal soils with root zone temperatures of 57°C. An analysis of 100 *D. lanuginosum* plants revealed that all were colonized by one dominant fungal endophyte *Curvularia protuberata*; however, when grass plants were grown without *Curvularia protuberata* in simulated geothermal conditions, the plants died (Redman et al., 2002). The extreme thermotolerance associated with this endophyte-plant symbiosis could be attributed to the fungal endophyte producing cell wall melanin that helps dissipate heat along the hyphae or forms a complex with oxygen radicals generated during heat stress (Redman et al., 2002; Verghese et al., 2012). Other soil fungi, such as *Trichoderma virens*, have been shown to enhance plant growth and help plants become more tolerant to extreme environmental conditions such as heavy metal stress through processes such as changes in valence and intracellular localization of the contaminant (Babu et al., 2014). Similarly, Ikram and others (2018) found the inoculation of heavy metal-rich soil with *Penicillium roqueforti* increased heavy metal tolerance and nutrient uptake, resulting in greater plant growth than wheat grown in control soil in the presence of heavy metals. Thus, fungi, particularly endophytes, use intracellular processes, nutrient acquisition, and reductions in reactive oxygen species to aid in a plant's ability to tolerate abiotic stresses.

1.1.2 Antagonistic interactions

Pathogenic fungi cause damage or death to host plants, with or without penetration of the plant cells. Biotrophic fungal pathogens are those that cannot survive without a living host plant and do not have a saprotrophic independent life stage (Pawlowski and Hartman, 2016). These biotrophic fungal pathogens have highly developed infection structures called appressoria for penetrating host plant cells and may also produce branched or unbranched haustoria within the plant cells for nutrient uptake. Necrotrophic fungal and oomycete plant pathogens rely on dead tissue to derive their nutrition (Pawlowski and Hartman, 2016). Necrotrophic fungal plant pathogens produce secondary metabolites with toxic properties, an example being deoxynivalenol (DON), a toxin produced by *Fusarium* and other fungi that induces cell death (Pusztahelyi et al., 2015). Necrotrophic fungal plant pathogens use various secretory pathways to cause localized cell death, resulting in localized necrosis, lesions, and in some cases, whole plant death (Pawlowski and Hartman, 2016). Certain environmental conditions, such as nitrogen limitation or soil flooding, can exacerbate the effects of these fungal pathogens (Pusztahelyi et al., 2015).

1.1.3 Dynamic interactions

Fungal–plant interactions can be dynamic, with endophytes sometimes transitioning from a commensal or mutualistic relationship to a pathogenic one under certain conditions. Various factors can impact the strength and nature of these interactions, including abiotic stresses, environmental factors, host range, host-specificity, tissue-specificity, and nutrient imbalances (Schulz and Boyle, 2005; Rai and Agarkar, 2016). Nutrient imbalances can lead to reactive oxidative stress and cause endophytes to shift from a neutral or mutualistic relationship to a pathogenic one. Environmental factors like temperature and humidity can also make host plants more susceptible to this shift (Freeman and Rodriguez, 1993; Rai and Agarkar, 2016). Plants exhibit a diverse array of interactions with fungal communities present in the soil, which span the full range of ecological possibilities, including competitive, exploitative, neutral, commensal, and mutualistic, and may vary across time, space, and edaphic conditions. Therefore, understanding how the interactions of root-associated fungal communities may be linked to soil management, plant health, and crop yield is of great interest.

2 Overview of methods for community analyses: studying the mycobiome

It is vital to acknowledge the tremendous knowledge debt owed to the pioneers of soil mycology, from Brefeld to Domsch, Gams, Warcup, and Barron, whose careful studies using microscopy and culturing provided direct evidence of the presence and activities of fungi in soil (Warcup and Talbot, 1965; Domsch et al., 2007). Although many fungi and other microbes cannot be cultured readily from soil, creative soil-baiting culture techniques have yielded bizarre and novel fungi that are rarely detected by molecular methods (Tzean and Barron, 1981; Barron, 1991). Overviews of culture-based methods for soil fungi have been published by Mueller et al. (2004) and Thorn et al. (2007).

The advent of high-throughput, next-generation sequencing (NGS) technologies has enabled researchers to determine the composition and functions of microbial communities associated with different crops. Metabarcoding is based upon DNA barcoding, which employs the use of a standardized, short, and informative region of DNA ('DNA barcode') for species identification (Hebert et al., 2003). Amplification and sequencing of a DNA barcode from the community of organisms in environmental DNA (eDNA) samples allows the detection and identification of both culturable and unculturable microbes or fungi from a diversity of habitats, including bulk soil, the rhizosphere, and plant tissues (Ursell et al., 2012). This has increased the understanding of how different factors affect microbial communities associated with host plants in unprecedented detail (Taylor et al., 2016). NGS can identify low abundance DNA found in community samples and produce species richness estimates at an affordable cost, and this quickly led to a rise in popularity when studying microbial ecology (Jumpponen et al., 2010; Tedersoo et al., 2015). DNA metabarcoding is a valuable tool that combines high NGS throughput with multiplexed mixed-species samples to determine their taxonomic identities in bulk (Caporaso et al., 2012; Taberlet et al., 2012).

The development of DNA barcoding has also encouraged extensive international efforts to build taxonomic reference libraries of many standardized regions. The process of metabarcoding includes the collection of an environmental sample, DNA extraction, target gene amplification, high-throughput NGS of the amplified community's DNA fragments, sequence processing, and statistical analyses (Caporaso et al., 2012; Taberlet et al., 2012; Lynch and Neufeld, 2015).

There are potential biases introduced during DNA extraction steps and PCR amplification that must be considered. Depending on the DNA extraction procedures there may be differences in community composition estimations or DNA quantities, with up to a 3x change in DNA quality using varying methods (Thakuria et al., 2008), both of which may affect gene copy number estimates (Feinstein et al., 2009) and lead to bias in PCR during metabarcoding (Pawluczyk et al., 2015). When isolating genomic DNAs from samples of roots or rhizosphere soil, a subsample of 100–200 mg is usually ample for use in a variety of available DNA kits (from Galenvs, Macherey-Nagel, Norgen, Omega, Qiagen, Zymo, or others). However, when dealing with bulk soils, particularly those with very low organic content or very high sand or clay fractions, it may be necessary to start from quantities as large as 2–5 g. In this case, options include large capacity commercial kits (e.g. Qiagen's DNeasy PowerMax or Norgen's Soil DNA MaxiKit) or starting with a manual extraction involving grinding in liquid nitrogen and alkaline lysis with protease, followed by cleaning the crude alcohol precipitate through much less expensive miniprep kits (e.g. ZymoBiomics DNA Miniprep Kit or Qiagen DNeasy PowerSoil Pro Kit). More efficient DNA extraction kits, including those based on magnetic bead technologies, may alleviate these problems. Distinguishing the active fraction of the mycobiome from transient or legacy members may require extraction from washed organic particles, removing transient spores and legacy DNAs bound to soil colloids (Lynch and Thorn, 2006). Soil samples in solution (before any physical or chemical lysis step) may also be stained with ethidium or propidium monoazide which, following photoactivation, damages DNAs outside of living cells. After removing the dye and continuing with the lysis and DNA extraction procedure, only DNAs from previously intact and viable cells will be PCR amplified (Emerson et al., 2017), but this 'viability PCR' technique does not eliminate viable but non-resident taxa.

The selection of appropriate metabarcoding primers is key to uncovering diversity and community composition in these studies (Taylor et al., 2016). An ideal set of forward and reverse primers must bind to highly conserved sites across the target group of organisms, must border a region with adequate variability to allow for species-specific identification, and produce an amplicon of an appropriate length for the intended NGS platform (Alberdi et al., 2018). Details of the selection of marker regions and primers for metabarcoding are discussed in Section 5.1 and details of sequencing platforms are discussed in Section 5.2.

The high-throughput sequencing of PCR-amplified metabarcoding primers has traditionally been followed by the construction of molecular operational taxonomic units (OTUs or MOTUs), which are clusters of reads that differ by less than a fixed dissimilarity threshold selected by the user, typically ranging from 95% to 99% similarity (Westcott and Schloss, 2015; Kopylova et al., 2016). An OTU table, which features the abundance of each identified OTU in each sample, serves as the basis for all future analyses, with each OTU treated as equivalent to a species. An attractive alternative is to use a method producing groups of sequences that are unique in length and composition (i.e. 100% identity), called amplified sequence variants (SVs or ASVs, Callahan et al., 2017). Although ASVs are meant to capture biological variation within species (Callahan et al., 2017), OTUs may greatly inflate estimates of richness compared to ASVs because of an overabundance of rare OTUs (Chiarello et al., 2022). Whereas methods based on OTUs report only a single sequence chosen as representative for each OTU (usually the most abundant in read count), ASV methods report each sequence variant detected, allowing direct comparisons between studies and datasets (Callahan et al., 2017). Excess (i.e. artifactually inflated) ASVs or OTUs can be clustered through sequence alignment and phylogenetic analysis to yield highly supported clades (clusters) at a level of sequence identity chosen to represent 'species-level clusters' for the fungi under analysis, noting that one threshold will not be suitable across the breadth of fungal phylogeny. Such manual, post-pipeline analysis, although labor intensive, can add valuable depth and taxonomic precision to a study.

The next step in the process is to use automated or manual methods to associate the most accurate taxon name with each OTU or ASV. The choice of method and the reference database used depends on the choice of metabarcoding marker gene: there are robust automated methods for identification of fungal ITS sequences using the UNITE_All reference database (Nilsson et al., 2019b; Rawson and Zahn, 2023) but less so for large or small ribosomal subunits or other marker genes. For LSU data, a phylogenetic binning approach was found to be more accurate than either the QIIME 2 or RDP classifiers (Eldred et al., 2021), and for other marker gene regions, it may be the only option. Phylogenetic binning requires the user to select authoritative reference sequences. Ideally, these are sequences derived from type strains of a species to represent a species name and of authoritatively identified, or type strain of the type species of a genus to represent generic names, type genus to represent a family, and so on. These are downloaded and aligned together with their OTUs or ASVs and then run using appropriate analyses to create phylogenetic trees in which eDNAs are 'binned' in well-supported clusters (clades) together with identified reference sequences (Eldred et al., 2021). The process of selecting trustworthy and accurately named reference sequences is made more challenging by misidentified sequences in GenBank (Bidartondo et al., 2008) as well as the historical practice in mycology of separately naming the sexual and asexual forms of the same taxa (Taylor, 2011). However, arriving at the correct name for sequences detected in eDNA studies helps move mycobiome research from studies of pattern to biological insights into ecological function. For a discussion of the current state of fungal identification, see Lücking et al. (2020).

3 Case study 1: root and rhizosphere mycobiomes in relation to yield in corn and in corn and winter wheat

A study of root mycobiome and soil physicochemical factors relating to yield in corn (maize) in southern Ontario found higher phylogenetic diversity associated with higher crop yield and indicator species that could be associated with high or low yield in single farm sites but not shared across multiple farms because of high inter-farm variation in community composition (Kandasamy et al., 2021a). Metabarcoding of the fungal communities of washed corn roots used primers for the ITS2 and D1 LSU regions, with the ITS2 primers 5.8S_Fun and ITS4_Fun (Taylor et al., 2016) and Ascomycota-selective primers to D1 LSU (LSUA, Asemaninejad et al., 2016) yielding complementary (partially overlapping) phylogenetic communities. Primers to the D1 LSU of Basidiomycota, Glomeromycota, basal fungi, and fungus-like organisms (LSUBG, Asemaninejad et al., 2016) performed poorly, yielding high numbers of non-target reads (plants, animals, and Rhizaria) and few target reads, while primers to the V4 region of SSU in Glomeromycota (AMV4.5N-F and AMDG-R, Sato et al., 2005) yielded few reads overall and few target OTUs. Fungal sequences were dominated by members of classes Sordariomycetes, Dothideomycetes, and Eurotiomycetes of the phylum Ascomycota (Fig. 2). Communities varied in a

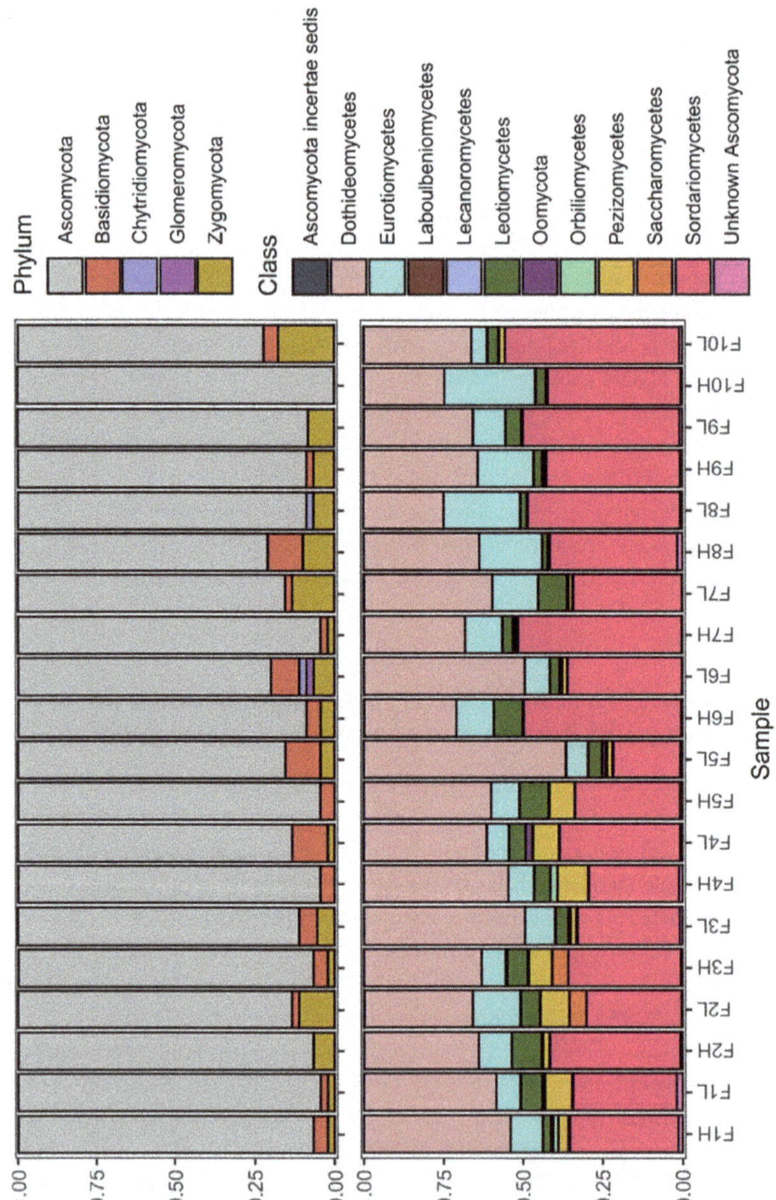

Figure 2 Relative abundances of corn root fungal communities across all farms. Relative abundance of fungal communities at the (A) phylum level (ITS2 general fungal primer), and the (B) class level (LSU Ascomycota specific primer), from Kandasamy et al. (2021a). H and L represent high-yielding and low-yielding sties, respectively.

site-specific manner; characters of soil texture and chemistry varied strongly both between and within farm sites. The ALDEx2 package in R was used as an ANOVA-like univariate comparison tool incorporating the Bayesian estimate of taxon abundance into a compositional framework, useful for the differentiation of OTU abundances between two conditions to identify taxa that are significantly associated with each condition (Fernandes et al., 2013), in this case high and low yield (Kandasamy et al., 2021a). Sequences identified as *Chalara fungorum*, *Fusarium oxysporum*, *Gibellulopsis* spp., *Mucor hiemalis*, *Neonectria fuckeliana*, *Penicillium janthinellum*, *Penicillium ochrochloron*, *P. paneum/chrysogenum*, *Setophoma terrestris*, *Talaromyces* sp., and *T. diversus* were found in washed corn roots from high-yielding sites from at least some farms in the study, but *Chalara fungorum*, *Penicillium ochrochloron*, *Setophoma terrestris*, and *T. diversus* were often abundant and found in both high- and low-yielding sites. No single soil physicochemical factor was highly explanatory of yield, but yield significantly varied inversely with pH, which in turn was partially positively correlated with cation exchange capacity (CEC), soil moisture, and organic matter and negatively with aluminum concentration. A clustered heatmap of Spearman correlation coefficients between ecologically important indicator taxa (as identified by ALDEx) and key soil variables provides insights into the ecology of the fungal taxa but does not provide any clear instructions to growers or anyone seeking to beneficially modify the soil mycobiome.

In a continuation of this study (Wallace, 2022), primers to the ITS2 and D1 LSU regions (LSUA and LSUBG) were used to amplify fungal DNA from rhizosphere soil and washed roots of winter wheat and corn from historically high and low yield sites within three fields that have a history of corn-soybean-wheat crop rotation. Significant differences in fungal species composition were found between each of the fields, requiring each field to be analyzed separately. As expected, root samples exhibited lower alpha diversity compared to rhizosphere samples, but there were no discernible trends between the alpha diversity of corn and wheat rhizosphere samples. Within a single field, the overall fungal community composition of samples from corn and wheat were significantly different from one another. Fungal community composition in corn and wheat rhizosphere soils was found to be significantly correlated with yield measured at harvest. The indicator species analysis of rhizosphere soils in each field revealed some key species that may be driving the differences in community composition of high and low yield sites, including *Pseudeurotium bakeri*, *Cryptococcus watticus*, *Solicoccozyma terricola*, and *Candida sake*, which were significantly associated with high yield sites, and *Leptodontidium camptobactrum* and *Soloacrosporiella*, which were significantly associated with low yield sites. Differences in fungal community composition of samples were correlated with soil physicochemical variables, including pH and concentrations of K, Mn, S, Fe, and Mg. Several soil physicochemical variables were significantly associated with yield of wheat and corn, with some soil characteristics differing between the two crops. Soil texture was an important parameter, along with pH, soil moisture, CEC, and concentrations of Al, Ca, Fe, H, Mg, Mn, and NO_3-N. Genera were identified as being significantly correlated with yield measured at harvest for both corn and wheat. Notably, *Fusicolla*, *Epicoccum*, and *Tetracladium* were positively correlated with yield, and *Plectosphaerella*, *Neonectria*, *Dactylonectria*, *Aphanomyces*, *Pythium*, *Corynespora*, and *Mrakia* were negatively correlated with yield. Of these, *Epicoccum* was strongly positively correlated with yield but negatively correlated with clay, soil moisture, pH, and concentrations of K and Mg, whereas *Plectosphaerella* was negatively correlated with yield and positively correlated with the same soil characters. *Sagenomella*, *Myrmecridium*, and *Soloacrosporiella*, all negatively (but not significantly) correlated with yield, were also strongly positively correlated with soil moisture, pH, and Mg concentration (Fig. 3). The results of this study emphasize the importance of comprehensive integrative analysis of soil fungi, soil characteristics, and yield.

The study by Wallace (2022) found that many of the soil physicochemical variables measured could serve as predictors of the separations of samples in the same field belonging to different crop and productivity sites based on fungal community composition. This was consistently observed with the ITS2, LSUA, and LSUBG primer sets. This suggests that these biotic and abiotic variables are highly different between sample types in each of the fields studied and implies that soil physicochemical variables are important predictors of differences between community composition and crop productivity (Chang et al., 2017). Although the fields studied employ similar cropping regimes and fertilizer, soil characteristics, which can vary in the same field, can greatly influence the microbial community (Tkacz and Poole, 2015). The variability in soil characteristics identified as significant are likely contributing to the differences in community composition that distinguish the sampling sites for high yielding and low yielding corn and high yielding and low yielding wheat. Several other studies have identified environmental factors as being important contributors to fungal diversity (Antoninka et al., 2015; Tedersoo et al., 2014; Tkacz et al., 2015; Xu et al., 2009). In one such study by Tedersoo et al. (2014), several climactic and edaphic factors were found to be predictors of soil fungi diversity and community composition, including mean annual precipitation, soil calcium concentration, soil pH, soil carbon/nitrogen ratio, and concentration of soil nutrients such as phosphorus. Soil physicochemical variables in fields and in site-specific locations within the fields should continue to be assessed and compared, as their associations with fungal communities may be paramount to understand differences in crop productivity in sites within the same field.

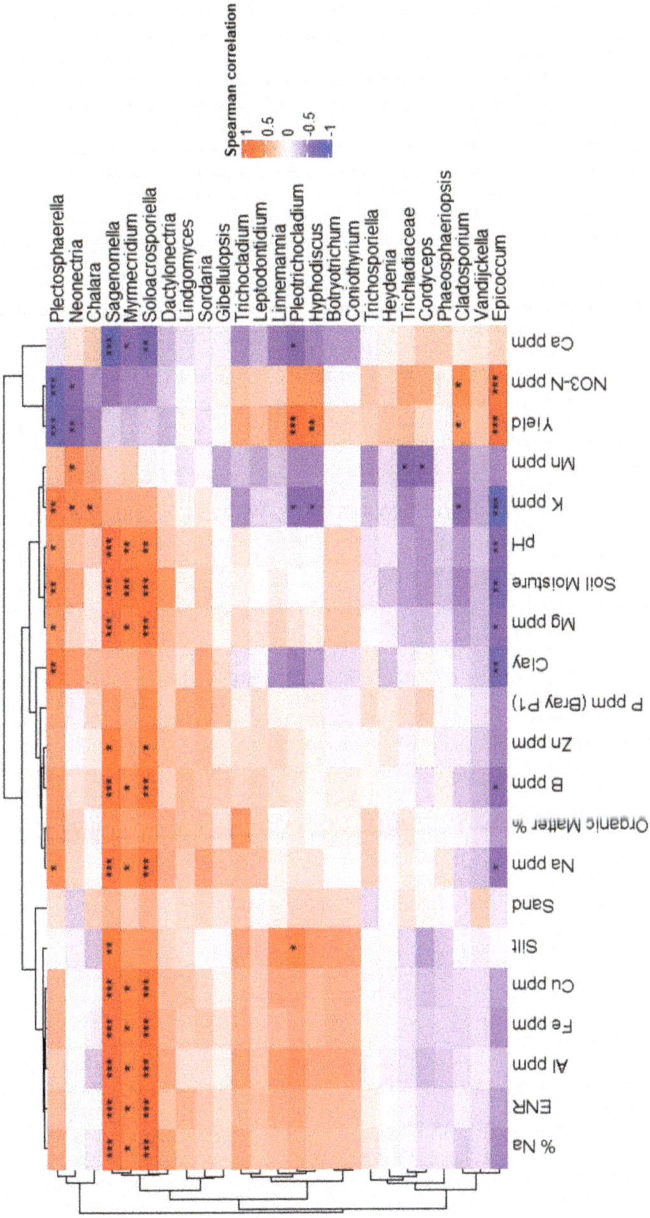

Figure 3 Correlation between field 1 key rhizosphere soil factors, including yield at harvest, and the top 25 most abundant taxa from corn and wheat rhizosphere samples from three fields amplified with ITS2 primers. Correlations were measured using Spearman's correlation coefficient. Positive correlations are displayed in red and negative correlations in blue. Color intensity is proportional to the correlation coefficients. Taxa and soil factors are clustered using the Spearman distance metric. If a *P*-value was less than 0.05, it was flagged with one star (*); less than 0.01, with 2 stars (**); less than 0.001, with three stars (***). Adjustment for multiple comparisons was done with the Benjamin and Hochberg procedure.

4 Case study 2: using fungi isolated from high-yielding soil to improve crop performance in low-yielding soils

The variation in crop yield within a single field is well known. In this study, fungal isolates associated with high-yielding sites within corn fields (Saeed Cheema, 2022) were used as inocula for corn seedlings grown in pots containing soil from a low-yielding site, in the hope of improving growth to the level seen in high-yielding soil. As a preliminary to this, various fungal inocula, including both putatively beneficial and pathogenic strains, were added in a barley carrier with corn seedlings grown in sterile vermiculite. Significantly greater corn shoot dry mass was obtained with the putatively beneficial isolate *Fusarium oxysporum* CM003 compared to the amended control (with barley carrier only), but corn grown with the putatively beneficial strains *Penicillium janthinellum* and *Trichoderma koningii* had significantly lower dry mass than corn in the unamended control, and *T. atroviride* (putatively beneficial), *F. oxysporum* CM004, and *Setophoma terrestris* (both putatively pathogenic) had no significant treatment effects (Fig. 4).

In the main experiment, corn seedlings were grown in soil from a high-yielding or low-yielding site from a cornfield, with the low-yielding soil uninoculated (control) or inoculated with various putatively beneficial or pathogenic fungal strains. The anticipated beneficial isolate, *F. oxysporum* (CM003) did not increase corn seedling growth in inoculated low yield soil to levels comparable to seedlings grown in high yield soil, nor were there any consistent, statistically significant difference in corn growth in any of the fungal inoculum treatments relative to controls (data not shown).

Genomic DNA was isolated from washed corn roots and used in metabarcoding analyses with primers to the ITS2 barcode region (5.8S_Fun

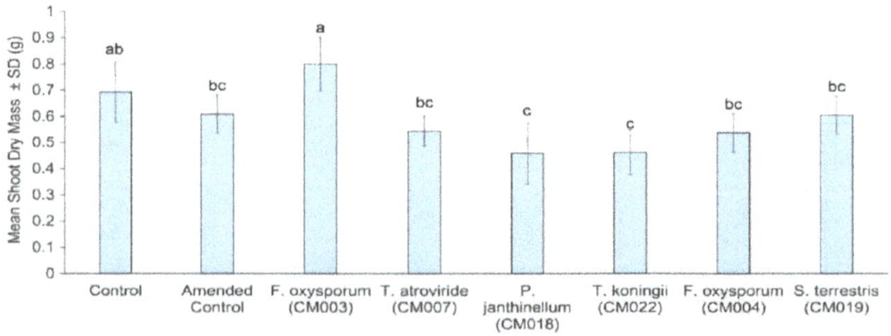

Figure 4 Mean shoot dry mass of corn grown 3 weeks in autoclaved vermiculite (control) or with autoclaved barley (amended control) or barley inoculated with various fungal isolates ($n = 6$). Data were analyzed using one-way ANOVA ($F_{(7, 40)} = 9.94$, $P < 0.001$), and Tukey's post-hoc test ($\alpha = 0.05$). Significance is denoted by lowercase letters above each bar. Treatments with the same letter are not considered significantly different.

and ITS4_Fun, Taylor et al., 2016). Several of the fungi used as primary inocula (*Trichoderma atroviride*, *T. koningii*, and *Penicillium janthinellum*) did not appear among the top ten fungi recovered from corn roots in these treatments, and *Setophoma terrestris* was much more abundant in uninoculated soils from both low-yield and high-yield sites than in the low yield soils inoculated with *Setophoma* (Fig. 5). Addition of autoclaved inoculum carrier (barley) to control soils greatly decreased the proportion of *Setophoma* and increased the proportion of *Actinomucor* recovered from roots of corn grown in these soils.

Principal coordinate analysis of the mycobiome communities in the various controls and treatments showed that the fungal community of high-yield soil was distinct from any of the low-yield soil controls, and none of the inoculated treatments shifted the community in the direction of amended or unamended high-yield soil (Fig. 6).

Soil chemical characters that were significantly correlated with the fungal mycobiome composition, according to Mantel tests using Pearson's correlation coefficient, were CEC, pH, and soil concentrations of aluminum, boron, calcium, manganese, potassium, sodium, and zinc, with high-yield soils characterized by higher values for Na and Zn, lower values for CEC and B, and high read counts for *Setophoma* and *Clohesyomyces*.

Overall, the results of case studies 1 and 2 suggest that both crops and the soil mycobiome respond to soil physicochemical characters and that, while the

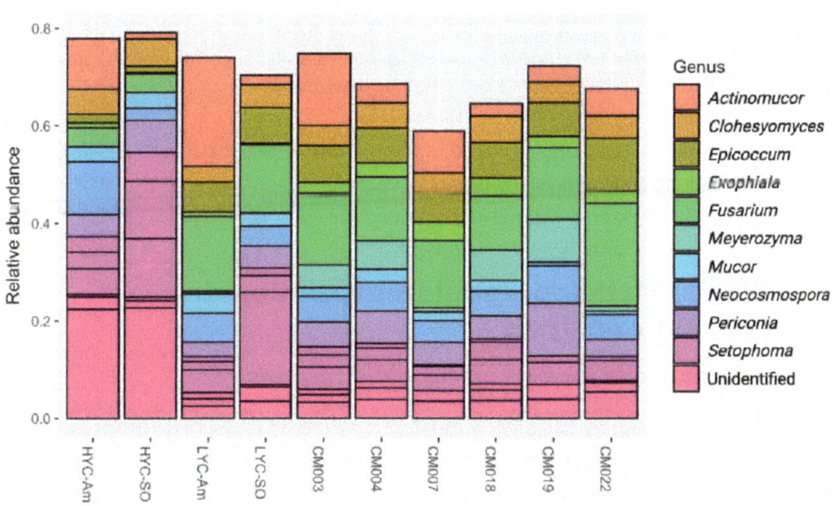

Figure 5 Relative abundance of the top ten genera recovered from washed roots of corn grown in high-yield (HY) or low-yield (LY) soils with varied fungal inocula, using metabarcoding with primers to the ITS2 region. SO = unamended control soils; Am = soils with autoclaved barley carrier; CM003 = *Fusarium oxysporum* (beneficial strain), CM004 = *F. oxysporum* (pathogenic strain), CM007 = *Trichoderma atroviride*, CM018 = *Penicillium janthinellum*, CM019 = *Setophoma terrestris*, and CM022 = *T. koningii*.

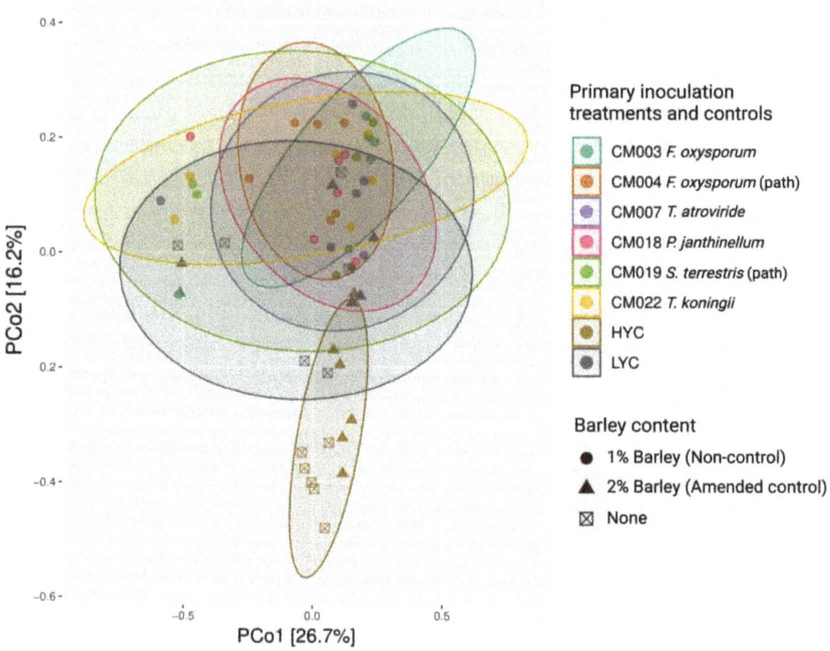

Figure 6 PcoA of the fungal communities recovered with ITS2 metabarcoding from washed roots of corn grown in high- (HYC) or low-yielding soils (LYC), or low-yielding soils with various fungal inoculants. Ellipses show 90% confidence area for each treatment. The fungal communities from high-yield control soils show much lower dispersion than those from control, amended, or inoculated low-yield soils, and the communities from unamended HYC are quite distinct from all other treatments.

composition of the root mycobiome is important in plant growth responses to soil conditions, it is not readily altered by simple inoculation of the soil in even the most controlled experimental conditions.

5 Detailed methodological considerations in mycobiome analyses

5.1 Considerations for barcoding gene regions

The best marker choice for eDNA metabarcoding is a short fragment that fits within the parameters of the intended sequencing platform, the dominant platform being Illumina MiSeq. At the time of writing, the latest Illumina MiSeq kit, the v3, allows for 2 × 300 nucleotide reads (forward and reverse), which can be paired or overlapped to allow for total amplicon lengths between 400 bp and 500 bp, including the primers with their barcodes and Illumina adaptors. Barcoding both forward and reverse primers ('double indexing') allows multiplexing of approximately 250 samples per run with sufficient read depth for each sample (Gloor et al., 2010; Kozich et al., 2013). Several long-read DNA sequencing technologies exist (e.g. PacBio and Oxford Nanopore) or are now entering the market (Eisenstein, 2023; Marx, 2023), which enable the use of longer barcoding regions, with some caveats discussed later.

Candidate gene markers require a central variable region flanked by two highly conserved regions in which, ideally, the variable region differentiates all species of a target group and the conserved regions on the ends of the target fragment are where PCR primers will anneal; these should not be shared with non-target taxa. Differentiating species using a gene marker requires that there is a barcoding gap – more interspecific variation than intraspecific variation. Additional thought must be given to genomic copy numbers for candidate genes. With single copy markers (typically protein-coding genes), any primer binding mismatches may result in an under-representation of species richness. Multiple-copy markers, such as nuclear ribosomal regions, are typically easier to amplify within a PCR, but may result in over-representation of species richness. Extrapolating from read numbers to template abundance and then to organism biomass or activity levels can be influenced by gene copy numbers per nucleus and density of nuclei relative to hyphal length (Ihrmark et al., 2012).

The region of the mitochondrial cytochrome *c* oxidase subunit 1 (COX1) is commonly used as an animal barcode (Andújar et al., 2018) but has been found to be a poor barcode locus for fungal community studies because it is difficult to amplify in fungi, often includes large introns, or lacks variability for many groups (Dentinger et al., 2011; Schoch et al., 2012). The internal transcribed spacer (ITS) region of the nuclear ribosomal DNA has been proposed as the barcode for molecular identification of fungi (Schoch et al., 2012). The ITS region, which comprises three subregions: ITS1, 5.8S, and ITS2, has many features that make it a strong candidate for a universal fungal DNA barcode region. These include its presence in all fungi, its variability within the ITS1 and ITS2 regions, a high number of copies per cell, and the near universality of its flanking regions (Scibetta et al., 2018). Internal transcribed spacer regions have been studied for several decades and are thus well represented in international reference databases (Schoch et al., 2012; Nilsson et al., 2019a,b). The entire ITS region has been proposed as a barcode region for metabarcoding studies and it is a good marker in lineages such as the Mucorales (Walther et al., 2013), but the length of this region (500–1000 bp) makes it unsuitable for many NGS platforms, including Illumina MiSeq (Scibetta et al., 2018). In addition, different fungal species have varying lengths of the ITS region, which causes biased amplification and inaccurate community results (Tedersoo et al., 2014; De Filippis et al., 2017). Although primer bias may always be an issue, the use of long-read technologies such as Oxford Nanopore or PacBio with circular consensus sequencing may permit highly informative analyses of the full-length ITS regions, although separation of the library into pools by amplicon length may be required to avoid length bias in the output (Tedersoo et al., 2018). Because the full fungal barcoding ITS region is too long for the commonly used Illumina technology, the ITS1 and ITS2 regions have most often been used as metabarcodes separately.

Anyone using ITS1, ITS2, or the full ITS region as a metabarcoding marker for fungi should consider the multiple issues with this region as caveats of their study (Kiss, 2012). These include possible heterogeneity between multiple copies of ITS within an organism (intraspecific variability) (O'Donnell and Cigelnik, 1997), gene heterozygosity not correlated with genomic relatedness (Tremble et al., 2020), copy number variability as related to the life strategy of a fungus (Kobayashi, 2011), length variation among target organisms that creates PCR biases (Ihrmark et al., 2012), and the outstanding problem that the most commonly used primers may not amplify all fungi (Pino-Bodas et al., 2013; Kijpornyongpan and Aime, 2016). Also, the ITS region does not differentiate closely related species within certain taxa (e.g. *Penicillium*, *Fusarium*, or polypores) or allow for the identification of unknown fungi at higher phylogenetic levels (Schoch et al., 2012; Lindahl et al., 2013; Öpik et al., 2014); even manual phylogenetic approaches fail because of the saturation of this region in deeper lineages. A comprehensive analysis on the advantages and disadvantages of the ITS region versus other nuclear and protein-coding regions for fungal eDNA metabarcoding can be found in Zelenka (2015). A global database of fungi records based on ITS metabarcoding studies is available at globalfungi.com (Větrovský et al., 2020).

Jumpponen and Jones (2009) were among the first to use existing ITS1 primers ITS1F (Gardes and Bruns, 1993) and ITS2 (White et al., 1990) on a 454 GS-FLX pyrosequencer to metabarcode phyllosphere fungi of a bur oak (*Quercus macrocarpa*). Alternative forward primers for the ITS1 region are ITS1Fngs, ITS5, and ITS1ngs, but all of these and ITS2 have mismatches with various groups of fungi, missing them altogether or creating a numerical bias against them in metabarcoding (Tedersoo et al., 2015). In addition, some fungi have an intron just downstream of primer ITS1F, making this primer unsuitable for these fungi (Tedersoo et al., 2015). A comparison between leaf-associated fungal communities amplified by the ITS1 and ITS2 rDNA regions have shown similar results in studies of community composition and structure (Bazzicalupo et al., 2013), and this was again confirmed by Blaalid et al. (2013). The ITS2 is considered more suitable for minibarcoding (shorter, platform-dependent barcoding regions) since it is less variable in length (Martin and Rygiewicz, 2005) resulting in fewer PCR cycles for the required product concentration (Ihrmark et al., 2012). However, Bazzicalupo et al. (2013) confirmed that ITS1 and ITS2 regions will invariably record different components of fungal communities.

Among the first primers of the ITS2 region used in metabarcoding were ITS3_KYO2 and ITS4_KYO3 (Toju et al., 2012), and these have been used successfully for root-associated fungi from agricultural soil rhizospheres (Asemaninejad et al., 2016; Kandasamy et al., 2021a, b), caterpillar gut microbiomes (Šigut et al., 2022), and many other applications. Other successful ITS2 primers developed for the Illumina include the 5.8S-Fun/ITS4-Fun (Taylor et al., 2016). Three primers developed by Ihrmark et al. (2012), fITS7, fITS9, and gITS7, outperformed ITS1-F/ITS4 (i.e. whole ITS amplicons) in both artificial communities and field samples since they produced shorter fragments and returned a more diverse amplicon community. The gITS7 primer returned a larger diversity of fungi but did amplify more plant sequences than the more exclusive fITS7 and fITS9 (Ihrmark et al., 2012). Taylor et al. (2016) noted that there may be low matches to Glomeromycota with fITS7, and fITS9 has strong secondary structure that affects amplification efficiency. The ITS2 subregion provides lower length variation and better resolution, resulting in less taxonomic bias than ITS1, and is the recommended target for fungal metabarcoding (Tedersoo et al., 2015; Nilsson et al., 2019b). In addition, primers targeting the ITS2 region are more selective and thereby advantageous when working with samples that include fungi as a minority, such as with soils and plant tissues (Taylor et al., 2016).

However, when using only ITS2 primers, several taxa have been found to be consistently under-represented (Asemaninejad et al., 2016), either because of primer mismatches or unusually long amplicons (as in Cantharellales). Several studies have highlighted the need for new targets and universal primer pairs as secondary DNA barcodes (Schoch et al., 2012). The D1 variable region of the large ribosomal subunit (LSU) (Hassouna et al., 1984) is useful in species-level identification and phylogenetic reconstruction of various fungal groups (Kurtzman and Robnett, 1998; Moncalvo et al., 2002). Combining ITS2 and D1 LSU primers recovers a greater diversity of fungi than with primers targeting only the ITS2 barcode (Asemaninejad et al., 2016; Poolman et al., 2021). In a study by Asemaninejad et al. (2016), D1 LSU primers recovered a greater number and phylogenetic diversity of sequences than ITS2 primers with approximately equivalent read depths. In the same study, OTUs belonging to 127 genera and 28 species were recovered using D1 LSU primers that were not recovered by ITS2 primers, and OTUs belonging to 10 unique genera and 16 species were recovered by ITS2 primers that were not recovered by LSU primers. Therefore, the combination of LSU and ITS2 primers used for metabarcoding analysis of fungal community data makes for a dataset that more accurately describes the community than either region alone.

Several other barcode genes have been used, with some being preferred for particular taxonomic groups. For instance, arbuscular mycorrhizal fungi (AMF, Glomeromycota) are better characterized by the V4 variable region of the 18S (SSU) rDNA than either of the ITS regions (Lumini et al., 2010; Kohout et al., 2014; Zelenka, 2015; Öpik and Davison, 2016; Thiéry et al., 2016). Specific primers include AML1 and AML2 (Lee et al., 2008), which were then improved upon by van Geel et al. (2014) by combining AML2 and NS31 (Simon et al., 1992; Helgason et al., 1998). The AMV4.5N-F/AMDG-R primer pair developed by Sato et al. (2005) were used to characterize AMF in various soil communities and continue to be used for new research (Ezeokoli et al., 2020; Higo et al., 2020; Wang et al., 2020; Kandasamy et al., 2021b). However, care must be taken when using the SSU variable regions as metabarcodes since the taxonomic accuracy can drop to <50% at the genus level (Edgar, 2018). Preference for the LSU D1–D2 region for metabarcoding in yeasts arose from the extensive use of this region for species identification which started in the early 1990s (Peterson and Kurtzman, 1991; Dulla et al., 2016; Boekhout et al., 2021) and because the ITS region was demonstrated to have high intraspecies variability in *Geotrichum candidum* (Alper et al., 2011). Other markers that have been used in specific groups include the mitochondrial 40S ribosomal protein S10 gene (rps10 gene) for oomycete community sequencing (Foster et al., 2022) and RNA polymerase II genes (rpb1 and rpb2; Matheny, 2005; Matheny et al., 2007) for lichen-forming Lecanoromycetes (Ascomycota) (Hofstetter et al., 2007) or general soil fungi (Větrovský et al., 2016). Use of alternate markers should be considered carefully since reference sequences for the basal Fungi, including Chytridiomycota, Zygomycota, and Glomeromycota are limited (Lutzoni et al., 2004). For a comprehensive list and map of primers for ITS, LSU, and SSU, please see https://unite.ut.ee/primers.php and for others mentioned in this chapter, please see Table 1.

5.2 Sequencing platforms for fungal metabarcoding

Illumina produced some of the first NGS platforms and continues to dominate the market by providing different machines that tackle a variety of sequencing methods for different applications such as DNA, RNA, methylation, and high-throughput sequencing (https://www.illumina.com/techniques/sequencing.html). Early players such as the Roche 454 pyrosequencing or Ion Torrent PGM proton detector sequencing platforms have gone by the wayside, both having high error rates and difficulties with homopolymer regions. The MiSeq continues to have the best cost-per-base performance despite the shorter read lengths (~200–300 nt) in comparison to Sanger chemistry (1000 nt). Much higher output options exist with Illumina, from HiSeq to NextSeq and NovaSeq systems, but all offer shorter reads (150 or 250 nt) than the maximum available on the MiSeq. These are all sequence-by-synthesis systems, detecting fluorescent nucleotides as they are added to the growing DNA molecule.

Newly emerging third-generation sequencing (TGS) technology such as the RSII, Sequel, Sequel II, IIe, and ONSO (Pacific Biosciences) and MinION, GridION, and PromethION (Oxford Nanopore) are defined by the detection

Table 1 Ribosomal RNA primers mentioned in this chapter

Primer name	Position in gene[d]	Sequence (5′ to 3′)	Product	Direction	Target	Remarks	Reference
SSU							
AML1	SSU 311	ATCAACTTTCGATGGTAGG ATAGA	SSU	F	Glomeromycota		Lee et al. (2008)
NS31	SSU 545	TTGGAGGGCAAGTCTG GTGCC	SSU	F	Fungi	Widely used	
AML2	SSU 1082	GAACCCAAACACTTTGG TTTCC	SSU	R	Glomeromycota		
AMV4.5N-F	SSU 622	AGCTCGTAGTTGAATTTCG	SSU	F	Glomeromycota		Sato et al. (2005)
AMDG-R	SSU 855	CCCAACTATCCCTATTAATCAT	SSU	R	Glomeromycota		
ITS							
ITS1-F	SSU 1732	CTTGGTCATTTAGAGGA AGTAA	ITS1	F	Fungi	Excludes basal fungi and *Tulasnella*	Gardes and Bruns (1993)
ITS2	5.8S 50	GCTGCGTTCTTCATCGATGC	ITS1	R	Universal	Excludes some Agaricales, Tremellales	White et al. (1990)
ITS3	5.8S 31	GCATCGATGAAGAACGCAGC	ITS2	F	Universal	Excludes Cantharellales	
ITS4	LSU 60	TCCTCCGCTTATTGATATGC	ITS2	R	Universal	Has central mismatches to some groups and Metazoa and Archaeorhizomycetes	
ITS3_KYO2	5.8S 36	GATGAAGAACGYAGYRAA	ITS2	F	Fungi	May exclude some chytrids, Glomeromycota, or other zygomycete groups[b].	Toju et al. (2012)
ITS4_KYO3	LSU 109	CTBTTVCCKCTTCACTCG	ITS2	R	Fungi		
5.8S-Fun[c]	5.8S 2	AACTTTYRRCAAYGGA TCWCT	ITS2	F	Fungi	Yields the full highly conserved 5.8S unit at 5′ end of amplicon with ITS4	Taylor et al. (2016)
ITS4-Fun[c]	LSU 61	AGCCTCCGCTTATTGATATGC TTAART	ITS2	R	Fungi	5′ 2 bases mismatched to many fungi	

(*Continued*)

Table 1 (*Continued*)

Primer name	Position in gene[d]	Sequence (5′ to 3′)	Product	Direction	Target	Remarks	Reference
fITS7[a]	5.8S 85	GTGARTCATCGAATCTTTG	ITS (5.8S)	F	Fungi; minus Glomeromycota	5′ mismatches to *Amanita*; excludes most *Penicillium*, Orbiliales and Mucorales	Ihrmark et al. (2012); gITS7 modified by RGT to broaden fungal matches
fITS9[a]	5.8S 42	GAACGCAGCRAAIIGYG	ITS (5.8S)	F	Fungi	5′ mismatches to *Amanita*; will still amplify Viridiplantae, algae and protists; strong 2° structure	
gITS7	5.8S 85	GyGARTCATCGARTCTTTG	ITS2	F	Universal	5′ mismatches to *Amanita*; mismatches some Saccharomycetales	
ITS4ngs	LSU 60	TCyTCCGCTTAnTGATATGC	ITS2	R	Universal	Modified by RGT to broaden fungal matches; works well with gITS7	White et al. (1990); Tedersoo et al. (2014)
				LSU			
LSU200-F (LSUBG-F)	LSU 87	AACKGCGAGTGAAGMGGGA	LSU	F	Fungi minus Ascomycota	Matches to many plants and unicellular eukaryotes; successful for Basidiomycota in conifer roots and peat; unsuccessful in corn roots[e]	Asemaninejad et al. (2016)
LSU481-R (LSUBG-R)	LSU 368	TCTTTCCCTCACGGTACTTG	LSU	R	Fungi minus Ascomycota		
LSU200A-F (LSUA-F)	LSU 87	AACKGCGAGTGAAGCRGBM	LSU	F	Ascomycota	Few off-target reads; successful in peat, conifer roots and corn roots[e]	
LSU476A-R (LSUA-R)	LSU 363	CSATCACTSTACTTGTKCGC	LSU	R	Ascomycota		

[a] Equimolar mix; I = inosine residues.
[b] See Asemaninejad et al. (2016).
[c] See primer design note in Taylor et al. (2016).
[d] Position of the first nucleotide (5') from the start of the gene region based on annotated *Saccharomyces cerevisiae* S288C (NC_001144.5).
[e] Asemaninejad et al. (2016); Poelman et al. (2021); Kandasamy et al. (2021a).

of single unmodified nucleotide additions using novel optical and electrical single-molecule techniques, which avoid PCR amplification and its inevitably associated biases (Steinbock and Radenovic, 2015). These long-read technologies are now entering the metabarcoding field (Tedersoo et al., 2020). In PacBio machines, amplicons are circularized, then read multiple times to generate high consensus accuracy. The PacBio RSII has 150 000 wells in a SMRT cell, but as few as 35 000–75 000 of these may produce successful single reads, resulting in 0.5–1 million bases per run (Rhoads and Au, 2015). Newer HiFi (highly accurate long read) sequencing options using the Sequel II and IIe systems released through PacBio can sequence the full-length 16S rDNA to generate 3.5 million >Q20 (1 error in 100) reads, to resolve up to 192 communities per SMRT Cell 8M, or perform metagenome profiling to generate 2.9 million >Q20 reads to profile up to 48 communities (Pacific Biosciences, n.d.). In both platforms, the read depth is 12–44x smaller than Illumina MiSeq but can provide slightly less accurate full 16S rDNA reads. Developments in long-read technology are occurring rapidly, with many new companies and products appearing on the market (Eisenstein, 2023).

The Oxford Nanopore MinION is a small, low-power device that is designed to perform *in situ* sequencing, as demonstrated by data from a rain forest (Quick et al., 2016), the Antarctic desert (Johnson et al., 2017), and the International Space Station (Castro-Wallace et al., 2017). MinION devices implement sequencing by shape and not sequencing by synthesis found in other platforms. Single amplicons in buffer solution are drawn through a Nanopore by a motor protein where a downstream electrode detects the changes in ion flow as differently shaped nucleotides pass through. The platform has been affected by low sequence quality (with error rates of 10–15%, or 1–5% in double-stranded sequencing methods) which makes it suited for resolving long repeats and bridging contigs in whole-genome sequencing applications, and less so for exploratory metabarcoding of natural communities (Loit et al., 2019; Bennett, 2021). Using the MinION with R9 chemistry in natural freshwater fungal communities, Bennett (2021) found that the portable device sequenced individual strands averaging 2 kbp without requiring amplification but it came at the cost of lower accuracy (~90%). In this study, the MinION was selected over the MiSeq in hopes that the higher error rates would not negate the phylogenetic resolution power of the full ITS sequences obtained. After quality filtering of 2D sequences, 3–15% of total reads passed (25 500–106 400 reads/sample), and there were also differences in throughput between samples, with only 322 711 reads per sample, which indicated that its use as a high-throughput ITS barcoding platform is not ideal (Bennett, 2021). Improvements to flow cell chemistry have helped with sequence quality; in a project sequencing the intergenic spacer (IGS) sequence between the tandem repeats of the RNA gene of fungal pathogens, using the R9.2 chemistry increased sequence quality to 98.3%, the R9.4.1 to 99.57%, and the R10.3 to 99.83% (Morrison et al., 2020). Unfortunately, the cost of consumables (barcoding kits and flow cells) makes the MinION more expensive and lower throughput for long-amplicon metabarcoding than PacBio. Again, rapid advances in technologies may change the comparisons reported here.

6 Future trends in research

Aside from the technological push to increase barcode lengths and sequencing depth, there is a parallel movement to better develop reference datasets from high-quality curated type and strain sequences. Within the fungi (of 14 582 NCBI genome assemblies), there are more representative Dikarya sequences of Ascomycota (11 789) and Basidiomycota (1967), but early diverging lineages – fungi *incertae sedis* – including Blastocladiomycota (10), Cryptomycota (3), Chytridiomycota (112), Microsporidia (99), Mucoromycota (302), Olpidiomycota (1), Sanchytriomycota (2), Zoopagomycota (210), and other cryptic species are often under-represented or missing entirely (Nephridiophagidae).[1]

One of the most popular references is SILVA (https://www.arb-silva.de/download/arb-files/) which contains curated and aligned rRNA reference sequences from the domains Bacteria, Archaea, and Eukaryota. SILVA release 138.1, as of August 2020, contains 2 225 272 high-quality SSU and 227 331 LSU sequences, with a respective increase of 134 072 and 28 488 sequences since its previous 132 iteration in December of 2017. In the current releases, Fungi comprise 10 273 (0.46%) and 5213 (2.29%) of all sequences, furthermore, Dikarya comprise 8588 (83.6%) and 4720 (90.5%) reads of Fungal SSU and LSU datasets, respectively. This leaves a substantial gap in knowledge for Fungi *incertae sedis*. Development of thorough reference datasets will continue to be the most limiting factor for metagenomic studies as technology allows for deeper sequencing of more low-abundant species in newer ecological niches. Currently, taxonomic identification relies on manually aligning query ASVs to references and analyzing tree topologies to identify low confidence matches or chimeric sequences that escaped detection. This process invariably takes up time prior to performing any basic taxonomic analyses. A final caution regards the many 'unidentified fungi' detected in environmental metabarcoding studies, of which a large proportion may not be Fungi at all (from personal experience and Rawson and Zahn, 2023). Investigators should always include sequences of non-fungal outgroups in the reference database(s) used for ASV or OTU identification to detect members of Rhizaria, Metazoa, or other non-fungal eukaryotes that are frequently misidentified as 'unidentified fungus' (Rawson and Zahn, 2023). This reiterates the continued need for improved,

1 As of May 2023; https://www.ncbi.nlm.nih.gov/assembly/?term=Fungi.

curated reference databases for identification of eDNA sequences, particularly from under-explored environments such as soil.

7 Conclusion

Root-associated fungi have huge impacts – positive and negative – on plant health and agricultural productivity. Disturbance, whether from agriculture, salinity, climate change, or other factors, can reduce the diversity of the soil fungal community, which in turn can lessen the ability of plants to cope with abiotic stressors and biotic stressors such as pests or pathogenic fungi. Metabarcoding has provided previously unknown dimensions about the diversity and identity of root-associated fungi, but what remains to be learned are functional properties of the fungal community or its individual members. Sustainable agriculture and healthy agroecosystems will depend upon healthy soil fungal communities and the scientific and technological abilities to manage beneficial and harmful soil fungi. In many good ways, the future is fungal.

8 Where to look for further information

8.1 Key resources

- Fungal Traits and FUNGuild: https://github.com/traitecoevo/fungaltraits and http://www.funguild.org/
- GlobalFungi database: https://globalfungi.com
- UNITE: https://unite.ut.ee/
- RDP: https://github.com/rdpstaff/classifier
- SILVA: https://www.arb-silva.de/download/arb-files/

8.2 Key references

For general eDNA methods, Taberlet et al. (2018); for fungal metabarcoding, Nilsson et al. (2019a), Tedersoo et al. (2019), Lahlali et al. (2021), and Tedersoo et al. (2022); for identification of fungi, including by eDNA sequences, see Lücking et al. (2020), Reynolds et al. (2022), and Vu et al. (2022); for assignment of ecological function, see Nguyen et al. (2016), Flores-Moreno et al. (2019), and Tanunchai et al. (2023).

9 References

Ai, C., Liang, G., Sun, J., Wang, X. and Zhou, W. (2012). Responses of extracellular enzyme activities and microbial community in both the rhizosphere and bulk soil to long-term fertilization practices in a fluvo-aquic soil, *Geoderma* 173–174, 330–338. doi: 10.1016/j.geoderma.2011.07.020.

Alberdi, A., Aizpurua, O., Gilbert, M. T. P. and Bohmann, K. (2018). Scrutinizing key steps for reliable metabarcoding of environmental samples, *Methods Ecol. Evol.* 9(1), 134–147. doi: 10.1111/2041-210X.12849.

Allen, M. F. (2007). Mycorrhizal fungi: highways for water and nutrients in arid soils, *Vadose Zone J.* 6(2), 291–297. doi: 10.2136/vzj2006.0068.

Almario, J., Jeena, G., Wunder, J., Langen, G., Zuccaro, A., Coupland, G. and Bucher, M. (2017). Root-associated fungal microbiota of nonmycorrhizal *Arabis alpina* and its contribution to plant phosphorus nutrition, *Proc. Natl. Acad. Sci. U. S. A.* 114(44), E9403–E9412.

Alper, I., Frenette, M. and Labrie, S. (2011). Ribosomal DNA polymorphisms in the yeast *Geotrichum candidum*, *Fungal Biol.* 115(12), 1259–1269. doi: 10.1016/j.funbio.2011.09.002.

Andújar, C., Arribas, P., Yu, D. W., Vogler, A. P. and Emerson, B. C. (2018). Why the COI barcode should be the community DNA metabarcode for the metazoa, *Mol. Ecol.* 27(20), 3968–3975.

Antoninka, A. J., Ritchie, M. E. and Johnson, N. C. (2015). The hidden Serengeti–mycorrhizal fungi respond to environmental gradients, *Pedobiologia* 58(5–6), 165–176. doi: 10.1016/j.pedobi.2015.08.001.

Asemaninejad, A., Weerasuriya, N., Gloor, G. B., Lindo, Z. and Thorn, R. G. (2016). New primers for discovering fungal diversity using nuclear larger ribosomal DNA, *PLoS ONE* 11(7), e0159043. doi: 10.1371/journal.pone.0159043.

Babu, A. G., Shim, J., Bang, K. S., Shea, P. J. and Oh, B. T. (2014). *Trichoderma virens* PDR-28: A heavy metal-tolerant and plant growth-promoting fungus for remediation and bioenergy crop production on mine tailing soil, *J. Environ. Manage.* 132, 129–134. doi: 10.1016/j.jenvman.2013.10.009.

Baron, N. C. and Rigobelo, E. C. (2022). Endophytic fungi: a tool for plant growth promotion and sustainable agriculture, *Mycology* 13(1), 39–55. doi: 10.1080/21501203.2021.1945699.

Barron, G. L. (1991). A new genus, *Rotiferophthora*, to accommodate the *Diheterospora*-like endoparasites of rotifers, *Can. J. Bot.* 69(3), 494–502. doi: 10.1139/b91-067.

Bazzicalupo, A. L., Bálint, M. and Schmitt, I. (2013). Comparison of ITS1 and ITS2 rDNA in 454 sequencing of hyperdiverse fungal communities, *Fungal Ecol.* 6(1), 102–109. doi: 10.1016/j.funeco.2012.09.003.

Bennett, D. (2021). *Development of Environmental Metabarcoding Analysis for Use in Ecological Studies of Aquatic Fungal and Oomycete Communities Using Nanopore Sequencing*, MSc Thesis, Central Washington University, Ellensburg, WA. Available at: https://digitalcommons.cwu.edu/etd/1550.

Berg, G., Rybakova, D., Fischer, D., Cernava, T., Vergès, M. C., Charles, T., Chen, X., Cocolin, L., Eversole, K., Corral, G. H., Kazou, M., Kinkel, L., Lange, L., Lima, N., Loy, A., Macklin, J. A., Maguin, E., Mauchline, T., McClure, R., Mitter, B., Ryan, M., Sarand, I., Smidt, H., Schelkle, B., Roume, H., Kiran, G. S., Selvin, J., Souza, R. S. C., van Overbeek, L., Singh, B. K., Wagner, M., Walsh, A., Sessitsch, A. and Schloter, M. (2020). Microbiome definition re-visited: old concepts and new challenges, *Microbiome* 8(1), 103. doi: 10.1186/s40168-020-00875-0.

Berg, G. and Smalla, K. (2009). Plant species and soil type cooperatively shape the structure and function of microbial communities in the rhizosphere, *FEMS Microbiol. Ecol.* 68(1), 1–13. doi: 10.1111/j.1574-6941.2009.00654.x.

Bidartondo, M. I., Bruns, T. D. and Blackwell, M. (2008). Preserving accuracy in GenBank, *Science* 319(5870), 1616.

Blaalid, R., Kumar, S., Nilsson, R. H., Abarenkov, K., Kirk, P. M. and Kauserud, H. (2013). ITS1 versus ITS2 as DNA metabarcodes for fungi, *Mol. Ecol. Resour.* 13(2), 218–224. doi: 10.1111/1755-0998.12065.

Boekhout, T., Aime, M. C., Begerow, D., Gabaldón, T., Heitman, J., Kemler, M., Khayhan, K., Lachance, M. A., Louis, E. J., Sun, S., Vu, D. and Yurkov, A. (2021). The evolving species concepts used for yeasts: from phenotypes and genomes to speciation networks, *Fungal Divers.* 109(1), 27–55. doi: 10.1007/s13225-021-00475-9.

Brundrett, M. C. (2004). Diversity and classification of mycorrhizal associations, *Biol. Rev. Camb. Philos. Soc.* 79(3), 473–495. doi: 10.1017/S1464793103006316.

Brundrett, M. C. (2009). Mycorrhizal associations and other means of nutrition of vascular plants: understanding the global diversity of host plants by resolving conflicting information and developing reliable means of diagnosis, *Plant Soil* 320(1–2), 37–77. doi: 10.1007/s11104-008-9877-9.

Brundrett, M. C. and Tedersoo, L. (2020). Resolving the mycorrhizal status of important northern hemisphere trees, *Plant Soil* 454(1–2), 3–34. doi: 10.1007/s11104-020-04627-9.

Buscot, F., Munch, J. C., Charcosset, J. Y., Gardes, M., Nehls, U. and Hampp, R. (2000). Recent advances in exploring physiology and biodiversity of ectomycorrhizas highlight the functioning of these symbioses in ecosystems, *FEMS Microbiol. Rev.* 24(5), 601–614. doi: 10.1111/j.1574-6976.2000.tb00561.x.

Callahan, B. J., McMurdie, P. J. and Holmes, S. P. (2017). Exact sequence variants should replace operational taxonomic units in marker-gene data analysis, *ISME J.* 11(12), 2639–2643. doi: 10.1038/ismej.2017.119.

Caporaso, J. G., Lauber, C. L., Walters, W. A., Berg-Lyons, D., Huntley, J., Fierer, N., Owens, S. M., Betley, J., Fraser, L., Bauer, M., Gormley, N., Gilbert, J. A., Smith, G. and Knight, R. (2012). Ultra-high-throughput microbial community analysis on the Illumina HiSeq and MiSeq platforms, *ISME J.* 6(8), 1621–1624. doi: 10.1038/ismej.2012.8.

Castro-Wallace, S. L., Chiu, C. Y., John, K. K., Stahl, S. E., Rubins, K. H., McIntyre, A. B. R., Dworkin, J. P., Lupisella, M. L., Smith, D. J., Botkin, D. J., Stephenson, T. A., Juul, S., Turner, D. J., Izquierdo, F., Federman, S., Stryke, D., Somasekar, S., Alexander, N., Yu, G., Mason, C. E. and Burton, A. S. (2017). Nanopore DNA sequencing and genome assembly on the International Space Station, *Sci. Rep.* 7(1), 18022. doi: 10.1038/s41598-017-18364-0.

Chang, H. X., Haudenshield, J. S., Bowen, C. R. and Hartman, G. L. (2017). Metagenome-wide association study and machine learning prediction of bulk soil microbiome and crop productivity, *Front. Microbiol.* 8, 519. doi: 10.3389/fmicb.2017.00519.

Chiarello, M., McCauley, M., Villéger, S. and Jackson, C. R. (2022). Ranking the biases: the choice of OTUs vs. ASVs in 16S rRNA amplicon data analysis has stronger effects on diversity measures than rarefaction and OTU identity threshold, *PLoS ONE* 17(2), e0264443. doi: 10.1371/journal.pone.0264443. PMID: 35202411; PMCID: PMC8870492.

Costa, R., Götz, M., Mrotzek, N., Lottmann, J., Berg, G. and Smalla, K. (2006). Effects of site and plant species on rhizosphere community structure as revealed by molecular analysis of microbial guilds, *FEMS Microbiol. Ecol.* 56(2), 236–249. doi: 10.1111/j.1574-6941.2005.00026.x.

DeAngelis, K. M., Brodie, E. L., DeSantis, T. Z., Andersen, G. L., Lindow, S. E. and Firestone, M. K. (2009). Selective progressive response of soil microbial community to wild oat roots, *ISME J.* 3(2), 168–178. doi: 10.1038/ismej.2008.103.

De Filippis, F., Laiola, M., Blaiotta, G. and Ercolini, D. (2017). Different amplicon targets for sequencing-based studies of fungal diversity, *Appl. Environ. Microbiol.* 83(17), e00905-17. doi: 10.1128/AEM.00905-17.

Dentinger, B. T. M., Didukh, M. Y. and Moncalvo, J. M. (2011). Comparing COI and ITS as DNA barcode markers for mushrooms and allies (Agaricomycotina), *PLoS ONE* 6(9), e25081. doi: 10.1371/journal.pone.0025081.

Dias, T., Dukes, A. and Antunes, P. M. (2015). Accounting for soil biotic effects on soil health and crop productivity in the design of crop rotations, *J. Sci. Food Agric.* 95(3), 447–454. doi: 10.1002/jsfa.6565.

Domsch, K. H., Gams, W. and Anderson, T.-H. (2007). *Compendium of Soil Fungi* (vol. 2). IHW-Verlag, Eching, Germany.

Dulla, E. L., Kathera, C., Gurijala, H. K., Mallakuntla, T. R., Srinivasan, P., Prasad, V., Mopati, R. D. and Jasti, P. K. (2016). Highlights of DNA Barcoding in identification of salient microorganisms like fungi, *J. Mycol. Med.* 26(4), 291–297. doi: 10.1016/j.mycmed.2016.05.006.

Edgar, R. C. (2018). Accuracy of taxonomy prediction for 16S rRNA and fungal ITS sequences, *PeerJ* 6, e4652. doi: 10.7717/peerj.4652.

Eisenstein, M. (2023). Navigating an increasingly crowded sequencing landscape, *Nature* 614(7949), 798–800.

Eldred, L. E., Thorn, R. G. and Smith, D. R. (2021). Simple matching using QIIME 2 and RDP reveals misidentified sequences and an under-representation of fungi in reference datasets, *Front. Genet.* 12, 768473, doi: 10.3389/fgene.2021.768473.

Emerson, J. B., Adams, R. I., Román, C. M. B., Brooks, B., Coil, D. A., Dahlhausen, K., Ganz, H. H., Hartmann, E. M., Hsu, T., Justice, N. B., Paulino-Lima, I. G., Luongo, J. C., Lymperopoulou, D. S., Gomez-Silvan, C., Rothschild-Mancinelli, B., Balk, M., Huttenhower, C., Nocker, A., Vaishampayan, P. and Rothschild, L. J. (2017). Schrödinger's microbes: tools for distinguishing the living from the dead in microbial ecosystems, *Microbiome* 5(1), 86. doi: 10.1186/s40168-017-0285-3.

Enebe, M. C. and Babalola, O. O. (2019). The impact of microbes in the orchestration of plants' resistance to biotic stress: a disease management approach, *Appl. Microbiol. Biotechnol.* 103(1), 9–25. doi: 10.1007/s00253-018-9433-3.

Ezeokoli, O. T., Mashigo, S. K., Maboeta, M. S., Bezuidenhout, C. C., Khasa, D. P. and Adeleke, R. A. (2020). Arbuscular mycorrhizal fungal community differentiation along a post-coal mining reclamation chronosequence in South Africa: a potential indicator of ecosystem recovery, *Appl. Soil Ecol.* 147, 103429. doi: 10.1016/j.apsoil.2019.103429.

Feinstein, L. M., Sul, W. J. and Blackwood, C. B. (2009). Assessment of bias associated with incomplete extraction of microbial DNA from soil, *Appl. Environ. Microbiol.* 75(16), 5428–5433. doi: 10.1128/AEM.00120-09. PMID: 19561189; PMCID: PMC2725469.

Fellbaum, C. R., Gachomo, E. W., Beesetty, Y., Choudhari, S., Strahan, G. D., Pfeffer, P. E., Kiers, E. T. and Bücking, H. (2012). Carbon availability triggers fungal nitrogen uptake and transport in arbuscular mycorrhizal symbiosis, *Proc. Natl. Acad. Sci. U. S. A.* 109(7), 2666–2671. doi: 10.1073/pnas.1118650109.

Fernandes, A. D., Macklaim, J. M., Linn, T. G., Reid, G. and Gloor, G. B. (2013). ANOVA-like differential expression (ALDEx) analysis for mixed population RNA-Seq, *PLoS ONE* 8(7), e67019. doi: 10.1371/journal.pone.0067019.

Flores-Moreno, H., Treseder, K. K., Cornwell, W. K.Maynard, D. S., Milo, A. M., Abarenkov, K., Afkhami, M. E., Aguilar-Trigueros, C. A., Bates, S., Bhatnagar, J. M., Busby, P. E., Christian, N., Crowther, T. W., Floudas, D., Gazis, R., Hibbett, D., Kennedy, P. F., Lindner, D. L., Nilsson, R. H., Powell, J., Schildhauer, M., Schilling, J. and Zanne, A. E. (2019). Fungaltraits aka funfun: a dynamic functional trait database for the world's fungi, *Dataset*. Available at: https://github.com/traitecoevo/fungaltraits.

Foster, Z. S. L., Albornoz, F. E., Fieland, V. J., Larsen, M. M., Jones, F. A., Tyler, B. M., Nguyen, H. D. T., Burgess, T. I., Riddell, C., Voglmayr, H., Martin, F. N. and Grünwald, N. J. (2022). A new oomycete metabarcoding method using the rps10 gene, *Phytobiomes J.* 6(3), 214–226. doi: 10.1094/PBIOMES-02-22-0009-R.

Freeman, S. and Rodriguez, R. J. (1993). Genetic conversion of a fungal plant pathogen to a non-pathogenic, endophytic mutualist, *Science* 260(5104), 75–78. doi: 10.1126/science.260.5104.75.

Garbeva, P., van Elsas, J. D. and van Veen, J. A. (2008). Rhizosphere microbial community and its response to plant species and soil history, *Plant Soil* 302(1–2), 19–32. doi: 10.1007/s11104-007-9432-0.

Gardes, M. and Bruns, T. D. (1993). ITS primers with enhanced specificity for basidiomycetes: application to the identification of mycorrhizae and rusts, *Mol. Ecol.* 2(2), 113–118.

Gloor, G. B., Hummelen, R., Macklaim, J. M., Dickson, R. J., Fernandes, A. D., MacPhee, R. and Reid, G. (2010). Microbiome profiling by Illumina sequencing of combinatorial sequence-tagged PCR products, *PLoS ONE* 5(10), e15406. doi: 10.1371/journal.pone.0015406.

Hassouna, N., Mithot, B. and Bachellerie, J. P. (1984). The complete nucleotide sequence of mouse 28S rRNA gene. Implications for the process of size increase of the large subunit rRNA in higher eukaryotes, *Nucl. Acids Res.* 12(8), 3563–3583. doi: 10.1093/nar/12.8.3563.

Hebert, P. D. N., Cywinska, A., Ball, S. L. and DeWaard, J. R. (2003). Biological identifications through DNA barcodes, *Proc. Biol. Sci.* 270(1512), 313–321. doi: 10.1098/rspb.2002.2218.

Helgason, T., Daniell, T. J., Husband, R., Fitter, A. H. and Young, J. P. (1998). Ploughing up the wood-wide web?, *Nature* 394(6692), 431. doi: 10.1038/28764.

Higo, M., Tatewaki, Y., Iida, K., Yokota, K. and Isobe, K. (2020). Amplicon sequencing analysis of arbuscular mycorrhizal fungal communities colonizing maize roots in different cover cropping and tillage systems, *Sci. Rep.* 10(1), 6039. doi: 10.1038/s41598-020-58942-3.

Hinsinger, P., Bengough, A. G., Vetterlein, D. and Young, I. M. (2009). Rhizosphere: biophysics, biogeochemistry and ecological relevance, *Plant Soil* 321(1–2), 117–152. doi: 10.1007/s11104-008-9885-9.

Hofstetter, V., Miadlikowska, J., Kauff, F. and Lutzoni, F. (2007). Phylogenetic comparison of protein-coding versus ribosomal RNA-coding sequence data: a case study of the Lecanoromycetes (Ascomycota), *Mol. Phylogenet. Evol.* 44(1), 412–426. doi: 10.1016/j.ympev.2006.10.016.

Ihrmark, K., Bodecker, I. T. M., Cruz-Martinez, K., Friberg, H., Kubartova, A., Schenck, J., Strid, Y., Stenlid, J., Brandström-Durling, M., Clemmensen, K. E. and Lindahl, B. D. (2012). New primers to amplify the fungal ITS2 region–evaluation by 454-sequencing of artificial and natural communities, *FEMS Microbiol. Ecol.* 82(3), 666–677.

Ikram, M., Ali, N. I., Jan, G., Jan, F. G., Rahman, I. U., Iqbal, A. and Hamayun, M. (2018). IAA producing fungal endophyte Penicillium roqueforti Thom., enhances stress tolerance and nutrients uptake in wheat plants grown on heavy metal contaminated soils, *PLoS ONE* 13(11), e0208150. doi: 10.1371/journal.pone.0208150.

Javot, H., Pumplin, N. and Harrison, M. J. (2007). Phosphate in the arbuscular mycorrhizal symbiosis: transport properties and regulatory roles, *Plant Cell Environ.* 30(3), 310–322.

Johnson, S. S., Zaikova, E., Goerlitz, D. S., Bai, Y. and Tighe, S. W. (2017). Real-time DNA qequencing in the Antarctic Dry Valleys using the Oxford nanopore sequencer, *J. Biomol. Tech.* 28(1), 2–7. doi: 10.7171/jbt.17-2801-009.

Jumpponen, A. and Jones, K. L. (2009). Massively parallel 454 sequencing indicates hyperdiverse fungal communities in temperate *Quercus macrocarpa* phyllosphere, *New Phytol.* 184(2), 438–448. doi: 10.1111/j.1469-8137.2009.02990.x.

Jumpponen, A., Jones, K. L. and Blair, J. (2010). Vertical distribution of fungal communities in tallgrass prairie soil, *Mycologia* 102(5), 1027–1041. doi: 10.3852/09-316.

Kandasamy, S., Weerasuriya, N., Subramanian, G., Thorn, R. G., Patterson, G., Ali, S. and Lazarovits, G. (2021a). Disentangling the association of corn root mycobiome with plant productivity and the importance of soil physicochemical balance in shaping their relationship, *Front. Sustain. Food Syst.* 5, 617332. doi: 10.3389/fsufs.2021.617332.

Kandasamy, S., Weerasuriya, N., White, J. F., Patterson, G. and Lazarovits, G. (2021b). Chapter 23. Soil's physical and nutritional balance is essential for establishing a healthy microbiome. In: White, J., Kumar, A. and Droby, S. (Eds). *Microbiome Stimulants for Crops*, Woodhead Publishing, Amsterdam, pp. 381–404.

Kijpornyongpan, T. and Aime, M. C. (2016). Rare or rarely detected? *Ceraceosorus guamensis* sp. nov.: a second described species of Ceraceosorales and the potential for underdetection of rare lineages with common sampling techniques, *Antonie Leeuwenhoek* 109(8), 1127–1139. doi: 10.1007/S10482-016-0715-4.

Kiss, L. (2012). Limits of nuclear ribosomal DNA internal transcribed spacer (ITS) sequences as species barcodes for Fungi, *Proc. Natl. Acad. Sci. U. S. A.* 109(27), E1811; author reply E1812.

Kobayashi, T. (2011). Regulation of ribosomal RNA gene copy number and its role in modulating genome integrity and evolutionary adaptability in yeast, *Cell. Mol. Life Sci.* 68(8), 1395–1403. doi: 10.1007/s00018-010-0613-2.

Kohout, P., Sudová, R., Janoušková, M., Čtvrtlíková, M., Hejda, M., Pánková, H., Slavíková, R., Štajerová, K., Vosátka, M. and Sýkorová, Z. (2014). Comparison of commonly used primer sets for evaluating arbuscular mycorrhizal fungal communities: is there a universal solution?, *Soil Biol. Biochem.* 68, 482–493. doi: 10.1016/j.soilbio.2013.08.027.

Kopylova, E., Navas-Molina, J. A., Mercier, C., Xu, Z. Z., Mahé, F., He, Y., Zhou, H. W., Rognes, T., Caporaso, J. G. and Knight, R. (2016). Open-source sequence clustering methods improve the state of the art, *MSystems* 1(1), e00003-15. doi: 10.1128/mSystems.00003-15.

Kozich, J. J., Westcott, S. L., Baxter, N. T., Highlander, S. K. and Schloss, P. D. (2013). Development of a dual-index sequencing strategy and curation pipeline for analyzing amplicon sequence data on the MiSeq Illumina sequencing platform, *Appl. Environ. Microbiol.* 79(17), 5112–5120. doi: 10.1128/AEM.01043-13.

Kurtzman, C. P. and Robnett, C. J. (1998). Identification and phylogeny of ascomycetous yeasts from analysis of nuclear large subunit (26S) ribosomal DNA partial sequences, *Antonie Leeuwenhoek* 73(4), 331–371. doi: 10.1023/A:1001761008817.

Lahlali, R., Ibrahim, D. S. S., Belabess, Z., Kadir Roni, M. Z., Radouane, N., Vicente, C. S. L., Menéndez, E., Mokrini, F., Barka, E. A., Galvão de Melo E Mota, M. and Peng, G. (2021). High-throughput molecular technologies for unraveling the mystery of soil microbial community: challenges and future prospects, *Heliyon* 7(10), e08142. doi: 10.1016/j.heliyon.2021.e08142.

Lee, J., Lee, S. and Young, J. P. (2008). Improved PCR primers for the detection and identification of arbuscular mycorrhizal fungi, *FEMS Microbiol. Ecol.* 65(2), 339–349. doi: 10.1111/j.1574-6941.2008.00531.x.

Lindahl, B. D., Nilsson, R. H., Tedersoo, L., Abarenkov, K., Carlsen, T., Kjøller, R., Kõljalg, U., Pennanen, T., Rosendahl, S., Stenlid, J. and Kauserud, H. (2013). Fungal community analysis by high-throughput sequencing of amplified markers – a user's guide, *New Phytol.* 199(1), 288–299. doi: 10.1111/nph.12243.

Liu, H., Li, J., Carvalhais, L. C., Percy, C. D., Prakash Verma, J., Schenk, P. M. and Singh, B. K. (2021). Evidence for the plant recruitment of beneficial microbes to suppress soil-borne pathogens, *New Phytol.* 229(5), 2873–2885. doi: 10.1111/nph.17057.

Loit, K., Adamson, K., Bahram, M., Puusepp, R., Anslan, S., Kiiker, R., Drenkhan, R. and Tedersoo, L. (2019). Relative performance of MinION (Oxford nanopore Technologies) versus Sequel (Pacific Biosciences) third-generation sequencing instruments in identification of agricultural and forest fungal pathogens, *Appl. Environ. Microbiol.* 85(21), e01368-19. doi: 10.1128/AEM.01368-19.

Lücking, R., Aime, M. C., Robbertse, B., Miller, A. N., Ariyawansa, H. A., Aoki, T., Cardinali, G., Crous, P. W., Druzhinina, I. S., Geiser, D. M., Hawksworth, D. L., Hyde, K. D., Irinyi, L., Jeewon, R., Johnston, P. R., Kirk, P. M., Malosso, E., May, T. W., Meyer, W., Öpik, M., Robert, V., Stadler, M., Thines, M., Vu, D., Yurkov, A. M., Zhang, N. and Schoch, C. L. (2020). Unambiguous identification of fungi: where do we stand and how accurate and precise is fungal DNA barcoding?, *IMA Fungus* 11, 14. doi: 10.1186/s43008-020-00033-z.

Lumini, E., Orgiazzi, A., Borriello, R., Bonfante, P. and Bianciotto, V. (2010). Disclosing arbuscular mycorrhizal fungal biodiversity in soil through a land-use gradient using a Pyrosequencing approach, *Environ. Microbiol.* 12(8), 2165–2179. doi: 10.1111/j.1462-2920.2009.02099.x.

Lutzoni, F., Kauff, F., Cox, C. J., McLaughlin, D., Celio, G., Dentinger, B., Padamsee, M., Hibbett, D., James, T. Y., Baloch, E., Grube, M., Reeb, V., Hofstetter, V., Schoch, C., Arnold, A. E., Miadlikowska, J., Spatafora, J., Johnson, D., Hambleton, S., Crockett, M., Shoemaker, R., Sung, G. H., Lücking, R., Lumbsch, T., O'Donnell, K., Binder, M., Diederich, P., Ertz, D., Gueidan, C., Hansen, K., Harris, R. C., Hosaka, K., Lim, Y. W., Matheny, B., Nishida, H., Pfister, D., Rogers, J., Rossman, A., Schmitt, I., Sipman, H., Stone, J., Sugiyama, J., Yahr, R. and Vilgalys, R. (2004). Assembling the fungal tree of life: progress, classification and evolution of subcellular traits, *Am. J. Bot.* 91(10), 1446–1480. doi: 10.3732/ajb.91.10.1446. PMID: 21652303.

Lynch, M. D. and Neufeld, J. D. (2015). Ecology and exploration of the rare biosphere, *Nat. Rev. Microbiol.* 13(4), 217–229. doi: 10.1038/nrmicro3400.

Lynch, M. D. and Thorn, R. G. (2006). Diversity of basidiomycetes in Michigan agricultural soils, *Appl. Environ. Microbiol.* 72(11), 7050–7056. doi: 10.1128/AEM.00826-06.

Márquez, L. M., Redman, R. S., Rodriguez, R. J. and Roossinck, M. J. (2007). A virus in a fungus in a plant: three-way symbiosis required for thermal tolerance, *Science* 315(5811), 513–515. doi: 10.1126/science.1136237. Erratum in: *Science* (2007) 316(5822), 201; author reply 201.

Martin, K. J. and Rygiewicz, P. T. (2005). Fungal-specific PCR primers developed for analysis of the ITS region of environmental DNA extracts, *BMC Microbiol.* 5, 28. doi: 10.1186/1471-2180-5-28.

Marx, V. (2023). Method of the year: long-read sequencing, *Nat. Meth.* 20(1), 6–11.

Matheny, P. B. (2005). Improving phylogenetic inference of mushrooms with RPB1 and RPB2 nucleotide sequences (*Inocybe*; Agaricales), *Mol. Phylogenet. Evol.* 35(1), 1–20.

Matheny, P. B., Wang, Z., Binder, M., Curtis, J. M., Lim, Y. W., Nilsson, R. H., Hughes, K. W., Hofstetter, V., Ammirati, J. F., Schoch, C. L., Langer, E., Langer, G., McLaughlin, D. J., Wilson, A. W., Frøslev, T., Ge, Z. W., Kerrigan, R. W., Slot, J. C., Yang, Z. L., Baroni, T. J., Fischer, M., Hosaka, K., Matsuura, K., Seidl, M. T., Vauras, J. and Hibbett, D. S. (2007). Contributions of rpb2 and tef1 to the phylogeny of mushrooms and allies (Basidiomycota, Fungi), *Mol. Phylogenet. Evol.* 43(2), 430–451. doi: 10.1016/j.ympev.2006.08.024.

Meena, M., Swapnil, P., Zehra, A., Aamir, M., Dubey, M. K., Goutam, J. and Upadhyay, R. S. (2017). Beneficial microbes for disease suppression and plant growth promotion. In: Singh, D. P., Singh, H. B. and Prabha, R. (Eds). *Plant-Microbe Interactions in Agro-Ecological Perspectives*, Springer Nature, Singapore, pp. 395–432.

Moncalvo, J. M., Vilgalys, R., Redhead, S. A., Johnson, J. E., James, T. Y., Catherine Aime, M., Hofstetter, V., Verduin, S. J., Larsson, E., Baroni, T. J., Greg Thorn, R., Jacobsson, S., Clémençon, H. and Miller, O. K. (2002). One hundred and seventeen clades of euagarics, *Mol. Phylogenet. Evol.* 23(3), 357–400. doi: 10.1016/S1055-7903(02)00027-1.

Morrison, G. A., Fu, J., Lee, G. C., Wiederhold, N. P., Cañete-Gibas, C. F., Bunnik, E. M. and Wickes, B. L. (2020). Nanopore sequencing of the fungal intergenic spacer sequence as a potential rapid diagnostic assay, *J. Clin. Microbiol.* 58(12), e01972-20. doi: 10.1128/JCM.01972-20.

Mueller, G. M., Bills, G. F. and Foster, M. S. (Eds). (2004). *Biodiversity of Fungi: Inventory and Monitoring Methods*, Elsevier Academic Press, Amsterdam.

Muneer, M. A., Huang, X., Hou, W., Zhang, Y., Cai, Y., Munir, M. Z., Wu, L. and Zheng, C. (2021). Response of fungal diversity, community composition and functions to nutrients management in red soil, *J. Fungi (Basel)* 7(7), 554. doi: 10.3390/jof7070554.

Nguyen, N. H., Song, Z., Bates, S. T., Branco, S., Tedersoo, L., Menke, J., Schilling, J. S. and Kennedy, P. G. (2016). FUNGuild: an open annotation tool for parsing fungal community datasets by ecological guild, *Fungal Ecol.* 20, 241–248.

Nilsson, R. H., Anslan, S., Bahram, M., Wurzbacher, C., Baldrian, P. and Tedersoo, L. (2019a). Mycobiome diversity: high-throughput sequencing and identification of fungi, *Nat. Rev. Microbiol.* 17(2), 95–109. doi: 10.1038/s41579-018-0116-y.

Nilsson, R. H., Larsson, K. H., Taylor, A. F. S., Bengtsson-Palme, J., Jeppesen, T. S., Schigel, D., Kennedy, P., Picard, K., Glöckner, F. O., Tedersoo, L., Saar, I., Kõljalg, U. and Abarenkov, K. (2019b). The UNITE database for molecular identification of fungi: handling dark taxa and parallel taxonomic classifications, *Nucl. Acids Res.* 47(D1), D259–D264. doi: 10.1093/nar/gky1022.

O'Donnell, K. and Cigelnik, E. (1997). Two divergent intragenomic rDNA ITS2 types within a monophyletic lineage of the fungus *Fusarium* are nonorthologous, *Mol. Phylogenet. Evol.* 7(1), 103–116. doi: 10.1006/mpev.1996.0376.

Öpik, M. and Davison, J. (2016). Uniting species- and community-oriented approaches to understand arbuscular mycorrhizal fungal diversity, *Fungal Ecol.* 24(B), 106–113. doi: 10.1016/j.funeco.2016.07.005.

Öpik, M., Davison, J., Moora, M. and Zobel, M. (2014). DNA-based detection and identification of Glomeromycota: the virtual taxonomy of environmental sequences, *Botany* 92(2), 135–147. doi: 10.1139/cjb-2013-0110.

Patel, J. S., Singh, A., Singh, H. B. and Sarma, B. K. (2015). Plant genotype, microbial recruitment and nutritional security, *Front. Plant Sci.* 6(August), 608. doi: 10.3389/fpls.2015.00608.

Pawlowski, M. L. and Hartman, G. L. (2016). Infection mechanisms and colonization patterns of fungi associated with soybean. In: Sultan, S. (Ed.), *Fungal Pathogenicity*. Tech Open, London, pp. 25–43. doi: 10.5772/62305.

Pawluczyk, M. L., Weiss, J., Links, M. G., Egaña Aranguren, M., Wilkinson, M. D. and Egea-Cortines, M. (2015). Quantitative evaluation of bias in PCR amplification and next-generation sequencing derived from metabarcoding samples, *Anal. Bioanal. Chem.* 407(7), 1841–1848. doi: 10.1007/s00216-014-8435-y.

Peterson, S. W. and Kurtzman, C. P. (1991). Ribosomal RNA sequence divergence among sibling species of yeasts, *Syst. Appl. Microbiol.* 14(2), 124–129. doi: 10.1016/S0723-2020(11)80289-4.

Petrini, O. (1996). Ecological and physiological aspect of host specificity in endophytic fungi. In: Redlin, S. C. and Carris, L. M. (Eds). *Endophytic Fungi in Grasses and Woody Plants*, APS Press, St. Paul, MN, pp. 87–100.

Philippot, L., Raaijmakers, J. M., Lemanceau, P. and Van Der Putten, W. H. (2013). Going back to the roots: the microbial ecology of the rhizosphere, *Nat. Rev. Microbiol.* 11(11), 789–799. doi: 10.1038/nrmicro3109.

Pino-Bodas, R., Martín, M. P., Burgaz, A. R. and Lumbsch, H. T. (2013). Species delimitation in *Cladonia* (Ascomycota): a challenge to the DNA barcoding philosophy, *Mol. Ecol. Resour.* 13(6), 1058–1068. doi: 10.1111/1755-0998.12086.

Pirozynski, K. A. and Malloch, D. W. (1975). The origin of land plants: a matter of mycotrophism, *Biosystems.* 6(3), 153–164. doi: 10.1016/0303-2647(75)90023-4.

Poelman, A., Weerasuriya, N., Vanderwolf, K. J., Malloch, D., McAlpine, D. F. and Thorn, R. G. (2021). Fungi associated with aeroponic roots in caves and mines of New Brunswick, *Fungal Ecol.* 52, 101074. doi: 10.1016/j.funeco.2021.101074.

Pusztahelyi, T., Holb, I. J. and Pócsi, I. (2015). Secondary metabolites in fungus-plant interactions, *Front. Plant Sci.* 6, 573. doi: 10.3389/fpls.2015.00573.

Quick, J., Loman, N. J., Duraffour, S., Simpson, J. T., Severi, E., Cowley, L., Bore, J. A., Koundouno, R., Dudas, G., Mikhail, A., Ouédraogo, N., Afrough, B., Bah, A., Baum, J. H., Becker-Ziaja, B., Boettcher, J. P., Cabeza-Cabrerizo, M., Camino-Sanchez, A., Carter, L. L., Doerrbecker, J., Enkirch, T., Dorival, I. G. G., Hetzelt, N., Hinzmann, J., Holm, T., Kafetzopoulou, L. E., Koropogui, M., Kosgey, A., Kuisma, E., Logue, C. H., Mazzarelli, A., Meisel, S., Mertens, M., Michel, J., Ngabo, D., Nitzsche, K., Pallash, E., Patrono, L. V., Portmann, J., Repits, J. G., Rickett, N. Y., Sachse, A., Singethan, K., Vitoriano, I., Yemanaberhan, R. L., Zekeng, E. G., Trina, R., Bello, A., Sall, A. A., Faye, O., Faye, O., Magassouba, N., Williams, C. V., Amburgey, V., Winona, L., Davis, E., Gerlach, J., Washington, F., Monteil, V., Jourdain, M., Bererd, M., Camara, A., Somlare, H., Camara, A., Gerard, M., Bado, G., Baillet, B., Delaune, D., Nebie, K. Y., Diarra, A., Savane, Y., Pallawo, R. B., Gutierrez, G. J., Milhano, N., Roger, I., Williams, C. J., Yattara, F., Lewandowski, K., Taylor, J., Rachwal, P., Turner, D., Pollakis, G., Hiscox, J. A., Matthews, D. A., O'Shea, M. K., Johnston, A. M., Wilson, D., Hutley, E., Smit, E., Di Caro, A., Woelfel, R., Stoecker, K., Fleischmann, E., Gabriel, M., Weller, S. A., Koivogui, L., Diallo, B., Keita, S., Rambaut, A., Formenty, P., Gunther, S. and Carroll, M. W. (2016).

Real-time, portable genome sequencing for Ebola surveillance, *Nature* 530(7589), 228–232. doi: 10.1038/nature16996.

Rai, M. and Agarkar, G. (2016). Plant-fungal interactions: what triggers the fungi to switch among lifestyles?, *Crit. Rev. Microbiol.* 42(3), 428–438. doi: 10.3109/1040841X.2014.958052.

Rashid, M. I., Mujawar, L. H., Shahzad, T., Almeelbi, T., Ismail, I. M. and Oves, M. (2016). Bacteria and fungi can contribute to nutrients bioavailability and aggregate formation in degraded soils, *Microbiol. Res.* 183, 26–41. doi: 10.1016/j.micres.2015.11.007.

Rawson, C. and Zahn, G. (2023). Inclusion of database outgroups reduces false positives in fungal metabarcoding taxonomic assignments, *Mycologia* 115(4), 571–577. doi: 10.1080/00275514.2023.2206931.

Ray, P., Lakshmanan, V., Labbé, J. L. and Craven, K. D. (2020). Microbe to microbiome: a paradigm shift in the application of microorganisms for sustainable agriculture, *Front. Microbiol.* 11, 622926. doi: 10.3389/fmicb.2020.622926.

Redman, R. S., Sheehan, K. B., Stout, R. G., Rodriguez, R. J. and Henson, J. M. (2002). Thermotolerance generated by plant/fungal symbiosis, *Science* 298(5598), 1581. doi: 10.1126/science.1072191.

Reynolds, N. K., Jusino, M. A., Stajich, J. E. and Smith, M. E. (2022). Understudied, underrepresented and unknown: methodological biases that limit detection of early diverging fungi from environmental samples, *Mol. Ecol. Resour.* 22(3), 1065–1085. doi: 10.1111/1755-0998.13540.

Rhoads, A. and Au, K. F. (2015). PacBio sequencing and its applications, *Genomics Proteomics Bioinformatics* 13(5), 278–289. doi: 10.1016/j.gpb.2015.08.002.

Rodriguez, R. J., Henson, J., Van Volkenburgh, E., Hoy, M., Wright, L., Beckwith, F., Kim, Y. O. and Redman, R. S. (2008). Stress tolerance in plants via habitat-adapted symbiosis, *ISME J.* 2(4), 404–416. doi: 10.1038/ismej.2007.106.

Saeed Cheema, N. F. (2022). *Manipulating the Root Mycobiome to Improve Plant Performance and Reduce Pathogen Pressure in Corn (Zea mays)*, MSc thesis, University of Western Ontario, London. Available at: https://ir.lib.uwo.ca/etd/8859.

Sato, K., Suyama, Y., Saito, M. and Sugawara, K. (2005). A new primer for discrimination of arbuscular mycorrhizal fungi with polymerase chain reaction-denature gradient gel electrophoresis, *Grassl. Sci.* 51(2), 179–181.

Schnitzer, S. A., Klironomos, J. N., Hillerislambers, J., Kinkel, L. L., Reich, P. B., Xiao, K., Rillig, M. C., Sikes, B. A., Callaway, R. M., Mangan, S. A., van Nes, E. H. and Scheffer, M. (2011). Soil microbes drive the classic plant diversity– productivity pattern, *Ecology* 92(2), 296–303. doi: 10.1890/10-0773.1.

Schoch, C. L., Seifert, K. A., Huhndorf, S., Robert, V., Spouge, J. L., Levesque, C. A., Chen, W., Fungal Barcoding Consortium and Fungal Barcoding Consortium Author List (2012). Nuclear ribosomal internal transcribed spacer (ITS) region as a universal DNA barcode marker for Fungi, *Proc. Natl. Acad. Sci. U. S. A.* 109(16), 6241–6246. doi: 10.1073/pnas.1117018109.

Schulz, B. and Boyle, C. (2005). The endophytic continuum, *Mycol. Res.* 109(6), 661–686. doi: 10.1017/S095375620500273X.

Scibetta, S., Schena, L., Abdelfattah, A., Pangallo, S. and Cacciola, S. O. (2018). Selection and experimental evaluation of universal primers to study the fungal microbiome of higher plants, *Phytobiomes J.* 2(4), 225–236. doi: 10.1094/PBIOMES-02-18-0009-R.

Šigut, M., Pyszko, P., Šigutová, H., Višňovská, D., Kostovčík, M., Kotásková, N., Dorňák, O., Kolařík, M. and Drozd, P. (2022). Fungi are more transient than bacteria in caterpillar gut microbiomes, *Sci. Rep.* 12(1), 15552. doi: 10.1038/s41598-022-19855-5.

Simon, L., Lalonde, M. and Bruns, T. D. (1992). Specific amplification of 18S fungal ribosomal genes from vesicular-arbuscular endomycorrhizal fungi colonizing roots, *Appl. Environ. Microbiol.* 58(1), 291–295. doi: 10.1128/aem.58.1.291-295.1992.

Smith, S. and Read, D. (2008). *Mycorrhizal Symbiosis*, Elsevier, Amsterdam. doi: 10.1016/B978-0-12-370526-6.X5001-6.

Southworth, D. (2012). *Biocomplexity of Plant-Fungal Interactions*, Wiley-Blackwell, Chichester.

Staskawicz, B. J. (2001). Genetics of plant-pathogen interactions specifying plant disease resistance, *Plant Physiol.* 125(1), 73–76. doi: 10.1104/pp.125.1.73.

Steinbock, L. J. and Radenovic, A. (2015). The emergence of nanopores in next-generation sequencing, *Nanotechnology* 26(7), 074003. doi: 10.1088/0957-4484/26/7/074003.

Taberlet, P., Bonin, A., Zinger, L. and Coissac, E. (2018). *Environmental DNA: for Biodiversity Research and Monitoring*, Oxford Academic, Oxford. doi: 10.1093/oso/9780198767220.001.0001.

Taberlet, P., Coissac, E., Pompanon, F., Brochmann, C. and Willerslev, E. (2012). Towards next-generation biodiversity assessment using DNA metabarcoding, *Mol. Ecol.* 21(8), 2045–2050. doi: 10.1111/j.1365-294X.2012.05470.x.

Tanunchai, B., Ji, L., Schroeter, S. A., Wahdan, S. F. M., Hossen, S., Delelegn, Y., Buscot, F., Lehnert, A. S., Alves, E. G., Hilke, I., Gleixner, G., Schulze, E. D., Noll, M. and Purahong, W. (2023). FungalTraits vs. FUNGuild: comparison of ecological functional assignments of leaf- and needle-associated fungi across 12 temperate tree species, *Microb. Ecol.* 85(2), 411–428. doi: 10.1007/s00248-022-01973-2.

Taylor, D. L., Walters, W. A., Lennon, N. J., Bochicchio, J., Krohn, A., Caporaso, J. G. and Pennanen, T. (2016). Accurate estimation of fungal diversity and abundance through improved lineage-specific primers optimized for Illumina amplicon sequencing, *Appl. Environ. Microbiol.* 82(24), 7217–7226. doi: 10.1128/AEM.02576-16.

Taylor, J. W. (2011). One fungus = one name: DNA and fungal nomenclature twenty years after PCR, *IMA Fungus* 2(2), 113–120. doi: 10.5598/imafungus.2011.02.02.01.

Tedersoo, L., Anslan, S., Bahram, M., Kõljalg, U. and Abarenkov, K. (2020). 'Identifying the 'unidentified' fungi: a global-scale long-read third-generation sequencing approach, *Fungal Divers.* 103(1), 273–293. doi: 10.1007/S13225-020-00456-4.

Tedersoo, L., Anslan, S., Bahram, M., Põlme, S., Riit, T., Liiv, I., Kõljalg, U., Kisand, V., Nilsson, H., Hildebrand, F., Bork, P. and Abarenkov, K. (2015). Shotgun metagenomes and multiple primer pair-barcode combinations of amplicons reveal biases in metabarcoding analyses of fungi, *MycoKeys* 10, 1–43. doi: 10.3897/mycokeys.10.4852.

Tedersoo, L., Bahram, M., Põlme, S., Kõljalg, U., Yorou, N. S., Wijesundera, R., Villarreal Ruiz, L., Vasco-Palacios, A. M., Thu, P. Q., Suija, A., Smith, M. E., Sharp, C., Saluveer, E., Saitta, A., Rosas, M., Riit, T., Ratkowsky, D., Pritsch, K., Põldmaa, K., Piepenbring, M., Phosri, C., Peterson, M., Parts, K., Pärtel, K., Otsing, E., Nouhra, E., Njouonkou, A. L., Nilsson, R. H., Morgado, L. N., Mayor, J., May, T. W., Majuakim, L., Lodge, D. J., Lee, S. S., Larsson, K. H., Kohout, P., Hosaka, K., Hiiesalu, I., Henkel, T. W., Harend, H., Guo, L. D., Greslebin, A., Grelet, G., Geml, J., Gates, G., Dunstan, W., Dunk, C., Drenkhan, R., Dearnaley, J., De Kesel, A., Dang, T., Chen, X., Buegger, F., Brearley, F. Q., Bonito, G., Anslan, S., Abell, S. and Abarenkov, K. (2014). Fungal biogeography. Global diversity and geography of soil fungi, *Science* 346(6213), 1256688, doi: 10.1126/science.1256688.

Tedersoo, L., Bahram, M., Zinger, L., Nilsson, R. H., Kennedy, P. G., Yang, T., Anslan, S. and Mikryukov, V. (2022). Best practices in metabarcoding of fungi: from experimental design to results, *Mol. Ecol.* 31(10), 2769–2795. doi: 10.1111/MEC.16460.

Tedersoo, L., Drenkhan, R., Anslan, S., Morales-Rodriguez, C. and Cleary, M. (2019). High-throughput identification and diagnostics of pathogens and pests: overview and practical recommendations, *Mol. Ecol. Resour.* 19(1), 47–76. doi: 10.1111/1755-0998.12959.

Tedersoo, L., Tooming-Klunderud, A. and Anslan, S. (2018). PacBio metabarcoding of Fungi and other eukaryotes: errors, biases and perspectives, *New Phytol.* 217(3), 1370–1385. doi: 10.1111/nph.14776. PMID: 28906012.

Thakuria, D., Schmidt, O., Mac Siúrtáin, M., Egan, D. and Doohan, F. M. (2008). Importance of DNA quality in comparative soil microbial community structure analyses, *Soil Biol. Biochem.* 40(6), 1390–1403.

Thorn, R. G., Scott, J. and Lachance, M. A. (2007). Methods for studying terrestrial fungal ecology and diversity. In: Reddy, C. A., Beveridge, T. J., Breznak, J. A., Marzluf, G. A., Schmidt, T. M. and Snyder, L. R. (Eds). *Methods for General and Molecular Microbiology* (3rd ed.), ASM Press, Washington, DC, pp. 929–951.

Thiéry, O., Vasar, M., Jairus, T., Davison, J., Roux, C., Kivistik, P. A., Metspalu, A., Milani, L., Saks, Ü., Moora, M., Zobel, M. and Öpik, M. (2016). Sequence variation in nuclear ribosomal small subunit, internal transcribed spacer and large subunit regions of *Rhizophagus irregularis* and *gigaspora margarita* is high and isolate-dependent, *Mol. Ecol.* 25(12), 2816–2832. doi: 10.1111/mec.13655.

Tkacz, A., Cheema, J., Chandra, G., Grant, A. and Poole, P. S. (2015). Stability and succession of the rhizosphere microbiota depends upon plant type and soil composition, *ISME J.* 9(11), 2349–2359. doi: 10.1038/ismej.2015.41.

Tkacz, A. and Poole, P. (2015). Role of root microbiota in plant productivity, *J. Exp. Bot.* 66(8), 2167–2175. doi: 10.1093/jxb/erv157.

Toju, H., Tanabe, A. S., Yamamoto, S. and Sato, H. (2012). High-coverage ITS primers for the DNA-based identification of ascomycetes and basidiomycetes in environmental samples, *PLoS ONE* 7(7), e40863. doi: 10.1371/journal.pone.0040863.

Tremble, K., Suz, L. M. and Dentinger, B. T. M. (2020). Lost in translation: population genomics and long-read sequencing reveals relaxation of concerted evolution of the ribosomal DNA cistron, *Mol. Phylogenet. Evol.* 148, 106804. doi: 10.1016/j.ympev.2020.106804.

Tzean, S. S. and Barron, G. L. (1981). A new and unusual fungus living in the gut of free-living nematodes, *Can. J. Bot.* 59(10), 1861–1866. doi: 10.1139/b81-246.

Ursell, L. K., Metcalf, J. L., Parfrey, L. W. and Knight, R. (2012). Defining the human microbiome, *Nutr. Rev.* 70 (Suppl. 1), S38–S44. doi: 10.1111/j.1753-4887.2012.00493.x.

Vandenkoornhuyse, P., Quaiser, A., Duhamel, M., Le Van, A. and Dufresne, A. (2015). The importance of the microbiome of the plant holobiont, *New Phytol.* 206(4), 1196–1206. doi: 10.1111/nph.13312.

Van Geel, M., Busschaert, P., Honnay, O. and Lievens, B. (2014). Evaluation of six primer pairs targeting the nuclear rRNA operon for characterization of arbuscular mycorrhizal fungal (AMF) communities using 454 Pyrosequencing, *J. Microbiol. Methods* 106, 93–100. doi: 10.1016/j.mimet.2014.08.006.

Verghese, J., Abrams, J., Wang, Y. and Morano, K. A. (2012). Biology of the heat shock response and protein chaperones: budding yeast (*Saccharomyces cerevisiae*) as a model system, *Microbiol. Mol. Biol. Rev.* 76(2), 115–158. doi: 10.1128/MMBR.05018-11.

Větrovský, T., Kolařík, M., Žifčáková, L., Zelenka, T. and Baldrian, P. (2016). The rpb2 gene represents a viable alternative molecular marker for the analysis of environmental fungal communities, *Mol. Ecol. Resour.* 16(2), 388–401. doi: 10.1111/1755-0998.12456.

Větrovský, T., Morais, D., Kohout, P., Lepinay, C., Algora, C., Awokunle Hollá, S., Bahnmann, B. D., Bílohnědá, K., Brabcová, V., D'Alò, F., Human, Z. R., Jomura, M., Kolařík, M., Kvasničková, J., Lladó, S., López-Mondéjar, R., Martinović, T., Mašínová, T., Meszárošová, L., Michalčíková, L., Michalová, T., Mundra, S., Navrátilová, D., Odriozola, I., Piché-Choquette, S., Štursová, M., Švec, K., Tláskal, V., Urbanová, M., Vlk, L., Voříšková, J., Žifčáková, L. and Baldrian, P. (2020). GlobalFungi, a global database of fungal occurrences from high-throughput-sequencing metabarcoding studies, *Sci. Data* 7(1), 228. doi: 10.1038/s41597-020-0567-7.

Vu, D., Nilsson, R. H. and Verkley, G. J. M. (2022). Dnabarcoder: an open-source software package for analysing and predicting DNA sequence similarity cutoffs for fungal sequence identification, *Mol. Ecol. Resour.* 22(7), 2793–2809. doi: 10.1111/1755-0998.13651.

Wallace, M. E. (2022). *Root-inhabiting and Rhizosphere Mycobiomes and Crop Yield of Corn and Wheat*, MSc thesis, University of Western Ontario, London. Available at: https://ir.lib.uwo.ca/etd/9028.

Walther, G., Pawłowska, J., Alastruey-Izquierdo, A., Wrzosek, M., Rodriguez-Tudela, J. L., Dolatabadi, S., Chakrabarti, A. and de Hoog, G. S. (2013). DNA barcoding in Mucorales: an inventory of biodiversity, *Persoonia* 30, 11–47. doi: 10.3767/003158513X665070.

Wang, F., Zhang, X., Zhang, S., Zhang, S. and Sun, Y. (2020). Interactions of microplastics and cadmium on plant growth and arbuscular mycorrhizal fungal communities in an agricultural soil, *Chemosphere* 254, 126791. doi: 10.1016/j.chemosphere.2020.126791.

Wang, Y., Jiao, P., Guo, W., Du, D., Hu, Y., Tan, X. and Liu, X. (2021). Changes in bulk and rhizosphere soil microbial diversity and composition along an age gradient of Chinese fir (Cunninghamia lanceolata) plantations in subtropical China, *Front. Microbiol.* 12, 777862. doi: 10.3389/fmicb.2021.777862.

Warcup, J. H. and Talbot, P. H. B. (1965). Ecology and identity of mycelia from soil III, *Trans. Br. Mycol. Soc.* 48(2), 249–259.

Westcott, S. L. and Schloss, P. D. (2015). De novo clustering methods outperform reference-based methods for assigning 16S rRNA gene sequences to operational taxonomic units, *PeerJ* 3, e1487. doi: 10.7717/peerj.1487.

White, J. F., Kingsley, K. L., Zhang, Q., Verma, R., Obi, N., Dvinskikh, S., Elmore, M. T., Verma, S. K., Gond, S. K. and Kowalski, K. P. (2019). Review: endophytic microbes and their potential applications in crop management, *Pest Manag. Sci.* 75(10), 2558–2565. doi: 10.1002/ps.5527.

White, T. J., Bruns, T., Lee, S. B. and Taylor, J. W. (1990). Amplification and direct sequencing of fungal ribosomal RNA genes for phylogenetics. In: Innis, M. A., Gelfand, D. H., Shinsky, J. J. and White, T. J. (Eds). *PCR Protocols: A Guide to Methods and Applications*, Academic Press, San Diego, CA, pp. 315–322.

Xu, Y., Wang, G., Jin, J., Liu, J., Zhang, Q. and Liu, X. (2009). Bacterial communities in soybean rhizosphere in response to soil type, soybean genotype and their growth stage, *Soil Biol. Biochem.* 41(5), 919–925. doi: 10.1016/j.soilbio.2008.10.027.

Zelenka, T. (2015). *Comparison of ITS nrDNA and Alternative Markers for Fungal Metabarcoding in Environmental Samples*, MSc Thesis, Charles University, Prague. Available at: https://dspace.cuni.cz/bitstream/handle/20.500.11956/2607/DPTX_2013_1_11310_0_420916_0_144059.pdf?sequence=1&isAllowed=y.

Part 3

Analysing structure and dynamics of soil microbiomes

Chapter 6

Advances in understanding microbial communities in the rhizosphere

R Regmi[†], Qi Yang[†] and Vadakattu V. S. R. Gupta, CSIRO Agriculture and Food, Australia

1 Introduction

Plant rhizosphere is a dynamic region governed by complex interactions between plants and organisms closely associated with roots (Walker et al., 2003). Root exudates, secretions from roots, rhizodeposits, and microbially modified products, dictate microbiome–plant interactions, benefiting plants, improving soil fertility, and enhancing toxic chemical breakdown (Vives-Peris et al., 2020). These interactions often benefit plants, improve soil fertility, and enhance the breakdown of toxic chemicals. Rhizosphere represents a thin layer of soil surrounding roots, i.e. the intersection between plant roots and the surrounding soil (~2 mm), harbours a dynamic and highly active microbial community. It can be viewed as a semicontinuous zone between the interior of the root (the endosphere), and the root surface commonly referred to as the rhizoplane (i.e. including root epidermal cells and the mucilage). The rhizosphere is also a crucial interface for plant nutrient acquisition, stress tolerance, and disease suppression, where soil-borne pathogens establish

† Equal first author

http://dx.doi.org/10.19103/AS.2024.0136.18

parasitic relationships with plants (Tracanna et al., 2021). Microbiomes in the rhizosphere play crucial roles in plant growth, and understanding and manipulating these cross-talks, facilitated by native and modified root exudates and gene products, could ensure food security and resource sustainability in a changing global environment (Wang et al., 2022). Also, the impact of climate change and associated abiotic stresses will have a significant impact on the below-ground plant traits, both in the rhizosphere and roots, a better understanding of the microbe–microbe and plant–microbe interactions under stress and the plasticity of their responses for different crops and in different soil environments would help in the development of adaptation and mitigation practices to climate change (George et al., 2024).

Rhizosphere organisms and plants interact, affecting plant function and growth. The rhizosphere may contain organisms that impact plant growth and productivity, and this dynamic interaction is influenced by co-evolutionary pressures (Broeckling et al., 2008). Root exudate is crucial for establishing specialized relationships between plants and their rhizosphere surroundings, but these relationships are not universal and are influenced by various factors. Among which plant-mediated factors such as species (Innes et al., 2004), genotypes within species (Cordovez et al., 2021), growth stages (Houlden et al., 2008), root architecture, and compartments (Rüger et al., 2023) have been reported to orchestrate the fate of microbiome assembly and functions. The rhizosphere is a dynamic system influenced by time, but studies have primarily focused on single time points and distance from roots, suggesting a need for a 'legacy' perspective in understanding its dynamics (Oliver et al., 2021). Wurst and Ohgushi (Wurst and Ohgushi, 2015) provide a generalized definition of 'legacy effects' as *'a specific case of long-term effects that persist after the biotic interaction that caused the effects cease.'* The rhizosphere legacy effects could be physical, chemical, and biological. The detail about this is described in a review article (Oliver et al., 2021). Root architecture, chemical carbon (C), and root exudates influence rhizosphere legacy, while plant-soil feedback is mediated by microbial legacies left by plants in the soil. Microbial soil legacies developed during previous crop plant establishment play a key role in shaping plant growth, even if those legacies have faded into the soil from the growth of current plant species.

Soil and plant factors not only affect microbiome assembly processes but also significantly influence functional capacity (Yan et al., 2017). Rhizosphere microbes exhibit enrichment of transporters, glycolysis, and hydrogen metabolism genes. Strong network correlations are observed among members of the bacterial community compared to soil. Selection is not random but species-specific, influenced by stochastic and deterministic factors like edaphic and plant factors (Yan et al., 2017). Plants exert strong selection pressure on rhizosphere microbiome through root exudates that lead to differential effects on host performance and root morphological traits through host-to-microbe interaction, microbe-to-microbe, and microbe-to-host feedback (Foster et al., 2017; Yue et al., 2023). Most of the research on rhizosphere microbial communities has concentrated on factors that determine their assemblage at various taxonomic levels and little is known about the relative contribution of plant and soil factors at different functional levels.

This chapter explores the dynamics of rhizosphere microbiomes, focusing on their composition and functional aspects. It evaluates host-specific selection, plant species, genotypes, domestication, and agricultural ecosystem management practices. The chapter also investigates the legacy of plant developmental stages and root architecture on microbiome assembly and functions. Moreover, it highlights the selective pressure on the rhizosphere microbiome for plant nutrition and health, focusing on factors driving soil-root-plant interactions. Finally, the chapter also discusses challenges in manipulating functional microbiomes and proposes methods to address these gaps.

2 Host selection of rhizosphere microbiome community

Rhizosphere microbiome assembly is influenced by plant species, genetic makeup, and domestication, which are key for plant health and crop productivity. Understanding these factors helps breeders create specific cultivars and agronomists create systems that promote beneficial microbial communities and biological functions, ultimately improving crop productivity. Comparing the recruitment and functions of microbiomes in different plant species and genotypes can help in this process (Yue et al., 2023). The rhizosphere–plant interactions modulated by these factors are diverse, and spatially and temporally dynamic (Pantigoso et al., 2022). This section explores the role of microbiome recruitment and functions in various plant species, genotypes within the same species, and domestication impact on food crops, aiding breeders in cultivar creation and agronomists in improving crop productivity.

2.1 Plant species and genotypes

Plant–microbiota interactions have evolved/developed over millennia with complex host-specific genetic and physiological factors driving the composition of microbial communities in the rhizosphere, endosphere, and phyllosphere environments. However, the mechanistic understanding of host selection of the bacterial community remains elusive most of the time. The plant species and genotypes within the same species exert profound selective pressure on the diversity, abundance, and activity of the microbiome (Berendsen et al., 2012). The plant genotypes could contribute 5–20% of the variation in the microbiome communities (Kroll et al., 2017; Emmett et al., 2018; Walters et al., 2018; Li et al., 2020; Tkacz et al., 2020; Valente et al., 2020; French et al., 2021). While comparing the bacterial communities associated with maize genotypes or other Poaceae, a significant correlation has been observed between the genetic distance of rhizosphere microbial communities and the phylogenetic distance of the host-plant genotype, indicating the influence of the evolutionary history of a given plant genotype on the choice of bacterial taxa and plants (Bouffaud et al., 2014). Moreover, root exudates of wheat, maize, canola, and medics were found to significantly affect the community structure when grown in the same soil signifying the effect of plant species on the overall microbiome structure (Haichar et al., 2008). In pearl millet, the rhizosphere soil physical properties and microbiome diversity have similarly been shown to be impacted by genetic traits (Ndour et al., 2017).

Genotype-specific variation of the root fungal communities has been reported to have a role in chickpea plant productivity (Bazghaleh et al., 2015). A comparison of root-associated fungal community structure for 13 chickpea cultivars with a wide range of phenotypes and genotypes based on internal transcribed spacer (ITS) and 18S rRNA gene markers using 454 amplicon pyrosequencing, showed that root-associated fungal community structure was greatly impacted by the varietal selection and the magnitude of the effect varied between the genotypes analyzed and the effects maintained over the years (Bazghaleh et al., 2015). The variation was also observed in both growth-promoting and inhibiting species, suggesting the possibility of improving crop productivity by manipulating the root fungal communities through chickpea genetic breeding techniques. In wheat, the ancient genotypes have been reported to have a higher ability to establish relationships with plant growth-promoting rhizobacteria (PGPR) than the modern cultivars suggesting an inability to maintain plant genes during breeding that promote PGPR interaction (Valente et al., 2020). While comparing the effects of major wheat cultivars developed during the last 50 years on bacterial and fungal communities' assembly, 12.8–20.6% of microbial β-diversity was explained by the plant traits of these cultivars. Wheat cultivars from the 1970s were enriched with a greater number of microbial taxa with the highest relative abundance as compared to new varieties. However, some of the bacterial taxa belonging to the *Nocardioides* and fungal phylum Glomeromycota were enriched in the cultivars from the 2000s. Interestingly, the root zone of new cultivars (2000s) exhibited the highest bacterial/fungal ratio with more phosphorous cycle-related bacterial *phoD*-genes than 50s and 60s varieties. These findings suggest, the genotype of ancient and modern cultivars strongly patterns, and this information could be used to develop microbiome-based breeding strategies (Lu et al., 2024). Moreover, the impact of genotype was more pronounced at the flowering stage and has been proposed to be the right growth stage to see the genotype effects. In barley and maize, the host genotype was found to account for 5–7%

of variation in the rhizosphere community assembly (Bulgarelli et al., 2015, Peiffer et al., 2013). In soybean, the inherent microbiome in different nutritional status soil and plant genotypes have combinedly manipulated the rhizosphere microbiome (Liu et al., 2019). The undomesticated variety *Glycine soja* exhibited higher rhizosphere bacterial diversity in comparison to the modern domesticated variety. Some of the taxa like *Rhizobium, Novosphingobium, Phenylbacterium, Streptomyces,* and *Nocaridioides* were highly enriched in the rhizosphere regardless of the soil types. The network analysis revealed the effect of soil types and genotype-specific preferences for key microbe–microbe interactions. The modulating role of the soil type and genotype emphasizes the importance of an integrated consideration of soil conditions and plant genetic variability for the soybean cultivar development through the breeding program. In barley, the microbiome assembly in the wild and domesticated varieties has been modulated through nitrogen availability (Terrazas et al., 2022). Nitrogen enrichment in barley microbiomes involves beneficial bacteria, recruitment influenced by host genotype. Nitrogen availability also affects taxonomic composition, with wild and domesticated genotypes maintaining functional signatures encoding different bacterial members. This information could be used to develop barley varieties with improved yields with reduced dependency on nitrogen fertilization. In *Brachypodium*, a significant correlation has been reported between plant trait variability and community structures (R^2=0.61, $p<0.001$) for bacteria and (R^2= 0.63, $p < 0.001$) for fungi. Furthermore, two distinct microbial sub-communities were predicted to have positive or negative associations with leaf greenness (Jacquiod et al., 2022). These results further echo the significance of the understanding of the association between plant genotypes/traits and microbiome, since not all community members are responsible for a selected trait.

2.2 Plant domestication

Rhizosphere microbiomes are coshaped by both host genotypes and domestication status. Plant domestication has a remarkable effect on the seed and root microbiome (Kim et al., 2020, Jonathan et al., 2017), especially on the rhizosphere microbiome composition (Sun et al., 2021). The rhizosphere microbiome plays an important role in crop performance and growth (Trivedi et al., 2020). Thus, dissecting the mechanisms of microbiome coexistence and plant–microbiome interactions induced by these effects at the root–soil interface offers new opportunities for using native root-associated populations to promote soil health and crop yield (Chen et al., 2021; Singh et al., 2020). For example, wheat genotypes have a pronounced effect on bacterial community diversity than fungal communities, while plant domestication has a stronger effect on fungal communities (Yue et al., 2023). Domestication has disrupted microbial associations, reduced keystone fungal taxa, and transformed the rhizosphere microbiome from a fungal-dominated to a bacterial-dominated one. In rice, domestication significantly decreased bacterial chemotaxis genes, resulting in a shift from a fungal-dominated to a bacterial-dominated microbiome (Sun et al., 2021). The differential selection of chemotaxis systems was, at least in part, due to changes in the metabolite profiles of rice roots affected by domestication. In soybean, irrespective of soil type, the non-domesticated species harboured higher rhizosphere bacterial diversity as compared to the domesticated genotypes (Liu et al., 2019).

2.3 Novel methods to manipulate host genetic factors for cultivar development

Research on rhizosphere microbiome assembly until now has shown differences in microbiome composition between cultivars or cultivars with no well-defined relationship to each other and limited mechanistic explanation where cultivars are chosen to represent a breadth of diversity within the host species (Clouse and Wagner, 2021). The effect of domestication or modern breeding has been addressed for some crop types with a limited understanding of plant host genetic components contributing to microbiome assembly. Most of these studies are less informative about the specific host traits/genes that could be targeted for their ability to improve the desirable microbiome. On the other hand, studies using mutant and transgenic lines have revealed key genes and traits that control microbiome recruitment and regulation of microbiomes (Vives-Peris et al., 2020). For example, transgenically induced production of glucosinolate defense chemicals in *Arabidopsis* changed the relative abundance of Rhizobiaceae and other potentially important bacteria, indicating that functional genetic modification of host plants can be used to develop plant-specific microbiomes (Bressan et al. 2009). Similarly, in a functional mutant study, the antimicrobial coumarin selectively inhibited fungal pathogens in the rhizosphere while promoting the colonization of the growth-promoting bacteria (Trivedi et al., 2020). Maize mutants lacking benzoxazinoids significantly altered rhizosphere bacterial and fungal communities, impacting growth and herbivore resistance in subsequent generations, even after winter dormancy (Hu et al., 2018). Another study on maize grown in nitrogen-depleted soils revealed carbohydrate-rich mucilage from areal roots harboured recruited microbial communities, including Bacteroidetes and beta/gamma Proteobacteria, which supplied 29–82% of plant nitrogen (Van Deynze et al., 2018).

Forward and reverse genetics approaches are used to study the complex relationship between plants and their microbiomes, with reverse genetics research focusing on changes in plant traits caused by genetic engineering

(Song et al., 2021a). Transitioning to gene-assisted ecological methods enhances sustainable agricultural and crop productivity by recruiting, studying, and utilizing the right microbiome consortium (Song et al., 2021b). Recent genetic screening and biochemical approaches have successfully identified driving factors for plant–microbe interactions, particularly single-microbe and plant interactions, determining plant resistance and susceptibility (Dangl and Jones, 2001). Similarly, rapid identification of genes and signals for symbiosis was achieved using the model legume plants and nitrogen-fixing rhizobia. In contrast to these studies, understanding plant phenotypes associated with a suite of microbiome, linked to growth promotion or induced systemic resistance, is challenging due to their variable effects and involvement of group/community of microbes (Song et al., 2021b). As a result, forward genetic screening for plant and bacterial genes that shape plant–microbiome interaction is challenging. Rescreening mutant collections in plants can help understand beneficial microbe enrichments and identify loci/genes, as demonstrated in Arabidopsis (Song et al., 2020). For reverse genetics, microbiome structure through amplicon sequencing could be linked to the functional process of the plants through host transcriptional profiling which in turn might help to unveil plant genes involved in the modulation of functional microbiomes. For example, in rice, the *NRT1.1B* gene locus has been shown to have a role in nitrogen homeostasis (Zhang et al., 2019). The reverse genetics approach has been used to demonstrate that chemical cues such as coumarins influence the rhizosphere microbiome, using SynCom and microbiome profiling on various coumarin biosynthetic enzyme mutants (Voges et al., 2019). By using transcriptional profiling, two genes have been reported to have a role in rhizobacteria-induced systematic resistance in *Arabidopsis* (Van der Ent et al., 2008). Genome-wide association studies reveal plant loci affecting defense and cell wall integrity contribute to microbiome community variation, while host genetic variation influences species richness in bacterial communities (Horton et al., 2014). Plants produce numerous metabolites, many of which are uncharacterized. Research has been focused on small molecules and metabolites that can affect the microbiome through recruitment or restriction (Huang et al., 2019). Sequencing plant mutants with metabolic biosynthesis defects, mutants can provide valuable insights into how genetic factors affect microbiome structure.

The structure of plant microbiota is often described as host-driven, especially in the rhizosphere, where microbial communities are formed mainly in response to variations in rhizodeposits and/or signal molecules (Pantigoso et al., 2022). Chemical cues from root exudates have been considered as a key driver of rhizosphere microbiome, recent research suggests that host and soil microbiomes may influence the development of a host-specific rhizosphere community through coevolution, with small noncoding RNAs called small RNAs, which act as key mediators in plant–microbe interactions and microbiota establishment (Ren et al., 2019). It has been discovered that rhizobial sRNAs regulate host genes involved in nodule initiation and development, with loss-of-function, over-expression, and mutations linked to root phenotypes like curling hairs and nodulation. This finding supports the potential of a novel way of improving plant–microbe interactions and functions through rhizosphere bacterial sRNAs (Ren et al., 2019). Recently, it has been shown that microbes-induced gene silencing in the rhizosphere using *Trichoderma* sRNAs can boost crop immunity by silencing essential genes in pathogenic fungi, suggesting sRNAs-based approaches as a potential alternative to microbial inoculants (Wen et al., 2023) (Figure 1).

2.4 Plant-rhizosphere microbiome interactions - functional perspective

The rhizosphere microbiome, influenced by factors like soil, plants, and plant–microbe communication, plays a multifunctional role at the rhizosphere–plant interface. Key functions include nutrient acquisition, stress tolerance, and pathogen suppression, driven by spatial and temporal variations in specific microbes/microbial groups (Figure 2) (Xun et al., 2021). Plant–microbe interactions as modulated by the plant type, edaphic (soil), and environmental factors are crucial in the expression of specific functions through the crop growth

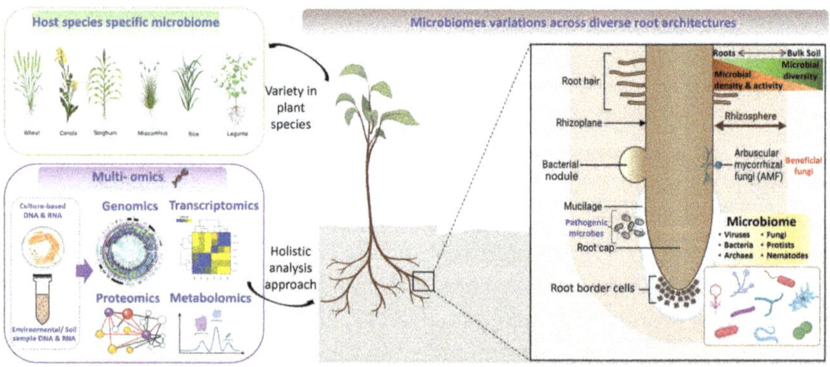

Figure 1 A graphical abstract showing the addressed questions in this chapter about the advances in understanding microbial communities in the rhizosphere. The subpanel (right) shows the distribution of life in the rhizosphere. The gradients in the diversity and abundance of microbiomes have opposite trends: while abundance and activity decrease from roots to the bulk soil, the diversity from roots to the bulk soil increases. A: higher density and lower diversity in the rhizosphere and reverse in bulk soil; B: the abundance of various microbial groups on rhizoplane, in the rhizosphere, in bulk soil; C: release of signalling compounds and attraction of rhizobia and other PGPR; D: border cells produced at root tips contributing to majority of root exudates and extracellular signals influencing microbial gene expression and growth and attraction of fungi, protozoa and nematodes.

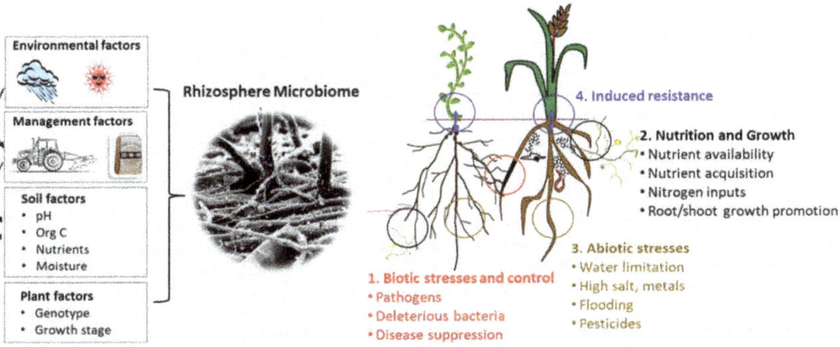

Figure 2 The different categories of abiotic and biotic processes related to plant health, nutrition, and growth are influenced by the plant-microbe interactions in the rhizosphere. Rhizosphere microbiomes are shaped by multiple plant-microbiome, microbe-microbe/microbe-biota interactions, and host plant exudation patterns. These effects are modulated by management (tillage, fertilizer), environmental factors, and host plants (Gupta and Sharma, 2021). Environmental factors and management practices can influence soil properties and in turn affect rhizosphere microbiomes.

period. PGPRs can enhance plant nutrient acquisition by improving soil nutrient availability or improving host plant root system architecture. For instance, the better nitrogen use efficiency of *Oryza sativa* sub sp. *indica* cultivars than *Oryza sativa* sub sp. *japonica* cultivars is due to the higher presence of N-cycling bacteria in the rhizosphere of *O. indica* cultivars, which is regulated by the host *NRT1.1B* gene (Zhang et al., 2019). During drought stress, *Actinobacteria* and *Streptomyces* are enriched (Narsing Rao et al., 2022) while salt stress enhances the enrichment of *Rhizobium* (Ventorino et al., 2012). These microbiomes regulate the flowering time to make plants more resistant to the stress. In pathogen suppression, microbiomes provide direct or indirect protection against soil-borne pathogens. In direct protection activity, rhizosphere microbiomes secrete antimicrobial compounds and compete for resources, and suppress pathogen colonization while in indirect protection activity, microbiomes elicit plant immune response thereby plants secrete defensive secondary metabolites and regulate growth-defense trade-off. Despite the multifunctional roles of rhizosphere microbiomes, current understanding of rhizosphere assembly patterns is still limited. Nevertheless, various assembly patterns of the rhizosphere microbiome have been proposed. A two-step selection process has been proposed in which at first, rhizodeposition fuels an initial substrate-driven community shift in the rhizosphere, and the host genotype further selects the endosphere (Yuan et al., 2018). Based on this, a three-step enrichment model was introduced to assemble the root-associated microbiome from the bulk soil to the root, introducing a rhizosphere microhabitat into the soil-root interface (Reinhold-Hurek et al., 2015). In addition, a multistep model suggests that rhizoplanes play a selective gating role in the assembly of root microbiomes from soil (Edwards et al., 2015). A 'boost-selection' model has recently been proposed to assemble the rhizosphere microbiome (Wang et al., 2020). However, these assembly models only considered the horizontal recruitment of root-associated microbiomes in a mass. In general, plants actively assemble their rhizosphere microbiome from surrounding microbial communities, including dominant horizontal transfer of bulk soil and minor vertical transfer via seeds (Berg and Raaijmakers, 2018). Recently, a new assembly model of 'functional compensation' has been proposed, that is, the rhizosphere microbiome is assembled to compensate for functional requirements of the host plant that cannot be fulfilled by itself. This assembly pattern involves both the recruitment of the soil microbiome from the outside and the release of the endosphere microbiome from the inside. Shao et al. (2021) observed massive compensatory colonization of the rhizosphere by the maize seed-derived phosphate-solubilizing bacterium *Burkholderia gladioli* when grown in P-limited soil. Therefore, plants recruit different communities of rhizosphere bacteria to fulfill the functional requirements of their fitness under different soil conditions.

Besides, the taxonomic diversity, the biological diversity of the microbial community dictated by the physiochemical and enzyme activities in the rhizosphere is important for maintaining the functioning of the soil ecosystem (Nsabimana et al., 2004). The large number of microbial species found in soil and the difficulties in their cultivation and species differentiation led to the development of approaches based on community rather than species biodiversity. One such technique involves community function analysis based on the use of carbon substrates (Degens and Harris, 1997). To gain a complete picture of the functional attributes of the rhizosphere microbiome, the relationship between the community composition and diversity should be deciphered. For example, enzyme activities showed a significant positive correlation with the total amount of microbial biomass in the rhizosphere during the growth phase of rice plants (Hussain et al., 2012). The bacterial biomass was reported to be higher in the rhizosphere of healthy plants compared to the Verticillium-wilted cotton plant, however, the fungal biomass was the opposite (Wei et al., 2021).

2.4.1 Rhizosphere microbiome and soil disease suppression

The rhizosphere is the hot spot where *in situ* microbiome provides resistance or tolerance to impacts by soil-borne pathogens. Resistant cultivars have been characterized by a higher abundance of known organisms with biocontrol potential as compared to susceptible varieties (Lazcano et al., 2021). These beneficial microbes can modulate pathogen dynamics and growth, plant-pathogen interactions, and/or increase the availability of nutrients such as nitrogen and phosphorus. Nutrient availability changes in the rhizosphere could occur through the release of organic chelates, enzymes (e.g. phosphatases), and/or increased microbial turnover (e.g. N mineralization compared to bulk soil (Finzi et al., 2015)). Cultivars with different degrees of resistance to soil-borne pathogens (Sánchez et al., 2016; Pastrana et al., 2017) suggest the interplay between roots and soil organisms during plant–pathogen interaction is the major driving factor for the disease outcome (Mendes et al., 2018; Hayden et al., 2018; Yao and Wu, 2010). Although the essential mechanisms involved in these interactions are not fully understood and may vary with the pathosystem, suppressive activity has been hypothesized to be complex in nature and results from a combination of the resident rhizosphere microbiome and the host plant. For example, rhizosphere bacterial community richness has been shown to modulate soil disease suppression by responding to soil abiotic properties and the abundance of bacterial groups, including the genera *Actinomyces*, *Bacillus*, and *Pseudomonas* (Yao and Wu, 2010). Also, different crop and soil management practices have been shown to impact suppressive capacities (Roget, 1995; Weller et al., 2002; Gupta et al., 2009, 2011). For example, crop rotational diversity has increased the suppressive capacities of some soil microbiomes against one pathogen (Peralta et al., 2018), while monoculture has been proven efficient to induce suppressiveness in other pathogens (Weller et al., 2002). Such variations suggest the disease outcomes driven by soil microbiomes are crop-specific selection of microbial communities in the rhizosphere and plant types–pathogen interactions specific and different individual microbial species or groups are involved in the suppression of different pathogens. For example, *Pseudomonas fluorescens* in the control of take-all disease of wheat (Raaijmakers and Weller, 2001). This leads to the thought-provoking question of the crop preference of this bacterium to maintain its threshold densities throughout the growing season (Schlatter et al., 2017). The capacity of crop species to select for particular genotypes has been strikingly proven in adjoining wheat and flax fields in Fargo, North Dakota. In comparison to the wheat species enriched in *P. brassicacearum*, the rhizosphere of a continuous flax-grown system since 1894 was enriched with F- and J-genotype traces (Landa et al., 2006). Therefore, disease suppression is not only the outcome of the presence of microbiomes but also their activity, diversity and abundance to make a significant impact on the host functions. At a coarse phylum level, Proteobacteria, Firmicutes, Acidobacteria, Actinobacteria, Bacteroidetes, and Cyanobacteria are common taxa present in most rhizospheres although different abundances across different plant types and environments occur as mentioned above. Relative abundances rather than the presence and absence of taxa have been reported to be the drivers of rhizosphere functions (Mendes et al., 2011). However, the variability in taxonomic difference might be due to the significant functional redundancy present in the soil microbiome, which can be considered as an ecological guarantee for the maintenance of function in case of possible induced disturbance or selective pressure exerted by above-ground activities (Chen et al., 2022). Another example of the importance of understanding the functional capacities of microbial communities and their management drivers is the observation of inhibition of expression of disease suppression against Rhizoctonia bare patch disease from the accumulation of high levels of mineral N prior to seedling growth (Gupta and Roget, 2011). The observation of a higher abundance of diazotrophic (*nif*H gene harbouring) communities in soils with high suppression capacity and bacterial species with higher suppression potential containing *nif*H gene suggest that diazotrophic functional expression is somehow linked to the suppression capacity. Also, the composition of diazotrophic communities was found to be significantly different between suppressive and non-suppressive soils (Gupta et al., 2014b). In view of the complex interactions observed, the driving force in shaping the rhizosphere microbiome should be studied in terms of its influence on functional potential in addition to taxonomy-based composition only.

Plants select a suite of microbiomes based on the functional traits that are important for interactions and may not be restricted to a specific taxonomic group (Yan et al., 2017). The enrichment of functional gene expression representing functional capacity further provides evidence that functional traits may be driving factors in the wide variation in taxonomic differentiation of rhizosphere composition across studies (Ofek-Lalzar et al., 2014). Plants also select a subset of microbiomes at different stages of development, according to the root types and sections for the specific functions which will be discussed in detail in the next sections.

3 Microbiome variations across diverse root architectures

The composition, diversity, and functionality of rhizosphere and root-associated microbial communities, vary between plant genotypes, both spatially and temporally, and is influenced by factors such as plant and root type including root architecture, exudation patterns, growth stage, plant health, and environmental conditions. For example, differentiated root exudation along a single root was observed, with the elongation zone behind the root tip being the primary source (Badri and Vivanco, 2009). Maize and rice studies (Gutjahr et al., 2015; Yu et al., 2018) revealed distinct transcriptomic profiles between axial and lateral roots, potentially influencing root exudation patterns. In tomatoes, microbial communities induced specific systemic changes in root exudation, with *Bacillus*-affiliated bacteria triggering the systemic exudation of acyl sugars as secondary metabolites (Korenblum et al., 2020). This led to alterations in both leaf and systemic root metabolomes and transcriptomes, suggesting glycosylated azelaic acid as a potential microbiome-induced signalling molecule. A recent review emphasized the link between biotic stress-induced root exudate composition and resultant microbial community changes suggesting that a better understanding of changes in root exudation when exposed to stress and the microbial community responses to it should help in the development of designer microbiome engineering to help plants cope with exposure to stress events (Sharma et al., 2023).

Root architectural phenotypes are crucial for crop breeding, but their impact on microbial associations is not fully understood. Variations in root architecture can impact root exudation, recruitment, function, and tissue decomposition (Kawasaki et al., 2016). Despite extensive research on plant genotypes' effects on microbiome assembly and functions, spatial variations within root systems, particularly across different root types, are often overlooked as most studies focus on the whole root system or a random subsampling.

In this section, we discuss responses in microbiome composition and activity by examining variations in diverse root architectures and plant growth stages. Understanding these variations can help link functionally important microbes to host and edaphic factors and develop options for modulating plant essential functions to enhance crop growth. This can be associated with differences in spatial and temporal aspects of different root types and their association functions, e.g. resource acquisition and anchorage, etc. which can be affected by environmental and host factors.

3.1 Root-type-dependent microbiome diversity

The abundance of individual microbial species and the overall community composition have been shown to vary among the root types within a plant species. The seminal and nodal root types have been particularly focused on as they vary in terms of their location, time of development, and nutrient uptake patterns. For example, in wheat, the seminal roots were found to support significantly larger populations of bacteria and fungi than nodal roots (Sivasithamparam et al., 1979; Kawasaki et al., 2021a&b). Similarly, in purple false broom (*Brachypodium distachyon*), seminal and the nodal roots hosted very different bacterial communities (Kawasaki et al., 2016). Seminal roots were dominated by bacterial order Oxalobacteraceae, whilst in the nodal roots, order Comamonadaceae was most prevalent, with the predicted microbial gene functions also being very different between the two root types. Furthermore, in *B. distachyon*, the bacterial and fungal communities that colonized the seminal, coleoptile nodal and leaf nodal roots showed small but significant differences in arbuscular mycorrhizal fungi (AMF) abundance (Donn et al., 2017). Additionally, microorganisms isolated from root tips and bases of roots showed different functional gene compositions (Kawasaki et al., 2016). Similarly, between axial (seminal or nodal) and lateral roots of maize, more than half of the identified fungal OTU did not overlap (Yu et al., 2018). A comprehensive analysis using group-specific amplicon sequencing has shown distinct spatial variation in the root microbiome colonizing the root systems of wheat and rice (Kawasaki et al., 2021b). The root type (seminal or nodal) was found to have a greater influence on microbiome structure than the exudates in the overall root system of rice (*Oryza sativa* L.) and wheat (*Triticum aestivum* L.). Soil type was found to additionally increase the variation between different root types, which was comparable with the variation detected between different plant species (Kawasaki et al., 2021a). In addition, different types of roots perform various processes to resist microbial infections or to acquire beneficial fungi. For instance, during the infection of take-all in wheat (*Gaeumannomyces graminis var. tritici*), reduction in the above-ground biomass was more pronounced on nodal roots infection than seminals (Sivasithamparam and Parker, 1978, Gilligan, 1980). Different parts of the wheat root system, i.e. seminal and crown roots, internodal space, and stem base, are infected by a number of soilborne fungal pathogens such as *Rhizoctonia solani, Bipolaris sorokiniana*, *Fusarium* species, *Curvularia* spp, *G. graminis* var. *tritici* diseases in wheat. The infection and impact of various diseases differ based on pathogen distribution, soil, and environmental factors. The new knowledge about the distinct variations in the spatial distribution of rhizosphere and root-associated microbiomes provides a new avenue for the development of designer management options for specific diseases.

3.2 Microbiome variability across plant root compartments

A significant variation in the composition and the functional capability of microorganisms associated with different root compartments has been reported in different crops wheat, barley, rice, and maize. In a study with wheat root tips (1–2 days old) and the root-base segments (1 week older), it was shown that they harboured different predominant bacteria clusters, each performing different functions, one related to acid production and the other associated with yellow pigmentation (Liljeroth et al., 1991). The compartmentation across the wheat root was the most significant factor explaining total community variation across six sample times (15.0–33.7% of total variation), whereas differences in crop development stage contributed by community succession (7.2-13.7%) and paddock location were lower (1.8–8.2%) with crop genotype showing no significant difference between two cultivars (0.6–1.3%) (Donn et al., 2015). Furthermore, the position along the roots (apex or base) had a greater influence on microbiome structure than the exudates in the root system of wheat and rice (Kawasaki et al., 2021b). Furthermore, the microbial communities were found to be significantly different between different positions along the root axes in wheat and rice (Kawasaki et al., 2021a). This study also identified bacterial and fungal taxa that were consistently more abundant in specific regions of the root system. Moreover, the comparison between bulk soil, rhizosphere, rhizoplane, and endosphere across six genotypes of two *Oryza* species showed clear distinction of bacterial community similarities according to compartment, irrespective of the soil type and host plant species, with plant genotype accounting for 2.4% only of the total variation (Edwards et al., 2015). In *B. distachyon*, closely related species to major cereal grains including wheat, maize, and rice, the bacterial community structure and functional gene content varied more between the tip and base of the same root type than between the different root types (nodal and seminal), suggesting a strong effect of root compartments on microbiome assembly and functions (Kawasaki et al., 2016). Barley (*Hordeum vulgare*)'s bacterial communities in the rhizosphere were substantially different in different root zones, including actively growing primary root tips, non-elongating secondary root tips, sites of lateral root emergence, and mature root sections 5 to 10 cm distal from the root tips (Yang and Crowley, 2000). More recently, the high variation in beta diversity at root tips and the root hair zone of maize (*Zea mays* L.) suggests a more stochastic community assembly (Rüger et al., 2021).

In addition to annual crop species, distinct variations in rhizosphere and root microbiomes including for total bacterial and fungal communities and for functional groups such as diazotrophs and nitrifying microbes, have been shown in a number of perennial grasses and trees (Chen et al. 2016; Beckers et al. 2017; Gupta et al. 2019a; Zhou et al., 2020a). For example, detailed descriptions of the ryegrass root-associated bacterial microbiomes across compartments (the bulk soil, 2-mm outer rhizosphere, rhizosphere, rhizoplane, and endosphere) were investigated using next-generation (NG) sequencing (Chen et al., 2016). Briefly, Increased Proteobacteria in the rhizosphere, rhizoplane, and endosphere, with dominant genera (*Methylobacterium*, *Rhizobium*, *Pseudomonas*, *Stenotrophomonas*, and *Enterobacter*) facilitating plant growth. Bacterial α-diversity decreased spatially from bulk soil to endosphere. A comparison of diazotrophic bacteria in the grass rhizosphere soils found similar abundances between the rhizosphere and roots of summer-active perennial grasses grown in South Australia, whereas their diversity was significantly higher in the rhizosphere soils (Gupta et al., 2014a). A diverse diazotrophic community was observed in both the below and above-ground plant parts but the composition was distinctly different for the different grass species (Gupta et al., 2019b; Roley et al., 2018). Also, in spite of plant species-based variation and that α-Proteobacteria were the dominant group of *nif*H-harbouring taxa both in the belowground and aboveground parts of the three grass species. The co-occurrence patterns within the *nif*H-harbouring community were compartmentalized to different parts of the plant, that is, each part of the plant represented a unique niche or microsite whose members interact to a larger degree with members within that niche (Gupta et al., 2019a). Similarly, for field-grown poplar trees (*Populus tremula* × *Populus alba*) using amplicon pyrosequencing (Beckers et al., 2017), the community structural variability of rhizosphere microbiomes was much lower than that of the endosphere microbiomes. These results confirmed microbiome niche differentiation at the rhizosphere soil–root interface and revealed additional fine-tuning in the endosphere microbiome within stem and leaf compartments, each representing a unique ecological niche for bacteria.

3.3 Microbiome dynamics across root development stages

Host developmental stages drive the differentiation in the ecological role of the rhizosphere microbiome. Therefore, studying rhizosphere and root microbiomes is essential to gain insights into community dynamics and changes in microbiome community structure over time and spatial context, especially in relation to plant growth and development. Baudoin et al. (2002) pioneered the concept, demonstrating that bacterial communities in different maize rhizosphere zones (ramification, root hair-elongation, apex) exhibited distinct functional abilities at 2 and 4 weeks after planting, with time influencing microbial pools more significantly than root-zone origins. In wheat, a stable biofilm covered 40% of the rhizoplane of young plants, with distinct clustering at the root cap and around root hairs (Watt et al., 2006). The elongation zone of growing roots (collected during the vegetative growth stage) had the least colonization but the highest Pseudomonas proportion, without detectable filamentous bacteria from the Actinobacteria group. In contrast, older roots (reproductive stage) or those in contact with dead roots had highly enriched (>50%) filamentous bacteria. Similarly, clear differences in community composition were demonstrated in wheat at four sampling times between rhizosphere and rhizoplane/endosphere communities, with lower diversity in the latter more tightly bound communities (Donn et al., 2015). Proteobacteria dominated in young plants, and Actinobacteria prevailed on senescent roots. The shifts were more pronounced in tightly bound bacterial communities (i.e. rhizoplane) over time compared to the rhizosphere and bulk soil. Similarly, in canola, crop phenology has been shown to impact the composition of soil bacterial and fungal communities due to the changes in the microenvironment provided by canola roots and their exudates (Gurusinghe et al., 2019). For example, Proteobacteria and Actinobacteria were generally found to be the dominant groups, however, the abundance of Firmicutes increased at flowering. Furthermore, the metagenomic analysis indicated that the assemblage of microbes from the associated bulk soil to the root compartment was based on the lifestyles of *Pteris vittata*, in particular the root–microbe interactions (Xiao et al., 2021).

In contrast, other studies indicated that the host genotypes select an overall core bacterial microbiome with growth-stage-related patterns and potential host-driven changes found to have significant effects at some stages only. For example, there were no significant differences in bacterial structure found in *Arabidopsis*'s rhizospheric bacterial community at vegetative, bolting, and flowering stages, but the microbial community at the seedling stage was distinct from the other developmental time points. The metatranscriptomics further revealed 81 unique transcripts significantly expressed at different developmental stages (Chaparro et al., 2014). Similarly, a 10-week period investigation of *Brassica napus* grown in Western Canada demonstrated that the genotypes play a more important role in microbial assembly than the growing stages (Taye et al., 2020).

3.4 Factors shaping distinct microbiomes across root architectures

Root exudates, entirely biochemical have been considered as the main host-based driving factor in the assembly, composition, and functions of the rhizosphere microbiome. Additionally, signal compounds and secondary metabolites released by plants and members of the soil microbiome also impact the composition of the rhizosphere microbiome. On the other hand, soil biotic (e.g. presence of specific microbes and some fungal pathogens) and abiotic (e.g. pH, nutrient status, and the presence of particular minerals and toxic metals) factors, environmental variables and management interventions and rhizosphere microbial products also exert profound effects on plant rhizosphere microbiomes and host functions (Gupta and Knox, 2010; Richardson et al., 2021; Gupta et al., 2023; Law et al., 2024). For example, management practices involving above-ground treatment of crops such as the application of herbicides (e.g. generally seen with herbicide-tolerant crops), other pesticides, and nutrients have been shown to modify bacterial and fungal composition in the rhizosphere which has been attributed partly to modifications to the quantity and quality of root exudates (Gupta and Knox, 2010).

In most agricultural soils, microbial activity is limited by the availability of labile C and/or energy, hence inputs of easily available organics from growing plants, decomposing residues, etc. increase the activity and abundance of microorganisms resulting in microbial hotspots (Figure 3). The rhizosphere environment is one such hotspot where root exudates, decomposable rhizodeposits, and the turnover of root hairs and older roots boost microbial activity and abundances of microflora and microfauna (Kuzyakov and Blagodatskaya 2015). In view of the root turnover through the different

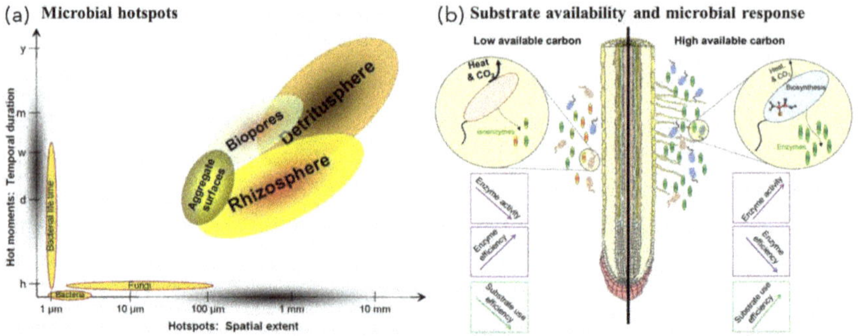

Figure 3 Schematic representation of microbial hotspots in soils including rhizosphere environments. (A) Spatial and temporal scales where the allocation of areas corresponds to the size and duration of microbial hotspots; logarithmic Y scale corresponds to h: hours, d: days, w: weeks, m: months, y: years and intensity of gray on shaded areas represents the spatial and temporal probability of microbial hotspots; (B) Microbial response to the input of root exudates in the rhizosphere. Substrate availability as affected by root hair proliferation and warming may control the optimal strategy of thermoregulation. For example, the production of multiple forms of key enzymes under low C availability reduces metabolic efficiency imposing a trade-off by reducing the allocation of available energy to growth whereas increased enzyme expression and shifting of resources to biosynthesis under high C input (Adapted from Kuzyakov and Blagodatskaya, 2015; Zhang et al., 2020).

stages of crop growth, the duration and magnitude of hotspot moment in the rhizosphere would be expected to vary, i.e. spatial and temporal effects of hotspots are modulated by the root architecture and the dynamics of root turnover. It is suggested that faster turnover and lower C use efficiency generally counterbalance with high C inputs. By combining soil zymography with enzyme kinetics measurements Zhang et al. community (2020) identified microbial hotspots in the rhizosphere *in situ* and quantified root hair effects on microbial growth strategy and functionality. They found that root exudate release, enzyme activities, and microbial substrate utilization in the rhizosphere were strongly influenced by root hair proliferation and environmental factors such as exposure to higher temperatures (e.g. warming). For example, under low available C conditions, shifts in gene expression and production of enzyme systems with a higher binding affinity and lower substrate use efficiency may occur, whereas, increasing C availability could result in increased enzyme activity and C resources shifted for the biosynthesis of new cellular material i.e. increased microbial biomass (Zhang et al., 2020; Figure 3). Such changes in the C and other inputs from roots are one of the primary factors regulating the temporal and spatial variations and succession changes in the phylogenetic and functional composition of rhizosphere microbial community structure and community adaptation (Donn et al., 2015; Zhalnina et al., 2018; Zhang et al., 2020). Since root exudate composition can vary between plant types, e.g.

grasses vs. legumes, stoichiometric ratios of exudates/root deposits can also affect microbial growth rates resulting in the mining from soil organic matter through priming effects (Kuzyakov and Blagodatskaya, 2015).

As discussed above, the recruitment of microorganisms and development of the root microbiome is largely driven by root exudates containing C, signalling molecules, and secondary metabolites, especially in the early stages of root colonization that signal microbe–microbe interactions (e.g. antagonism, quorum sensing, formation of biofilms, surface adhesion) and plant–microbe interactions (recognition, infection, receptors and secretory systems), triggering plant hormone pathways and host-induced plant immune responses. The rhizosphere signalling dictated by root exudates has been extensively reviewed since the early 1990s to unveil the mechanistic effects of host factors on rhizosphere microbiome assembly and functions (Arshad and Frankenberger, 1997; Bais et al., 2006; Watt et al., 2006; Faure et al., 2009; Swamy et al., 2016; Sasse et al., 2018; Richardson et al., 2021; Pantigoso et al., 2022; Singh et al., 2022; Sharma et al., 2023). The co-evolution of plants and microbes underpins important ecological functions and understanding the nature of signalling mechanisms between them has been an important research area.

In general, rhizospheric signalling mechanisms can be categorized into three major types (i) plant to microorganisms signalling through low molecular weight molecules secreted by plants; (ii) interspecies and intraspecies microbial signalling, chiefly through quorum-sensing molecules; and (iii) microorganisms responding to plant signalling by microbially produced compounds (Singh et al., 2022). Recent metabolomic studies have found differential root exudation composition and patterns in various environments and plant developmental stages, as well as consumption of specific exometabolites by rhizosphere microbes (Sasse et al., 2018). Genome-wide association studies and other comparative genomic approaches have suggested a link between plant genetic factors, exudation profiles, and microbiome compositions. A number of reviews have been published addressing root exudation patterns and the assembly, and macroecological patterns of root microbiome (Brunel et al., 2020; Cordovez et al., 2019; Fitzpatrick et al., 2020; Munoz-Ucros et al., 2021).

As discussed in previous sections of this chapter, it is evident that root type and localization have a significant influence on the diversity of their microbiomes, and such differences could be driven by variations in root structure and the quantity and composition of root exudates (Gupta and Sharma, 2021). For example, organic acid composition was found to be of particular importance to influence microbiome assemblage on roots. The release of malate by *Arabidopsis* roots has been shown to mediate the recruitment of a biofilm-forming beneficial *Bacillus subtilis* (Rudrappa et al., 2008). The primary composition of exudates, such as simple hexose sugars, amino sugars, and acids and organic acids, influences microbial growth through processes that include chemotaxis, attachment, and metabolic stimulation through preferential C use (Jones et al., 2009). Moreover, variation in exudate composition in the elongation zone and around the root apex has consequently been shown to contribute to substantial differences in root colonization and community structure (Dennis et al., 2010; Kawasaki et al., 2016; Zhalnina et al., 2018). Consistently, it is proposed that the identification and selection of genotypes with distinct patterns of root exudate, or specific alteration of exudation profiles, may thus provide an opportunity for the selection of more beneficial microbial communities (Quiza et al., 2015; Zhalnina et al., 2018). More comprehensively, Kawasaki et al. (2018) have shown that malate efflux occurs from both lateral and axial (seminal and nodal) root tips, whereas citrate efflux occurs from the axial roots and not from the laterals. The exudate compounds (released via aluminium-activated malate transporter and the multidrug and toxic compound extrusion pathways), separately and in combination, significantly altered microbiome composition throughout the root system of rice and wheat, although the root type (seminal or nodal), position along the roots (apex or base), and soil type had a greater influence on microbiome structure than the exudates (Kawasaki et al., 2021a). In fact, some other studies have proved that seminal and nodal roots have distinct transcriptomes (Tai et al., 2015) and proteomes (Liu et al., 2020), which means they are likely to release different compounds into the rhizosphere. *In vitro* studies further highlighted the importance of exudate composition in shaping community structure utilizing artificial rhizosphere solutions to assess changes in community composition (Baudoin et al., 2003; Shi et al., 2011). Another mechanism by which plants can influence rhizosphere microbiome composition is through the production of border cells which can alter the chemical and physical properties of the local environment. For example, border cell-associated gene expression and specific proteins and signal molecules produced by border cells can dramatically alter the direction of root growth in soil, behaviour of soilborne microflora, both beneficial and pathogenic microbes and has been suggested to neutralize abiotic and biotic threats to the new root tissue, i.e. function in defence of the root tip as it grows through the soil (Gunawardena et al., 2005; Zhu et al., 2003). Border cell production has been shown to vary significantly between crops and among cultivars of wheat and cotton (Zhu et al., 2003; Knox et al., 2006; Curlango-Rivera et al., 2013). However, a comparison of border cell numbers between cotton varieties differing in disease resistance against soilborne fungal diseases showed a moderate relationship ($r^2 = 0.65$) with resistance ranks (Knox et al., 2006). In spite of such evidence, the true consequences of variations in border cell production in terms of effects on rhizosphere microbiome community structure and plant fitness are yet to be fully understood.

The root exudation could also shape distinctive bacterial communities as well as their functionality in relation to different plant development stages.

For example, a shift from sugars to amino acids and phenolic compounds was revealed from *Arabidopsis* root exudation as roots developed (Chaparro et al., 2013). These changes correlated with shifts in functional gene composition (e.g. secondary metabolite biosynthesis genes), as shown by mRNA sequencing. The study on slender wild oats (*Avena barbata*) reinforced this correlation, linking temporal changes in root exudates to microbial substrate preferences, particularly a rise in Proteobacteria (Zhalnina et al., 2018). Overall, these findings emphasize a strong link between rhizosphere development, community composition, and functional gene presence, driven by root exudate composition. In addition to root exudate-mediated modulation, recent evidence suggests that microbial–plant interactions could be modulated through the migration of small non-coding RNAs (sRNAs) between rhizosphere microbes and hosts, altering the plant phenotype (Regmi et al., 2022).

3.5 Plant root exudate effects on pathogen-plant interactions and plant health

Bacteria have evolved diverse mechanisms to compete in various ecological niches, which play a fundamental role in shaping the composition of the rhizosphere and root microbiomes, consequently influencing the host plant health (Chepsergon and Moleleki, 2023). As discussed before, exudates play a critical role in shaping the rhizosphere microbiome and inducing variations among root types, compartments, and developmental stages. The quality and quantity of various root exudates with different compositions secreted from different locations of the root are mainly determined by root development processes, e.g. branching and root system architecture. The resulting microbial community variations can again in turn act as biotic factors to further modulate specific rhizospheric microbiota by altering the root exudates (Sharma et al., 2023), as well as promote plant health. For example, plants encountering *Pseudomonas syringae* pv tomato (*Pst*) (biotic factor) infection exhibited a significant increase in the amount of amino acids and nucleotides, and long-chain organic acids and simultaneously decreased the amount of sugars, alcohols, and short-chain organic acids in the root exudates, which were employed as signalling molecules (Li et al., 2021; Wen et al., 2021). These modifications in the root exudate composition changed the rhizosphere microbiome structure by recruiting beneficial rhizobacterium *Bacillus subtilis*, which elicited disease resistance characteristics in *Arabidopsis thaliana* (Lakshmanan et al., 2012; Chen et al., 2013; Dastogeer et al., 2020). Another example of this type of interaction is from tomato roots. Strigolactone (SL) (signalling molecule) synthesis was found to increase in the roots of tomato plants undergoing infestation of parasitic nematode *Meloidogyne incognita* (biotic factor), augmenting the root-AMF interactions for better resilience of tomato plants (Xu et al., 2018). In addition to that, the secretion of specific molecules (signalling molecules) like benzonitrile, benzothiazole, dimethyl trisulfide, formic acid, and terpene-like compounds from the roots of tomato plants infected with *Fusarium oxysporum* (biotic factor) was found to recruit *Bacillus* sp. in the rhizosphere (Gulati et al., 2020). Thus, under the influence of various biotic factors, plants release a diverse range of signalling molecules, ultimately shaping the rhizomicrobiota. Biotic factors influencing the rhizospheric microbial community may be due to changes in primary and secondary metabolite concentration gradients along the root and rhizosphere. The altered rhizosphere microbiomes can in turn affect the plant root exudation pattern.

4 Microbiome communication in rhizosphere

While the general effects of variations in quantitative and qualitative aspects of root exudates and rhizodeposition on rhizosphere microbiome among different plant types and within a plant type occur temporally, there is a need for a better understanding of specific factors associated with communications between plant host and microbiome. This is necessary in order to manipulate desired members or groups of microbes for designing ecofriendly strategies for improved plant nutrition, growth, and health.

4.1 Within the microbiome

The rhizosphere microbiome harbors many complex interacting organisms that drive functional activities that have a positive impact on plant health. Microbial co-occurrence networks can unravel complex relationships within the microbiome and offer insight into community structure (Fuhrman, 2009; Faust and Raes, 2012; de Vries et al., 2018; Ramirez et al., 2018; Ling et al., 2022). For example, highly modular but unstable bacterial networks in the rhizosphere were revealed by investigating 557 pairs of published 16S rDNA amplicon sequences from the bulk soils and rhizosphere in different ecosystems, which reflect the interactions and adaptations of microorganisms to dynamic conditions (Ling et al., 2022). Network analysis has been found particularly useful in recent years to understand how microbe–microbe associations change in response to environmental parameters (Barberán et al., 2012; Ma et al., 2016; Morriën et al., 2017; Ofaim et al., 2017; de Vries et al., 2018; Leff et al., 2018; Zhou et al., 2021; Ling et al., 2022). For example, a study addressing the question of how co-occurrence networks respond to disturbances (e.g. climate extremes) in grassland mesocosms, indicated that drought promotes destabilizing properties in soil bacterial, but not fungal, co-occurrence networks and that changes in bacterial community link more strongly to soil functioning during recovery than do changes in fungal communities (de Vries et al., 2018). In microbial networks, quorum sensing (QS) coordinates microbe activities through signalling molecules like N-acyl-l-homoserine lactones (AHLs) (Hassani et al., 2019). This process involves interactions among various microorganisms, influencing adaptation and community dynamics. AHLs produced by bacteria diffuse into the rhizosphere, where plants produce metabolites to mimic their actions. Similarly, canavanine, in root exudates of some legumes (Worst et al., 2015), is used to select beneficial rhizobia and remove other bacteria genera from the rhizosphere (Swamy et al., 2016). Another mimic, catechin, secreted by *Combretum albiflarum*, disrupts *Pseudomonas aeruginosa* PAO1 signalling, acting as the plant's defense against pathogens (Vandeputte et al., 2010). Some plant growth-promoting microorganisms (PGPM) use this mechanism to disrupt pathogen signalling and virulence gene expression (reviewed in Olanrewaju and Babalola (2022)). Moreover, bacteria have evolved diverse mechanisms that enable them to compete in different ecological niches, for example, bacterial competition in the rhizosphere also influences microbial composition, subsequently influencing the host plant health (reviewed in Chepsergon and Moleleki, (2023)). These interactions involve exploitative competition, where microbes compete for limited resources and interference competition, achieved through the production of diffusible molecules or secretion of effector proteins to antagonize competitors.

Keystone taxa, i.e. taxa that have a large influence on the community (Berry and Widder, 2014; Shi et al., 2016; Banerjee et al., 2018; Banerjee et al., 2019), can be identified by statistically analyzing network scores. Despite being numerically inconspicuous, keystone taxa confer greater biotic connectivity to the community and thus can be indicators of community shifts and compositional turnover (Herren and McMahon, 2018). The impact of abiotic factors and host genotypes on the plant microbiome is found to be facilitated via keystone taxa (Agler et al., 2016), and the root microbial network complexity is linked to plant survival (Duran et al., 2018). Agricultural intensification may alter the structure of the root microbial network and the abundance of keystone taxa, which in turn may have implications for crop performance (Wittwer et al., 2017; Toju et al., 2018).

In addition, keystone species could be grouped into microbial generalists and specialists that are thought to have distinct impacts on community dynamics. A recent study identified the ecological characteristics of microbial generalists and specialists across over 3,000 farmland soil samples from eleven countries that encompassed seven climate types (Xu et al., 2022). Examining keystone species of generalist and specialist groups in network topology, both contribute to species interactions in distinct ways. Special keystone species show a stronger deterministic influence, indicating lower tolerance to environmental changes. General keystone species, on the other hand, exhibit traits associated with stochastic processes, suggesting a crucial role in maintaining community stability amid environmental disturbances.

4.2 Across food web

Soil food webs play a crucial role in many soils through essential plant nutrition and health-related functions supporting agriculture production and sustaining ecosystem health (Gupta et al., 2019b). This intricate system comprises diverse organisms, including nematodes, bacteria, archaea, protozoa, fungi, and invertebrates, engaging in interactions like predation, parasitism, mutualism, or competition. These interactions, be they direct or indirect, antagonistic or beneficial, contribute to the formation of a functional ecosystem. The composition and structure of food webs are influenced by factors such as abiotic conditions (Garbeva et al., 2008), soil management (Xiong et al., 2018), plant species, community identity (Garbeva et al., 2008; Ramirez et al., 2018), plant growth stage (Roley, 2021), and various other elements. Particularly noteworthy is the active role of plants in selecting microbial community members, and recruiting beneficial organisms upon pathogen attack (Berendsen et al., 2012). In addition, microbe–fauna interactions (e.g. predation by protozoa, nematodes, etc.) play an important role in nutrient (e.g. N, P, S) mineralization and supply in the rhizosphere environment thereby affecting plant nutrition (Gupta and Sharma, 2021). Also, functionally significant feedbacks have been suggested to exist between consumers of rhizosphere microorganisms (i.e. fauna) and plant roots (Bonkowski et al., 2009). The dynamics of microbial predators' interactions have been shown to vary with soil type; for instance, protists could be more significant bacterial predators than nematodes in clay-rich soils, while the opposite appears true for very sandy soils (Rønn et al., 2012). Interactions between fungi and nematodes play a crucial role in maintaining a stable food web by providing essential ecosystem services such as nutrient mineralization and acquisition and soil disease suppression (Guan et al., 2018). Protists can also enhance plant protection by feeding on pathogenic organisms and through changing bacterial community dynamics. For example, bacteria-protozoa and fungi-mycophagous protist interactions in the rhizosphere can affect plant nutrition, soilborne disease incidence, and plant health. This can happen through changes in the microbiome composition and abundances of beneficial microbes or microbial groups and/or effects on pathogen growth in the rhizosphere (Gupta et al., 1999; Müller et al., 2013). Protists exhibit species-specific effects on bacterial communities, with soil type influencing microbial trophic interactions. A preferential grazing by protozoa can result in changes in rhizosphere microbial community composition in turn influencing plant growth and health (Rønn et al., 2002; Jousset et al., 2017). The microbial species responsible for soil disease suppression employ various mechanisms, including competition for space and resources, nutrients, antibiosis, hyper-parasitism, and induced systemic resistance (Pushpavathi et al., 2016; Zamioudis et al., 2014). Therefore, interactions within the soil food web can have a significant impact on the health and resilience of plants against soil-borne diseases (Guan et al., 2018).

Rhizosphere food webs are often not universally identical and factors modulating the variations in food webs and their impacts on overall farming systems might be specific depending on abiotic conditions (e.g. aggregation and pore network structure), soil management, plant species, and microbiome community (Geisen et al., 2021). Recent application of advanced 'omics' technologies has improved our understanding of both the microbial and some of the faunal communities in terms of their composition, community structure, functions, and interactions within the soil food web. For example, the application of metatranscriptomics techniques has provided a more comprehensive picture of active soil protist communities including in the plant rhizosphere (Geisen et al., 2015; Fujino et al., 2023). Similarly, group-specific amplicon sequencing of switchgrass rhizosphere soils revealed that most protists play varied ecological roles across plant growth stages, i.e. play diverse protist roles, including plant pathogenic and omnivorous types (Ceja-Navarro et al., 2021). Also, from the co-occurrence network analysis, rhizosphere communities were found to be more complex than those in the bulk soils. Both environmental filtering and plant control were found to drive protist community properties. Understanding protist patterns in rhizosphere soil is crucial for grasping the root-microbe-soil system ecology. In addition, Ren et al. (2023) found increased predatory protists and decreased protistan phototrophs with organic fertilization, which suggested that long-term organic fertilization primarily influences fungal diseases through protist predation, offering insights for biocontrol strategies against fungal infections. Moreover, *18S rRNA* gene-based amplicon data illustrated the chemical fertilizer application was associated with lower protist richness and diversity as opposed to the application of biofertilizer (Guo et al., 2018). These results suggest that the application of organic fertilizers and biofertilizers can affect the soil biota community at higher trophic levels (protists) with potential consequences for plant health and productivity. Overall, the improved understanding of the responses in the active populations, and community structure of different members of the rhizosphere food web suggests new and innovative options to manipulate total or specific groups of functional microbiomes for plant growth promotion, improved plant nutrition, and health as part of sustainable agricultural systems.

5 Integrative *omics* and other techniques to understand the rhizosphere

Recent advances in *omics* tools and techniques have expanded our understanding of microbiome assembly, although pinpointing functional drivers associated with the host phenotype remains challenging, especially for complex systems like the rhizosphere. Furthermore, rhizosphere microbiome research faces challenges due to bias in sample collection and limitations in bioinformatics analysis, which can over or underrepresent microbiome assembly and functional interpretation due to technical barriers. Additionally, the different types of *omics* techniques, e.g. genomics, transcriptomics, epigenomics, proteomics, and metabolomics, are generally used independently to study specific aspects of microbiome–host interactions. However, individual *omics* analyses are limited to correlations, so there is a need to collect multiple *omics* datasets (integrative *omics*) simultaneously to gain deeper insights into the regulatory networks. Combining multiple techniques would allow a better understanding of the mechanistic aspects of desirable traits and functions. Therefore, integrative multi-*omics*, incorporating cross-cutting technologies, can effectively address challenges by accurately capturing microbiome dynamics within communities and their interactions with host plants and the environment (Sergaki et al., 2018; Chetty and Blekhman, 2024). Also, imaging techniques such as fluorescence *in situ* hybridization (FISH), mass spectrometry imaging (MSI) methods like Stable Isotope probing (SIP) and nanoscale secondary mass spectrometry (NanoSIMS) and Raman spectroscopy have helped to understand the spatial dimension of the microbiome distribution and for scaling up the knowledge from microscale (e.g. from specific parts of roots to entire root system and larger scale) (Law et al., 2024). Bandara et al. (2021), for *Zea mays* rhizosphere, used a resin embedding and sectioning method that facilitates simultaneous correlative characterization i.e. microbial identification and chemical analysis at micro- to nano-meter scale spatial resolution. Some examples of new findings of the rhizosphere microbiome studies conducted by multi-omics techniques are shown in Table 1. For example, data from genomics and transcriptomics analysis for the same samples through co-isolation (Hurt et al., 2001) facilitates linking community structure to the transient functions, cross-linking of the microbial cells to assemble high-quality metagenomes and better resolution of functional genes and associate mobile genetic elements with their hosts (Wu et al., 2023). Spatial profiling using *omics* and microscopy facilitates the deconvolution of host and microbiome interactions (Saarenpää et al., 2023; Cao et al., 2023). Similarly, molecular interactions among the members of rhizosphere microbiome and plant host–microbiome interactions may also be explored by studying non-coding RNAs like small RNAs (RNAs) to gain a better understanding of the gene regulatory networks in rhizosphere communities (Ren et al., 2019; Regmi et al., 2022) (Figure 4). For example, recent research suggests that plants and rhizosphere microbes could also communicate via small non-coding RNAs (sRNAs). These molecules could move within and sometimes across cells and regulate their targets by sequence complementarity affecting specific functional traits (Regmi et al., 2022). For example, rhizobial tRNA-derived sRNAs regulate host modulation, while cross-kingdom RNA interference is implicated

Table 1 Examples of new findings about the rhizosphere microbiome and their role in plant health and nutrition-related functions using advanced integrative *omics* techniques.

	Aims	Omics approach	Key findings	Reference
1	Exploration of the relationships between the dynamics of exudate composition and the growth of specific soil microbes and whether those relationships can be predicted and generalized	(Comparative) genomic (*16S rRNA* gene-based amplicon sequencing); (exo) -Metabolomics	Dynamic patterns in root exudate composition, driven by plant developmental processes, interact with microbial metabolite preferences, shaping microbial community assembly in the rhizosphere	Zhalnina et al., 2018
2	Characterization of biosynthetic genes in plant genomes that are critical for shaping root microbial community	Metabolomics; Genomics (*16S rRNA* gene-based amplicon sequencing)	New information that assists the engineering root microbiota and further paves the way for investigating the functions of root microbiota in plant growth and health	Huang et al., 2019
3	Identification of signal molecules produced by bacteria and their hosts involved in rhizobial infection and root nodule formation in legumes	Genomics; Transcriptomics	Demonstrated that rhizobial tRNA-derived small RNAs are signal molecules regulating plant nodulation	Ren et al., 2019
4	Quantification of the changes in the microbial network shaped by diverse farming methods for designing environmentally friendly agroecosystems	Ionomics; Metabolomics; Genomics (*16S rRNA* gene-based amplicon sequencing)	Identified organic nitrogen as a crucial factor boosting crop yield during soil solarization, providing a potential solution to make crop production more sustainable	Ichihashi et al., 2020
5	Assessment of the role of iron competition within the microbiome for plant protection for a better understanding of how the microbiome and the resulting ecological interactions affect pathogen growth and disease outbreak	Genomics (*16S rRNA* gene-based amplicon sequencing); together with culture-based method, *in vitro* and in planta bioassays	Rhizosphere microbiome members with growth-promoting siderophores, the secreted iron-scavenging molecules, although competitively inferior, facilitated plant infection by the pathogen, offering the potential for microbiome engineering and pathogen control	Gu et al., 2020

(*Continued*)

Table 1 (*Continued*)

	Aims	Omics approach	Key findings	Reference
6	Testing the hypothesis that rhizosphere microbiota induces several host systemic signals that trigger distal metabolic reprogramming and consequently differential root exudation	Metabolomics; Genomics (*16S rRNA* gene-based amplicon sequencing)	Rhizosphere microbiome (tomato) affects the chemical composition of root exudation through a systemic root–root signalling mechanism, and the microbiome-reprogrammed systemic root exudation promotes soil conditioning.	Korenblum et al., 2020
7	Exploration of structural differences of bacterial communities in different root microenvironments under desert leguminous plant shrubs	Genomics (*16S rRNA* gene-based amplicon sequencing)	Diversity and structural composition of bacterial communities have significant hierarchical differences across four rhizo-compartment types, and soil physicochemical factors have a significant impact on driving the differentiation	Zhou et al., 2020b
8	Testing the hypothesis that the success in artificial selection of microbiota depends on the stabilization of community structure	Genomics (*16S rRNA* gene- and ITS region-based amplicon sequencing)	Identified a remarkable correlation between the variability in plant traits and selected microbiota structures, revealing two distinct sub-communities associated with high or low leaf greenness, whose abundance was significantly steered by directional selection.	Jacquiod et al., 2022
9	Demonstration of plant-antagonistic *Pseudomonas protegens* evolving into mutualists in rhizosphere of *Arabidopsis thaliana*	Genomics; together culture-based method, *in vitro* assay	*P. protegens* rapidly diversifies into mutualist genotypes that have positive effects on plant growth and competitive advantage over ancestral genotypes in the rhizosphere	Li et al., 2021
10	Characterization of the microbiome assembly of *Artemisia annua*, the only plant origin of artemisinin for industrial production, on root microbiomes	Metabolomics; Genomics (*16S rRNA* gene-based amplicon sequencing)	Demonstrated *A. annua* shapes root-associated microbial communities, promoting plant growth and enhancing inter-kingdom co-occurrence networks in the rhizosphere	Shi et al., 2022

11	Comparison of the links between total and active microbial diversity and soil ecosystem functions in the rhizosphere soil in grasslands	Genomics (*16S rRNA* gene-based amplicon sequencing); Proteomics; Metabolomics	Distinct compositions of active and total microbial communities in *Trifolium pratense* rhizosphere reflect correlations with soil functions and microbial richness, which highlights soil microbes' adaptability in rhizosphere	Wahdan et al., 2021
12	Characterization of root-associated environments, especially the interaction between biological and chemical components, in a field condition	Transcriptomics; Genomics (*16S rRNA* gene-based amplicon sequencing); Metabolomics (mineral)	Spatiotemporal dynamics of respective factors revealed a significant interaction between the characteristic bacterial community and the mineral environment in the rhizosphere	Yamazaki et al., 2021
13	Examination of how plants impact the rhizosphere bacterial community and recruit beneficial microorganisms to suppress soilborne fungal pathogens and promote plant growth	Genomics (*16S rRNA* gene-based amplicon sequencing); together culture-based method	Successive wheat plantings and pathogen infection can shape the rhizosphere microbial communities and specifically accumulate a group of beneficial microbes, which may offer the potential for addressing agronomic concerns associated with plant diseases and crop productivity.	Yin et al., 2021
14	A deeper understanding of the molecular mechanism of the rhizosphere–plant interaction	Metabolomics; Genomics (together with molecular imaging and biochemistry)	Uncovered a common mechanism by which fungi can gain access to phenazine-replete environments and provides a tractable model system, in particular revealing how rhizosphere microbial communities and in plant infection sites negotiate community membership via a chemical dialectic	Dahlstrom and Newman, 2022

(*Continued*)

Table 1 (*Continued*)

	Aims	Omics approach	Key findings	Reference
15	Identification of the rhizosphere microbes that actively consume plant-derived carbon	Isotope probing, quantitative PCR, marker gene sequencing, and shotgun metagenomic sequencing	Highlights the important roles and the differential contributions of rhizosphere microbes to belowground carbon dynamics, building a more nuanced understanding of the complexity of plant–microbe interactions in the rhizosphere	Fan et al., 2022
16	Testing if heterogeneity in plant-microbiome composition and function can merge out of an initially homogeneous species pool.	Genomics (*16S rRNA* gene-based amplicon sequencing); together culture-based method	Demonstrated that an initially homogenized soil can rapidly diverge into rhizosphere microbiomes varying in their ability to promote plant protection, suggesting that early-life interventions may have significant effects on later microbiome states and highlighting an opportunity for microbiome diagnostics and plant disease prevention	Gu et al., 2022
17	Bridging the gap between general rhizobacterial assemblages and functions associated with community-wide dormancy capacity, heterotrophic strategies, and individual nutrient cycling processes.	Genomics (*16S rRNA* gene-based amplicon sequencing)	The rhizosphere selects microorganisms, forming a seed bank and reducing diversity. Dormancy involves toxin–antitoxin systems, and sporulation is common in bulk soils. Genes related to organic compound conversion, N_2-fixation, and denitrification are enriched, while nitrification genes are depleted	Ling et al., 2022
18	A deeper understanding of basic principles governing metabolite use relevant to the plant–microbe interactions in the rhizosphere	Metabolomics	The chemical ecology of coumarins and phenazines affects iron acquisition by pseudomonads, suggesting a mechanism by which root exudates may tune the utility of microbial secondary metabolites	McRose et al., 2023

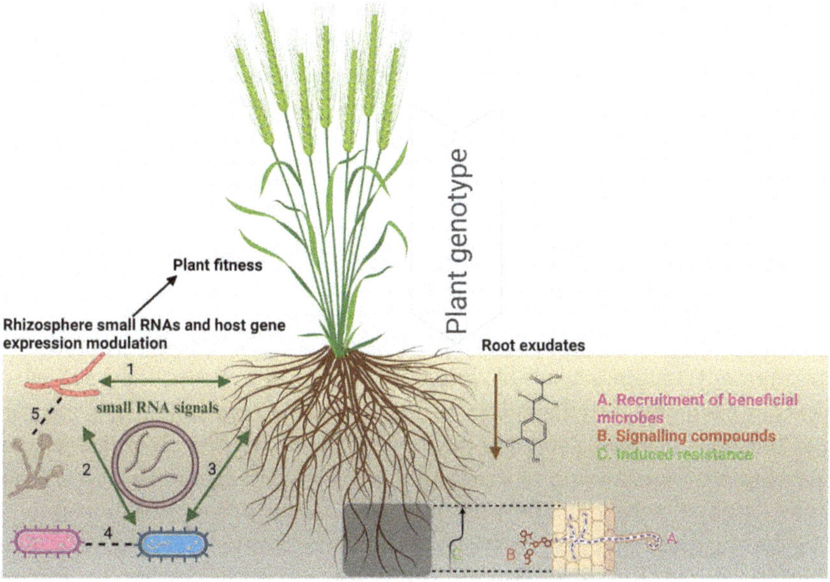

Figure 4 A conceptual diagram showing the potential roles of small non-coding RNAs (sRNAs) in influencing rhizosphere plant-microbe and microbe-microbe interactions. Rhizosphere-plant sRNA communication could happen in different ways: (1) between fungi and plant roots, (2) between fungi and bacteria, (3) between bacteria and plants (Ren et al., 2019). Additionally, sRNA signalling could occur across different (4) bacterial and (5) fungal species in the rhizosphere (Wen et al., 2023). Figure not to scale.

in the Arbuscular mycorrhizal association between *Gigaspora margarita* and its plant host (Silvestri et al., 2020). By understanding sRNA communication in the rhizosphere, microbes-induced gene silencing strategies could be achieved by modulating sRNA-target combinations to boost crop immunity (Wen et al., 2023). Although, rhizosphere sRNA research is in its infancy, early evidence shows promise for its potential use to manipulate rhizosphere plant–microbe and plant–pathogen interactions to specifically target gene functions related to plant health, nutrition, and tolerance to abiotic stresses. For example, engineered complementary dsRNAs in *Trichoderma harzianium* were reported to silence pathogenicity factors of soil-borne pathogens *Verticillium dahliae* and *Fusarium oxysporum*, reducing their virulence and increasing plant immunity of cotton and rice, respectively (Wen et al., 2023). This technology could therefore help avoid the need for genetically modified crops. Although its technical application is at the neonate stage, microbial sRNAs' ability to alter host gene expression and phenotypes presents promising opportunities for agricultural and therapeutic interventions, demonstrating the potential of microbiome-based alternatives to conventional agrichemicals and antibiotics. *In situ* manipulation of designer microbes could be the next step in this field, however technical difficulties for the effective delivery of sRNAs need to be addressed. Overall, this type of information about the composition of communities, expression of functional specific genes/processes, and the metabolic regulators of functions not only helps to understand the drivers of microbiome assembly but also their roles in plant health and fitness, enabling the development of intervention options for sustainable farming systems.

6 Development of designer plant-rhizosphere microbiome combinations

Lately, microbiome research has focused on developing sustainable management strategies by employing the beneficial properties of the soil–plant microbiome interactions. Based on the discussion in previous sections, the path to achieving functionally suitable designer rhizosphere microbiomes requires an in-depth understanding of the spatial and temporal dynamics of microbial composition and the identification of factors that regulate their functionality. It is now accepted that perturbations of the rhizosphere microbiome can alter plant fitness either directly or indirectly by altering their microenvironment leading to changes in soil fertility (Foo et al., 2017; Mahmud et al., 2021). Different strategies have been suggested to modify the composition of total microbial community composition or alter individual microbes or microbial groups, thereby enhancing the functional activities of beneficial microbes (Song et al., 2021a).

6.1 Above- and belowground manipulation – biologicals and other amendments

Both phyllosphere and rhizosphere microbial communities can be modified over short and longer periods of time by adopting different strategies such as microbiome transplantation (Raaijmakers and Mazzola, 2016), introduction of single or SynComs (de Souza et al., 2020), soil amendments (Andreo-Jimenez et al., 2021), agronomic management including above-ground treatment of plants (Ney et al., 2018; Gupta and Knox, 2010). These manipulations directly or indirectly enhance plant growth either by inducing the activity of beneficial microbes or making favourable environments for plants by modifying the surrounding environment.

6.2 Plant breeding-based modification of plant-associated microbiota

Host-mediated recruitment and activation of specific beneficial microbes is one approach that could be used to manipulate beneficial soil–plant microbiome interactions. Beneficial microbiomes can be modulated over longer periods by selecting the plant lines that facilitate the establishment of a selective rhizosphere (Ryan et al., 2009). Breeding programmes must be devised to rationally manipulate desirable characteristics to obtain beneficial microorganisms ultimately suppressing the biotic and abiotic stresses (Nerva et al., 2022). However, the continuous selection of genomic plant traits, may also negatively impact the interactions among plants and beneficial microorganisms, as a result, modern varieties are less resilient to stresses (Chen et al., 2015). For example, long-term nitrogen fertilization in legumes results in the recruitment of less functional rhizobacteria, providing reduced benefits to the host (Weese et al., 2015). Moreover, in soybeans, domestication and breeding have changed the natural legume defences mechanism ultimately reducing the agricultural benefits of symbiosis (Kiers et al., 2007). Similarly, an investigation of phylogenetically diverse herbaceous crops shows that wild varieties recruit more beneficial microbes while domesticated crops enhance the assembly of harmful microbiota (Martín-Robles et al., 2020). These findings suggest negligence including beneficial microbes–plant interactions in breeding programs that often only address commercially demanded traits.

6.3 Vertically transmitted microbiota

Plant–microbe interactions can create phenotypes without changing the genomic information of plants (Wei and Jousset, 2017). The genotypic characteristics of the parent plant can be changed by microbial transfer, which can be transmitted vertically to the next plant generation, and the desired plant phenotype can be maintained by plant breeding methods. For example, the introduction of beneficial bacteria at the flowering of progeny seeds has been shown to modifies plant microbiomes and to retain desirable plant traits (Mitter et al., 2017). In an aromatic clonal *Glechoma hederacea* plant, symbiotic bacteria and fungi were transmitted to its progeny while archaea were not passed. However, the heritable communities exhibited lower richness, suggesting filtration during transmission (Vannier et al., 2018). By understanding the complexities of the rhizosphere, the most favourable conditions for plant growth can be maintained by modifying the soil, breeding or creating better plants, and manipulating plant–microorganism interactions.

7 Conclusion

Plant–microbe interactions in the rhizosphere of crop plants play a key role in the development of sustainable agricultural practices and promoting resilient farming systems with the microbiome performing multiple ecological services such as nutrient acquisition, abiotic and biotic stress tolerance and addressing climate change issues. For this, a thorough understanding of factors governing the assembly and functional capacity of the rhizosphere microbiome is necessary, including host plant-mediated factors and microbe-related determinants. Currently, rhizosphere sampling strategies are often insufficient due to low resolution and multiple confounding factors including variations in microbiome composition and functionality among plant genotypes and root types within the same plant. Longitudinal separation of developmentally diverse root zones in single root types can detect development-dependent microbial gradients. To tease this apart, fine-scale isolation of rhizosphere and next-generation sequencing should be applied to target plant–microbe interactions at the cellular level. Considering these points, there is a need to better understand factors that affect microbiome assembly and functions, including host genetics, plant species, domestication, root morphology, and microbial interactions. Further elucidation of rhizosphere microbiome-plant interdependencies that mediate plant health with a focus on the spatial and temporal impacts of microbial assembly and their role in executing specific plant functions will ultimately lead to greater opportunity for the development of more sustainable farming systems.

8 Acknowledgements

The preparation of this manuscript was supported through funding from CSIRO Business Unit Agriculture & Food, Microbiomes For One Systems Health – Future Science Platform, Australia. The authors declare no conflict of interest.

9 References

Agler, M. T., et al. 2016. Microbial hub taxa link host and abiotic factors to plant microbiome variation. *PLoS Biology*, 14, e1002352.

Andreo-Jimenez, B., et al. 2021. Chitin- and keratin-rich soil amendments suppress *Rhizoctonia solani* disease via changes to the soil microbial community. *Applied and Environmental Microbiology*, 87, e00318-21.

Arshad, M. & Frankenberger, W. T. 1997. Plant growth-regulating substances in the rhizosphere: microbial production and functions. In: Sparks, D. L. (ed.) *Advances in Agronomy*. New York: Academic Press.

Badri, D. V. & Vivanco, J. M. 2009. Regulation and function of root exudates. *Plant, Cell & Environment*, 32, 666–681.

Bais, H. P., et al. 2006. The role of root exudates in ehizosphere interactions with plants and other organisms. *Annual Review of Plant Biology*, 57, 233–266.

Bandara C. D., et al. 2021. Microbial identification, high-resolution microscopy and spectrometry of the rhizosphere in its native spatial context. *Frontiers in Plant Science*, 12, 668929.

Banerjee, S., et al. 2018. Keystone taxa as drivers of microbiome structure and functioning. *Nature Reviews Microbiology*, 16, 567–576.

Banerjee, S., et al. 2019. Agricultural intensification reduces microbial network complexity and the abundance of keystone taxa in roots. *The Isme Journal*, 13, 1722–1736.

Barberán, A., et al. 2012. Using network analysis to explore co-occurrence patterns in soil microbial communities. *The Isme Journal*, 6, 343–351.

Baudoin, E., et al. 2002. Impact of growth stage on the bacterial community structure along maize roots, as determined by metabolic and genetic fingerprinting. *Applied Soil Ecology*, 19, 135–145.

Baudoin, E., et al. 2003. Impact of artificial root exudates on the bacterial community structure in bulk soil and maize rhizosphere. *Soil Biology and Biochemistry*, 35, 1183–1192.

Bazghaleh, N., et al. 2015. Genotype-specific variation in the structure of root fungal communities is related to chickpea plant productivity. *Applied and Environmental Microbiology*, 81, 2368–2377.

Beckers, B., et al. 2017. Structural variability and niche differentiation in the rhizosphere and endosphere bacterial microbiome of field-grown poplar trees. *Microbiome*, 5, 25.

Berendsen, R. L., et al. 2012. The rhizosphere microbiome and plant health. *Trends in Plant Science*, 17, 478–486.

Berg, G. & Raaijmakers, J. M. 2018. Saving seed microbiomes. *The Isme Journal*, 12, 1167–1170.

Berry, D. & Widder, S. 2014. Deciphering microbial interactions and detecting keystone species with co-occurrence networks. *Frontiers in Microbiology*, 5, 90985.

Bonkowski, M., et al. 2009. Rhizosphere fauna: the functional and structural diversity of intimate interactions of soil fauna with plant roots. *Plant and Soil*, 321, 231–233.

Bouffaud, M.-L., et al. 2014. Root microbiome relates to plant host evolution in maize and other Poaceae. *Environmental Microbiology*, 16, 2804–2814.

Broeckling, C. D., et al. 2008. Root exudates regulate soil fungal community composition and diversity. *Applied and Environmental Microbiology*, 74, 738–744.

Bressan, M., et al. 2009. Exogenous glucosinolate produced by *Arabidopsis thaliana* has an impact on microbes in the rhizosphere and plant roots. *The Isme Journal*, 3, 1243–1257.

Brunel, C., et al. 2020. Towards unraveling macroecological patterns in rhizosphere microbiomes. *Trends in Plant Science*, 25, 1017–1029.

Bulgarelli, D., et al. 2015. Structure and function of the bacterial root microbiota in wild and domesticated barley. *Cell Host Microbe*, 17, 392–403.

Cao, Z., et al. 2023. Spatial profiling of microbial communities by sequential Fish with error-robust encoding. *Nature Communications*, 14, 1477.

Ceja-Navarro, J. A., et al. 2021. Protist diversity and community complexity in the rhizosphere of switchgrass are dynamic as plants develop. *Microbiome*, 9, 96.

Chaparro, J. M., et al. 2013. Root exudation of phytochemicals in arabidopsis follows specific patterns that are developmentally programmed and correlate with soil microbial functions. *PLoS One*, 8, e55731.

Chaparro, J. M., et al. 2014. Rhizosphere microbiome assemblage is affected by plant development. *The Isme Journal*, 8, 790–803.

Chen, H., et al. 2022. Functional redundancy in soil microbial community based on metagenomics across the globe. *Frontiers in Microbiology*, 13, 878978

Chen, L., et al. 2016. Structural and functional differentiation of the root-associated bacterial microbiomes of perennial ryegrass. *Soil Biology and Biochemistry*, 98, 1–10.

Chen, Q.-L., et al. 2021. Potential of indigenous crop microbiomes for sustainable agriculture. *Nature Food*, 2, 233–240.

Chen, Y., et al. 2013. Biocontrol of tomato wilt disease by *Bacillus subtilis* isolates from natural environments depends on conserved genes mediating biofilm formation. *Environmental Microbiology*, 15, 848–864.

Chen, Y. H., et al. 2015. Crop domestication and its impact on naturally selected trophic interactions. *Annual Review of Entomology*, 60, 35–58.

Chepsergon, J. & Moleleki, L. N. 2023. Rhizosphere bacterial interactions and impact on plant health. *Current Opinion in Microbiology*, 73, 102297.

Chetty, A. & Blekhman, R. 2024. Multi-omic approaches for host-microbiome data integration. *Gut Microbes*, 16, 2297860.

Clouse, K. M. & Wagner, M. R. 2021. Plant genetics as a tool for manipulating crop microbiomes: opportunities and challenges. *Frontiers in Bioengineering and Biotechnology*, 9, 567548

Cordovez, V., et al. 2019. Ecology and evolution of plant microbiomes. *Annual Review of Microbiology*, 73, 69–88.

Cordovez, V., et al. 2021. Successive plant growth amplifies genotype-specific assembly of the tomato rhizosphere microbiome. *Science of the Total Environment*, 772, 144825.

Curlango-Rivera, G., et al. 2013. Intraspecies variation in cotton border cell production: Rhizopshere microbiome implications. *American Journal of Botany*, 100, 1706–1712.

Dahlstrom, K. M. & Newman, D. K. 2022. Soil bacteria protect fungi from phenazines by acting as toxin sponges. *Current Biology*, 32, 275–288.e5.

Dangl, J. L. & Jones, J. D. G. 2001. Plant pathogens and integrated defence responses to infection. *Nature*, 411, 826–833.

Dastogeer, K. M. G., et al. 2020. Plant microbiome–an account of the factors that shape community composition and diversity. *Current Plant Biology*, 23, 100161.

De Souza, R. S. C., et al. 2020. From microbiome to traits: designing synthetic microbial communities for improved crop resiliency. *Frontiers in Plant Science*, 11, 1179.

De Vries, F. T., et al. 2018. Soil bacterial networks are less stable under drought than fungal networks. *Nature Communications*, 9, 3033.

Degens, B. P. & Harris, J. A. 1997. Development of a physiological approach to measuring the catabolic diversity of soil microbial communities. *Soil Biology and Biochemistry*, 29, 1309–1320.

Dennis, P. G., et al. 2010. Are root exudates more important than other sources of rhizodeposits in structuring rhizosphere bacterial communities? *Fems Microbiology Ecology*, 72, 313–327.

Donn, S., et al. 2015. Evolution of bacterial communities in the wheat crop rhizosphere. *Environmental Microbiology*, 17, 610–621.

Donn, S., et al. 2017. Root type is not an important driver of mycorrhizal colonisation in Brachypodium distachyon. *Pedobiologia*, 65, 5–15.

Durán, P., et al. 2018. Microbial interkingdom interactions in roots promote *Arabidopsis* survival. *Cell*, 175, 973–983.e14.

Edwards, J., et al. 2015. Structure, variation, and assembly of the root-associated microbiomes of rice. *Proceedings of the National Academy of Sciences*, 112, E911–E920.

Emmett, B. D., et al. 2018. Eighty years of maize breeding alters plant nitrogen acquisition but not rhizosphere bacterial community composition. *Plant and Soil*, 431, 53–69.

Fan, K., et al. 2022. Identification of the rhizosphere microbes that actively consume plant-derived carbon. *Soil Biology and Biochemistry*, 166, 108577.

Faure, D., et al. 2009. Molecular communication in the rhizosphere. *Plant and Soil*, 321, 279–303.

Faust, K. & Raes, J. 2012. Microbial interactions: from networks to models. *Nature Reviews Microbiology*, 10, 538–550.

Finzi, A. C., et al. 2015. Rhizosphere processes are quantitatively important components of terrestrial carbon and nutrient cycles. *Global Change Biology*, 21, 2082–2094.

Fitzpatrick, C. R., et al. 2020. The plant microbiome: from ecology to reductionism and beyond. *Annual Review of Microbiology*, 74, 81–100.

Foo, J. L., et al. 2017. Microbiome engineering: Current applications and its future. *Biotechnology Journal*, 12, 1600099.

Foster, K. R., et al. 2017. The evolution of the host microbiome as an ecosystem on a leash. *Nature*, 548, 43–51.

French, E., et al. 2021. Emerging strategies for precision microbiome management in diverse agroecosystems. *Nature Plants*, 7, 256–267.

Fuhrman, J. A. 2009. Microbial community structure and its functional implications. *Nature*, 459, 193–199.

Fujino, M., et al. 2023. Protists modulate active bacterial community composition in paddy field soils. *Biology and Fertility of Soils*, 59, 709–721.

Garbeva, P., et al. 2008. Rhizosphere microbial community and its response to plant species and soil history. *Plant and Soil*, 302, 19–32.

Geisen S., et al. 2015. Metatranscriptomic census of active protists in soils. The *Isme Journal*, 9, 2178–2190.

Geisen, S. & Quist, C. W. 2021. Microbial–Faunal Interactions in the Rhizosphere. In: Gupta, V.V.S.R. & Sharma, A.K. (eds.) *Rhizosphere Biology: Interactions Between Microbes and Plants*. Singapore: Springer.

George, T. S., et al. 2024. Bottom-up perspective-The role of roots and rhizosphere in climate change adaptation and mitigation in agroecosystems. *Plant and Soil*, 500, 297-323. doi:10.1007/s11104-024-06626-6.

Gilligan, C. 1980. Colonization of lateral, seminal and adventitious roots of wheat by the take-all fungus, *Gaeumannomyces graminis* var. *tritici*. *The Journal of Agricultural Science*, 94, 325–329.

Gu, S., et al. 2020. Competition for iron drives phytopathogen control by natural rhizosphere microbiomes. *Nature Microbiology*, 5, 1002–1010.

Gu, Y., et al. 2022. Small changes in rhizosphere microbiome composition predict disease outcomes earlier than pathogen density variations. *The Isme Journal*, 16, 2448–2456.

Guan, P., et al. 2018. Soil microbial food web channels associated with biological soil crusts in desertification restoration: The carbon flow from microbes to nematodes. *Soil Biology and Biochemistry*, 116, 82–90.

Gulati, S., et al. 2020. A non-invasive soil-based setup to study tomato root volatiles released by healthy and infected roots. *Scientific Reports*, 10, 12704.

Gunawardena, U., et al. 2005. Tissue-specific localization of pea root infection by *Nectria haematococca*. mechanisms and consequences. *Plant Physiology*, 137, 1363–1374.

Guo, S., et al. 2018. Continuous application of different fertilizers induces distinct bulk and rhizosphere soil protist communities. *European Journal of Soil Biology*, 88, 8–14.

Gupta, H., et al. 2023. Unearthing the modern trends and concepts of rhizosphere microbiome in relation to plant productivity. In: Chhabra, S., Prasad, R., Maddela,

N. R. & Tuteja, N. (eds.) *Plant Microbiome for Plant Productivity and Sustainable Agriculture*. Singapore: Springer.

Gupta, V. V. S. R. & Knox, O. G. G. 2010. How best can we design rhizosphere plant-microbe interactions for the benefit of plant growth? In: Gupta V. V. S. R., & Ryder M. M., Radcliffe J. (eds.) *The Rovira Rhizosphere Symposium – Celebrating 50 Years of Rhizosphere Research*, pp. 11–24. Deakin: The Crawford Fund.

Gupta, V. V. S. R. & Roget, D.K. 2011. Biological suppression of Rhizoctonia disease lin wheat – effect of carbon and available nitrogen in soil. In: *New Frontiers in Plant Pathology for Asia and Oceania, ACPP APPS 2011*, p. 33. Darwin, Australia.

Gupta, V. V. S. R. & Sharma, A. K. 2021. *Rhizosphere Biology: Interactions Between Microbes and Plants*, p. 356. Singapore: Springer.

Gupta, V. V. S. R., et al. 1999. Effects of microfuana and mesofauna on *Rhizoctonia solani* in a South Australian soil. In: Magarey, R.C. (ed.) *Proceedings of the First Australasian SoilBorne Disease Symposium*, pp. 134–136. Brisbane: BSES.

Gupta, V. V. S. R., et al. 2009. Soil type and rotation effects on the suppression of rhizoctonia bare patch disease in wheat. In: *5th Australasian SoilBorne Disease Symposium – Extended Abstracts*, pp. 85–87, Thredbo, NSW, Australia.

Gupta, V. V. S. R., et al. 2011. Principles and management of soil biological factors for sustainable rainfed farming systems. In: Tow, P., Cooper, I., Partridge, I., Birch, C. (eds.) *Rainfed Farming Systems*, pp. 149–184. New York: Springer Science and Business Media.

Gupta, V. V. S. R., et al. 2014a. Nitrogen cycling in summer active perennial grass systems in South Australia: non-symbiotic nitrogen fixation. *Crop & Pasture Science*, 65, 1044–1056.

Gupta, V. V. S. R., et al. 2014b. Diversity of nifH-harbouring bacteria and non-symbiotic N_2 fixation in Australian cropping soils. In: Gupta V. V. S. R. & Unkovich M., Kaiser B. N. (eds.) *The 17th Australian Nitrogen Fixation Proceedings*, pp. 40–41. Urrbrae: Australian Society for Nitrogen Fixation.

Gupta, V. V. S. R., et al. 2019a. Harnessing the benefits of soil biology in conservation agriculture. In: Pratley, J. & Kirkegaard, J (eds.) *Australian Agriculture in 2020: From Conservation to Automation*, pp. 237–253. Wagga Wagga: Agronomy Australia and Charles Sturt University, Australia.

Gupta, V. V. S. R., et al. 2019b. Diazotroph diversity and nitrogenfixation in summer active perennial grasses in a mediterranean region agricultural soil. *Frontiers in Molecular Biosciences*, 6, 115.

Gurusinghe, S., et al. 2019. Technologies for the selection, culture and metabolic profiling of unique rhizosphere microorganisms for natural product discovery. *Molecules*, 24, 1955.

Gutjahr, C., et al. 2015. Transcriptome diversity among rice root types during asymbiosis and interaction with arbuscular mycorrhizal fungi. *Proceedings of the National Academy of Sciences*, 112, 6754–6759.

Haichar, F. E. Z., et al. 2008. Plant host habitat and root exudates shape soil bacterial community structure. *The Isme Journal*, 2, 1221–1230.

Hassani, M. A., et al. 2019. Interactions and coadaptation in plant metaorganisms. *Annual Review of Phytopathology*, 57, 483–503.

Hayden, H. L., et al. 2018. Comparative metatranscriptomics in disease suppressive and non-suppressive soils for *Rhizoctonia solani* AG8. *Frontiers in Microbiology*, 9, 1–19.

Herren, C. M. & Mcmahon, K. D. 2018. Keystone taxa predict compositional change in microbial communities. *Environmental Microbiology*, 20, 2207–2217.

Horton, M. W., et al. 2014. Genome-wide association study of *Arabidopsis thaliana* leaf microbial community. *Nature Communications*, 5, 5320.

Houlden, A., et al. 2008. Influence of plant developmental stage on microbial community structure and activity in the rhizosphere of three field crops. *Fems Microbiology Ecology*, 65, 193–201.

Hu, L., et al. 2018. Root exudate metabolites drive plant-soil feedbacks on growth and defense by shaping the rhizosphere microbiota. *Nature Communications*, 9, 2738.

Huang, A. C., et al. 2019. A specialized metabolic network selectively modulates *Arabidopsis* root microbiota. *Science*, 364, eaau6389.

Hurt, R. A., et al. 2001. Simultaneous recovery of Rna and Dna from soils and sediments. *Applied and Environmental Microbiology*, 67, 4495–503.

Hussain, Q., et al. 2012. Microbial community dynamics and function associated with rhizosphere over periods of rice growth. *Plant, Soil and Environment*, 58, 55–61.

Ichihashi, Y., et al. 2020. Multi-omics analysis on an agroecosystem reveals the significant role of organic nitrogen to increase agricultural crop yield. *Proceedings of the National Academy of Sciences*, 117, 14552–14560.

Innes, L., et al. 2004. The impacts of individual plant species on rhizosphere microbial communities in soils of different fertility. *Biology and Fertility of Soils*, 40, 7–13.

Jacquiod, S., et al. 2022. Artificial selection of stable rhizosphere microbiota leads to heritable plant phenotype changes. *Ecology Letters*, 25, 189–201.

Jonathan, W. L., et al. 2017. Plant domestication and the assembly of bacterial and fungal communities associated with strains of the common sunflower, *Helianthus annuus*. *The New Phytologist*, 214, 412–423.

Jones, D. L., et al. 2009. Carbon flow in the rhizosphere: carbon trading at the soil–root interface. *Plant and Soil*, 321, 5–33.

Jousset, A., et al. 2017. Where less may be more: how the rare biosphere pulls ecosystems strings. *The Isme Journal*, 11, 853–862.

Kawasaki, A., et al. 2016. Microbiome and exudates of the root and rhizosphere of brachypodium distachyon, a model for wheat. *PLoS One*, 11, e0164533.

Kawasaki, A., et al. 2018. A sterile hydroponic system for characterising root exudates from specific root types and whole-root systems of large crop plants. *Plant Methods*, 14, 114.

Kawasaki, A., et al. 2021a. Manipulating exudate composition from root apices shapes the microbiome throughout the root system. *Plant Physiology*, 187, 2279–2295.

Kawasaki, A., et al. 2021b. The microbiomes on the roots of wheat (*Triticum aestivum* L.) and rice (*Oryza sativa* L.) exhibit significant differences in structure between root types and along root axes. *Functional Plant Biology*, 48, 871–888.

Kiers, E. T., et al. 2007. Human selection and the relaxation of legume defences against ineffective rhizobia. *Proceedings of the Royal Society B: Biological Sciences*, 274, 3119–3126.

Kim, H., et al. 2020. Domestication of Oryza species eco-evolutionarily shapes bacterial and fungal communities in rice seed. *Microbiome*, 8, 20.

Knox, O. G. G., et al. 2006. Constitutive expression of Cry proteins in roots and border cells of transgenic cotton. *Euphytica*, 154, 83–90.

Korenblum, E., et al. 2020. Rhizosphere microbiome mediates systemic root metabolite exudation by root-to-root signaling. *Proceedings of the National Academy of Sciences*, 117, 3874–3883.

Kroll, S., et al. 2017. Genomic dissection of host–microbe and microbe–microbe interactions for advanced plant breeding. *Current Opinion in Plant Biology*, 36, 71–78.

Kuzyakov, Y. & Blagodatskaya, E. 2015. Microbial hotspots and hot moments in soil: concept & review. *Soil Biology and Biochemistry*, 83, 184–199.

Lakshmanan, V., et al. 2012. Microbe-associated molecular patterns-triggered root responses mediate beneficial rhizobacterial recruitment in *Arabidopsis*. *Plant Physiology*, 160, 1642–1661.

Landa, B. B., et al. 2006. Enrichment and genotypic diversity of phlD-containing fluorescent *Pseudomonas* spp. in two soils after a century of wheat and flax monoculture. *Fems Microbiology Ecology*, 55, 351–368.

Law, S. R., et al. 2024. Life at the borderlands: microbiomes of interfaces critical to One Health. *Fems Microbiology Reviews*, 48, fuae008.

Lazcano, C., et al. 2021. The rhizosphere microbiome plays a role in the resistance to soil-borne pathogens and nutrient uptake of strawberry cultivars under field conditions. *Scientific Reports*, 11, 3188.

Leff, J. W., et al. 2018. Predicting the structure of soil communities from plant community taxonomy, phylogeny, and traits. *The Isme Journal*, 12, 1794–1805.

Li, E., et al. 2021. Rapid evolution of bacterial mutualism in the plant rhizosphere. *Nature Communications*, 12, 3829.

Li, Z., et al. 2020. Effect of wheat cultivars with different resistance to *Fusarium* head blight on rhizosphere *Fusarium graminearum* abundance and microbial community composition. *Plant and Soil*, 448, 383–397.

Liljeroth, E., et al. 1991. Changes in bacterial populations along roots of wheat (*Triticum aestivum* L.) seedlings. *Biology and Fertility of Soils*, 10, 276–280.

Ling, N., et al. 2022. Rhizosphere bacteriome structure and functions. *Nature Communications*, 13, 836.

Liu, F., et al. 2019. Soil indigenous microbiome and plant genotypes cooperatively modify soybean rhizosphere microbiome assembly. *BMC Microbiology*, 19, 201.

Liu, Z., et al. 2020. Seminal and nodal roots of barley differ in anatomy, proteome and nitrate uptake capacity. *Plant and Cell Physiology*, 61, 1297–1308.

Lu, J., et al. 2024. Wheat cultivar replacement drives soil microbiome and microbial cooccurrence patterns. *Agriculture, Ecosystems & Environment*, 360, 108774.

Ma, B., et al. 2016. Geographic patterns of co-occurrence network topological features for soil microbiota at continental scale in eastern China. *The Isme Journal*, 10, 1891–1901.

Mahmud, K., et al. 2021. Rhizosphere microbiome manipulation for sustainable crop production. *Current Plant Biology*, 27, 100210.

Martín-Robles, N., et al. 2020. Crops and their wild progenitors recruit beneficial and detrimental soil biota in opposing ways. *Plant and Soil*, 456, 159–173.

McRose, D. L., et al. 2023. The chemical ecology of coumarins and phenazines affects iron acquisition by pseudomonads. *Proceedings of the National Academy of Sciences*, 120, e2217951120.

Mendes, L. W., et al. 2018. Influence of resistance breeding in common bean on rhizosphere microbiome composition and function. *The Isme Journal*, 12, 212–224.

Mendes, R., et al. 2011. Deciphering the rhizosphere microbiome for disease-suppressive bacteria. *Science*, 332, 1097–100.

Mitter, B., et al. 2017. A new approach to modify plant microbiomes and traits by introducing beneficial bacteria at flowering into progeny seeds. *Frontiers in Microbiology*, 8, 11.

Morriën, E., et al. 2017. Soil networks become more connected and take up more carbon as nature restoration progresses. *Nature Communications*, 8, 14349.

Müller, M. S., et al. 2013. Protozoa drive the dynamics of culturable biocontrol bacterial communities. *PLoS One*, 8, e66200.

Munoz-Ucros, J., et al. 2021. Spatiotemporal patterns of rhizosphere microbiome assembly: from ecological theory to agricultural application. *Journal of Applied Ecology*, 58, 894–904.

Narsing Rao, M. P., et al. 2022. Actinobacteria-plant interactions in alleviating abiotic stress. *Plants*, 11, 2976.

Ndour, P. M. S., et al. 2017. Pearl millet genetic traits shape rhizobacterial diversity and modulate rhizosphere aggregation. *Frontiers in Plant Science*, 8, 1288.

Nerva, L., et al. 2022. Breeding toward improved ecological plant–microbiome interactions. *Trends in Plant Science*, 27, 1134–1143.

Ney, L., et al. 2018. Examining trophic-level nematode community structure and nitrogen mineralization to assess local effective microorganisms' role in nitrogen availability of swine effluent to forage crops. *Applied Soil Ecology*, 130, 209–218.

Nsabimana, D., et al. 2004. Size, activity and catabolic diversity of the soil microbial biomass as affected by land use. *Applied Soil Ecology*, 26, 81–92.

Ofaim, S., et al. 2017. Analysis of microbial functions in the rhizosphere using a metabolic-network based framework for metagenomics interpretation. *Frontiers in Microbiology*, 8, 1606.

Ofek-Lalzar, M., et al. 2014. Niche and host-associated functional signatures of the root surface microbiome. *Nature Communications*, 5, 4950.

Olanrewaju, O. S. & Babalola, O. O. 2022. The rhizosphere microbial complex in plant health: a review of interaction dynamics. *Journal of Integrative Agriculture*, 21, 2168–2182.

Oliver, I. C., et al. 2021. Rhizosphere legacy: plant root interactions with the soil and its biome. In: Gupta, V. V. S. R. & Sharma, A. K. (eds.) *Rhizosphere Biology: Interactions Between Microbes and Plants*. Singapore: Springer.

Pantigoso, H. A., et al. 2022. The rhizosphere microbiome: plant–microbial interactions for resource acquisition. *Journal of Applied Microbiology*, 133, 2864–2876.

Pastrana, A. M., et al. 2017. Potential inoculum sources and incidence of strawberry soilborne pathogens in spain. *Plant Disease*, 101, 751–760.

Peiffer, J. A., et al. 2013. Diversity and heritability of the maize rhizosphere microbiome under field conditions. *Proceedings of the National Academy of Sciences*, 110, 6548–6553.

Peralta, A. L., et al. 2018. Crop rotational diversity increases disease suppressive capacity of soil microbiomes. *Ecosphere*, 9, e02235.

Pushpavathi, Y., et al. 2016. Biological control of *Fusarium* wilt disease in banana with emphasis on *Trichoderma* spp. and *Pseudomonas* spp.. *Plant Archives*, 16(1), 51–55.

Quiza, L., et al. 2015. Harnessing phytomicrobiome signaling for rhizosphere microbiome engineering. *Frontiers in Plant Science*, 6, 507.

Raaijmakers, J. M. & Mazzola, M. 2016. Soil immune responses. *Science*, 352, 1392–1393.

Raaijmakers, J. M. & Weller, D. M. 2001. Exploiting genotypic diversity of 2,4-diacetylphloroglucinol-producing *Pseudomonas* spp.: characterization of superior root-colonizing *P. fluorescens* strain Q8r1-96. *Applied and Environmental Microbiology*, 67, 2545–2554.

Ramirez, K. S., et al. 2018. Network analyses can advance above-belowground ecology. *Trends in Plant Science*, 23, 759–768.

Regmi, R., et al. 2022. Do small RNAs unlock the below ground microbiome-plant interaction mystery? *Frontiers in Molecular Biosciences*, 9, 1017392.

Reinhold-Hurek, B., et al. 2015. Roots shaping their microbiome: global hotspots for microbial activity. *Annual Review of Phytopathology*, 53, 403–424.

Ren, B., et al. 2019. Rhizobial tRNA-derived small RNAs are signal molecules regulating plant nodulation. *Science*, 365, 919–922.

Ren, P., et al. 2023. Predatory protists play predominant roles in suppressing soil-borne fungal pathogens under organic fertilization regimes. *Science of the Total Environment*, 863, 160986.

Richardson, A. E., et al. 2021. Root microbiome structure and microbial succession in the rhizosphere. In: Gupta, V. V. S. R. & Sharma, A. K. (eds.) *Rhizosphere Biology: Interactions Between Microbes and Plants.* Singapore: Springer.

Roget, D. K. 1995. Decline in root rot (*Rhizoctonia solani* AG-8) in wheat in a tillage and rotation experiment at Avon, South Australia. *Australian Journal of Experimental Agriculture*, 35, 1009–1013.

Roley, S. S. 2021. Diazotrophic nitrogen fixation in the rhizosphere and endosphere. In: Gupta, V. V. S. R. & Sharma, A. K. (eds.) *Rhizosphere Biology: Interactions Between Microbes and Plants.* Singapore: Springer.

Roley, S. S., et al. 2018. Associative nitrogen fixation (Anf) in switchgrass (*Panicum virgatum*) across a nitrogen input gradient. *PLoS One*, 13, e0197320.

Rønn R., et al. 2002. Impact of protozoan grazing on bacterial community structure in soil microcosms. *Applied and Environmental Microbiology*, 68, 6094–6105.

Rønn, R., et al. 2012. Interactions between bacteria, protozoa and nematodes in soil. *Acta Protozoologica*, 51, 223–235.

Rudrappa, T., et al. 2008. Root-secreted malic acid recruits beneficial soil bacteria. *Plant Physiology*, 148, 1547–1556.

Rüger, L., et al. 2021. Assembly patterns of the rhizosphere microbiome along the longitudinal root axis of maize (*Zea mays* L.). *Frontiers in Microbiology*, 12, 614501.

Rüger, L., et al. 2023. Responses of root architecture and the rhizosphere microbiome assembly of maize (*Zea mays* L.) to a soil texture gradient. *Soil Biology and Biochemistry*, 181, 109026.

Ryan, P. R., et al. 2009. Rhizosphere engineering and management for sustainable agriculture. *Plant and Soil*, 321, 363–383.

Saarenpää, S., et al. 2023. Spatial metatranscriptomics resolves host–bacteria–fungi interactomes. *Nature Biotechnology*. doi:10.1038/s41587-023-01979-2.

Sánchez, S., et al. 2016. Susceptibility of strawberry cultivars to root and crown rot caused by *Macrophomina phaseolina*. *Journal of Berry Research*, 6, 345–354.

Sasse, J., et al. 2018. Feed your friends: do plant exudates shape the root microbiome? *Trends in Plant Science*, 23, 25–41.

Schlatter, D., et al. 2017. Disease suppressive soils: new insights from the soil microbiome. *Phytopathology*, 107, 1284–1297.

Sergaki, C., et al. 2018. Challenges and approaches in microbiome research: from fundamental to applied. *Frontiers in Plant Science*, 9, 1205.

Shao, J., et al. 2021. Rhizosphere microbiome assembly involves seed-borne bacteria in compensatory phosphate solubilization. *Soil Biology and Biochemistry*, 159, 108273.

Sharma, I., et al. 2023. Biotic stress-induced changes in root exudation confer plant stress tolerance by altering rhizospheric microbial community. *Frontiers in Plant Science*, 14, 1132824.

Shi, S., et al. 2016. The interconnected rhizosphere: high network complexity dominates rhizosphere assemblages. *Ecology Letters*, 19, 926–936.

Shi, S., et al. 2011. Effects of selected root exudate components on soil bacterial communities. *Fems Microbiology Ecology*, 77, 600–610.

Shi, Y., et al. 2022. Assembly of rhizosphere microbial communities in *Artemisia annua*: recruitment of plant growth-promoting microorganisms and inter-kingdom interactions between bacteria and fungi. *Plant and Soil*, 470, 127–139.

Silvestri A., et al. 2020. Different genetic sources contribute to the small Rna population in the arbuscular mycorrhizal fungus *Gigaspora margarita*. *Frontiers in Microbiology*, 11, 395.

Singh, B. K., et al. 2020. Crop microbiome and sustainable agriculture. *Nature Reviews Microbiology*, 18, 601–602.

Singh, K., et al. 2022. Unraveling the secrets of rhizobacteria signaling in rhizosphere. *Rhizosphere*, 21, 100484.

Sivasithamparam, K. & Parker, C. A. 1978. Effect of infection of seminal and nodal roots by the take-all fungus on tiller numbers and shoot weight of wheat. *Soil Biology and Biochemistry*, 10, 365–368.

Sivasithamparam, K., et al. 1979. Rhizosphere micro-organisms of seminal and nodal roots of wheat grown in pots. *Soil Biology and Biochemistry*, 11, 155–160.

Song, C., et al. 2021a. Designing a home for beneficial plant microbiomes. *Current Opinion in Plant Biology*, 62, 102025.

Song, S., et al. 2021b. Mechanisms in plant–microbiome interactions: lessons from model systems. *Current Opinion in Plant Biology*, 62, 102003

Song, Y., et al. 2020. Loss of a plant receptor kinase recruits beneficial rhizosphere-associated *Pseudomonas*. *bioRxiv*, 2020.11.02.364109.

Sun, Y., et al. 2021. Rice domestication influences the composition and function of the rhizosphere bacterial chemotaxis systems. *Plant and Soil*, 466, 81–99.

Swamy, M. K., et al. 2016. Root exudates and their molecular interactions with rhizospheric microbes. In: Hakeem, K. R. & Akhtar, M. S. (eds.) *Plant, Soil and Microbes: Volume 2: Mechanisms and Molecular Interactions*. pp. 59–77. Cham: Springer International Publishing.

Tai, H., et al. 2015. Transcriptomic and anatomical complexity of primary, seminal, and crown roots highlight root type-specific functional diversity in maize (*Zea mays* L.). *Journal of Experimental Botany*, 67, 1123–1135.

Taye, Z. M., et al. 2020. Core and differentially abundant bacterial taxa in the rhizosphere of field grown *Brassica napus* genotypes: implications for canola breeding. *Frontiers in Microbiology*, 10, 3007.

Terrazas, R. A., et al. 2022. defining composition and function of the rhizosphere microbiota of barley genotypes exposed to growth-limiting nitrogen supplies. *mSystems*, 7, e00934-22.

Tkacz, A., et al. 2020. Agricultural selection of wheat has been shaped by plant-microbe interactions. *Frontiers in Microbiology*, 11, 132.

Toju, H., et al. 2018. Network hubs in root-associated fungal metacommunities. *Microbiome*, 6, 116.

Tracanna, V., et al. 2021. Dissecting disease-suppressive rhizosphere microbiomes by functional amplicon sequencing and 10× metagenomics. *mSystems*, 6, e0111620.

Trivedi, P., et al. 2020. Plant–microbiome interactions: from community assembly to plant health. *Nature Reviews Microbiology*, 18, 607–621.

Valente, J., et al. 2020. Ancient wheat varieties have a higher ability to interact with plant growth-promoting rhizobacteria. *Plant, Cell & Environment*, 43, 246–260.

Van Der Ent, S., et al. 2008. MYB72 is required in early signaling steps of rhizobacteria-induced systemic resistance in *Arabidopsis*? *Plant Physiology*, 146, 1293–1304.

Van Deynze, A., et al. 2018. Nitrogen fixation in a landrace of maize is supported by a mucilage-associated diazotrophic microbiota. *PLoS Biology*, 16, e2006352.

Vandeputte, O. M., et al. 2010. Identification of catechin as one of the flavonoids from *Combretum albiflorum* bark extract that reduces the production of quorum-sensing-controlled virulence factors in *Pseudomonas aeruginosa* PAO1. *Applied and Environmental Microbiology*, 76, 243–253.

Vannier, N., et al. 2018. A microorganisms' journey between plant generations. *Microbiome*, 6, 79.

Ventorino, V., et al. 2012. Response to salinity stress of *Rhizobium leguminosarum* bv. *viciae* strains in the presence of different legume host plants. *Annals of Microbiology*, 62, 811–823.

Vives-Peris, V., et al. 2020. Root exudates: from plant to rhizosphere and beyond. *Plant Cell Reports*, 39, 3–17.

Voges, M. J. E. E. E., et al. 2019. Plant-derived coumarins shape the composition of an *Arabidopsis* synthetic root microbiome. *Proceedings of the National Academy of Sciences*, 116, 12558–12565.

Wahdan, S. F. M., et al. 2021. Targeting the active rhizosphere microbiome of *Trifolium pratense* in grassland evidences a stronger-than-expected belowground biodiversity-ecosystem functioning link. *Frontiers in Microbiology*, 12, 629169.

Walker, T. S., et al. 2003. Root exudation and rhizosphere biology. *Plant Physiology*, 132, 44–51.

Walters, W. A., et al. 2018. Large-scale replicated field study of maize rhizosphere identifies heritable microbes. *Proceedings of the National Academy of Sciences*, 115, 7368–7373.

Wang, L., et al. 2022. Ensuring future food security and resource sustainability: insights into the rhizosphere. *iScience*, 25, 104168.

Wang, X., et al. 2020. An amplification-selection model for quantified rhizosphere microbiota assembly. *Science Bulletin*, 65, 983–986.

Watt, M., et al. 2006. Rates of root and organism growth, soil conditions, and temporal and spatial development of the rhizosphere. *Annals of Botany*, 97, 839–855.

Weese, D. J., et al. 2015. Long-term nitrogen addition causes the evolution of less-cooperative mutualists. *Evolution*, 69, 631–642.

Wei, F., et al. 2021. Composition of rhizosphere microbial communities associated with healthy and *Verticillium* wilt diseased cotton plants. *Frontiers in Microbiology*, 12, 618169.

Wei, Z. & Jousset, A. 2017. plant breeding goes microbial. *Trends in Plant Science*, 22, 555–558.

Weller, D. M., et al. 2002. Microbial populations responsible for specific soil suppressiveness to plant pathogens. *Annual Review of Phytopathology*, 40, 309–48.

Wen, H.-G., et al. 2023. Microbe-induced gene silencing boosts crop protection against soil-borne fungal pathogens. *Nature Plants*, 9, 1409–1418.

Wen, T., Zhao, M., Yuan, J., Kowalchuk, G. A. & Shen, Q. 2021. Root exudates mediate plant defense against foliar pathogens by recruiting beneficial microbes. *Soil Ecology Letters*, 3, 42–51.

Wittwer, R. A., et al. 2017. Cover crops support ecological intensification of arable cropping systems. *Scientific Reports*, 7, 41911.

Worst, E. G., et al. 2015. Cell-free expression with the toxic amino acid canavanine. *Bioorganic & Medicinal Chemistry Letters*, 25, 3658–3660.

Wu, R., et al. 2023. Hi-C metagenome sequencing reveals soil phage–host interactions. *Nature Communications*, 14, 7666.

Wurst, S. & Ohgushi, T. 2015. Do plant- and soil-mediated legacy effects impact future biotic interactions? *Functional Ecology*, 29, 1373–1382.

Xiao, E., et al. 2021. Root microbiome assembly of As-hyperaccumulator *Pteris vittata* and its efficacy in arsenic requisition. *Environmental Microbiology*, 23, 1959–1971.

Xiong, W., et al. 2018. Soil protist communities form a dynamic hub in the soil microbiome. *The Isme Journal*, 12, 634–638.

Xu, Q., et al. 2022. Microbial generalists and specialists differently contribute to the community diversity in farmland soils. *Journal of Advanced Research*, 40, 17–27.

Xu, X., et al. 2018. Strigolactones positively regulate defense against root-knot nematodes in tomato. *Journal of Experimental Botany*, 70, 1325–1337.

Xun, W., et al. 2021. Rhizosphere microbiome: functional compensatory assembly for plant fitness. *Computational and Structural Biotechnology Journal*, 19, 5487–5493.

Yamazaki, S., et al. 2021. Field multi-omics analysis reveals a close association between bacterial communities and mineral properties in the soybean rhizosphere. *Scientific Reports*, 11, 8878.

Yan, Y., et al. 2017. Functional traits dominate the diversity-related selection of bacterial communities in the rhizosphere. *The Isme Journal*, 11, 56–66.

Yang, C.-H. & Crowley, D. E. 2000. Rhizosphere microbial community structure in relation to root location and plant iron nutritional status. *Applied and Environmental Microbiology*, 66, 345–351.

Yao, H. & Wu, F. 2010. Soil microbial community structure in cucumber rhizosphere of different resistance cultivars to *Fusarium* wilt. *Fems Microbiology Ecology*, 72, 456–463.

Yin, C., et al. 2021. Rhizosphere community selection reveals bacteria associated with reduced root disease. *Microbiome*, 9, 86.

Yu, P., et al. 2018. Root type and soil phosphate determine the taxonomic landscape of colonizing fungi and the transcriptome of field-grown maize roots. *New Phytologist*, 217, 1240–1253.

Yuan, J., et al. 2018. Root exudates drive the soil-borne legacy of aboveground pathogen infection. *Microbiome*, 6, 156.

Yue, H., et al. 2023. Plant domestication shapes rhizosphere microbiome assembly and metabolic functions. *Microbiome*, 11, 70.

Zamioudis, C., et al. 2014. β-Glucosidase BGLU42 is a MYB72-dependent key regulator of rhizobacteria-induced systemic resistance and modulates iron deficiency responses in *Arabidopsis* roots. *New Phytologist*, 204, 368–379.

Zhalnina, K., et al. 2018. Dynamic root exudate chemistry and microbial substrate preferences drive patterns in rhizosphere microbial community assembly. *Nature Microbiology*, 3, 470–480.

Zhang, J., et al. 2019. NRT1.1B is associated with root microbiota composition and nitrogen use in field-grown rice. *Nature Biotechnology*, 37, 676–684.

Zhang, Z., et al. 2020. Rhizosphere hotspots: root hairs and warming control microbial efficiency, carbon utilization and energy production. *Soil Biology and Biochemistry*, 148, 107872.

Zhou, Y., et al. 2020a. Diversity and structural differences of bacterial microbial communities in rhizocompartments of desert leguminous plants. *PLoS One*, 15, e0241057.

Zhou, Y., et al. 2020b. Nitrifying microbes in the rhizosphere of perennial grasses are modified by biological nitrification inhibition. *Microorganisms*, 29, 1687.

Zhou, Y., et al. 2021. Warming reshaped the microbial hierarchical interactions. *Global Change Biology*, 27, 6331–6347.

Zhu, M., et al. 2003. Inhibition of growth and development of root border cells in wheat by Al. *Physiologia Plantarum*, 117, 359–367.

Chapter 7

Nematodes and their trophic interactions in the soil microbiome

Liliane Ruess, Humboldt-Universität zu Berlin, Germany

1 Introduction

1.1 Nematodes in the soil microbiome

The soil microbiome is teeming with diverse microorganisms and fauna. These interact with each other, forming complex and dynamic interrelationships. The soil microbiome includes all soil type-dependent microorganisms, including microfaunal grazers, nematodes and protists (Costa et al., 2018). However, microbial metazoa are rarely considered as a key component in the soil microbiome, and most research focuses on prokaryotic taxa (Bik, 2019). This is mainly due to the complex taxonomy of this microfauna as well as the lack of published molecular data for barcoding (di Montanara et al., 2022). For example, in nematodes, most of the available genome data sets are on parasitic taxa and the model species *Caenorhabditis elegans* (Zhang et al., 2017; Qing et al., 2020; Powers et al., 2021). Free-living nematodes, which predominate in the soil microbiome, are understudied: they are in a 'Goldilocks zone of neglect' (Bik, 2019).

http://dx.doi.org/10.19103/AS.2024.0136.30

To achieve a holistic view of microbiome processes, soil nematodes should be embraced as a key element in microbial ecology. The present chapter therefore introduces the major functions of free-living as well as plant parasitic nematodes in the soil and rhizosphere environment, with a focus on arable systems. It addresses their role as central components of food webs and as biological indicators for soil and plant health (Section 2). Further, it highlights the importance of root-feeding taxa for plant–microbe interactions in the rhizosphere (Section 3). Finally, nematodes as vectors for human pathogens among microbial hotspots and the potential for foodborne diseases are discussed (Section 4).

1.2 The 'good guys' - free-living nematodes

Nematodes are the most abundant and diverse multicellular soil organisms, with numerous species and millions of individuals, e.g. up to 200 species m^{-2} in one site and trophic groups at all food web levels (Yeates et al., 1993; Yeates, 2010). Globally it has been estimated that there is a total nematode biomass of approximately 0.3 giga tonnes in surface soils (van den Hoogen, et al., 2019). Within the soil microbiome, bacterial and fungal grazers regulate community structure and activity, which, in turn, affects nutrient mineralisation by their microbial prey (Whalen et al., 2013; Richter et al., 2019; Mielke, et al., 2022). Although, when compared with microorganisms, nematodes represent only a small amount of biomass in this micro-consortium, they have a key position in energy flow and nutrient cycling (e.g. Nieminen, 2008; Neher, 2010). Moreover, nematode life history traits and feeding relationships provide a framework to understand food web disturbance, structure and diversity as well as processes such as carbon pathways (Bongers, 1990; Ferris et al., 2001; Ruess and Ferris, 2004). This concept of 'nematode fauna analysis' has proven useful as a tool for relating soil and plant health to land use and arable management (Mulder and Maas, 2017).

1.3 The 'bad guys' - plant parasitic nematodes

In contrast to free-living nematodes, the role of plant parasites is well known in agriculture. Nematodes are among the most widespread root feeders: they break down plant cells, disrupt physiological processes and modify immune responses (Gillet et al., 2017; Liu and Park, 2018; Siddique et al., 2022). This damage is of great agricultural and economic importance, and the annual loss is estimated at ~14% of the world's crop production (Nicol et al., 2011). The primary effects of root feeders are on plant resource allocation pattern and tissue nutrient concentrations (Maboreke et al., 2017; Gilarte et al., 2020; Bell et al., 2021). Secondary effects are nematode-induced 'leakage' of plant cell content, increasing carbon (C) translocation to the rhizosphere, which stimulates the associated microbiome (Yeates et al., 1999; Poll et al., 2007). Through these diverse synergistic and antagonistic effects, nematodes alter microbial community patterns, C fluxes and root signalling pathways and, in turn, the organisation of the rhizosphere microbiome.

1.4 Nematodes and food-borne diseases

A growing concern in agriculture is outbreaks of food-borne illness linked to consumption of fresh or partially processed products in both industrialised and developing countries. Human pathogens enter arable soil mainly via recycled waste water in surface irrigation or via animal manure used as fertiliser (Roth et al.; 2015; Iwu and Okoh, 2019). Bacteria such as *Escherichia coli* and *Salmonella enterica* can then persist in the soil for long periods of time (Tran et al., 2020; Ramos et al., 2021). However, these pathogens have to be transported between microbial hotspots, e.g. from manure patches to the rhizosphere microbiome, in order to contaminate crops. A neglected aspect of this dissemination is soil-inhabiting fauna. As the most predominant metazoan in soil, nematodes have a considerable potential as a vector for pathogenic bacteria (Nykyri et al., 2014; Diaz and Restif, 2014; Kroupitski, et al., 2015). This may be particularly important in sustainable agriculture, with greater application of animal manure as an organic soil amendment. However, there is still a lack of knowledge about the role of nematodes (as well as other soil fauna) as vectors for human pathogens.

2 Nematodes as key biota in the soil micro-food web

2.1 Functional diversity and role of nematodes in the micro-food web

Nematodes dominate all soil ecosystems in terms of biomass and abundance due to their multiple adaptations to differing environmental conditions (Andrássy, 2005). Their high trophic diversity encompasses functional groups at each level of the soil food web (Fig. 1). Nematode diets range from bacteria, fungi and algae to other microfauna, predominantly other nematodes (Yeates et al., 1993). On the other hand, nematodes often become prey of soil mesofauna such as mites and springtails (Chernova et al., 2007; Walter and Proctor, 2013; Rueda-Ramírez et al., 2023). Through these multiple trophic interactions, nematodes connect microbial and faunal soil food web compartments, i.e. they couple the soil microbiome with higher trophic levels. They are an essential link in a multichannel food web across different size classes of soil-associated biota (Potapov, 2022).

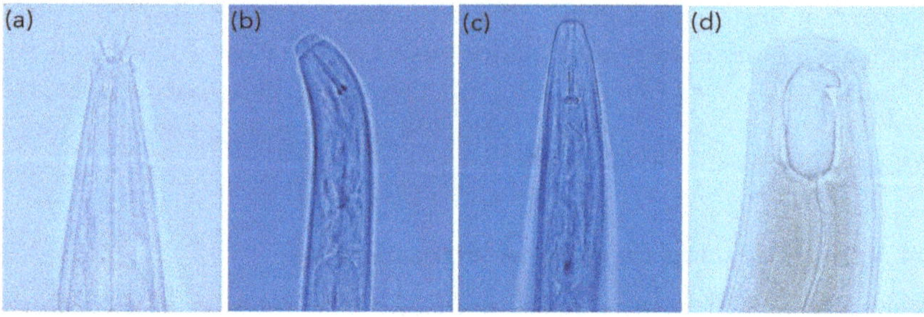

Figure 1 Common nematode trophic groups. (a) *Acrobeles* – bacterial feeder with an unarmed mouth cavity, the lip appendages (probolae) likely serve to sort bacteria; (b) *Aphelenchoides* – fungal feeder with a stylet with delicate knobs to pierce fungal hyphae; (c) *Merlinius* – root feeder with a strong stylet with prominent knobs used to rupture the plant cell wall and suck cell content. *Prionchulus* – predator with a barrel-shaped oral cavity bearing a dorsal tooth and opposite a ridge with rasp teeth.

Due to this high complexity in trophic relationships, functional groups are formed, including species with a similar trophic level and feeding pattern and a comparable function in the food web (Ferris and Tuomisto, 2015; Heijboer et al., 2017). Nematode functional groups are established in two major soil food chains (Crotty et al., 2014; Lazarova et al., 2021):

- The herbivore ('green') chain; and
- The detritivore ('brown') chain.

Living root tissue opens the herbivore C and energy channels by serving as a resource for root feeders, while bacterial and fungal feeders occupy the detrital decomposition pathway. Via cross-feeding, omnivores and predators can trigger energy and nutrient flows between both food chains (Wolkovich, 2016). Nematodes not only have a central role in the carbon and energy flux in the microbiomes of soil (brown chain) and rhizosphere (green chain) but also connect the fluxes between these two fundamental pathways.

Nematodes can both contribute to and detract from agricultural production systems given their diverse functional roles. In the area of plant production, bacterial feeders can act as vectors for both beneficial and harmful bacteria in the rhizosphere microbiome (Nykyri et al., 2014; Martins et al., 2022). Their preference for grazing on plant pathogens can also promote competing beneficial microbiota and vice versa (Bonkowski et al., 2009; Rønn et al., 2013). While root feeders are a severe plant pest, omnivores and predators can function as their control agents. The impact of nematode functional diversity thereby is not restricted to the microbiome itself. It acts at different scales, from the soil and rhizosphere microbiome to single plants and on to cultivated fields (Lazarova et al., 2021).

2.2 Trophic interactions shaping microbiome communities and processes

The structure and function of the micro-food web play a key role in the composition and activity of microbial assemblages and, in turn, soil nutrient cycling, which is crucial for plant production (De Vries and Wallenstein, 2017; Fierer, 2017; Jiang et al., 2022). Microfaunal grazers contribute to the microbial loop in the rhizosphere microbiome, a tri-tropic interaction that makes nutrients available to plants (Bonkowski et al., 2009; Zheng et al., 2022). Nematodes and protozoa are 'the bacterial predators' in the soil microbiome, and their predation pressure is a driving force in bacterial evolution. Soil bacteria have developed diverse defence strategies to survive or escape from their predators. The strategies include physical adaptations, e.g. to prevent contact, recognise and attack (e.g. engulfing predators), and chemical defence such as antibiotics (Martins et al., 2022). The impact of microfaunal grazing on microbial community structure as well as mineralisation processes in the soil microbiome is discussed comprehensively elsewhere in this book (see chapter on protists). Some aspects specific to nematodes are summarised later.

The global abundance of bacterial-feeding nematodes has been estimated as high as 1.9×10^{20} Ind. g^{-1} soil dry weight, distributed in 43 families comprising at least 129 species (van den Hoogen et al., 2019). Dunger (2008) investigated the nematode population in an afforested dune soil and found 6.9×10^{6} Ind. M^{-2} nematodes residing at a 0–25 cm depth. Approximately 42% of these nematodes were bacterial feeders, consuming approximately 16.6 g of bacterial cells and 5.3 g of protein per year. These trophic interactions directly affect microbial activity and C release from the microbial biomass (Richter et al., 2019). Nematode grazing activity further accounts for an additional release of ammonium, estimated as high as 32–38% of annual nitrogen (N) mineralisation in an arable land (Whalen et al., 2013).

However, bacterial feeders do not just drive microbiome community structure and function. Fungal feeders also impact on fungal and bacterial α-diversity and stimulate C and N cycling (Kane et al., 2023). In a $^{13}CO_2$-labelling field study, Pausch et al. (2016) showed that, despite a fungal C stock being less than half of that of bacteria, the C transfer from fungi via fungal-feeding nematodes into higher trophic levels of the fungal decomposition channel far exceeded that of the bacterial pathway. Recent micro-food web studies suggested that the transformation of exogenous C from bacteria to bacterial feeders accelerates the turnover of the C pool, while the C flux from fungi to fungal-feeders and then to omnivores and predators contributes to exogenous C sequestration (Zhang et al., 2023). This highlights the role of the functional diversity of nematodes in shaping microbiome community structure and processes.

2.3 Ecological tools to assess soil conditions and plant health

The high abundance and diversity of nematodes, multiple trophic links and impact on microbiome processes make them good indicators of soil and food web conditions (Ferris, 2010a; Ferris and Tuomisto, 2015; Mulder and Maas, 2017). Morphological traits and functional roles in nematodes show phylogenetic clustering and some genetic conservation at the genus level (Ross et al., 2022), paving the way for a framework of functional indices. These consider nematode population dynamics based on the family structure in relation to:

- soil ecosystem disturbance (Maturity Index; Bongers, 1990);
- nutrient availability (Enrichment Index; Ferris et al., 2001); and
- structure and function of the micro-food web (Structure Index; Ferris et al., 2001).

More details on different nematode indices are presented in Table 1. The indices are widely used as ecological tools to determine soil health and identify sustainable management practices (e.g. Lu et al., 2020; Melakeberhan et al., 2021; Biswal, 2022).

Functional indices as ecological tools are very useful in assessing the impact of environmental changes on soil conditions, but they do not show the magnitude of ecosystem response. To account for this, Ferris (2010b) developed 'metabolic footprints', including nematode biomass (i.e. weight) and metabolic activity (i.e. respiration). This concept combines form and function by taking advantage of the vermiform body-shape and standardised morphometrics in nematodes. This trait-based approach was tested across 200 sites (arable fields, managed grassland and wooded areas) as a functional descriptor of land use (Mulder and Maas, 2017). Using conversion factors for C content, C pools and fluxes in different parts of the microbiome were quantified, ranging from top soil to subsoil micro-food webs (Pausch et al., 2018). An overview of the different footprints and their indicator values are presented in Table 1.

The use of nematodes as biological indicators in soil systems has been successfully used to relate soil and food web conditions to grassland, arable and forest habitats, as shown in a meta-analysis across 83 sites (Ruess, 2003). More recently, nematode community analysis-based models have been used to identify sustainable soil health management. For example, Melakeberhan et al. (2021) designed models to assess the impact of more sustainable and integrated arable practices. A comprehensive review of nematodes as soil indicators is provided by Biswal (2022), and further case studies are summarised in Table 1. Nematode community analyses have great potential to support the

Published by Burleigh Dodds Science Publishing Limited, 2025.

Table 1 Nematode functional indices and metabolic footprints as ecological tools according to Bongers (1990), Ferris et al. (2001) and Ferris (2010). Given are the indicated soil (microbiome) and micro-food web conditions as well as related case studies in agricultural systems

Ecological tool	Soil and/or food web conditions	Case studies in arable land inferring the impact of
Functional index Maturity index (MI)	Disturbance	Manure application (Roth et al., 2015; Yang et al., 2016); tillage (Neher et al., 2019; Bongiorno, 2020); pesticides (Neher et al., 2011; Gupta et al., 2019); sewage sludge (Georgieva et al., 2002; Johansen et al., 2023)
Channel index (CI)	Major decomposition pathways	Irrigation (Ferris et al., 2001; Ewald et al., 2022); plant residues (Scharroba et al., 2012; Glavatska et al., 2017; Maina et al., 2021); tillage and crop rotation (Zhang et al., 2015)
Enrichment index (EI)	Nutrient (N) enrichment	Fertilisation (Ferris and Matute, 2003; Pan et al., 2020; Hu et al., 2022); litter amendment (Scharroba et al., 2012); cover crops (Ferris et al., 2001; Djigal et al., 2012)
Structure index (SI)	Food web structure and diversity	N addition (Zhou et al., 2022); intensive cultivation (Karuri, 2023), tillage (Zhang et al., 2015; Zhong et al., 2017)
Metabolic footprint		
Enrichment footprint	Responsiveness to resource pulses	Fertilisation (Pan et al., 2020); crop growth (Yang et al., 2020; Ewald et al., 2022)
Structure footprint	Impact on higher trophic levels	Crop growth (Yang et al., 2020; Ewald et al., 2022); continuous cropping (Yang et al., 2022)
Functional footprint	Overall food web response	Tillage (Zhang et al., 2015; Pothula et al., 2022; Schmidt et al., 2022); fertiliser type (Maina et al., 2021), crop rotation (Zhang et al., 2015; Zhang et al., 2016)
Herbivore, bacterial and fungal footprint	C and energy flux in the green and brown food chain	Litter amendment (Ewald et al., 2020); fertilisation (Pan et al., 2020); crop presence (Ewald et al., 2020; Karuri, 2023), pest management (Yang et al., 2022)
Composite footprint	Entire metabolic response of the population	Intercrop (Leiririo et al., 2022); crop growth (Yang et al., 2020)

maintenance of environmental quality and sustainable production practices in the future.

3 Plant parasites and microbiomes

3.1 Modification of the soil microbiome by plant parasitic nematodes

Microbiomes play a vital role in plant fitness and health, including those in bulk soil and those in the rhizosphere (Fierer, 2017; O´Brien et al., 2021). The composition and dynamics of microbial communities in these small-scale habitats are controlled by multiple abiotic and biotic factors. In particular the rhizosphere microbiome is characterised by a multitude of chemical signals from plants, bacteria and nematodes that shape environmental conditions and influence which organisms will flourish (Berg et al., 2014; Lakshmanan et al., 2014; Markalanda et al., 2022). Plant–microbiome interactions are therefore an important factor in sustainable agriculture (Gupta et al., 2021; Omotayo and Babalola, 2021).

Root-feeding nematodes have evolved several functional groups differing in parasitic lifestyle (Yeates et al., 1993), including:

- migratory and sedentary ectoparasites; and
- migratory and sedentary endoparasites.

Migratory ectoparasites have the most primitive mode of parasitism. They remain outside the root and use their stylet to feed on epidermal cells or root hairs. Sedentary ectoparasites also remain outside the root, feeding from a single plant cell for a prolonged period of time. Migratory endoparasites enter the root and periodically feed as they migrate, while sedentary endoparasites inhabit galls or cysts and establish specialised feeding sites at the vascular bundle. All functional groups induce leakage of root cell content, i.e. sugars, amino acids and carbolic acids, into the rhizosphere environment as shown by ^{13}C- or ^{14}C-labelling experiments (Denton et al., 1998; Yeates et al., 1999; Poll et al., 2007; Maboreke et al., 2017). Herbivore nematodes can thus shift microbial diversity and community structure (Maboreke et al., 2017; Malacrinò et al., 2021).

A special position in these interactions is held by root-knot nematodes. The rhizosphere microbiome undergoes major modifications when roots are infected with these parasites. Changes can be caused by the disruption of secondary metabolite synthesis, e.g. flavonoids (Skider et al., 2022) or negative interference with rhizobia mutualism (Wood et al., 2018). Root-knot nematodes also enhance bacteria that contain plant cell wall-degrading enzymes, thereby promoting nematode root colonisation (Lu et al., 2023). The microbiome community on the surface of root-knots is very specific, suggesting adaptions in functional capabilities to inhabit this 'gallobiome' (Haase et al., 2007; Masson et al., 2020). More details on these multitrophic interactions are provided in the case study presented in Section 5.

Published by Burleigh Dodds Science Publishing Limited, 2025.

3.2 *Turning the tables: soil microbiome effects on plant parasitic nematodes*

Microbiomes in the soil and plant rhizosphere protect crops against multiple stressors including phytopathogens such as root-feeding nematodes. This impact can be mediated via antagonistic bacteria as well as growth-promoting bacteria (Silva et al., 2018; Dukare et al., 2022) or mutualistic rhizosphere interactions (Markalanda et al., 2022). Plant secondary metabolites and systemic resistance are also important (Thahira et al., 2022). Details on these effects are discussed elsewhere in this book. In a nutshell, the different plant–microbiome interactions are key for plant fitness and resistance against plant parasitic nematodes.

Most is known about root-knot nematodes of the genus *Meloidogyne* (see review by Ciancio et al., 2016). *Meloidogyne*-based disease complexes harbour specific bacterial communities. These pathosystems can devastate important crops such as coffee and tomato (Lamelas et al., 2020). The soil microbiome can decrease root invasion of second-stage juveniles (J2) and progeny in *Meloidogyne hapla*. The former is related to plant-mediated non-direct effects of the microbiome, while the latter is caused by a direct negative impact of microbiome taxa on nematode reproduction (Topalović et al., 2020). Specific microorganisms are known as antagonists, e.g. *Bacillus amyloliquefaciens*, producing acetophenone and killing 99% of *M. incognita* J2 within 24 h (Zhao et al., 2023). Arbuscular mycorrhizal fungi in combination with *Pseudomonas fluorescens* can also reduce root infestation by *M. javanica* (Sharma et al., 2021).

Microbiome biota therefore have great potential in ecofriendly nematode pest-management. Specialised pathogens such as the parasitic bacterium *Pasteuria penetrans* or the nematode-trapping fungus *Pochonia chlamydosporia* are commonly used (Ciancio et al., 2016). Other microbial taxa and free-living nematodes are less studied as biocontrol tools. In the case of the ectoparasite *Tylenchorhynchus*, a study from Piśkiewicz et al. (2009) revealed a preference for nematode migration towards roots without microbial predators. Castillo et al. (2017) investigated different farms growing potatoes and found that soil with a high abundance of *Bacillus*, *Arthrobacter* and *Lysobacter*, and a nematode community comprising at least 30% fungal feeders, had the lowest populations of the migratory endoparasite *Pratylenchus neglectus*. In addition to antagonistic interactions in the rhizosphere microbiome, the soil microbiome can affect plant parasitic nematodes while migrating between crop plants. The question of whether and how free-living nematodes influence this parasitic interaction remains a task for future research.

3.3 The nematode gut microbiome - gateway to a new understanding of the host-parasite relationship?

As previously stated, the impact of root-feeding nematodes on plant health is affected by the soil microbiome. However, very little is known about the influence of the nematode gut microbiome on this host-parasite interaction. Investigations of the model nematode *C. elegans* showed that the worms' gut microbiota were assembled from the surrounding microbial environment (Berg et al., 2016; Clark and Hodgkin, 2016). Thereafter, composition is shaped by the nematode host, predominantly by immune, xenobiotic and metabolic signalling pathways (Zhang et al., 2021). Ecological network analyses revealed strong bacteria-to-bacteria interactions, with distinct community shifts in competitive species (Johnke et al., 2020). Overall, the assembly of the nematode gut microbiome is non-random, and its composition is different from that in the soil (Zheng et al., 2020).

This native gut microbiome influences nematode fitness, which, in turn, can have implications for nematode parasitism with respect to plants. Gut bacteria promote nematode life history traits, e.g. body size as well as energy allocation from lipid storage to reproduction (Lo et al., 2022). Gut microbes enhance nematode fitness under stresses such as high temperature, changes in osmolarity or exposure to chemicals (Dirksen et al., 2016; Lee et al., 2020). It has been shown recently that the composition of the gut microbiome can foster the development of the parasitic bacterium *Pasteuria penetrans* in the root-knot nematode *M. javanica*, offering potential for biocontrol (Lindo-Seminario et al., 2022).

The nematode gut microbiome is also shaped by agricultural management, particularly fertilisation practice. Zheng et al. (2020) reported that fertiliser type (none, inorganic or mixed) had a greater influence than fertilisation duration (5, 8, or 10 years). For example, the use of inorganic fertilisers significantly decreased the diversity of gut microorganisms, while the use of organic fertilisers had no impact on composition (Zheng et al., 2019). This may be important in the dissemination of antibiotic resistance genes that organic fertilisers, in particular, spread. Shared antibiotic resistance genes among soil and nematodes after the application of slurry (pig manure with antibiotics) indicate horizontal gene transfer (Zheng et al., 2019). Nematode gut microbiomes might therefore serve as a refuge or vector for antibiotic resistance genes and enhance dispersal of resistomes in the soil ecosystem. Taken together, with the effects on nematode fitness, parasite gut microbiomes can significantly influence host relationships.

Published by Burleigh Dodds Science Publishing Limited, 2025.

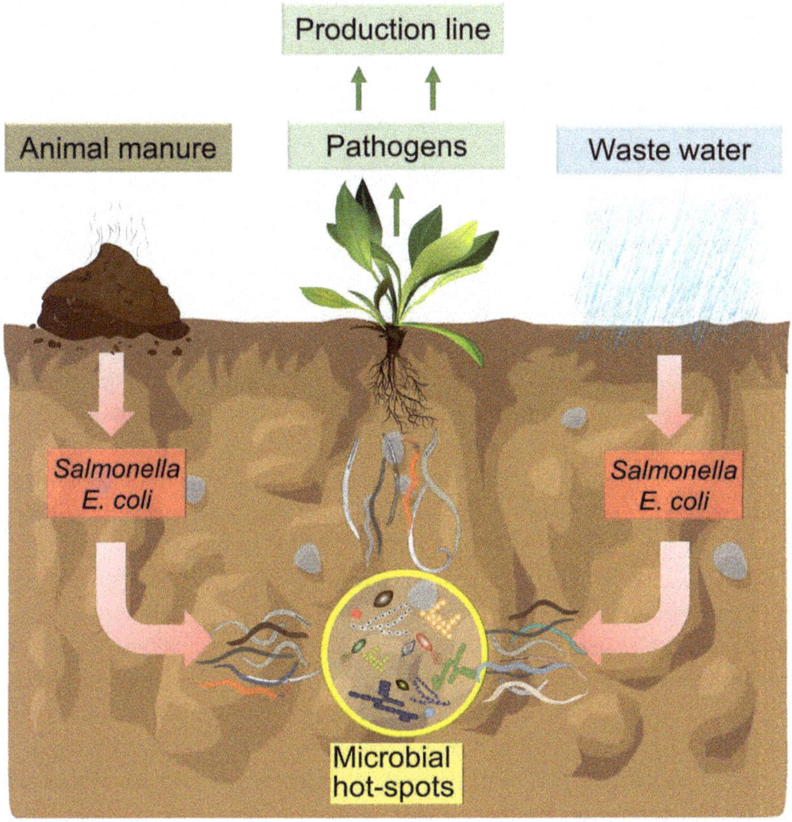

Figure 2 Root-knot nematodes and their multitrophic interactions in the rhizosphere. The second-stage juveniles penetrate near the root tip, resulting in the *leakage* of plant metabolites that promote microorganisms. *Root-knots* with mature females represent an active metabolic sink for plant carbon, reducing its availability for microorganisms. The altered exudation pattern on the surface of root-knots fosters fungi over bacteria. Phytohormones as host response to nematode infestation cause *local and systemic changes in root hair morphology*, i.e. an increase in length. These microscale modifications increase sugar exudation into the rhizosphere. Both the metabolic and morphological changes have feedback mechanisms (red arrows) on microbiome processes and thus nutrient dynamics and plant production.

4 Nematodes and food-borne diseases

4.1 The soil microbiome as a source for human pathogens

Soil microbiomes are a cornerstone of the one health concept. This includes beneficial microorganisms, plant and human pathogens, and overall microbial diversity (Banerjee and van der Heijden, 2023). Over the last decade, the soil microbiome in arable systems has developed as a 'battleground' for human pathogens (Overbeek van et al., 2014). A major hazard in fresh produce is shiga-toxin producing strains of *Escherichia coli* and *Salmonella enterica* (Fig. 2). As a nutrient-rich microbial hotspot, the plant rhizosphere is a reservoir for opportunistic pathogens such as *Enterobacter* and *Staphylococcus* (Berg et al., 2005; Mendes et al., 2013). A major concern is that these pathogens often acquire antimicrobial resistance, further aggravating risks to human and environmental health (Iwu and Okoh, 2019).

Fruits and vegetables grown on open fields can become contaminated with human pathogenic microorganisms before and during harvest. Major transmission routes are contaminated irrigation water (e.g. recycled wastewater) and raw or improperly composted animal manure as an organic amendment (Nicholson et al., 2005; Franz and Bruggen van , 2008; Roth et al., 2015) (Fig. 2). Once released into the environment, these human pathogens can persist in the soil for months (Jamieson et al., 2002; Guan and Holley, 2003; Iwu and Okoh, 2019). The most important environmental drivers for their persistence are soil conditions (temperature, moisture, pH and type) and arable management such as type or method of incorporation of manure (Overbeek van et al., 2014; Iwu and Okoh, 2019; Tran et al., 2020).

Complex biotic interactions in the microbiome are critical for the survival of human pathogens in the soil environment. Indigenous microbial communities in soil and the rhizosphere provide a strong barrier against allochthonous taxa. However, many human pathogenic bacteria are highly competitive and good colonisers. They accumulate nutrients competitively and produce antimicrobial metabolites, allowing proliferation in the presence of autochthonous taxa (Mendes et al., 2013). The high nutrient content of manure-amended soil in particular promotes successful establishment (van Overbeek et al., 2010). Soil bacterial diversity has been shown to be negatively correlated with pathogen survival (Elsas van et al., 2012), highlighting the importance of a diverse soil microbiome for food safety.

4.2 Nematodes as vectors for bacterial pathogens

As dominant soil bacteria predators, bacterial feeders are key to disease management for both plant and human health (Martins et al., 2022; Nykyri et al., 2014). It has long been known that they harbour, protect and disperse bacteria (e.g. Wasilewska and Webster, 1975; Sutherland and Webster, 1993; Anderson et al., 2003). They engulf bacteria via unarmed stoma (Fig. 1) and defecate 30–60% of ingested cells under viable conditions (Freckman and Caswell, 1985).

Interactions of *Escherichia coli* and *Salmonella enterica* with the model nematode *C. elegans* have been investigated the most (Diaz and Restif, 2014). Key aspects reported were:

- nematode attraction by and migration to bacteria (Caldwell et al., 2003; Kenney et al., 2006);
- bacterial ingestion and protection against sanitisers (Caldwell et al., 2003; Park et al., 2014);
- bacterial persistence in the gut (Kenney et al., 2005); and
- shedding of viable cells (Anderson et al., 2006).

Other taxa with known vector potential for human pathogens are the thermotolerant *Diploscapter*, found in compost, sewage and agricultural soil (Gibbs et al., 2005), and the opportunist *Acrobeloides buetschlii*, which is widespread across many soils (Kroupitski et al., 2015; Hemmerling et al., 2023). By moving between bacterial hotspots, e.g. soil and rhizosphere microbiome, bacterial feeders have a great vector potential for pathogen transfer from soil to crop roots (Fig. 2).

In contrast, plant parasites carry a mouth stylet and cannot ingest bacteria undamaged (Fig. 1). They are nevertheless vectors and transport bacteria externally on their cuticle (Topolavić et al., 2019). This attachment is mediated by surface epitopes or specific carbohydrates on their surface coat (Spiegel and McClure, 1995; Davies and Curtis, 2011). The bacterial community carried can be diverse, with, e.g., 24 bacterial strains isolated from the pine wood nematode *Bursaphelenchus xylophilus* (Zhao et al., 2003). Injuries caused by parasitic nematodes also serve as entry sites for human pathogens (Tyler and Triplett, 2008).

Using plant parasitic nematodes as a vehicle, pathogenic bacteria can access roots (Hallman et al., 1998; Maghodia et al., 2008). Under appropriate conditions, they grow in the plant tissue as in the case for *Salmonella* in lettuce and *Escherichia coli* in maize, resulting in subsequent dissemination to aerial parts of the plant (Bernstein et al., 2007a, b). On the other hand, Fornefeld et al. (2018) did not find that *M. hapla* and *Pratylenchus crenatus* improved internalisation of lettuce by *Salmonella* Typhimurium. Overall, plant parasites can assist pathogen invasion by providing wounds for pathogen entry and by actively transferring bacterial cells on their external surface into the plant (Fig. 2). However, their vector function is not as well documented as that for bacterial feeders, and further studies are needed.

5 Case study: root parasites shape the microbiome and rhizosphere processes of their host plant

The nature and outcome of the multitrophic interactions in the rhizosphere are vital for plant fitness and health. Plant parasitic nematodes have severe negative impacts on crops. Only two nematode genera, *Meloidogyne* and *Heterodera*, are responsible for ~10% reduction of the global food production (Nicol et al., 2011), which calls for effective and sustainable pest management. The primary concern is direct damage to specific plant hosts, i.e. the nematode's removal of carbon and energy, resulting in lower crop yield. Common solutions include crop rotation and the use of tolerant or resistant plants. Other more recent approaches include molecular breeding to improve plant defence (Liu and Park, 2018) or rhizosphere engineering to promote plant growth (Hakim et al., 2021) to develop a more eco-friendly, sustainable agriculture. However, a big gap in knowledge exists on the interplay between plant parasitic nematodes, microbiome community dynamics and host-plant responses.

As mentioned in the previous sections (see Section 3.1), feeding by nematodes leads to the release of plant metabolites in the rhizosphere, which promotes microbial communities. However, this goes beyond just increasing microbial substrates (e.g. sugars) in the rhizosphere. An illustration of this is a greenhouse experiment on barley (*Hordeum vulgare*) infected with the root-knot

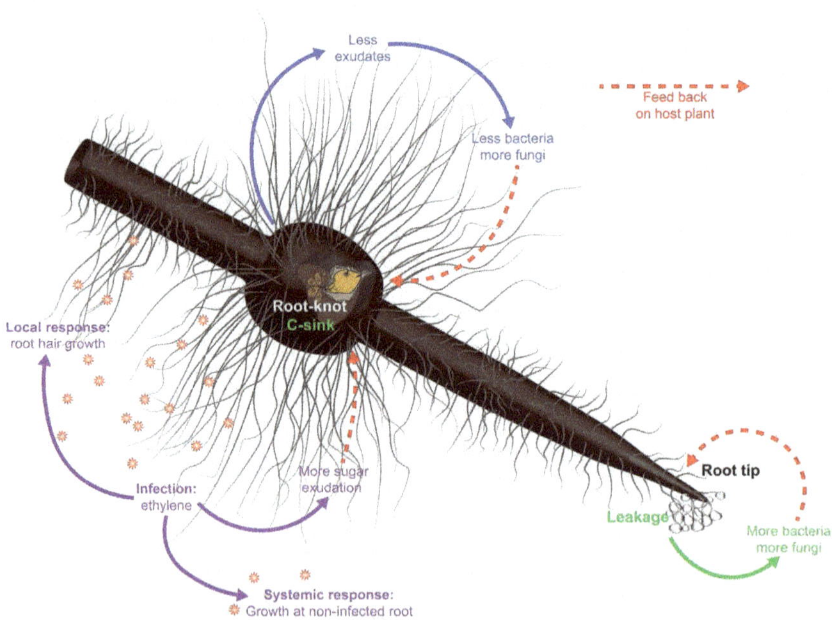

Figure 3 Nematodes as vectors for food-borne diseases. Human pathogens such as the shiga-toxin-producing strains of *Escherichia coli* and *Salmonella enterica* are introduced into the soil by agricultural practice, primarily inadequately treated animal manure and waste water. Nematodes share their environment with bacteria and can harbour, protect and disperse them. By moving between bacterial hotspots, they transport human pathogens into the rhizosphere (bacterial feeders) or even into the plant (root feeders), eventually entering the food production line.

Published by Burleigh Dodds Science Publishing Limited, 2025.

nematode *M. incognita*, and root hair morphology and rhizodeposition were investigated (Haase et al., 2007) as well as the rhizosphere microbiome, plant C allocation and biomass (Poll et al., 2007). The experiment studied a whole parasitic cycle, from the entry of second-stage juveniles at the root tip to feeding site establishment at the vascular cylinder and to the formation of root-knots by the host plant. Infestation rates of barley were kept below severe damage to allow the assessment of direct and indirect effects. The entire cycle of interactions between nematodes, microorganisms and plants is presented in Fig. 3.

Pulse-labelling barley with $^{13}CO_2$ showed that infestation with *M. incognita* enhanced plant C translocation to the rhizosphere, which altered microbial community dynamics (Poll et al., 2007). Barley root biomass and C content in shoots increased. However, with increasing densities of *M. incognita*, less plant C was sequestered by microorganisms and bacterial and fungal biomass decreased, indicating the shift from a mutualistic relationship to a parasitic interaction at higher infection rates. This highlights the positive impact of nematode-induced leakage on the rhizosphere microbiome at low levels of root herbivory (Fig. 3), as reported in other studies (Denton et al., 1998; Yeates et al., 1999; Maboreke et al., 2017).

This general pattern was modulated by the nematode life cycle. Positive responses of bacterial and fungal biomass occurred at the point when nematodes entered the root. Moreover, the composition of the soil microbiome changes with time and shifted in favour of fungi after root-knot development. The root exudation pattern also mirrored the life cycle of *M. incognita*. Enhanced release of total sugars and amino acids as well as carboxylic acid fumarate from roots occurred only during nematode invasion, i.e. with mechanical penetration that causes leakage (Haase et al., 2007). After establishment inside roots, nematodes acted as a strong C sink (Fig. 3). ^{13}C labelling revealed that recent photo-assimilates were transferred to parasites and that availability of plant C for rhizosphere microorganisms was reduced (Poll et al., 2007).

Finally, the presence of *M. incognita* also had a striking impact on root hair morphology. Root infestation reduced root hair numbers but stimulated root hair length (Haase et al., 2007) (Fig. 3). Such changes are often related to phytohormone signalling, e.g. via ethylene, which is triggered by local wounding. However, as elongation of root hairs was apparent in both infected and uninfected roots of the same barley plant, nematodes had a systemic impact. The enlarged root surface further resulted in a higher sugar release across the entire root system. These microscale morphological modifications were connected to the nematode's life cycle and diminished after feeding site establishment.

Root-knots formed a specific microhabitat within barley roots. Compared with uninfected and infected root zones, the release rates of total amino acids were significantly reduced, and carboxylates malate and fumarate were absent even at root-knots (Fig. 3). This altered exudation pattern increased the biomass of saprotrophic fungi, while bacteria declined, particularly gram-positive taxa. Overall, this suggests competitive consumption of plant metabolites by the nematode in the root-knot, as reported by Gommers and Dropkin (1977), resulting in an adapted microbiome, the so-called 'gallobiome' (Masson et al., 2020).

In summary, the infection of crops by root-feeding nematodes leads to changes in root morphology, exudation pattern and the microbiome, which can all have feedback mechanisms affecting plant fitness and health (Fig. 3). The results from this study offer a basis for more sustainable crop management. The anatomical traits reported have the potential for the inhibition of similar events in plant molecular breeding, while the biochemical traits triggering microbiome interactions can be used to foster beneficial bacteria in rhizosphere engineering. Given the considerable economic impact of root parasitic nematodes, the response pattern of the plant and microbiome should be pooled to develop management measures for sustainable crop production.

6 Conclusion

Maintaining agricultural production on a sustainable basis requires conservation of soil biodiversity and ecosystem services. Nematodes play a central role in soil food webs and have important effects on processes such as the mineralisation of nutrients or the flow of C and energy from microorganisms to the meso- and macrofauna. These multiple functions make nematodes well-suited indicators to assess the impact of agricultural management, e.g. the shift from conventional to organic farming. Nematode indicators can track microbiome disturbance, nutrient enrichment, decomposition channels and food web conditions. Nematodes thus provide a holistic measure of soil conditions, supporting research into approaches for improved and sustainable crop production.

On the other hand, root-feeding nematodes are serious plant pests with major effects on plant health and yield. Managing these impacts on crop production in a sustainable manner is a challenge for agriculture. However, indirect effects on the rhizosphere microbiome as well as the synergistic effects on plant hormone balance offer the potential for biotic control. Microbiome-based agronomic approaches, viewing the plant and its rhizosphere microbiome as a holobiont, co-evolving with root parasites, may promote more sustainable management of plant diseases.

Finally, specific agricultural practices can foster food-borne diseases. This applies particularly to organic farming with the application of manure and the use of recycled or waste water for irrigation in arid areas. The latter is likely to increase with climate change. The role of nematodes as vectors for human pathogens has received little attention, although they influence pathogen presence in the rhizosphere of crops and also in the crops themselves. Integration of these biotic interactions into a 'One Health' framework would lead to more efficient pathogen regulation strategies and, in turn, healthy fresh produce in sustainable agriculture.

7 Future trends

The framework for micro-food web diagnostics offered by nematode faunal analysis is a valuable tool to understand the structure and function of the soil microbiome. It links species, morphological traits and measured or inferred functional roles, therewith reflecting the multifunctionality of soil food webs (Potapov, 2022). However, the potential of nematodes as bioindicators is complicated by the enormous diversity of the phylum. A future trend is to match genetic inter-relatedness and functional trait expression, so that soil conditions and health can be assigned without the need for exact taxonomic identification. Initial attempts to link functional traits to novel sequences of nematodes have been made by Ross et al. (2022).

Over the last decade, improving the soil microbiome for sustaining ecosystem services has become a popular area of research (Chernov and Semenov, 2021). Biofertiliser and bio-pesticide products are used in organic and low-input farming to augment the soil microbial community (Rengalakshmi et al., 2018; Cornell et al., 2021). These bio-inputs play an integral role in maintaining soil quality. A future option is to exploit interactions with soil fauna. For instance, the fungal-feeding nematode *Aphelenchus avenae* has been shown to promote soil microbial activity, thereby alleviating damping-off disease caused by *Pythium ultimum* in tomato (Zhang et al., 2023). Climate change also affects belowground food chains. Thakur et al. (2021) showed that warming altered the dynamics in the rhizosphere microbiome differently, i.e. extreme heat events were detrimental for microorganisms, while their microfaunal grazers exhibited thermal acclimation. Future effective microbiome management therefore should include soil fauna to achieve healthy and sustainable farming systems.

Recently, conservation of soil mite predators using free-living nematodes as prey has been suggested as a way to improve agricultural ecosystem performance and sustainability (Azevedo et al., 2019; Rueda-Ramírez et al., 2023). Many predatory mites are generalists and feed on a diverse diet, making them excellent candidates for biological control programmes. Free-living nematodes serve as prey for many soil predatory mites (Heidemann et al., 2014a,b). For some mite species, they are an essential food in order to lay eggs (Walter et al., 1987) or undergo complete development into adults (Rueda-Ramírez et al., 2019), while other species show increased fecundity with a nematode diet

(Moreira et al., 2015). Free-living nematodes can thus harness the soil food web for biological control of root-knot nematodes or other soil-borne pests.

8 Acknowledgements

I am very grateful to Dr. Diana Rueda-Ramírez for transforming the scientific graphics into appealing illustrations. I also thank Martin Ewald MSc for the beautiful microscopic nematode pictures.

9 Where to look for further information

A standard introduction to nematodes in soil ecosystems is D.W. Freckman, *Nematodes in Soil Ecosystems* (1982, University of Texas Press). Good single sources on nematode management in sustainable agriculture are:

- Meghvansi, M. K. and Chaudhary, K. K. (2022). *Sustainable Management of Nematodes in Agriculture, Vol.1: Organic Management*. Springer, Germany; 425 pp.
- Subbotin, S. A. and Chitambar, J. J. (2019). *Plant Parasitic Nematodes in Sustainable Agriculture of North America, Vol.2: Northeastern, Midwestern and Southern USA*. Springer, Germany; 457 pp.
- Grewal, P. S., Ehlers, R. U. and Shapiro-Ilan, D. I. (2005). *Nematodes as Biocontrol Agents*. CABI, Wallingford, UK; 489 pp.

The international nematological societies inform about recent advances in nematology in both fundamental and economic aspects. They support networking among scientists and promote and extend knowledge in all areas of nematology:

- Society of Nematologists (SON): https://www.nematologists.org/.
- European Society of Nematologists (ESN): https://www.esn-online.org/.
- Organisation of Nematologists of Tropical America (ONTA): https://ontaweb.org/.

There are a number of current online sources with useful information on nematodes:

- Nemaplex is a virtual encyclopaedia on nematodes (mainly soil and plant taxa) developed and maintained by Howard Ferris. The website contains information on biology, ecology, indices and various resources, e.g. collections, methods, teaching (http://nemaplex.ucdavis.edu).
- WormBook is a collection of original, peer-reviewed chapters covering topics related to the biology of Caenorhabditis elegans and other

nematodes. It also contains a collection of protocols on methods (http://www.wormbook.org/).

- WormBase provides the research community with current and accessible information concerning the genetics, genomics and biology of *Caenorhabditis elegans* and related nematodes. This international consortium of biologists and computer scientists was founded in 2000 (https://wormbase.org).

10 References

Anderson, G. L., Caldwell, K. N., Beuchat, L. R. and Williams, P. L. (2003). Interaction of a free-living soil nematode, *Caenorhabditis elegans*, with surrogates of foodborne pathogenic bacteria. *J. Food Prot.* 66(9), 1543–1549.

Anderson, G. L., Kenny, S. J., Millner, P. D., Beuchat, L. R. and Williams, P. L. (2006). Shedding of foodborne pathogens by *Caenorhabditis elegans* in compost-amended and unamended soil. *Food Microbiol.* 23(2), 146–153.

Andrássy, I. (2005). Free-living nematodes of Hungary (Nematoda errantia). In: Csuzdi, C. and Mahunka, S. (Eds). *Pedozoologica Hungarica 3*, Hungarian Natural History Museum, Budapest, Hungary, 518 pp.

Azevedo, L. H., Leite, L. G., Chacon-Orozco, J. G., Moreira, M. F. P., Ferreira, M. P., Cano, L. M., Borges, V., Rueda-Ramírez, D., de Moraes, G. J. and Palevsky, E. (2019). Free living nematodes as alternative prey for soil predatory mites: an interdisciplinary case study of conservation biological control. *Biol. Control* 132, 128–134.

Neher, D., Bao, Y. and Chen, S. (2011). Effect of soil disturbance and biocides on nematode communities and extracellular enzyme activity in soybean cyst nematode suppressive soil. *Nematology* 13(6), 687–699.

Bell, C. A., Magjourilou, E., Urwin, P. E. and Field, K. J. (2021). The influence of competing root symbionts on below-ground plant resource allocation. *Ecol. Evol.* 11(7), 2997–3003.

Berg, G., Eberl, L. and Hartmann, A. (2005). The rhizosphere as a reservoir for opportunistic human pathogenic bacteria. *Environ. Microbiol.* 7(11), 1673–1685.

Berg, G., Grube, M., Schloter, M. and Smalla, K. (2014). Unravelling the plant microbiome: looking back and future perspectives. *Front. Microbiol.* 5, 148.

Berg, M., Stenuit, B., Ho, J., Parke, C., Knight, M., Alvarez-Cohen, L. and Shapira, M. (2016). Assembly of the *Caenorhabditis elegans* gut microbiome from diverse soil microbial environments. *ISME J.* 10, 1998–2009.

Bernstein, S., Sela, S. and Neder-Lavon, S. (2007a). Effect of irrigation regimes on persistence of *Salmonella enterica* serovar Newport in small experimental pots designed for plant cultivation. *Irrig. Sci.* 70, 471–475.

Bernstein, N., Sela, S., Pinto, R. and Loffe, M. (2007b). Evidence for internalization of *Escherichia coli* into the aerial parts of maize via the root system. *J. Food Prot.* 70(2), 471–475.

Bik, H. M. (2019). Microbial metazoan are microbes too. *mSystems* 4(3), e00109-19.

Biswal, D. (2022). Nematodes as ghosts of land use past: elucidating the roles of soil nematode community studies as indicators of soil health and land management practices. *Appl. Biochem. Biotechnol.* 194(5), 2357–2417.

Bnaerjee, S. and van der Heijden, M. G. A. (2023). Soil microbiomes and one health. *Nat. Rev. Microbiol.* 21(1), 6–20.

Bongers, T. (1990). The maturity index: an ecological measure of environmental disturbance based on nematode species composition. *Oecologia* 83(1), 14–19.

Bongiorno, G. (2020). Novel soil quality indicators for the evaluation of agricultural management practices: a biological perspective. *Front. Agric. Sci. Eng.* 7(3), 257–274.

Bonkowski, M., Villenave, C. and Griffiths, B. (2009). Rhizosphere fauna: the functional and structural diversity of intimate interactions of soil fauna with plant roots. *Plant Soil* 321(1–2), 213–233.

Caldwell, K. N., Adler, B. B., Anderson, G. L., Williams, P. L. and Beuchat, L. R. (2003). Ingestion of *Salmonella enterica* serotype Poona by a free-living nematode, *Caenorhabditis elegans*, and protection against inactivation by produce sanitizers. *Appl. Environ. Microbiol.* 69(7), 4103–4110.

Castillo, J. D., Vivanco, J. M. and Manter, D. K. (2017). Bacterial microbiome and nematode occurrence in different potato agricultural soils. *Microb. Ecol.* 74(4), 888–900.

Chernov, T. I. and Semenov, M. V. (2021). Management of soil microbial communities: opportunities and prospects (a Review) (a review). *Eurasian Soil Sc.* 54(12), 1888–1902.

Chernova, N. M., Bokova, A. I., Varshav, E. V., Goloshchapova, N. P. and Savenkova, Y. Y. (2007). Zoophagy in Collembola. *Entomol. Rev.* 87(7), 799–811.

Ciancio, A., Colagiero, M., Pentimone, I. and Rosso, L. (2016). Soil microbial communities and their potential for root-knot nematode management. *Environ. Eng. Manag. J.* 15(8), 1833–1839.

Clark, L. C. and Hodgkin, J. (2016). *Caenorhabditis* microbiota: worm guts get populated. *BMC Biol.* 14, 37.

Cornell, C., Kokkoris, V., Richards, A., Horst, C., Rosa, D., Bennet, J. A. and Hart, M. M. (2021). Do bioinoculants affect resident microbial communities? A metanalysis. *Front. Agronom.* 3, 753474.

Costa, J., Oliveira, R. S., Tiago, I., Ma, Y., Galhano, C., Freitas, H. and Castro, P. (2018). Soil microorganisms. In: Sáchnez-Moreiras, A. M. and Reigosa, M. J. (Eds). *Advances in Plant Ecophysiology Techniques*, Springer Nature, Germany, pp. 457–482.

Crotty, F. V., Blackshaw, R. P., Adl, S. M., Inger, R. and Murray, P. J. (2014). Divergence of feeding channels within the soil food web determined by ecosystem type. *Ecol. Evol.* 4(1), 1–13.

Davies, K. G. and Curtis, R. H. (2011). Cuticule surface coat of plant-parasitic nematodes. *Annu. Rev. Phytopathol.* 49, 135–156.

De Vries, F. T. and Wallenstein, M. D. (2017). Below-ground connections underlying above-ground food production: a framework for optimising ecological connections in the rhizosphere. *J. Ecol.* 105(4), 913–920.

Di Montanara, A. C., Baldrighi, E., Franzo, A., Catani, L., Grassi, E., Sandulli, R. and Semprucci, F. (2022). Free-living nematode research: state of the art, prospects, and future directions. A bibliometric analysis approach. *Ecol. Inform.* 72, 101891.

Diaz, S. A. and Restif, O. (2014). Spread and transmission of bacterial pathogens in experimental populations of the nematode *Caenorhabditis elegans*. *Appl. Environ. Microbiol.* 80(17), 5411–5418.

Dirksen, P., Marsh, S. A., Braker, I., Heitland, N., Wagner, S., Nakad, R., Mader, S., Petersen, C., Kowallik, V., Rosenstiel, P., Félix, M. A. and Schulenburg, H. (2016). The native microbiome of the nematode *Caenorhabditis elegans*: gateway to a new host-microbiome model. *BMC Biol.* 14, 38.

Djigal, D., Chabrier, C., Duyck, P.-F., Achard, R., Quénéhervé, P. and Tixier, P. (2012). Cover crops alter the soil nematode food web in banana agroecosystems. *Soil Biol. Biochem.* 48, 142–150.

Dukare, A., Mhatre, P., Maheshwari, H. S., Bagul, S., Manjuantha, B. S., Khade, Y. and Kamble, U. (2022). *Delineation of Mechanistic Approaches of Rhizosphere Microorganisms Facilitated Plant Health and Resistance under Challenging Conditions*, Biotech Publishing (vol. 12), article: 57.

Dunger, W. (2008). *Tiere im Boden* (4th edn.), Westarp-Wissenschaftsverlag, Germany.

Elsas, van, J. D., Chiurazzi, M., Mallon, C. A., Elhottova, D., Kristufek, V. and Salles, J. F. (2012). Microbial diversity determines the invasion of soil by a bacterial pathogen. *Proc. Natl. Acad. Sci. U. S. A.* 109(4), 1159–1164.

Ewald, M., Glavatska, O. and Ruess, L. (2020). Effects of resource manipulation on nematode community structure and metabolic footprints in an arable soil across time and depth. *Nematology* 22(9), 1025–1043.

Ewald, M., Rusch, D., Rißmann, C., Trost, B., Theuerl, S. and Ruess, L. (2022). Effects of irrigation and fertilization practice on soil nematode communities in arable land. *Appl. Soil Ecol.* 177, 104546.

Ferris, H. (2010a). Contribution of nematodes to the structure and function of the soil food web. *J. Nematol.* 42(1), 63–67.

Ferris, H. (2010b). Form and function: metabolic footprints of nematodes in the soil food web. *Eur. J. Soil Biol.* 46(2), 97–104.

Ferris, H., Bongers, T. and Goede, de, R. G. M. (2001). A framework for soil food web diagnostics: extension of the nematode faunal analysis concept. *Appl. Soil Ecol.* 18(1), 13–29.

Ferris, H. and Matute, M. M. (2003). Structural and functional succession in the nematode fauna of a soil food web. *Appl. Soil Ecol.* 23(2), 93–110.

Ferris, H. and Tuomisto, H. (2015). Unearthing the role of biological diversity in soil health. *Soil Biol. Biochem.* 85, 101–109.

Fierer, N. (2017). Embracing the unknown: disentangling the complexities of the soil microbiome. *Nat. Rev. Microbiol.* 15(10), 579–590.

Fornefeld, E., Baklawa, M., Hallmann, J., Schikora, A. and Smalla, K. (2018). Sewage sludge amendment and inoculation with plant-parasitic nematodes do not facilitate the internalization of *Salmonella* Typhimurium LT2 in lettuce plants. *Food Microbiol.* 71, 111–119.

Franz, E. and Bruggen, van, A. H. (2008). Ecology of *E. coli* O157:H7 and *Salmonella enterica* in the primary vegetable production chain. *Crit. Rev. Microbiol.* 34(3–4), 143–161.

Freckman, D. W. and Caswell, E. P. (1985). The ecology of nematodes in agroecosystems. *Annu. Rev. Phytopathol.* 23(1), 275–296.

Georgieva, S. S., McGrath, S. P., Hooper, D. J. and Chambers, B. S. (2002). Nematode communities under stress: the long-term effects of heavy metals in soil treated with sewage sludge. *Appl. Soil Ecol.* 20(1), 27–42.

Gibbs, D. S., Anderson, G. L., Beuchat, L. R., Carta, L. K. and Williams, P. L. (2005). Potential role of *Diploscapter* sp. strain LKC25, a bacterivorous nematode from soil, as a vector of food-borne pathogenic bacteria to preharvest fruits and vegetables. *Appl. Environ. Microbiol.* 71(5), 2433–2437.

Gilarte, P., Plett, J., Pendall, E., Carrillo, Y. and Nielsen, U. N. (2020). Direct and indirect trophic interactions of soil nematodes impact chickpea and oat nutrition. *Plant Soil* 457(1–2), 255–268.

Gillet, F., Bournaud, C., de Souza Júnior, J. D. A. and Fatima Grossi-de-Sa, M. (2017). Plant-parasitic nematodes: towards understanding molecular players in stress responses. *Ann. Bot.* 199, 775–789.

Glavatska, O., Müller, K., Butenschoen, O., Schmalwasser, A., Kandeler, E., Scheu, S., Totsche, K. U. and Ruess, L. (2017). Disentangling the root- and detritus-based food chain in the micro food web of an arable soil by plant removal. *PLOS ONE* 12(7), e0180264.

Gommers, F. J. and Dropkin, V. H. (1977). Quantitative histochemistry of nematode induced transfer cells. *Phytopathol.* 67, 869–873.

Guan, T. Y. and Holley, R. A. (2003). Pathogen survival in swine manure environments and transmission of human enteric illness - a review. *J. Environ. Qual.* 32(2), 383–392.

Gupta, D., Bhandari, S. and Bhusal, D. R. (2019). Variation of nematode indices under contrasting pest management practices in a tomato growing agro-ecosystem. *Heliyon* 5(10), e02621.

Gupta, R., Anand, G., Gaur, R. and Yadav, D. (2021). Plant-microbiome interactions for sustainable agriculture: a review. *Physiol. Mol. Biol. Plants* 27(1), 165–179.

Haase, S., Ruess, L., Neumann, G., Marhan, S. and Kandeler, E. (2007). Low-level herbivory by root-knot nematodes (*Meloidogyne incognita*) modifies root hair morphology and rhizodeposition in host plants (*Hordeum vulgare*). *Plant Soil* 301(1–2), 151–164.

Hakim, S., Naqqash, T., Nawaz, M. S., Laraib, I., Siddique, M. J., Zia, R., Mirza, M. S. and Imran, A. (2021). Rhizosphere engineering with plant growth-promoting microorganisms for agriculture and ecological sustainability. *Front. Sustain. Food Syst.* 5, 617157.

Hallman, J., Quadt-Hallman, A., Rodríguez-Kábana, R. and Kloepper, J. W. (1998). Interaction between *Meloidogyne incognita* and endophytic bacteria in cotton and cucumber. *Soil Biol. Biochem.* 30(7), 925–937.

Heidemann, K., Hennies, A., Schakowske, J., Blumenberg, L., Ruess, L., Scheu, S. and Maraun, M. (2014a). Free-living nematodes as prey for higher trophic levels of forest soil food webs. *Oikos* 123(10), 1199–1211.

Heidemann, K., Ruess, L., Scheu, S. and Maraun, M. (2014b). Nematode consumption by mite communities varies in different forest microhabitats as indicated by molecular gut content analysis. *Exp. Appl. Acarol.* 64(1), 49–60.

Heijboer, A., Ruess, L., Traugott, M., Jousset, A. and de Ruiter, P. C. (2017). Empirical methods of identifying and quantifying trophic interactions for constructing soil food web models. In: de Ruiter, P. C., McCann, K. S. and Wolters, V. (Eds), *Adaptive Food Webs -Stability and Transitions of Real and Model Ecosystems*, Jackson College, Moore, Cambridge University Press, UK, pp. 257–285.

Hemmerling, C., Ackermann, M. and Ruess, L. (2023). Food choice between the human pathogen Salmonella enterica and common soil bacteria is related to nematode life traits. *Appl. Soil Ecol.* 192.

van den Hoogen, J., Geisen, S., Routh, D., Ferris, H., Traunspurger, W., Wardle, D. A., de Goede, R. G. M., Adams, B. J., Ahmad, W., Andriuzzi, W. S., Bardgett, R. D., Bonkowski, M., Campos-Herrera, R., Cares, J. E., Caruso, T., de Brito Caixeta, L., Chen, X., Costa, S. R., Creamer, R., Mauro da Cunha Castro, J., Dam, M., Djigal, D., Escuer, M., Griffiths, B. S., Gutiérrez, C., Hohberg, K., Kalinkina, D., Kardol, P., Kergunteuil, A., Korthals,

G., Krashevska, V., Kudrin, A. A., Li, Q., Liang, W., Magilton, M., Marais, M., Martín, J. A. R., Matveeva, E., Mayad, E. H., Mulder, C., Mullin, P., Neilson, R., Nguyen, T. A. D., Nielsen, U. N., Okada, H., Rius, J. E. P., Pan, K., Peneva, V., Pellissier, L., Carlos Pereira da Silva, J., Pitteloud, C., Powers, T. O., Powers, K., Quist, C. W., Rasmann, S., Moreno, S. S., Scheu, S., Setälä, H., Sushchuk, A., Tiunov, A. V., Trap, J., van der Putten, W., Vestergård, M., Villenave, C., Waeyenberge, L., Wall, D. H., Wilschut, R., Wright, D. G., Yang, J. I. and Crowther, T. W. (2019). Soil nematode abundance and functional group composition at a global scale. *Nature* 572(7768), 194–198.

Hu, H., Yao, Z., Gao, S. and Zhu, B. (2022). Nematode response to long-term fertilization in purple soil. *Biodivers. Sci.* 30(12), 2022189.

Iwu, C. D. and Okoh, A. I. (2019). Preharvest transmission routes of fresh produce associated bacterial pathogens with outbreak potentials: a review. *Int. J. Environ. Res. Public Health* 16(22), 4407.

Jamieson, R. C., Gordon, R. J., Sharples, K. E., Stratton, G. W. and Madani, A. (2002). Movement and persistence of faecal bacteria in agricultural soils and subsurface drainage water: a review. *Can. Biosys Eng.* 44, 1–9.

Jiang, Y., Wang, Z., Liu, Y., Wang, Y., Wang, Q. and Liu, T. (2022). Nematodes and their bacterial prey improve phosphorous acquisition by wheat. *New Phytol.* 237, 974–986.

Johansen, J. L., Dam, M., Kudjordjie, E. N., Santos, S. S., Palmqvist, A., Magid, J. and Vestergård, M. (2023). Effects of long-term fertilization with contemporary Danish human urine, composted household waste and sewage sludge on soil nematode abundance and community structure. *Sci. Total Environ.* 860, 160485.

Johnke, J., Dirksen, P. and Schulenburg, H. (2020). Community assembly of the native *C. elegans* microbiome is influenced by time, substrate and individual bacterial taxa. *Environ. Microbiol.* 22(4), 1265–1279.

Kane, J. L., Kotcon, J. B., Freedman, Z. B. and Morrissey, E. M. (2023). Fungivorous nematodes drive microbial diversity and carbon cycling in soil. *Ecology* 104(1), e3844.

Karuri, H. (2023). Nematode community response to intensive tomato production in the tropics. *Rhizosphere* 25, 100681.

Kenney, S. J., Anderson, G. L., Williams, P. L., Millner, P. D. and Beuchat, L. R. (2005). Persistence of *Escherichia coli* O157:H7, *Salmonella* Newport, and *Salmonella* Poona in the gut of a free-living nematode, *Caenorhabditis elegans*, and transmission to progeny and uninfected nematodes. *Int. J. Food Microbiol.* 101(2), 227–236.

Kenney, S. J., Anderson, G. L., Williams, P. L., Millner, P. D. and Beuchat, L. R. (2006). Migration of *Caenorhabditis elegans* to manure and manure compost and potential for transport of *Salmonella* Newport to fruits and vegetables. *Int. J. Food Microbiol.* 106(1), 61–68.

Kroupitski, Y., Pinto, R., Bucki, P., Belausov, E., Ruess, L., Spiegel, Y. and Sela, S. (2015). *Acrobeloides buetschlii* as a potential vector for enteric pathogens. *Nematology* 17(4), 447–457.

Lakshmanan, V., Selvaraj, G. and Bais, H. P. (2014). Functional soil microbiome: belowground solutions to an aboveground problem. *Plant Physiol.* 166(2), 689–700.

Lamelas, A., Desgarennes, D., López-Lima, D., Villain, L., Alomos-Sánchez, A., Artacho, A., Latorre, A., Moya, A. and Carrión, G. (2020). The bacterial microbiome of *Meloidogyne*-based disease complex in coffee and tomato. *Front. Plant Sci.* 11, 136.

Lazarova, S., Coyne, D., Rodríguez, G., Peteira, B. and Ciancio, A. (2021). Functional diversity of soil nematodes in relation to the impact of agriculture - a review. *Diversity* 13(2), 64.

Lee, S., Kin, Y. and Choi, J. (2020). Effect of soil microbial feeding on the gut microbiome and cadmium toxicity in *Caenorhabditis elegans*. *Ecotox. Env. Saf.* 187, 109777.

Leiririo, J. L., Karuri, H. and Nyaga, J. M. (2022). Nematode-based soil food web condition in mung bean under semi-arid conditions. *J. Agric. Food Res.* 10, 100465.

Lindo-Seminario, D., Mendez-Farroñan, S., Canta-Ventura, J., Córdova-Campos, J., Condemarin-Montealegre, C., Gutiérrez-Calle, S., Morales-Pizarro, A. and Mialhe-Matonnier, E. (2022). Endosymbiont microbiota of *Meloidogyne javanica* adult females infected by *Pasteuria penetrans*. *Trop. Subtrop. Agroecosys* 25, 132.

Liu, W. and Park, S.-W. (2018). Underground mystery: interactions between plant roots and parasitic nematodes. *Curr. Plant Biol.* 15, 25–29.

Lo, W. S., Han, Z., Witte, H., Rösler, W. and Sommer, R. J. (2022). Synergistic interaction of gut microbiota enhances the growth of nematode through neuroendocrine signaling. *Curr. Biol.* 32(9), 2037–2050.e4.

Lu, P., Shi, H., Tao, J., Jin, J., Wang, S., Zheng, Q., Liu, P., Xiang, B., Chen, Q., Xu, Y., Li, Z., Tan, J. and Cao, P. (2023). Metagenomic insights into the changes in the rhizosphere microbial community caused by the root-knot nematode *Meloidogyne incognita* in tabaco. *Environ. Res.* 216(4), 114848.

Lu, Q., Liu, T., Wang, N., Dou, Z., Wang, K. and Zuo, Y. (2020). A review of soil nematodes as biological indicators for the assessment of soil health. *Front. Agr. Sci. Eng.* 7(3), 275–281.

Maboreke, H. R., Graf, M., Grams, T. E. E., Herrmann, S., Scheu, S. and Ruess, L. (2017). Multitrophic interactions in the rhizosphere of a temperate forest tree affect plant carbon flow into the belowground food web. *Soil Biol. Biochem.* 115, 526–536.

Maghodia, A. B., Spiegel, Y. and Sela, S. (2008). Interaction between *Escherichia coli* and the plant-parasitic nematode *Meloidogyne javanica*. *J. Appl. Microbiol.* 105(6), 1810–1816.

Maina, S., Karuri, H. and Ng´endo, R. N. (2021). Free-living nematode assemblages associated with maize residues and their ecological significance. *J. Nematol.* 53, e2021-038.

Malacrinò, A., Wang, M., Caul, S., Karley, A. J. and Bennett, A. E. (2021). Herbivory shapes the rhizosphere bacterial microbiota in potato plants. *Environ. Microbiol. Rep.* 13(6), 805–811.

Markalanda, S. H., McFadden, C. J., Cassidy, S. T. and Wood, C. W. (2022). The soil microbiome increases plant survival and modifies interactions with root endosymbionts in the field. *Ecol. Evol.* 12(1), e8283.

Martins, S. J., Taerum, S. J., Triplett, L., Emerson, J. B., Zasada, I., Toledo, de, B. F., Kovac, J., Martin, K. and Bull, C. T. (2022). Predators of soil bacteria in plant and human health. *Phytobiomes J.* 6(3), 184–200.

Masson, A.-S., Bich, H. H., Simonin, M., Thi Nguyen, H., Czernic, P., Moulin, L. and Bellafiore, S. (2020). Deep modifications of the microbiome of rice roots infected by the parasitic nematode *Meloidogyne graminicola* in highly infested fields in Vietnam. *FEMS Microbiol. Ecol.* 26, fiaa099.

Melakeberhan, H., Bonito, G. and Kravchenko, A. N. (2021). Application of nematode community analyses-based models towards identifying sustainable soil health management outcomes: a review of the concepts. *Soil Syst.* 5(2), 32.

Mendes, R., Garbeva, P. and Raaijmakers, J. M. (2013). The rhizosphere microbiome: significance of plant beneficial, plant pathogenic and human pathogenic microorganisms. *FEMS Microbiol. Rev.* 37(5), 634–663.

Mielke, L., Taubert, M., Cesarz, S., Ruess, L., Küsel, K., Gleixner, G. and Lange, M. (2022). Nematode grazing increases the allocation of plant-derived carbon to soil bacteria and saprophytic fungi, and activates bacterial species of the rhizosphere. *Pedobiologia* 90, 150787.

Moreira, G. F., de Morais, M. R., Busoli, A. C. and de Moraes, G. J. (2015). Life cycle of *Cosmolaelaps jaboticabalensis* (Acari: Mesostigmata: Laelapidae) on *Frankliniella occidentalis* (Thysanoptera: Thripidae) and two factitious food sources. *Exp. Appl. Acarol.* 65(2), 219–226.

Mulder, C. and Maas, R. (2017). Unifying the functional diversity in natural and cultivated soil using the overall body-mass distribution of nematodes. *BMC Ecol.* 17(1), 36.

Neher, D. A. (2010). Ecology of plant and free-living nematodes in natural and agricultural soil. *Annu. Rev. Phytopathol.* 48, 371–394.

Neher, D. A., Nishanthan, T., Grabau, Z. J. and Chen, S. Y. (2019). Crop rotation and tillage affect nematode communities more than biocides in monoculture soybean. *Appl. Soil Ecol.* 140, 89–97.

Nicholson, F. A., Groves, S. J. and Chambers, B. J. (2005). Pathogen survival during livestock manure storage and following land application. *Bioresour. Technol.* 96(2), 135–143.

Nicol, J. M. and Turner, S. J., Coyne, L. D., den Nijs, L., Hockland, S., Maafi, Z. T. (2011). Current nematode threats to world agriculture. In: Jones, J., Gheysen, G., and Fenoll, C. (Eds.), *Genomics and Molecular Genetics of Plant-Nematode Interaction*, Springer, Germany, 2011, pp. 21–43.

Nieminen, J. K. (2008). Soil animals and ecosystem processes: how much odes nutrient cycling explain? *Pedobiologia* 51(5–6), 367–373.

Nykyri, J., Fang, X., Dorati, F., Bakr, R., Pasanen, M., Niemi, O., Palva, E. T., Jackson, R. W. and Pirhonen, M. (2014). Evidence that nematodes may vector the soft rot-causing enterobacterial phytopahogens. *Plant Pathol.* 63(4), 747–757.

O'Brien, A. M., Ginnan, N. A., Rebolleda-Gómez, M. and Wagner, M. R. (2021). Microbial effects on plant phenology and fitness. *Am. J. Bot.* 108(10), 1824–1837.

Omotayo, O. P. and Babalola, O. O. (2021). Resident rhizosphere microbiome's ecological dynamics and conservation: towards achieving the envisioned sustainable development goals, a review. *Int. Soil Water Cons. Res* 9, 127–142.

Overbeek, van, L. S., Franz, E., Semenov, A. V., De Vos, O. J. and Bruggen, van, A. H. C. (2010). The effect of the native bacterial community structure on the predictability of *E. coli* O157:H7 survival in manure-amended soil. *Lett. Appl. Microbiol.* 50(4), 425–430.

Overbeek, van, L. S., Doorn, van, J., Wichers, J. H., Amerongen, van, A., Roermund, van, H. J. W. and Willemsen, P. T. J. (2014). The arable ecosystem as battleground for emergence of new human pathogens. *Front. Microbiol.* 5, 104.

Pan, F., Han, X., Li, N., Yan, J. and Xu, Y. (2020). Effect of organic amendment amount on soil nematode community structure and metabolic footprints in soybean phase of a soybean-maize rotation on mollisols. *Pedosphere* 30(4), 544–554.

Park, M. R., Oh, S., Yun, H. S., Kim, S. H., Ko, Y. H., Ryu, J. H., Rhee, M. S., Shin, O. S. and Kim, Y. (2014). Enhanced protection of pathogenic *Escherichia coli* ingested by a soil nematode *Caenorhabditis elegans* against sanitizer treatments. *Biosci. Biotechnol. Biochem.* 78(11), 1917–1922.

Pausch, J., Kramer, S., Scharroba, A., Scheunemann, N., Butenschoen, O., Kandeler, E., Marhan, S., Riederer, M., Scheu, S., Kuzyakov, Y. and Ruess, L. (2016). Small but active – pool size does not matter for carbon incorporation in belowground food webs. *Funct. Ecol.* 30(3), 479–489.

Pausch, J., Hüninghaus, M., Kramer, S., Scharroba, A., Scheunemann, N., Butenschoen,O., Marhan, S., Bonkowski, M., Kandeler, E., Scheu, S., Kuzyakov, Y. and Ruess, L. (2018). Carbon budgets of top- and subsoil food webs in an arable system. *Pedobiologia* 69, 29–33.

Piśkiewicz, A. M., Millano, de, M. J. K., Duyts, H. and Putten, van der, W. H. (2009). Plant ectoparasitic nematodes prefer roots without their microbial enemies. *Plant Soil* 316(1–2), 277–284.

Poll, J., Marhan, S., Haase, S., Hallmann, J., Kandeler, E. and Ruess, L. (2007). Low amounts of herbivory by root-knot nematodes affect microbial community dynamics and carbon allocation in the rhizosphere. *FEMS Microbiol. Ecol.* 62(3), 268–279.

Potapov, A. M. (2022). Multifunctionality of belowground food webs: resource, size and spatial energy channels. *Biol. Rev. Camb. Philos. Soc.* 97(4), 1691–1711.

Pothula, S. K., Gray, P. and Bernard, E. C. (2022). Increasing levels of physical disturbance affect soil nematode community composition in a previously undisturbed ecosystem. *J. Nematol.* 54, 10.2478.

Powers, T. O., Harris, T. S., Higgins, R. S., Mullin, P. G. and Powers, K. S. (2021). Nematode biodiversity assessments need vouchered databases: a bold reference library for plant-parasitic nematodes in the superfamily Criconematoidea. *Genome* 64(3), 232–241.

Qing, X., Wang, M., Karssen, G., Bucki, P., Bert, W. and Braun-Miyara, S. (2020). PPNID: a reference database and molecular identification pipeline for plant-parasitic nematodes. *Bioinformatics* 36(4), 1052–1056.

Ramos, T. D. M., Jay-Russel, M. T., Millner, P. D., Baron, J. N., Stover, J., Pagliari, P., Hutchinson, M., Lilley, J., Rowley, N., Haghani, V., Aminabadi, P., Kenney, A., Hashem, F., Martínez-López, B., Bihn, E. A., Clements, D. P., Shade, J. B., Sciligo, A. R. and Pires, A. F. A. (2021). Survival and persistence of foodborne pathogens in manure-amended soils and prevalence on fresh produce in certified organic farms: a multi-regional baseline analysis. *Front. Sustain. Food Syst.* 5, 67476.

Rengalakshmi, R. M. M., Prabavathy, V. R., Prabavathy, V. R., Jegan, S. and Selvamukilan, B. (2018). Building bioeconomy in agriculture: harnessing soil microbes for sustaining ecosystem services. In: *Towards a Sustainable Bioeconomy: Principles, Challenges and Perspectives*, Springer, Germany, pp. 261–277.

Richter, A., Kern, T., Wolf, S., Struck, U. and Ruess, L. (2019). Trophic and non-trophic interactions in binary links affect carbon flow in the soil micro-food web. *Soil Biol. Biochem.* 135, 239–247.

Rønn, R., Vestergård, M. and Ekelund, F. (2013). Interactions between bacteria, protozoa and nematodes in soil. *Acta Protozool.* 51, 223–235.

Ross, G. M., Berg, M. P., Salomon, S. and Nielsen, U. N. (2022). Phylogenies of traits and functions in soil invertebrate assemblages. *Austral. Ecol.* 47(3), 465–481.

Roth, E., Samara, N., Ackermann, M., Seiml-Buchinger, R., Saleh, A. and Ruess, L. (2015). Fertilization and irrigation practice as source of microorganisms and the impact on nematodes as their potential vectors. *Appl. Soil Ecol.* 90, 68–77.

Rueda-Ramírez, D., Rios-Malaver, D., Varela-Ramírez, A. and Moraes, de, G. J. (2019). Biology and predation capacity of *Parasitus bituberosus* (Acari: Mesostigmata:

Parasitidae) on *Frankliniella occidentalis* (Thysanoptera: Thripidae) and free-living nematodes as complementary diet. *Pest Manag. Sci.* 75(7), 1819–1830.

Rueda-Ramírez, D., Palevsky, E. and Ruess, L. (2023). Soil Nematodes as a means of conservation of soil predatory mites for biocontrol. *Agronomy* 13(1), 32.

Ruess, L. (2003). Nematode soil faunal analysis of decomposition pathways in different ecosystems. *Nematology* 5(2), 179–181.

Ruess, L. and Ferris, H. (2004). Decomposition pathways and successional changes. *Nematol. Monogr. Perspect.* 2, 547–556.

Scharroba, A., Dibbern, D., Hünninghaus, M., Kramer, S., Moll, S., Butenschoen, O., Bonkowski, M., Buscot, F., Kandeler, E., Koller, R., Krüger, D., Lueders, T., Scheu, S. and Ruess, L. (2012). Effects of resource availability and quality on the structure of the micro-food web of an arable soil across depth. *Soil Biol. Biochem.* 50, 111–119.

Schmidt, J. H., Theisgen, L. V., Finckh, M. R. and Šišić, A. (2022). Increased resilience of peas toward root rot pathogens can be predicted by the nematode metabolic footprint. *Front. Sustain. Food Syst.* 6, 881520.

Sharma, M., Saini, I., Kaushik, P., Aldawsari, M. M., Balawi, T. A. and Alam, P. (2021). Mycorrhiza fungi and *Pseudomonas fluorescens* application reduces root-knot nematode (*Meloidogyne javanica*) infestation in eggplant. *Saudi J. Biol. Sci.* 28(7), 3685–3691.

Siddique, S., Coomer, A., Baum, T. and Williamson, V. M. (2022). Recognition and response in plant-nematode interactions. *Annu. Rev. Phytopathol.* 60, 143–162.

Silva, J. C. P. D., Mederios, F. H. V. D. and Campos, V. P. (2018). Building soil suppressiveness against plant-parasitic nematodes. Biocont. *Sci. Techn.* 28, 423–445.

Skider, M. M., Vestergård, M., Kyndt, T., Topalović, O., Kudjordjie, E. N. and Nicolaisen, M. (2022.). Genetic disruption of *Arabidopsis* secondary metabolite synthesis leads to microbiome-mediated modulation of nematode invasion. *ISME J.* 16(9), 2230–2241.

Spiegel, Y. and McClure, M. A. (1995). The surface coat of plant parasitic nematodes: chemical composition, origin and biological role - a review. *J. Nematol.* 27(2), 127–134.

Sutherland, J. R. and Webster, J. M. (1993). Nematode pests of forest trees. In: Evans, K., Trudgill, D. L. and Webster, J. M. (Eds). *Plant Parasitic Nematodes in Temperate Agriculture*, CAB International, Wallingford UK, pp. 351–380.

Thahira, T., Qadeer, K., Sosso, J. and Zakeel, M. C. M. (2022). Role of plant microbiome in crop protection. In: Soni, R., Suyal, D. C. and Goel, R. (Eds). *Plant Protection: From Chemicals to Biologicals*, De Gruyter, Germany, pp. 573–600.

Thakur, M. P., Putten, van der, W. H., Apon, F., Angelini, E., Vreš, B. and Geisen, S. (2021). Resilience of rhizosphere microbial predators and their prey communities after extreme heat events. *Funct. Ecol.* 35(1), 216–225.

Topalović, O., Heuer, H., Reineke, A., Zinkernagel, J. and Hallmann, J. (2020). Antagonistic role of the microbiome from a *Meloidogyne hapla* - suppressive soil against species of plant-parasitic nematodes of different life strategies. *Nematology* 22(1), 75–86.

Topolavić, O., Elhady, A., Hallmann, J., Richert-Pöggeler, K. R. and Heuer, H. (2019). Bacteria isolated from the cuticule of plant-parasitic nematodes attached to and antagonized the root-knot nematode *Meloidogyne hapla*. *Sci. Rep.* 9(1), 11477.

Tran, D. T. Q., Bradbury, M. I., Ogtrop, van, F. F. V., Bozkurt, H., Jones, B. J. and McConchie, R. (2020). Environmental drivers for persistence of *Escherichia coli* and *Salmonella* in manure-amended soil: a meta-analysis. *J. Food Prot.* 83(7), 1268–1277.

Tyler, H. L. and Triplett, E. W. (2008). Plants as a habitat for beneficial and/or human pathogenic bacteria. *Annu. Rev. Phytopathol.* 46(1), 53–73.

Walter, D. E. and Proctor, H. C. (2013). *Mites: Ecology, Evolution & Behaviour* (2nd edn.), Springer, The Netherlands. 494 pp.
Wasilewska, L. and Webster, J. M. (1975). Free-living nematodes as disease factors of man and his crops. *Int. J. Environ. Stud.* 7(3), 201–204.
Whalen, J. K., Kernecker, M. L., Thomas, B. W., Sachdeva, V. and Ngosong, C. (2013). Soil food web controls on nitrogen mineralization are influences by agricultural practices in humid temperate climates. *CAB Rev.* 8, 1–18.
Wolkovich, E. M. (2016). Reticulated channels in soil food webs. *Soil Biol. Biochem.* 102, 18–21.
Wood, C. W., Pilkington, B. L., Vaidya, P., Biel, C. and Stinchcombe, J. R. (2018). Genetic conflict with a parasitic nematode disrupts the legume-rhizobia mutualism. *Evol. Lett.* 2(3), 233–245.
Yang, B., Chen, Q., Liu, X., Chen, F., Liang, Y., Qiang, W., He, L. and Ge, F. (2020). Effects of pest management practices on soil nematode abundance, diversity, metabolic footprint and community composition under paddy rice fields. *Front. Plant Sci.* 11, 88.
Yang, L., Zhang, F., Luo, Y. and Tang, P. (2022). Continuous cotton cropping affects soil micro-food web. *Appl. Soil Ecol.* 171, 104304.
Yang, Y. R., Li, X. G., Zhou, Z. G., Zhang, T. L. and Wang, X. X. (2016). Differential responses of soil nematode community to pig manure application levels in ferric acrisols. *Sci. Rep.* 6, 35334.
Yeates, G. W. (2010). Nematodes in ecological webs. In: *Encyclopedia of Life Sciences*, John Wiley & Sons, Hoboken, pp. 1–11.
Yeates, G. W., Bongers, T., Goede, de, R. G. M., Freckman, D. W. and Georgieva, S. S. (1993). Feeding habits in soil nematode families and genera - an outline for soil ecologists. *J. Nematol.* 25(3), 315–331.
Denton, C. S., Mercer, C. F., Yeates, G. W. and Saggar, S. (1998). Impact of clover cyst nematode (*Heterodera trifolii*) infection on soil microbial activity in the rhizosphere of white clover (*Trifolium repens*) - a pulse-labelling experiment. *Nematologica* 44(1), 81–90.
Yeates, G. W., Saggar, S., Hedley, C. B. and Mercer, C. F. (1999). Increase in ^{14}C-carbon translocation to the soil microbial biomass when five species of plant-parasitic nematodes infect root of white clover. *Nematology* 1(3), 295–300.
Zhang, F., Berg, M., Dierking, K., Félix, M. A., Shapira, M., Samuel, B. S. and Schulenburg, H. (2017). *Caenorhabditis elegans* as model for microbiome research. *Front. Microbiol.* 8, 485.
Zhang, F., Weckhorst, J. L., Assié, A., Hosea, C., Ayoub, C. A., Khodakova, A. S., Loeza Cabrera, M. L., Vidal Vilchis, D., Félix, M. A. and Samuel, B. S. (2021). Natural genetic variation drives microbiome selection in the *Caenorhabditis elegans* gut. *Curr. Biol.* 31(12), 2603–2618.e9.
Zhang, P., Zhang, W. and Hu, S. (2023). Fungivorous nematode *Aphelenchus avenae* and Collembola *Hypogastrura perplexa* alleviate damping-off disease caused by *Pythium ultimum* in tomato. *Plant Soil* 482(1–2), 175–189.
Zhang, S., Zeng, H.-C. and Jin, Z.-Q. (2016). Response of soil nematode community composition and diversity to different crop rotations and tillage in the tropics. *Appl. Soil Ecol.* 107, 134–143.

Published by Burleigh Dodds Science Publishing Limited, 2025.

Zhang, Z., Zhang, X., Jhao, J., Zhang, X. and Liang, W. (2015). Tillage and rotation effects on community composition and metabolic footprints of soil nematodes in a black soil. *Eur. J. Soil Biol.* 66, 40–48.

Zhang, Z., Li, B., Wang, H., Zhang, X., Cui, S., An, T., Fu, S., Mahamood, M., Zhang, D. and Wang, J. (2023). The fungal feeding channel of the soil micro-food web contributes to the transformation of exogenous C into soil C – a ^{13}C labelling experiment. *Land Degrad. Dev.* 34(2), 466–477.

Zhao, B. G., Wang, H. L., Han, S. F. and Han, Z. M. (2003). Distribution and pathogenicity of bacteria species carried by *Bursaphelenchus xylophilus* in China. *Nematology* 5(6), 899–906.

Zhao, X., Lin, C., Tan, J., Yang, P., Wang, R. and Qi, G. (2023). Changes in rhizosphere microbiome and metabolites in *Meloidogyne incognita* infested soil. *Plant Soil* 483(1–2), 331–353.

Zheng, F., Zhu, D., Giles, M., Daniell, T., Neilson, R., Zhu, Y. G. and Yang, X. R. (2019). Mineral and organic fertilization alters the microbiome of a soil nematode *Dorylaimus stagnalis* and its resistome. *Sci. Total Environ.* 680, 70–78.

Zheng, F., Zhu, D., Chen, Q. L., Bi, Q. F., Yang, X. R., O'Connor, P. and Zhu, Y. G. (2020). The driving factors of nematode gut microbiota under long term fertilization. *FEMS Microbiol. Ecol.* 96(4), fiaa037.

Zheng, J., Dini-Andreote, F., Luan, L., Geisen, S., Xue, J., Li, H., Sun, B. and Jiang, Y. (2022). Nematode predation and competitive interactions affect microbe-mediated phosphorus dynamics. *mBio* 13(3), e0329321.

Zhong, S., Zeng, H.-C. and Jin, Z.-Q. (2017). Influences of different tillage and residue management systems on soil nematode community composition and diversity in the tropics. *Soil Biol. Biochem.* 107, 234–243.

Zhou, J., Lu, M., Sheng, X. and Wu, J. (2022). Effects of nitrogen addition and warming on nematode ecological indices: a meta-analysis. *Eurp J. Soil Biol.* 110, 103407.

Chapter 8

Advances in understanding soil microbiomes in ecosystem functions across trophic chains

Karoline Jetter, University of Ulm, Germany; and Patrick Schäfer, Justus Liebig University, Germany

1 Introduction

Soil is the basis for all agricultural development and one of, if not, the most important resources to feed humanity. While being the foundation of our well-being, soil microbiomes, defined as a consortium of all bacteria and fungi living in soil environments, host about 15% of global biomass (Bar-On et al., 2018). In addition to bringing in a huge biomass, soil is one of the most diverse biomes on our planet by hosting the most heterogeneous and diverse microbial communities (Xiong et al., 2021) and harboring around 25% of Earth's terrestrial species (Guerra et al., 2021). Soil microbiomes involve microbes with gene functions responsible for decomposition, climate regulation, plant growth promotion and stress tolerance, pest and disease control, degradation of pollutants and nutrient cycling and therefore display key players in ecosystem functioning (Hartmann and Six, 2023). The interaction between soil

http://dx.doi.org/10.19103/AS.2024.0136.31

microbes and plants (i.e. plant–soil–feedback, the process of plants altering the biotic and abiotic soil properties which in turn alters the ability of plants to grow in this soil in the future) is one of the most important interactions in agroecosystems as it has the potential to support, e.g. plant growth and stress tolerance that are key to securing food supply for a growing world population, especially under changing climates. Considering plant microbiomes, soil-derived microorganisms are mainly associated with the rhizosphere and the roots themselves (Zarraonaindia et al., 2015). Microbial biomass and activity are up to 20 times higher in surroundings where roots and fungal hyphae are growing compared to bulk soil (Finzi et al., 2015; Kuzyakov and Blagodatskaya, 2015). The soil microbiome is therefore considered as a second genome that plants access and add to their functional diversity (Mendes et al., 2013).

In addition to being essential for the soil–plant interaction, soil microbial communities and their functions are, within the frame of the One Health concept, interconnected to all organisms living in the ecosystem and with that to ecosystem health. Understanding which factors affect the general soil microbiome functioning and what consequences a changing soil microbiome has for its essential services in ecosystem functioning is key to keeping a balanced One Health equilibrium with a healthy environment and healthy humanity (Banerjee and van der Heijden, 2023). In this review, we will discuss the effect of different abiotic and biotic factors on soil microbiome diversity and its functions from the perspective of complex ecosystems. This will include experimental setups to study ecosystem functions in soil–plant feedback in natural and controlled lab-scale environments.

2 Ecosystems in the One Health concept

An ecosystem describes the functional interrelatedness of all living organisms with the physical environment that they live in. Originally coined by the English botanist Sir Arthur Tansley in 1935, the term 'ecosystem' includes all elements, no matter whether they are chemical, physical, or biological, that can impact an organism under study (Golley, 1984). Since then ecosystems and their functions are studied intensively from all possible perspectives. By including not only animals, plants, and the physical environment like soil and water but also humans and by taking the assumption that health and well-being of humans are linked to the health of the ecosystem, the One Health concept emphasizes on the interaction and dependences of different partners in an ecosystem (Banerjee and van der Heijden, 2023). Microbes, although being the smallest partners in an ecosystem and not getting major attention in the context of One Health, play a pivotal role in ecosystem health. By interconnecting all components of the One Health concept and contributing to many ecosystem functions, the health of an ecosystem relies on the contribution of microbial communities (Berg et al., 2020).

3 Impact of abiotic factors on soil microbiomes

Soil is present in almost every ecosystem and subjected to several abiotic and biotic stresses that can change the composition of microbiomes and, eventually, its functions (reviewed in Fierer (2017)). Based on the physico-chemical properties of soils, it is difficult to generalize the impact of abiotic factors and climate change on soil microbial communities. While the impact of abiotic factors has been reviewed recently (Fierer, 2017; Jansson and Hofmockel, 2020), we will focus on some critical factors to explain how these changes in soil microbial communities can affect ecosystem functioning.

3.1 Impact of abiotic soil properties

Soil is not a homogeneous compartment but very diverse in its structure, pore size, permeability, mineral-to-organic matter composition, water content, and pH. The community composition of soil microbiota varies a lot in relation to these factors. A well-developed soil structure with a good drainage helps plant roots and fungi to properly establish and build a well-connected network, thereby allowing sufficient water and oxygen supply for microbes at different soil levels. This habitat structure raises microbial activities in the soil and supports nutrient cycling and supply to plants and other organisms (Fig. 3 in Hartmann & Six, 2023).

Several studies name soil pH to be the driving force determining soil microbial community composition (Fierer and Jackson, 2006; Lauber et al., 2009). About 43–85% of the variation in microbial community abundances involved in N-cycling at the landscape scale can be explained by environmental parameters, with pH being the most pressing one (Bru et al., 2011). Soils of low pH harbor more homogeneous communities that are dominated by *Acidobacteria*, whereas soils with a higher pH show a more diverse microbial population (Griffiths et al., 2011; Mukherjee et al., 2014; Tan et al., 2020). Bacterial diversity was found to be significantly higher in arable soils at pH 8 than at pH 4, whereas fungi were only slightly affected (Rousk et al., 2010). This might be explained by a higher availability of carbon and nitrogen substrate that, in turn, influences soil microbiomes (Kemmitt et al., 2006). Microbes rely on the availability of nutrients and hence the distribution of macro- and micronutrients plays a pivotal role in the community distribution of soil microbes. As main macronutrients, nitrogen availability (Cederlund et al., 2014) and soil carbon content (Sul et al., 2013) were found to predict soil microbial communities and distribution of microbes was further associated with several micronutrients, such as available iron, copper, zinc, and manganese (Peng et al., 2022).

3.2 Impact of climatic factors

Generally, a good supply of water in a well-structured soil promotes higher microbial activity, and soil moisture is one of the major factors influencing microbial diversity (Brockett et al., 2012; Frindte et al., 2019) with higher humidity of ecosystems (e.g. tropical rainforest) positively correlating with higher microbial biomass (Serna-Chavez et al., 2013). However, comparing different agricultural sites in China, Tan et al. (2020) found a negative correlation between mean annual precipitation and Shannon-index, describing (microbial) diversity within a sample based on species abundances and evenness (Shannon & Weaver, 1949). A fast increase in soil moisture, e.g. due to heavy rains or flooding on dry soil implies high-energy costs based on regulating osmotic stress to soil microbes and therefore might affect cell viability and integrity resulting in a release of mineralizable carbon (Halverson et al., 2000).

Temperature as another major climatic factor highly impacts the microbial composition in soil (Frindte et al., 2019). In an artificial soil warming experiment over 2 years, Nottingham et al. (2022) found a significant reduction of fungal and bacterial diversity causing large shifts in microbial community composition. In addition, a study on maize soils found higher temperatures to be negatively correlated with Shannon-index (Tan et al., 2020). In grasslands, experimental warming of soils for several years not only increased microbial biomass (Zhang et al., 2014) but also resulted in a shift in microbial community composition toward Gram-positive bacteria (Melillo et al., 2017) and a decreased fungal-to-bacterial ratio (Melillo et al., 2017; Zhang et al., 2014). Given that ecosystem functioning in soils is usually dependent on certain key taxa and complex interactions between components of the soil food web (Bardgett, 2002), this decline in soil microbial diversity can heavily impact ecosystem functioning.

By linking high temperature and low soil moisture, drought challenges soil microbial communities resulting in reduced microbial diversity (McHugh et al., 2017). With a lower water availability in the soil and therefore loss of connection between pores, resources are ill-distributed. Drought is therefore considered to reduce microbial functions that are important for ecosystems as previously reviewed in Schimel (2018) and Tecon and Or (2017). This might lead to reduced productivity under drought conditions or reduced carbon loss through suppressed respiration. However, it is difficult to make a general statement as soil properties, e.g. porosity and soil texture, strongly impact the accessibility of water and nutrients and therefore responses of soil microbial communities to drought (Patel et al., 2021). For instance, when looking at these factors in the long term, soil microbial communities might have the ability to adapt to slow changes in climatic conditions (Jansson and Hofmockel, 2018). A gradual change in climatic conditions might allow soil microbes to select heat- or stress-resistant genes or favor-respective species. A study on grassland and forest soils showed that historical exposure to changes in soil water content altered the ability of soil microbes to respond to new changes in soil moisture (Waldrop & Firestone, 2006) suggesting the existence of a microbiome-derived memory of ecosystems to operate under changing environmental conditions.

4 Impact of agricultural land use on soil microbiomes

Soils with a high biodiversity, especially high microbial diversity allow the different organisms to perform essential soil functions. A high biodiversity is therefore important to sustain human health (Wall et al., 2015). The cultivation of land for agricultural purposes has increased tremendously over the last decades in line with a growing global population and demand for food. The intensification of food production and agricultural land use has led to land conversion, soil perturbation, over-fertilization, intensive irrigation, and enhanced application of pesticides. These trends have led to soil degradation, environmental pollution, eutrophication, and greenhouse gas emissions (Matson et al., 1997). Intensive agriculture was further found to impact the resistance and resilience of the soil food web to drought (de Vries et al., 2012). To build up sustainable land use strategies and secure crop production, it is essential to understand the different responses of soil microbial communities to different stressors connected to such an intensive agricultural land use.

4.1 Impact of soil perturbation

Soil microbiomes are highly dependent on soil structure and soil properties (König et al., 2020) and any soil disturbance can have strong impacts on the functioning of soil microbiomes in soil organic matter (SOM) decomposition, carbon sequestration, and nutrient turnover. This can further feedback on soil–plant interactions (Six et al., 2006). The trend to larger farm sites requires larger and most importantly heavier machines and equipment for efficient tillage, sowing or harvest, leading to soil compaction in intensive agricultural land use. This compaction of soil structure can lead to a change in microbial community structure (Longepierre et al., 2021). It can further affect soil health and function and with that reduce productivity and ecosystem health and function (Hartmann and Six, 2023). Tillage exposes soils from deeper soil layers to the surface, increases macroaggregate turnover and by that disrupts the existing pore network, and increases SOM decomposition with a reduction in soil organic carbon stocks (Six et al., 2002). These changes in soil structure also heavily disrupt the hyphal network of fungi (Young and Ritz, 2000). Comparing different tillage intensities, soil organic carbon content in the upper soil layers was found to be higher in non-tillage as compared to conventional tillage soil (Hernanz et al., 2002). In addition, changes in soil density, macroaggregate structure, and pore connectivity can affect water drainage, nutrient availability, and oxygen penetration (Schäffer et al., 2008). As these factors are essential for microbial soil communities (discussed later), any changes in soil structure can harm microbial life in the soil.

4.2 Impact of fertilization

Soil nutrient content affects soil microbiota. In this respect, abiotic factors like soil moisture (as discussed above) in addition to mineral and organic fertilizers often composed of manure or slurry from cattle, pigs, or horses or digestates of biogas plants is especially relevant in agricultural fields. Several studies have shown short-term effects of different organic and mineral fertilizers on microbial biomass, community structure, and enzymatic activities (Guo et al., 2011; Lazcano et al., 2013). However, as fertilization in agricultural soils is usually not necessarily a short-term application but rather a long-term treatment, the focus shifts to understanding the effects of long-term application of fertilizers.

Long-term fertilization has a more persistent effect on microbial diversity and activity while organic farming increased microbial richness, decreased evenness, and shifted the structure of soil microbiota as compared to conventionally managed soils with mineral fertilization (Giacometti et al., 2014; Hartmann et al., 2015). Based on 107 datasets of 64 long-term experiments to determine the effect of mineral fertilizers, Geisseler and Scow (2014) showed that in most studies microbial biomass increased due to an increase in soil pH. In a 2-year field experiment, nitrogen application decreased the fungal/bacterial ratio whereas bacterial activity increased (de Vries et al., 2006) and long-term addition of nitrogen disrupts multipartite synergies between arbuscular mycorrhizal fungi and soil microbial communities. These synergies usually enhance plant and fungal nitrogen uptake from organic matter and a disruption leads to a reduced mycorrhizal nitrogen acquisition from organic matter and has therefore a tremendous impact on nutrient cycling and ecosystem functioning (Hestrin et al., 2019).

4.3 Impact of plants

Plants can benefit from soil microbes by receiving nutrients or other factors supporting plant growth and development. It is therefore not further surprising that plants shape the community composition of the soil microbiome thereby recruiting beneficial microbes from the soil that colonize the root endosphere (interior root part) or rhizosphere (immediate area surrounding roots). The root morphology and root exudates assist in this recruitment process (Rolfe et al., 2019; Saleem et al., 2018). In grassland, interactions between above- and belowground ecosystem compartments are so strong that plant species diversity could predict beta diversity of respective soil microbial communities (Prober et al., 2015). Certain mycorrhizal fungi, plant pathogens, or nitrogen-fixing bacteria (e.g. nodulating rhizobia) were found to associate preferentially with certain plant species and genotypes (Berg and Smalla, 2009; Hartmann et al., 2015; Lundberg et al., 2012). By changing several abiotic and biotic factors like oxygen pressure and carbon and nitrogen availability, plant roots also influence nitrogen conversion of soil microorganisms. Roots consume oxygen and thereby increase the anaerobic capacities of soils. In addition to a higher carbon availability (e.g. root exudates) in the rhizosphere, it promotes denitrification (Henry et al., 2008; Philippot et al., 2009, 2013). As microbial community structures differ between soil types, such rhizosphere effects can vary between soil type and plant species (Schreiter et al., 2014).

4.4 Impact of antibiotics and antimicrobial resistance genes

Antibiotics are one of the revolutionizing findings in modern medicine. Soil microbiomes are an important source of antibiotics. Certain habitats such as the nutrient-rich rhizosphere represent not only a microbiologically highly diverse but also competitive environment, where microbes produce and release antibiotics to control microbial competitors (Abrudan et al., 2015).

In some exceptional cases, antibiotics are directly applied to plants due to the lack of alternatives and the highly destructive nature of certain plant quarantine diseases (e.g. against fire blight in apple or pear orchards (Psallidas & Tsiantos, 2000)). By irrigation and drainage, these antibiotics can leach into the soil and reach the soil microbiota. Certain antibiotics such as sulfonamides can move into lower soil levels and ground water, but most antibiotic compounds remain in the upper layers of surface soil (Aust et al., 2010; Blackwell et al., 2009; Ostermann et al., 2013) where they can interact with other organisms interacting with the soil microbiome and soil compounds.

Livestock are often treated with antibiotics used in human therapeutics. Using manure and slurry for fertilization in the field can lead to an enrichment of antimicrobial resistance genes and bacteria in the soil (Heuer et al., 2011). Depending on the pharmacokinetics of antibiotics, antibiotic compounds can persist in the soil from months to years (Aust et al., 2010; Ostermann et al., 2013; Rosendahl et al., 2012). Repeated application of manure containing antimicrobial resistant agents as fertilizer might therefore lead to an accumulation of taxa containing possible pathogens and reducing other microbes contributing to a healthy soil microbiome (Ding et al., 2014). Manure is a potential source for antimicrobial resistance genes on mobile genetic elements (Binh et al., 2008; Wolters et al., 2014). Mobile elements such as broad-host range plasmids (Heuer et al., 2011) are a natural, evolution-based mechanism facilitating horizontal gene transfer (reviewed in Jechalke et al., 2014) that can equip soil bacteria, well adapted to the soil environment, with antimicrobial resistance genes. In fact, the transfer of antimicrobial resistance genes is enhanced by manure fertilization (Götz and Smalla, 1997; Smalla et al., 2000). This might lead to the persistence and spreading of resistance genes in the environment. An accumulation of resistance genes might sustainably affect the structure of soil microbiomes and consequently even translate in an altered overall ecosystem functioning.

5 Microbial communities in ecosystem functioning: plant-soil interaction

Soil microorganisms drive organic matter decomposition, humus formation, and nutrient transformation; processes most critical for crop productivity (Singh and Gupta, 2018; Van Der Heijden et al., 2008). Plants recruit their endosphere and rhizosphere microbiomes from the soil and rely on it for their development and reproduction even under changing environments. Prokaryotic fractions of rhizosphere communities are dominated by *Proteobacteria*, *Actinobacteria, Acteriodetes*, *Firmicutes*, and *Acidobacteria*, while in fungal phyla *Ascomycota* and *Basidisomycota* are the most common taxa (Mohanram and Kumar, 2019). Although arbuscular fungi (phylum Glomeromycota) and ectomycorrhizal fungi are well-studied, these groups represent only minor diversity among fungal communities from roots (Vandenkoornhuyse et al., 2002). Soil microbes can promote plant growth indirectly by fixing essential macro- (e.g. N, P, and K) and micronutrients (e.g. Fe) for the plant in the soil, either symbiotically in the endosphere of the roots, in a free-living lifestyle in the rhizosphere, or directly by producing plant growth promoting factors and plant hormones. Plant growth is further promoted by inducing stress resistance against abiotic stresses to the plant, e.g. the climate change factors drought, flooding, salinity, and heavy metal pollution as well as biotic plant stressors like herbivores (Mendes et al., 2013). To attract and interact with such beneficial microbes, plant roots excrete exudates such as sugars, organic acids, amino acids, and phenolics (Compant et al., 2019). This high nutrient environment makes the rhizosphere a quite competitive habitat and explains the distinct composition of microbial community in the rhizosphere as compared to the more diverse bulk soil community (García-Salamanca et al., 2013). Furthermore, above- and belowground plant microbiomes differ significantly, although they all share a certain amount of soil-derived microbes. It indicates, however, soil as the main source and reservoir for plant microbiomes (Edwards et al., 2015; Rochefort et al., 2021).

Because of their significance in plant and ecosystem resilience under challenging environments, much attention was given to reveal how plant microbes impact plant traits related to nutrient acquisition, drought tolerance, disease resistance, and plant fitness (Philippot et al., 2013; Trivedi et al., 2020). Similarly, a focus was given on microbiome-based mechanisms that explain the outbreak of diseases. A dysbiosis, as an imbalance in microbiome structure, usually disturbs and reduces the biodiversity of microbial communities in the soil with potentially negative effects on plant fitness and ecosystem functions (Wagg et al., 2014). In the most extreme scenario, plant diseases occur as a consequence of a massive expansion of pathogens in a microbial community. Dysbiosis can be the result or the origin of a loss of soil biodiversity, which is in turn often accompanied by a reduction of plant diversity, productivity, nutrient retention, and belowground carbon allocation (Wagg et al., 2014). In the following sections, we will discuss these soil–plant microbiome interactions and their role in ecosystem functioning in the trophic chain and nutrient cycling.

5.1 Nutrient cycling

For all developmental processes from germination to reproduction, plants not only need sunlight, water, and carbon dioxide but are heterotrophs for macro- and micronutrients. For most vascular plants, these nutrients are taken up via roots from the soil. As discussed above, soil microorganisms are the main players in nutrient cycling and supply to different organisms associated with the soil. In addition, microbes produce hormones, e.g. used in quorum sensing but also with functions in plant growth promotion (Eichmann et al., 2021). Plant growth can be induced by the increase of nutrient availability or production of plant growth-promoting factors such as hormones that stimulate root growth or change the root system architecture for improved nutrient uptake.

5.1.1 Nitrogen

Nitrogen is one of the most limiting factors in plant growth and plants rely on the uptake of nitrogen from the soil environment as they cannot access atmospheric nitrogen. Nitrogen is usually present as nitrate, ammonium, and organic nitrogen in the soil, and soil associate bacteria can metabolize different forms of nitrogen (Näsholm et al., 2009). Plant roots, however, prefer inorganic nitrogen (i.e. nitrate and ammonium) rather than organic nitrogen which might affect the efficiency of nitrogen uptake (Bloom, 2015).

Nitrogen is available in the form of different nitrogen compounds that are the results of several reactions carried out by a complex network of microorganisms in the soil. These processes can roughly be separated into nitrogen fixation, nitrogen assimilation, ammonification, nitrification, denitrification, and anaerobic ammonia oxidation (Kuypers et al., 2018). Some of the intermediate products like ammonium resulting from nitrogen fixation and mineralization can directly be taken up by the plant or quickly processed to plant-available nitrate.

Arbuscular mycorrhizal fungi can acquire a substantial quantity of mineral nitrogen from organic and inorganic nitrogen sources and transfer it to the host plants (Hodge and Storer, 2014). Plants have therefore evolved mechanisms based on the release of the hormone strigolactone to attract mycorrhizal fungi and establish symbioses (Aliche et al., 2020; Kodama et al., 2022).

Acquiring nitrogen from organic or inorganic nitrogen sources is by far not restricted to arbuscular mycorrhizal fungi. Hestrin et al. (2019) showed that multipartite synergies between arbuscular mycorrhizal fungi and soil microbial communities enhance plant and fungal nitrogen acquisition from organic matter and microbial acquisition of plant photosynthates. Mycorrhizal fungi promote nitrogen acquisition but have limited ability to access organic nitrogen, while soil microbial communities mineralize organic nitrogen into bioavailable forms resulting in an up to ten-fold increase in nitrogen uptake.

Legumes further release flavonoids to attract N_2-fixing, nodulating rhizobia (Hassan and Mathesius, 2012). These symbioses can provide more than 300 kg of nitrogen per hectare per year (Bardgett and Van Der Putten, 2014; Fierer, 2017; Richardson et al., 2009).

5.1.2 Phosphorus

Phosphorus is another critical macronutrient for plants and other organisms. Although soils generally contain sufficient amounts of total phosphorus, these phosphorus resources occur predominantly as inorganic fractions that are adsorbed to soil mineral surfaces or organic matter (Richardson and Simpson, 2011). Only a small portion, namely orthophosphate anions (hydrogen phosphate (HPO_4^{2-}) and dihydrogen phosphate ($H_2PO_4^{1-}$)) are directly available for plants. Hence, plants usually depend on their microbial partners including mycorrhizal and endophytic fungi for phosphorus mobilization and supply (Bodenhausen et al., 2019; Jorquera et al., 2008). The underlying processes mediated by microorganisms can be grouped into three different mechanisms. First, they promote root growth and the resulting changes in root system architecture improve phosphorus uptake (Richardson and Simpson, 2011). Second, microbes can enhance phosphorus availability in the soil through mineralization and solubilization and, third, through assimilation and precipitation processes (Richardson et al., 2009). Arbuscular mycorrhizal fungi, for instance, are able to take up solubilized phosphorus from the environment and to translocate it directly to colonized plants subsequently (Kobae, 2019).

In an extensive review of data from 399 papers published on a nationwide field survey in China, Li et al. (2021) found phosphate-solubilizing bacteria (PSB) and fungi to include *Bacillus*, *Pseudomonas*, *Enterobacter*, *Burkholderia*, *Penicillium*, and *Aspergillus*. Additionally, to these genera, a study on Argentinian acidic soil added bacteria from the genera *Pantoea*, *Acinetobacter*, and *Exiguobacterium* to the genera containing PSB. Some of these isolated strains showed high phosphorus-solubilizing activity in *in vitro* assays and promoted plant growth with higher P and N contents after inoculation (Collavino et al., 2010). A number of these fungi are endophytes with the ability to transfer phosphorus to their host plants to promote growth under low P conditions (Almario et al., 2017; Bakshi et al., 2015; Hiruma et al., 2016).

5.2 Plant growth-promoting factors – abiotic and biotic stress tolerance

Plant growth can not only be promoted by bacteria indirectly by transforming nutrients like, e.g. nitrogen or phosphorus into available forms for uptake by the plant but can also be promoted directly by the production of phytohormones and other signals. Plant fitness can further be enhanced by microbes promoting stress tolerance against abiotic stressors or by supporting plant defense against pathogens or herbivores (Trivedi et al., 2020). Different hormones are involved in plant–microbiome interactions and microbes and plants use the same hormone for different purposes. As hormones and their role in plant–microbiome interactions have been reviewed very detail recently in Eichmann et al. (2021), we will not elaborate on these interactions in detail, but we will give a short summary of how plant–microbiome interactions can promote plant growth by other mechanisms than nutrient cycling as described above.

Microbes can produce signaling compounds, i.e. plant hormones, which can enhance lateral root growth and branching and promote the development of root hairs for better nutrient acquisition (e.g. reaching microbiota-rich top soil levels) and water uptake from the soil. Nitrogen limitation induces primary root growth, enhances lateral root density and elongation, and root hair formation to help plants to overcome low phosphorus and nitrogen levels (Jia and von Wirén, 2020).

To cope with drought or floods, plants take advantage of their microbial partners and select a stress-resistance-promoting microbiome (Fitzpatrick et al., 2019; Naylor and Coleman-Derr, 2018; Xu et al., 2018). The plant hormone abscisic acid is involved in adaptation to abiotic stress, particularly in drought and salinity stress, and plays a pivotal role in modulating the plant immune signaling network (Asselbergh et al., 2008). It dampens the immune response and therefore facilitates shifts in the endophytic community composition (Fitzpatrick et al., 2018).

Further, microbial activity can enhance plant fitness and survival by supporting the plant in its defense mechanisms against biotic stressors like pathogens or herbivores. As a first line of defense, plants recruit beneficial soil microorganisms against soil-borne pathogens by releasing plant root exudates (Mendes et al., 2011). This defense mechanism is based on the production of certain compounds like antifungal metabolites (Cha et al., 2016; Mendes et al., 2011) or volatiles (Carrión et al., 2018; Hol et al., 2015). Endophytic microbes can create a second layer of defense by enriching selected microbes that can produce enzymes and secondary metabolites against pathogens. Here, for instance, volatile compounds have been found to protect plants against insects (Hammerbacher et al., 2019; Sharifi et al., 2018). In addition, induced systemic resistance (ISR) can be promoted by the rhizosphere microbiome (Pieterse et al., 2014). As a result, plant infection can cause differences in root exudates that subsequently allow for establishing of further selected ISR-inducing bacteria. The promotion of jasmonic acid signaling by root-associated bacteria and fungi, e.g. also affects aboveground herbivory (Yuan et al., 2018).

Taken together, plants can defend themselves in several ways against several stressors by recruiting different beneficial microbes using root exudates and signaling compounds like phytohormones. Soil microorganisms play a pivotal role in this recruiting of beneficial microbes as they serve as a reservoir of microbes available to the plant (Edwards et al., 2015; Rochefort et al., 2021).

6 Soil microbiomes and ecosystem function

Ecosystems, as defined above, include all different organisms that interact with each other and their environment and all physical, chemical, and biological interrelations above- and belowground (Golley, 1984). Hence, in an ecosystem, there are not only plants that interact with soil microorganisms. Geophagy, so the consumption of soil, is very common in different animal species (Mahaney and Krishnamani, 2003) and soil microbes end up in animal guts with the potential to establish and contribute to the gut microbiome. By applying manure and slurry of livestock to the fields as fertilizer, gut microbiomes of farm animals again contribute to soil microbiomes. Accordingly, the eco-holobiont concept suggests a circular microbial loop which is formed by biotic (plants and animals) and environmental (e.g. soil, water, and air) microbiomes which can impact the assembly of microbiomes of all organisms within an ecosystem (Singh et al., 2020). To understand how changes in the composition and diversity of soil microbiomes affect the interacting partners within an ecosystem, in terms of ecosystem services, one needs to take several layers of an ecosystem and trophic layers into account.

Several studies have already shown effects on biodiversity and connected changes on ecosystem functions; however, they mostly focus on a few taxa and trophic networks. Assessing dynamics and effects of different biotic and abiotic factors on all these layers of an ecosystem is very complex and involves a lot of different compartments and species and, connected to that, different expertise. Therefore, cooperation platforms combining different research topics are needed. Cooperation platforms can contain a network for exchange of ideas and connect groups working on different topics and different sites like the Asia-Pacific Biodiversity Observation Network (Takeuchi et al., 2021). The Biodiversity Exploratories (www.biodiversity-exploratories.de, funded by the German Research Foundation, DFG) in Germany take the concept of a cooperation platform even further and support a joint effort to understand effects of land use on functional biodiversity on different levels of ecosystems and to assess the impact of biodiversity changes on ecosystem processes (Fischer et al., 2010). By combining around 40 different projects involving more than 250 researchers working on standardized forest and grassland plots subjected to different forest and agricultural land use practices, this research platform covers almost all aspects of biodiversity including not only individual species but whole groups of organisms from different trophic levels. Unraveling the impact of different environmental factors on soil microbiomes and their relations to other compartments of an ecosystem using such a joint research platform is highly needed to understand the effects of changing microbial diversity on ecosystem services (Fig. 1).

7 Systems to study ecosystem services across trophic chains in a controlled environment: mesocosms

Assessing ecosystem functions and effects of the natural environment on ecosystem compartments in the field are the first step to understand how different layers of an ecosystem interact and affect each other's ability to contribute to a healthy ecosystem and its services. In a second step, to study, for instance, the effects of future climatic conditions, land use intensification or diversity loss on ecosystem functions, controlled systems in a lab environment are a good tool, given that they can be manipulated to the interests of the study and the according ecosystem parts. Several studies use microcosm or mesocosm setups to assess different aspects of ecosystems in controlled environments to analyze the effect of individual factors on soil microbiome.

A broad range of factors, like the effects of several climatic factors (Rillig et al., 2019), the effect of invasive plant species (Stefanowicz et al., 2019), the effect of silver nanoparticles (Peixoto et al., 2020), or the effect of soil fauna diversity (Postma-Blaauw et al., 2006) on soil microbiome have been studied in such setups. By using large pots or tubes, most of these studies only include specific layers of an ecosystem and therefore focus only on specific interactions, e.g. soil–plant or soil–earthworm interactions. To include several

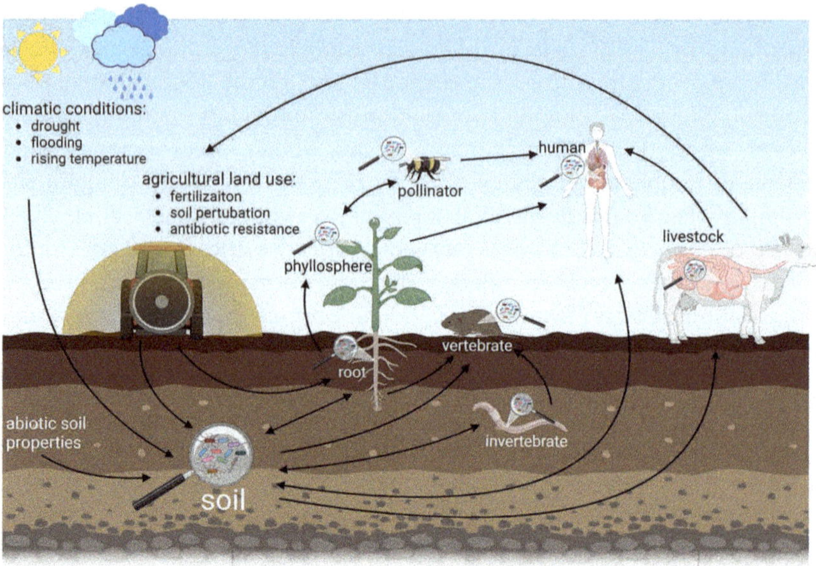

Figure 1 Biotic and abiotic factors affect soil microbial community composition and interrelations of soil microbiomes to other representative compartments within an ecosystem. The concept of the eco-holobiont suggests a circular microbial loop which is formed by biotic (plants and animals) and environmental (soil, water, air) microbiomes which can impact the assembly of microbiomes of all organisms within an ecosystem (Singh et al., 2020). Understanding how different factors alter soil microbial communities and their function is essential to unravel impacts on interrelations within the ecosystem and with that ecosystem services. Source: Figure created with BioRender.com.

layers of trophic interactions within an ecosystem, more complex setups are needed.

The iDiv Ecotron setup, led by the iDiv (German Center for integrative Biodiversity research) and Helmholtz Center for environmental research (UFZ) in Germany, builds an excellent research platform to study complex ecosystems and their functions under almost natural conditions by controlling for environmental conditions (Türke et al., 2017). This setup allows indoor multitrophic diversity research by manipulating above- and belowground trophic chains and studying the effects on diverse ecosystem functions, e.g. the effect of artificial light at night on soil communities and soil function (Cesarz et al., 2023) or the effects of aboveground predators on aboveground–belowground interactions and connected ecosystem functions (Schmidt et al., 2021).

The EcoUnits can be equipped with a broad range of ecosystem types, plant, animal, and bacterial species and a broad number of individuals thereby allowing biodiversity research from different biomes and assessing different facets of life. By controlling light, temperature, irrigation, fertilization, and species composition, the ecosystems can be manipulated to the specific factors affecting ecosystem services under study. For further information and visualization of the EcoUnits, please refer to Türke et al. (2017) and Schmidt et al. (2021).

Studies that focus on microbial biodiversity in different layers of an ecosystem, including plants, soil invertebrates, insects, and secondary consumers within a trophic chain of the ecosystem, are still lacking. Using a similar setup as the Ecotron would reveal a broader understanding of how microbiomes associated with different compartments of the trophic chain are regulated by a changing environment. Manipulation of different factors could help assessing changes and possible effects on ecosystem services.

8 Conclusion

Based on the One Health concept, which not only integrates animals, plants, and the physical environment but also humans, the interactions between different organisms and layers of the ecosystem are important. Accordingly, human well-being strongly correlates with the health of an ecosystem and its biotic and abiotic environment. Understanding the effects of different stressors on soil microbiomes and their functions within an ecosystem is necessary to maintain a sustainable and healthy environment. In this review, we discussed effects of different abiotic and biotic factors including those relevant in climate change and agricultural land use on soil microbiome diversity and its functions in complex ecosystems. In this respect, ecosystem functions of soil in plant growth promotion and stress tolerance reflects the interconnectedness of different organisms and the abiotic environment within an ecosystem.

To investigate ecosystem services and functions, different experimental setups need to be considered as provided in experimental systems such as the Biodiversity Exploratories, where different layers and various aspects of an ecosystem can be studied in a natural context. To reproduce and control such ecosystems and layers even more, we present the concept of indoor micro- or mesocosms, like the iDiv Ecotron, where different components and species of an ecosystem can be reconstructed under controlled conditions in the lab. These indoor ecosystems provide analytical platforms for further research on the role of microbiome diversity in the function of natural ecosystems and to identify the function of different ecosystem components in establishing a healthy and balanced One Health equilibrium.

9 Where to look for further information

Literature recommendations concerning the topics covered in this chapter are Fierer (2017) covering abiotic factors that drive soil microbial community composition, Hartmann and Six (2023) covering soil microbiomes in agroecosystems, Jechalke et al. (2014), who review the fate and effects of veterinary antibiotics in soil and Banerjee, and van der Heijden (2023), who describe soil microbiomes in the One Health concept. Concerning plant–soil interactions, Trivedi et al. (2020) describe several plant microbiome interactions and Eichmann et al. (2021) review plant hormones as communicators between plants and microbes.

10 References

Abrudan MI, Smakman F, Grimbergen AJ, et al. (2015) Socially mediated induction and suppression of antibiosis during bacterial coexistence. *Proceedings of the National Academy of Sciences of the United States of America* 112(35): 11054–11059.

Aliche EB, Screpanti C, De Mesmaeker A, et al. (2020) Science and application of strigolactones. *New Phytologist* 227: 1001–1011.

Almario J, Jeena G, Wunder J, et al. (2017) Root-associated fungal microbiota of nonmycorrhizal Arabis alpina and its contribution to plant phosphorus nutrition. *Proceedings of the National Academy of Sciences of the United States of America* 114(44): E9403–E9412.

Asselbergh B, De Vleesschauwer D, Höfte M (2008) Global switches and fine-tuning-ABA modulates plant pathogen defense. *709 MPMI* 21(6): 709–719.

Aust MO, Thiele-Bruhn S, Seeger J, et al. (2010) Sulfonamides leach from sandy loam soils under common agricultural practice. *Water, Air, and Soil Pollution* 211(1–4): 143–156.

Bakshi M, Vahabi K, Bhattacharya S, et al. (2015) WRKY6 restricts Piriformospora indica-stimulated and phosphate-induced root development in Arabidopsis. *BMC Plant Biology* 15(1): 305.

Banerjee S and van der Heijden MGA (2023) Soil microbiomes and one health. *Nature Reviews Microbiology* 21(1): 6–20.

Bardgett RD and Van Der Putten WH (2014) Belowground biodiversity and ecosystem functioning. *Nature* 515(7528): 505–511.

Bardgett RD (2002) Causes and consequences of biological diversity in soil. *Zoology* 105: 367–374.

Bar-On YM, Phillips R and Milo R (2018) The biomass distribution on Earth. *Proceedings of the National Academy of Sciences of the United States of America* 115(25): 6506–6511.

Berg G and Smalla K (2009) Plant species and soil type cooperatively shape the structure and function of microbial communities in the rhizosphere. *FEMS Microbiology Ecology* 68(1): 1–13.

Berg G, Rybakova D, Fischer D, et al. (2020) Microbiome definition re-visited: old concepts and new challenges. *Microbiome* 8(1): 103.

Binh CTT, Heuer H, Kaupenjohann M, et al. (2008) Piggery manure used for soil fertilization is a reservoir for transferable antibiotic resistance plasmids. *FEMS Microbiology Ecology* 66: 25–37.

Blackwell PA, Kay P, Ashauer R, et al. (2009) Effects of agricultural conditions on the leaching behaviour of veterinary antibiotics in soils. *Chemosphere* 75(1): 13–19.

Bloom AJ (2015) The increasing importance of distinguishing among plant nitrogen sources. *Current Opinion in Plant Biology* 25: 10–16.

Bodenhausen N, Somerville V, Desirò A, et al. (2019) Petunia- And Arabidopsis-specific root microbiota responses to phosphate supplementation. *Phytobiomes Journal* 3(2): 112–124.

Brockett BFT, Prescott CE and Grayston SJ (2012) Soil moisture is the major factor influencing microbial community structure and enzyme activities across seven biogeoclimatic zones in western Canada. *Soil Biology and Biochemistry* 44(1): 9–20.

Bru D, Ramette A, Saby NPA, et al. (2011) Determinants of the distribution of nitrogen-cycling microbial communities at the landscape scale. *ISME Journal* 5(3): 532–542.

Carrión VJ, Cordovez V, Tyc O, et al. (2018) Involvement of Burkholderiaceae and sulfurous volatiles in disease-suppressive soils. *ISME Journal* 12(9): 2307–2321.

Cederlund H, Wessén E, Enwall K, et al. (2014) Soil carbon quality and nitrogen fertilization structure bacterial communities with predictable responses of major bacterial phyla. *Applied Soil Ecology* 84: 62–68.

Cesarz S, Eisenhauer N, Bucher SF, et al. (2023) Artificial light at night (ALAN) causes shifts in soil communities and functions. *Philosophical transactions of the Royal Society of London. Series B, Biological sciences* 378(1892): 20220366.

Cha JY, Han S, Hong HJ, et al. (2016) Microbial and biochemical basis of a Fusarium wilt-suppressive soil. *ISME Journal* 10(1): 119–129.

Collavino MM, Sansberro PA, Mroginski LA, et al. (2010) Comparison of in vitro solubilization activity of diverse phosphate-solubilizing bacteria native to acid soil and their ability to promote Phaseolus vulgaris growth. *Biology and Fertility of Soils* 46(7): 727–738.

Compant S, Samad A, Faist H, et al. (2019) A review on the plant microbiome: ecology, functions, and emerging trends in microbial application. *Journal of Advanced Research* 19: 29–37.

de Vries FT, Hoffland E, van Eekeren N, et al. (2006) Fungal/bacterial ratios in grasslands with contrasting nitrogen management. *Soil Biology and Biochemistry* 38(8): 2092–2103.

de Vries FT, Liiri ME, Bjørnlund L, et al. (2012) Land use alters the resistance and resilience of soil food webs to drought. *Nature Climate Change* 2(4): 276–280.

Ding GC, Radl V, Schloter-Hai B, et al. (2014) Dynamics of soil bacterial communities in response to repeated application of manure containing sulfadiazine. *PLoS One* 9(3): e92958.

Edwards J, Johnson C, Santos-Medellín C, et al. (2015) Structure, variation, and assembly of the root-associated microbiomes of rice. *Proceedings of the National Academy of Sciences of the United States of America* 112(8): E911–E920.

Eichmann R, Richards L and Schäfer P (2021) Hormones as go-betweens in plant microbiome assembly. *Plant Journal* 105(2): 518–541.

Fierer N (2017) Embracing the unknown: disentangling the complexities of the soil microbiome. *Nature Reviews Microbiology* 15(10): 579–590.

Fierer N and Jackson RB (2006) The diversity and biogeography of soil bacterial communities. *Proceedings of the National Academy of Science* 103(3): 626–631.

Finzi AC, Abramoff RZ, Spiller KS, et al. (2015) Rhizosphere processes are quantitatively important components of terrestrial carbon and nutrient cycles. *Global Change Biology* 21(5): 2082–2094.

Fischer M, Bossdorf O, Gockel S, et al. (2010) Implementing large-scale and long-term functional biodiversity research: the biodiversity exploratories. *Basic and Applied Ecology* 11(6): 473–485.

Fitzpatrick CR, Copeland J, Wang PW, et al. (2018) Assembly and ecological function of the root microbiome across angiosperm plant species. *Proceedings of the National Academy of Sciences of the United States of America* 115: E1157–E1165.

Fitzpatrick CR, Mustafa Z and Viliunas J (2019) Soil microbes alter plant fitness under competition and drought. *Journal of Evolutionary Biology* 32(5): 438–450.

Frindte K, Pape R, Werner K, et al. (2019) Temperature and soil moisture control microbial community composition in an arctic–alpine ecosystem along elevational and micro-topographic gradients. *ISME Journal* 13(8): 2031–2043.

García-Salamanca A, Molina-Henares MA, van Dillewijn P, et al. (2013) Bacterial diversity in the rhizosphere of maize and the surrounding carbonate-rich bulk soil. *Microbial Biotechnology* 6(1): 36–44.

Geisseler D and Scow KM (2014) Long-term effects of mineral fertilizers on soil microorganisms: a review. *Soil Biology and Biochemistry*. 75: 54–63.

Giacometti C, Cavani L, Baldoni G, et al. (2014) Microplate-scale fluorometric soil enzyme assays as tools to assess soil quality in a long-term agricultural field experiment. *Applied Soil Ecology* 75: 80–85.

Golley, Frank B (1984) Historical Origins of the Ecosystem Concept in Biology. The ecosystem Concept in Anthropology. Edited by E.F. Moran. Pp. 33–49. Washington DC: AAAS.

Götz A and Smalla K (1997) Manure enhances plasmid mobilization and survival of pseudomonas putida introduced into field soil. *Applied and Environmental Microbiology* 63(5): 1980–1986.

Griffiths RI, Thomson BC, James P, et al. (2011) The bacterial biogeography of British soils. *Environmental Microbiology* 13(6): 1642–1654.

Guerra CA, Bardgett RD, Caon L, et al. (2021) Tracking, targeting, and conserving soil biodiversity: a monitoring and indicator system can inform policy. *Science* 371(6526): 239–241.

Guo P, Wang C, Jia Y, et al. (2011) Responses of soil microbial biomass and enzymatic activities to fertilizations of mixed inorganic and organic nitrogen at a subtropical forest in East China. *Plant and Soil* 338(1): 355–366.

Halverson LJ, Jones TM and Firestone MK (2000) Release of intracellular solutes by four soil bacteria exposed to dilution stress. *Soil Science Society of America Journal* 64(5): 1630–1637.

Hammerbacher A, Coutinho TA and Gershenzon J (2019) Roles of plant volatiles in defence against microbial pathogens and microbial exploitation of volatiles. *Plant Cell and Environment*.

Hartmann M and Six J (2023) Soil structure and microbiome functions in agroecosystems. *Nature Reviews Earth and Environment*.

Hartmann M, Frey B, Mayer J, et al. (2015) Distinct soil microbial diversity under long-term organic and conventional farming. *ISME Journal* 9(5): 1177–1194.

Hassan S and Mathesius U (2012) The role of flavonoids in root-rhizosphere signalling: Opportunities and challenges for improving plant-microbe interactions. *Journal of Experimental Botany*.

Henry S, Texier S, Hallet S, et al. (2008) Disentangling the rhizosphere effect on nitrate reducers and denitrifiers: Insight into the role of root exudates. *Environmental Microbiology* 10(11): 3082–3092.

Hernanz JL, López R, Navarrete L, et al. (2002) Long-term effects of tillage systems and rotations on soil structural stability and organic carbon stratification in semiarid central Spain. *Soil & Tillage Research*.

Hestrin R, Hammer EC, Mueller CW, et al. (2019) Synergies between mycorrhizal fungi and soil microbial communities increase plant nitrogen acquisition. *Communications Biology* 2(1).

Heuer H, Schmitt H and Smalla K (2011) Antibiotic resistance gene spread due to manure application on agricultural fields. *Current Opinion in Microbiology*.

Hiruma K, Gerlach N, Sacristán S, et al. (2016) Root Endophyte Colletotrichum tofieldiae Confers Plant Fitness Benefits that Are Phosphate Status Dependent. *Cell* 165(2): 464–474.

Hodge A and Storer K (2014) Arbuscular mycorrhiza and nitrogen: Implications for individual plants through to ecosystems. *Plant and Soil* 386(1–2): 1–19.

Hol WHG, Garbeva P, Hordijk C, et al. (2015) Non-random species loss in bacterial communities reduces antifungal volatile production. *Ecology* 96(8): 2042–2048.

Jansson JK and Hofmockel KS (2018) The soil microbiome – from metagenomics to metaphenomics. *Current Opinion in Microbiology* 43: 162–168.

Jansson JK and Hofmockel KS (2020) Soil microbiomes and climate change. *Nature Reviews Microbiology* 18: 35–46.

Jechalke S, Heuer H, Siemens J, et al. (2014) Fate and effects of veterinary antibiotics in soil. *Trends in Microbiology* 22(9): 536–545.

Jia Z and von Wirén N (2020) Signaling pathways underlying nitrogen-dependent changes in root system architecture: From model to crop species. *Journal of Experimental Botany* 71(15): 4393–4404.

Jorquera MA, Hernández MT, Rengel Z, et al. (2008) Isolation of culturable phosphobacteria with both phytate-mineralization and phosphate-solubilization activity from the rhizosphere of plants grown in a volcanic soil. *Biology and Fertility of Soils* 44(8): 1025–1034.

Kemmitt SJ, Wright D, Goulding KWT, et al. (2006) pH regulation of carbon and nitrogen dynamics in two agricultural soils. *Soil Biology and Biochemistry* 38(5): 898–911.

Kobae Y (2019) Dynamic phosphate uptake in arbuscular mycorrhizal roots under field conditions. *Frontiers in Environmental Science*.

Kodama K, Rich MK, Yoda A, et al. (2022) An ancestral function of strigolactones as symbiotic rhizosphere signals. *Nature Communications* 13(1).

König S, Vogel HJ, Harms H, et al. (2020) Physical, Chemical and Biological Effects on Soil Bacterial Dynamics in Microscale Models. *Frontiers in Ecology and Evolution*.

Kuypers MMM, Marchant HK and Kartal B (2018) The microbial nitrogen-cycling network. *Nature Reviews Microbiology* 16(5): 263–276.

Kuzyakov Y and Blagodatskaya E (2015) Microbial hotspots and hot moments in soil: Concept & review. *Soil Biology and Biochemistry* 83: 184–199.

Lauber CL, Hamady M, Knight R, et al. (2009) Pyrosequencing-based assessment of soil pH as a predictor of soil bacterial community structure at the continental scale. *Applied and Environmental Microbiology* 75(15): 5111–5120.

Lazcano C, Gómez-Brandón M, Revilla P, et al. (2013) Short-term effects of organic and inorganic fertilizers on soil microbial community structure and function: A field study with sweet corn. *Biology and Fertility of Soils* 49(6): 723–733.

Li J tian, Lu J li, Wang H yu, et al. (2021) A comprehensive synthesis unveils the mysteries of phosphate-solubilizing microbes. *Biological Reviews* 96(6): 2771–2793.

Longepierre M, Widmer F, Keller T, et al. (2021) Limited resilience of the soil microbiome to mechanical compaction within four growing seasons of agricultural management. *ISME Communications* 1(1).

Lundberg DS, Lebeis SL, Paredes SH, et al. (2012) Defining the core Arabidopsis thaliana root microbiome. *Nature* 488(7409): 86–90.

Mahaney WC and Krishnamani R (2003) Understanding Geophagy in Animals: Standard Procedures for Sampling soils. *Journal of Chemical Ecology* 29(7): 1503–1523.

Matson PA, Parton WJ, Power AG, et al. (1997) Agricultural Intensification and Ecosystem Properties. *57. R. J. Hobbs and L. Atkins, Aust. J. Ecol* 379: 504–509.

McHugh TA, Compson Z, Gestel N van, et al. (2017) Climate controls prokaryotic community composition in desert soils of the southwestern United States. *FEMS Microbiology Ecology* 93(10).

Melillo JM, Frey SD, Deangelis KM, et al. (2017) Long-term pattern and magnitude of soil carbon feedback to the climate system in a warming world. *Science* 358: 101–105.

Mendes R, Kruijt M, de Bruijn I, et al. (2011) Deciphering the rhizosphere Microbiome for Disease-Suppressive Bacteria. *Science* 332(6033): 1097–1100.

Mendes R, Garbeva P and Raaijmakers JM (2013) The rhizosphere microbiome: Significance of plant beneficial, plant pathogenic, and human pathogenic microorganisms. *FEMS Microbiology Reviews* 37(5): 634–663.

Mohanram S and Kumar P (2019) Rhizosphere microbiome: revisiting the synergy of plant-microbe interactions. *Annals of Microbiology*.

Mukherjee S, Juottonen H, Siivonen P, et al. (2014) Spatial patterns of microbial diversity and activity in an aged creosote-contaminated site. *ISME Journal* 8(10): 2131–2142.

Näsholm T, Kielland K and Ganeteg U (2009) Uptake of organic nitrogen by plants. *New Phytologist*.

Naylor D and Coleman-Derr D (2018) Drought stress and root-associated bacterial communities. *Frontiers in Plant Science*.

Nottingham AT, Scott JJ, Saltonstall K, et al. (2022) Microbial diversity declines in warmed tropical soil and respiration rise exceed predictions as communities adapt. *Nature Microbiology* 7(10): 1650–1660.

Ostermann A, Siemens J, Welp G, et al. (2013) Leaching of veterinary antibiotics in calcareous Chinese croplands. *Chemosphere* 91(7): 928–934.

Patel KF, Fansler SJ, Campbell TP, et al. (2021) Soil texture and environmental conditions influence the biogeochemical responses of soils to drought and flooding. *Communications Earth and Environment*.

Peixoto S, Khodaparast Z, Cornelis G, et al. (2020) Impact of Ag2S NPs on soil bacterial community – A terrestrial mesocosm approach. *Ecotoxicology and Environmental Safety*.

Peng Z, Liang C, Gao M, et al. (2022) The neglected role of micronutrients in predicting soil microbial structure. *npj Biofilms and Microbiomes* 8(1).

Philippot L, Hallin S, Börjesson G, et al. (2009) Biochemical cycling in the rhizosphere having an impact on global change. *Plant and Soil* 321(1–2): 61–81.

Philippot L, Raaijmakers JM, Lemanceau P, et al. (2013) Going back to the roots: The microbial ecology of the rhizosphere. *Nature Reviews Microbiology* 11(11): 789–799.

Pieterse CMJ, Zamioudis C, Berendsen RL, et al. (2014) Induced systemic resistance by beneficial microbes. *Annual Review of Phytopathology*: 347–375.

Postma-Blaauw MB, Bloem J, Faber JH, et al. (2006) Earthworm species composition affects the soil bacterial community and net nitrogen mineralization. *Pedobiologia* 50(3): 243–256.

Prober SM, Leff JW, Bates ST, et al. (2015) Plant diversity predicts beta but not alpha diversity of soil microbes across grasslands worldwide. *Ecology Letters* 18(1): 85–95.

Psallidas PG and Tsiantos J (2000) Chemical control of fire blight. In *Fire Blight, the Disease and its Causative Agent, Erwinia amylovora*, pp. 199–234, Edited by Vanneste J. L. Wallingford, UK & New York: CABI Publishing.

Richardson AE and Simpson RJ (2011) Soil microorganisms mediating phosphorus availability. *Plant Physiology* 156(3): 989–996.

Richardson AE, Barea JM, McNeill AM, et al. (2009) Acquisition of phosphorus and nitrogen in the rhizosphere and plant growth promotion by microorganisms. *Plant and Soil*.

Rillig MC, Ryo M, Lehmann A, et al. (2019) The role of multiple global change factors in driving soil functions and microbial biodiversity. *Science* 366: 886–890.

Rochefort A, Simonin M, Marais C, et al. (2021) Transmission of Seed and Soil Microbiota to Seedling. *mSystems* 6(3).

Rolfe SA, Griffiths J and Ton J (2019) Crying out for help with root exudates: adaptive mechanisms by which stressed plants assemble health-promoting soil microbiomes. *Current Opinion in Microbiology*.

Rosendahl I, Siemens J, Kindler R, et al. (2012) Persistence of the Fluoroquinolone Antibiotic Difloxacin in Soil and Lacking Effects on Nitrogen Turnover. *Journal of Environmental Quality* 41(4): 1275–1283.

Rousk J, Bååth E, Brookes PC, et al. (2010) Soil bacterial and fungal communities across a pH gradient in an arable soil. *ISME Journal* 4(10): 1340–1351.

Saleem M, Law AD, Sahib MR, et al. (2018) Impact of root system architecture on rhizosphere and root microbiome. *Rhizosphere* 6: 47–51.

Schäffer B, Stauber M, Mueller TL, et al. (2008) Soil and macro-pores under uniaxial compression. I. Mechanical stability of repacked soil and deformation of different types of macro-pores. *Geoderma* 146(1–2): 183–191.

Schimel JP (2018) Life in dry soils: effects of drought on soil microbial communities and processes. *Annual Review of Ecology, Evolution, and Systematics* 12: 409–432.

Schmidt A, Hines J, Türke M, et al. (2021) The iDiv ecotron–a flexible research platform for multitrophic biodiversity research. *Ecology and Evolution* 11(21): 15174–15190.

Schreiter S, Ding GC, Heuer H, et al. (2014) Effect of the soil type on the microbiome in the rhizosphere of field-grown lettuce. *Frontiers in Microbiology* 5(144):144.

Serna-Chavez HM, Fierer N and Van Bodegom PM (2013) Global drivers and patterns of microbial abundance in soil. *Global Ecology and Biogeography* 22(10): 1162–1172.

Shannon, Claude E., & Weaver, Warren (1949) The mathematical theory of communication, by CE Shannon (and recent contributions to the mathematical theory of communication), W. Weaver. University of illinois Press.

Sharifi R, Lee SM and Ryu CM (2018) Microbe-induced plant volatiles. *New Phytologist* 220(3), 684–691.

Singh BK, Liu H and Trivedi P (2020) Eco-holobiont: a new concept to identify drivers of host-associated microorganisms. *Environmental Microbiology* 22(2): 564–567.

Singh JS and Gupta VK (2018) Soil microbial biomass: A key soil driver in management of ecosystem functioning. *Science of the Total Environment* 634: 497–500.

Six J, Conant RT, Paul EA, et al. (2002) Stabilization mechanisms of soil organic matter: Implications for C-saturation of soils. *Plant and Soil* 241: 155–176.

Six J, Frey SD, Thiet RK, et al. (2006) Bacterial and fungal contributions to carbon sequestration in agroecosystems. *Soil Science Society of America Journal* 70(2): 555–569.

Smalla K, Heuer H, Götz A, et al. (2000) Exogenous Isolation of Antibiotic Resistance Plasmids from Piggery Manure Slurries Reveals a High Prevalence and Diversity of IncQ-Like Plasmids. *Applied and Environmental Microbiology* 66(11): 4854–62.

Stefanowicz AM, Stanek M, Majewska ML, et al. (2019) Invasive plant species identity affects soil microbial communities in a mesocosm experiment. *Applied Soil Ecology* 136: 168–177.

Sul WJ, Asuming-Brempong S, Wang Q, et al. (2013) Tropical agricultural land management influences on soil microbial communities through its effect on soil organic carbon. *Soil Biology and Biochemistry* 65: 33–38.

Takeuchi Y, Muraoka H, Yamakita T, et al. (2021) The Asia-Pacific Biodiversity Observation Network: 10-year achievements and new strategies to 2030. *Ecological Research* 36(2): 232–257.

Tan W, Wang J, Bai W, et al. (2020) Soil bacterial diversity correlates with precipitation and soil pH in long-term maize cropping systems. *Scientific Reports* 10(1): Article 6012.

Tecon R and Or D (2017) Biophysical processes supporting the diversity of microbial life in soil. *FEMS Microbiology Reviews* 41(5): 599–623.

Trivedi P, Leach JE, Tringe SG, et al. (2020) Plant–microbiome interactions: from community assembly to plant health. *Nature Reviews Microbiology* 24: e592–e602.

Türke M, Feldmann R, Fürst B, et al. (2017) Multitrophische Biodiversitätsmanipulation unter kontrollierten Umweltbedingungen im iDiv Ecotron.

Vandenkoornhuyse P, Baldauf SL, Leyval C, et al. (2002) Extensive fungal diversity in plant roots. *Science* 295(5562): 2051.

Van Der Heijden MGA, Bardgett RD and Van Straalen NM (2008) The unseen majority: soil microbes as drivers of plant diversity and productivity in terrestrial ecosystems. *Ecology Letters* 11(3): 296–310.

Wagg C, Bender SF, Widmer F, et al. (2014) Soil biodiversity and soil community composition determine ecosystem multifunctionality. *Proceedings of the National Academy of Sciences of the United States of America* 111(14): 5266–5270.

Waldrop MP and Firestone MK (2006) Response of microbial community composition and function to soil climate change. *Microbial Ecology* 52(4): 716–724.

Wall DH, Nielsen UN and Six J (2015) Soil biodiversity and human health. *Nature* 528(7580): 69–76.

Wolters B, Kyselková M, Krögerrecklenfort E, et al. (2014) Transferable antibiotic resistance plasmids 1 from biogas plant digestates often belong to the IncP-1ε subgroup. *Frontiers in Microbiology* 5: 1–11.

Xiong W, Jousset A, Li R, et al. (2021) A global overview of the trophic structure within microbiomes across ecosystems. *Environment International* 151: 106438.

Xu L, Naylor D, Dong Z, et al. (2018) Drought delays development of the sorghum root microbiome and enriches for monoderm bacteria. *Proceedings of the National Academy of Sciences of the United States of America* 115(18): E4284–E4293.

Young IM and Ritz K (2000) Tillage, habitat space and function of soil microbes. *Soil & Tillage Research* 53: 201–213.

Yuan J, Zhao J, Wen T, et al. (2018) Root exudates drive the soil-borne legacy of aboveground pathogen infection. *Microbiome* 6(1): 1–12.

Zarraonaindia I, Owens SM, Weisenhorn P, et al. (2015) The soil microbiome influences grapevine-associated microbiota. *mBio* 6(2): 1–10.

Zhang B, Chen S, He X, et al. (2014) Responses of soil microbial communities to experimental warming in alpine grasslands on the Qinghai-Tibet Plateau. *PLoS One* 9(8).

Part 4

Soil microbiomes and ecosystem services

Chapter 9

Advances in understanding the role of soil microbiomes in carbon cycling in soil

Alain F. Plante, University of Pennsylvania, USA; and Jennifer Pett-Ridge, Lawrence Livermore National Laboratory, USA

1 Introduction

Microorganisms drive carbon transformation in soils, the largest active terrestrial carbon reservoir, and they shape both soil carbon loss and persistence, the dynamic process by which organic molecules are retained over long timescales. The diverse microorganisms that comprise soil microbiomes execute a myriad of biogeochemical functions. At large scales, decomposition of organic matter is one of the most important ecological functions of soil microbiomes because it closes the loops of many terrestrial biogeochemical cycles, including carbon, which begin with fixation and photosynthesis to bring elements in mineral form into organic forms in the biosphere. In this chapter, the author focuses primarily on catabolic heterotrophic respiration, which results in soil organic matter decomposition and carbon mineralization. The microbial decomposition of organic matter and its conversion to CO_2 and other metabolites is a biogeochemical function shared among a wide range of phylogenic groups (e.g. the orange ring in Fig. 1). Because such a wide phylogenetic range of soil microbiomes participates in carbon cycling in soil, the diversity of organisms is sometimes of secondary importance compared to the total number, mass or volume of microbes, and their rates of activity. For this

http://dx.doi.org/10.19103/AS.2024.0136.21

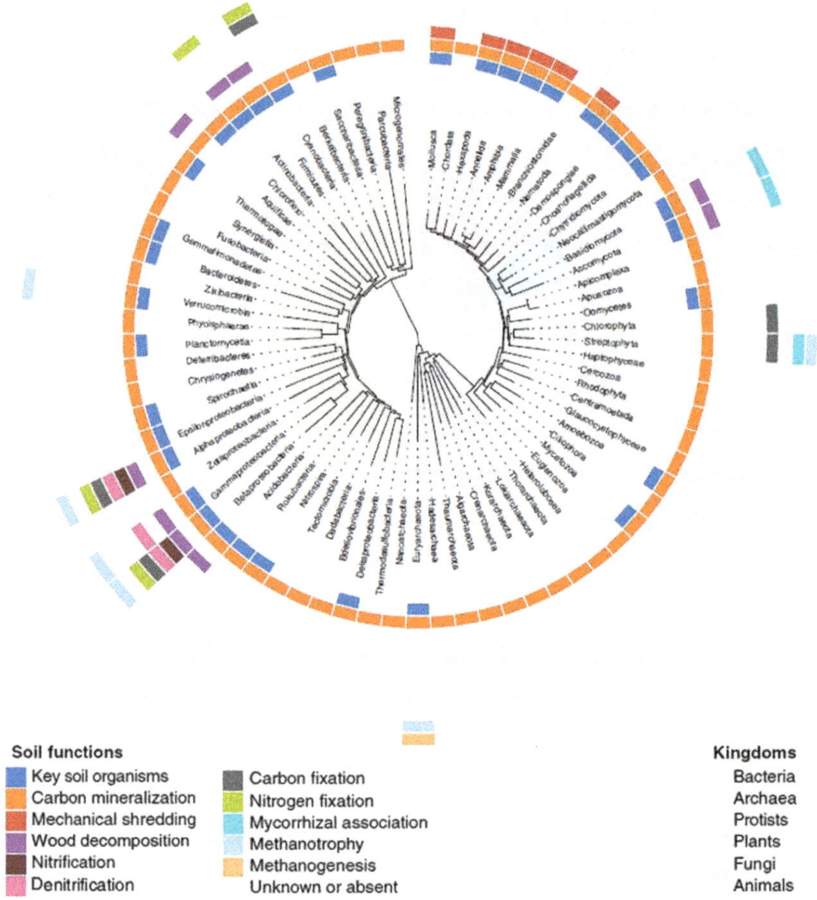

Figure 1 The distribution of common soil microbiome functions (outer rings) across phylogenetic classes of soil organisms (inner ring and tree). The orange blocks in the outer rings illustrate the wide range of phylogenetic groups involved in soil carbon cycling. Source: Crowther et al. (2019), with permission.

reason, soil microbiologists and biogeochemists refer to the soil microbiomes as 'soil microbial biomass' when speaking from the functional biogeochemical context. The term 'soil microbial biomass' was first coined by David Jenkinson in a 1966 paper in the then *Journal of Soil Science* (later to become the *European Journal of Soil Science*) titled 'Studies on the decomposition of plant materials in soil. II. Partial sterilization of soil and the soil microbial biomass' (Jenkinson, 1966). Even then, before the -omics revolution, the focus of soil microbiology was the identification of organisms responsible for specific soil processes.

The concept of quantifying properties of the soil microbial population as a whole was unusual. However, the methods developed by Jenkinson, Powlson, Brookes, Vance, and others proved revolutionary (Tate, 2017). The lens created by the soil microbial biomass approach linked soil microbiology to biogeochemistry and illustrated the varied roles of soil microbiomes in organic matter decomposition and formation in soils.

Our understanding of the storage and cycling of carbon in soil has undergone a recent paradigm shift (Schmidt et al., 2011). Once, the dominant view was that the selective preservation of remnants of plant materials in various stages of decomposition contributed to the abiotic synthesis of complex molecules (i.e. humic and fulvic acids and humin) that constituted the majority of persistent soil organic matter. The new paradigm places the soil microbiome as the central agent in both decomposition of incoming organic matter (and thus carbon cycling) as well direct contributors to persistent organic matter (and thus carbon storage). As such, soil microbiome contributions to biochemical processes extend beyond the metabolic processes of the living biomass to dead biomass that serves as both a substrate for subsequent microbial growth and metabolism, as well as the formation and accumulation of persistent soil organic carbon residues (Sokol et al., 2022).

2 Dead soil microbiomes and organic carbon storage

In recent decades, there has been a growing appreciation for the role of soil microbiomes as our understanding of the processes involved in the formation of soil organic matter has evolved. Soil microbiomes have been predominantly viewed as passive agents of catabolic transformation of incoming organic materials (see Section 3), but recent evidence suggests that incorporation of microbial biomass components into soil organic matter is substantial. Microbial inputs may be more important to the formation of persistent mineral-associated soil organic matter than previously thought under the traditional paradigm, particularly if microbial inputs have a greater propensity to be stabilized than are plant-derived inputs (Liang et al., 2020). The mechanism by which microbially synthesized organic matter is transformed into stabilized or persistent soil organic matter was conceptualized by Liang et al. (2017) as the 'microbial carbon pump' (MCP), where generations of microorganisms cycle energy and matter (e.g. carbon), store them in their biomass via anabolism, and turnover organic substrates via cell uptake–biosynthesis–growth–death, resulting in the deposition of microbial-derived compounds in the soil (Fig. 2). While short-term direct microbial contributions to soil organic matter are minimal because the living microbial biomass generally makes up less than 5% of the total soil carbon pool, this does not reflect the total amount of carbon that has, at some point, cycled through the soil microbiome, or the accumulation of dead microbial

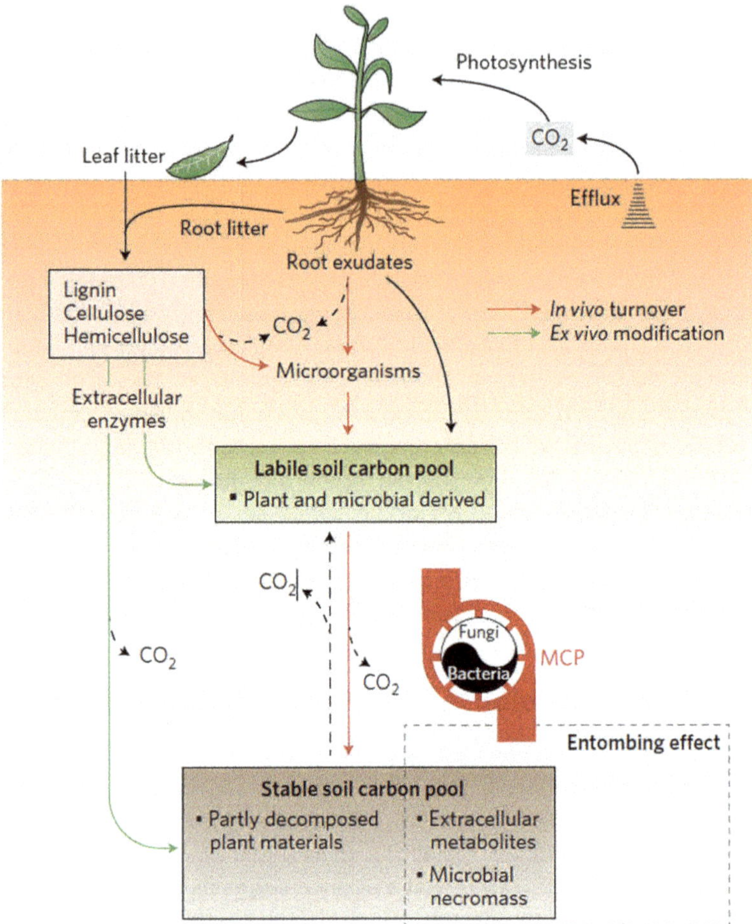

Figure 2 Microbial metabolic processes involved in the conversion of primary production inputs to soil organic matter. *In vivo* transformations involving microbial assimilation of organic materials (red lines) are coupled to the soil MCP to demonstrate that microbial necromass and metabolites are the precursors to persistent soil C. Source: Liang et al. (2017), with permission.

material over longer time scales. The accurate accounting of plant versus microbially derived organic matter is important because the two pathways imply different controls on soil organic matter formation and persistence. On the one hand, plant-dominated soil organic matter would imply formation controlled by microbial traits regulating extracellular depolymerization or persistence of plant-derived compounds independent of microbial transformation. On the other hand, microbially dominated soil organic matter would imply formation regulated by microbial turnover and transformation of plant material by assimilation and anabolism, in addition to extracellular depolymerization.

Current estimates for contributions to soil organic matter composition indicate that microbial necromass is a direct source of 30–80% of soil organic matter, while plant-derived material can comprise 20–70% of soil organic matter (Angst et al., 2021; Khan et al., 2016; Liang et al., 2019; Simpson et al., 2007). The majority of the evidence for the microbial contributions to soil organic matter is qualitative or from semi-quantitative biomarker and molecular fingerprinting, specifically in the non-particulate (plant dominated) mineral associated pool. The first line of evidence comes from stoichiometry, specifically carbon to nutrient ratios. Soil organic matter stoichiometry is typically nutrient enriched, and thus more similar to microbial stoichiometry than plant material stoichiometry, with reported C:N:P stoichiometry of 72:61:1 (Cleveland and Liptzin, 2007), 52:5:1 (Kirkby et al., 2011), or 61:7:1 (Tipping et al., 2016). Other lines of qualitative evidence include isotopic (e.g. $\delta^{13}C$ and $\delta^{15}N$) and advanced molecular (e.g. nuclear magnetic resonance spectroscopy) analyses demonstrating strong similarities between soil organic matter and microbial biomass, and imaging showing the close association of microbial cells with mineral surfaces (Whalen et al., 2022).

Analyses of specific indicator compounds such as proteins, DNA/RNA, or cell well and membrane constituents such as phospholipid fatty acid and fatty acid methyl esters have identified and quantified microbially derived compounds in soil organic matter. The analysis of amino sugars in particular has been used to determine microbial contributions. Amino sugars, which occur as key components of microbial cell walls while not being synthesized by plants to any significant degree, have emerged as the most widespread biomarker approach.

To assess the microbial contribution to soil organic matter in global temperate agricultural, grassland and forest ecosystems, Liang et al. (2019) used published amino sugar biomarker data combined with stoichiometry and scaling calculations. Previous work has suggested that soil fungi produce the majority of glucosamine while bacteria produce muramic acid. As illustrated in Fig. 3, bacterial necromass carbon was estimated as the measured muramic acid concentration times 45 (a conversion factor from muramic acid to bacterial necromass C) and bacterial necromass nitrogen was estimated as the measured muramic acid concentration times 6.67 (a similar conversion factor). Similarly, fungal necromass carbon was calculated by estimating fungal glucosamine by subtracting bacterial glucosamine from total glucosamine, assuming that muramic acid and glucosamine on average occur at a 1-to-2 molar ratio in bacterial cells, and multiplying the concentration of fungal glucosamine by a conversion factor of 9. Summing the bacterial and fungal contributions, the study found that microbial necromass contributed 56% to total soil carbon in temperate agricultural soils (Fig. 4a). Using the same calculation methods, but a different published data set, Wang et al. (2021) found that microbial necromass

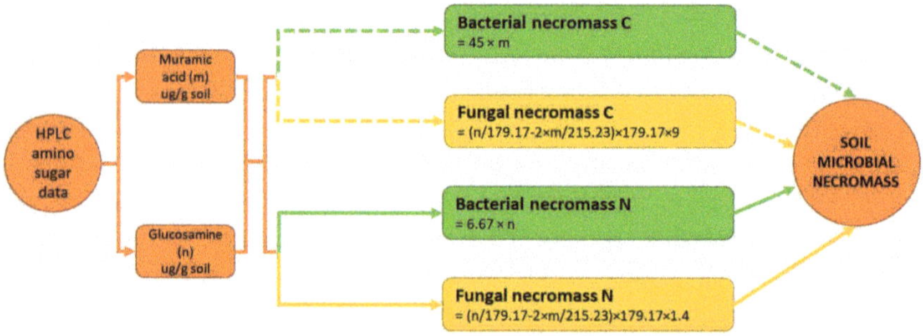

Figure 3 Framework for the quantification of microbial necromass contribution to soil organic matter using measured concentrations of amino sugars, where MurA represents muramic acid and GluN represents glucosamine. Source: Liang et al. (2019), with permission.

C contributed 51% in cropland topsoils (Fig. 4b). They also found that the contribution of microbial necromass to soil organic carbon decreased in croplands (from 51% to 24%), and that fungal necromass carbon had consistently greater contributions than did bacterial necromass. In a similar meta-analysis using the same measurements and calculations, but focused more specifically on agricultural soils, Zhou et al. (2023) examined how cropland management practices affected microbial necromass accumulation. They found that nitrogen fertilization, cover cropping, reduced tillage, manuring, and straw amendment all increased total microbial necromass (Fig. 4c), and that the necromass contribution per unit of added carbon ranged from 45% to 52%. These findings suggest that even in organic matter depleted and intensively managed soils, microbial necromass plays an important role in organic matter storage, and thus potentially in soil health (see Section 5).

While the evidence for the contribution of dead microbial material to persistent soil organic matter has been sufficiently compelling to initiate a paradigm shift in our understanding of the mechanisms of soil organic matter formation and stabilization, it remains contentious and incomplete (Whalen et al., 2022). Potential errors introduced by the approach using amino sugars include the assumption that all microbial necromass carbon is retained in soils to the same extent and at the same rate, conversion factors account for only the cellular debris contributions without including extracellular products of microbes, and finally plant derived contributions are assumed to be the difference between total and microbial derived organic matter rather than independent quantification. More generally, there are substantial challenges to ascribing a plant or microbial origin to many compounds found in soil organic matter. For instance, while lignin is exclusively plant derived, many phenolic and aromatic compounds that are chemically similar may derive from either plant or microbial sources. Similarly, galactose and short-chain lipids are assumed to

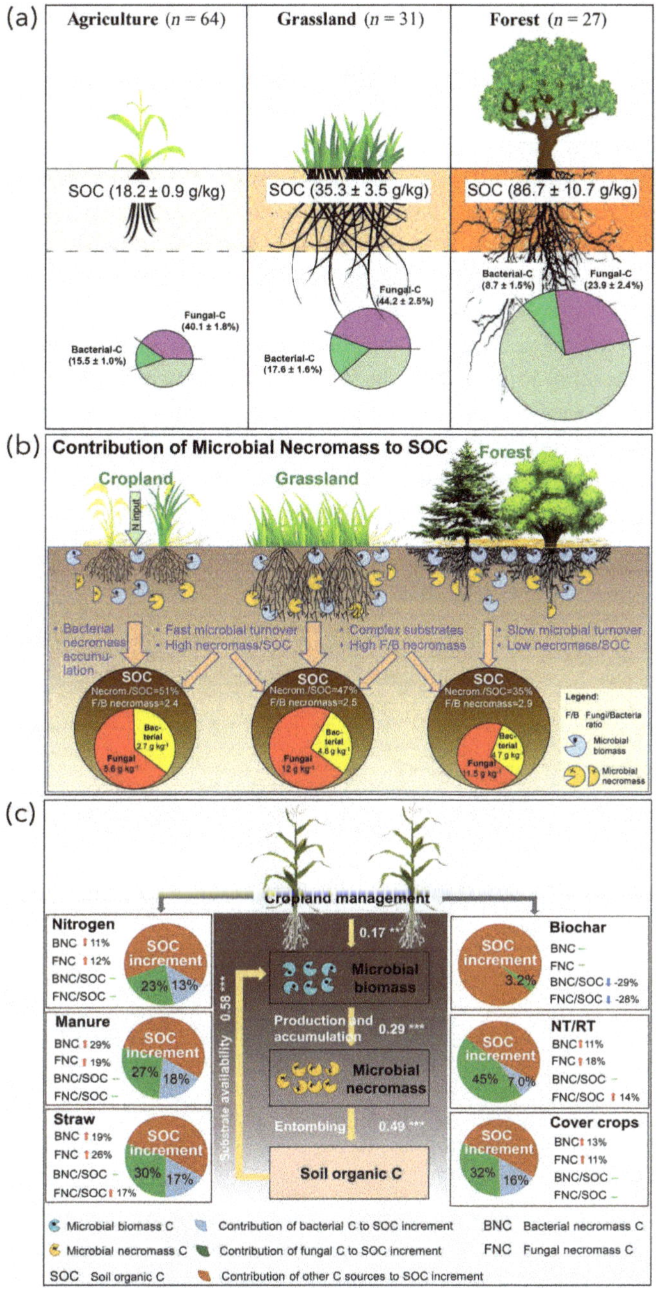

Figure 4 Estimates of microbial necromass contributions to soil organic matter across ecosystem types. Sources: (a) Liang et al. (2019), (b) Wang et al. (2021), and (c) Zhou et al. (2023), with permissions.

be primarily of microbial origin, but may be sourced from plant root exudates (Fig. 5).

In addition to uncertainty about the quantification of microbial versus plant derived contributions to soil organic matter storage and cycling, there are open questions about microbial death rates in soil and the degree to which the mechanisms by which microbes actually die matters to the rate and extent to which microbiomes contribute to organic matter persistence (Camenzind et al., 2023; Sokol et al., 2022). The mechanisms of microbial death in soil vary from predation by grazing protists or bacteriovores, cell lysis due to viral infection or osmotic shock, cell desiccation, and anti-microbial chemicals (Fig. 6). Successional processes that result in changes at the population and community levels may alter the dominant functional traits of the microbiomes, which in turn will have consequences on soil carbon cycling by the surviving microbes (see Section 3). Microbial death pathways may also have differential effects on the fate of dead microbial cell materials, as necromass may become more or less available for decomposition by the surviving organisms. For instance, Sokol et al.

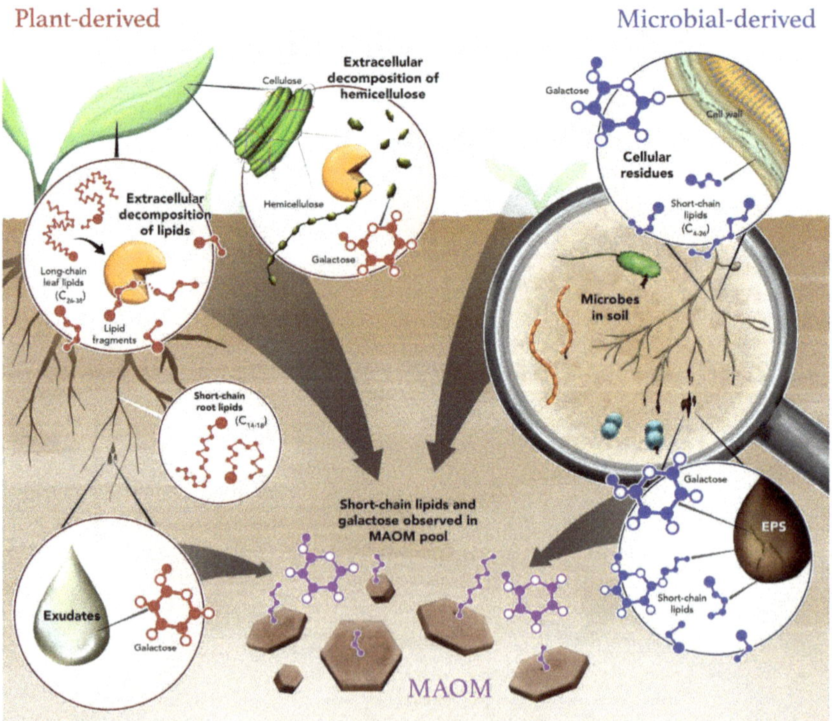

Figure 5 Challenges to the attribution of molecular markers to plant- (red) or microbial- (blue) derived sources. Galactose and short-chain lipids are typically assumed to be microbial derived; however, both can also derive from plant materials, creating the potential for mixed (purple) pools. Source: Whalen et al. (2022), with permission.

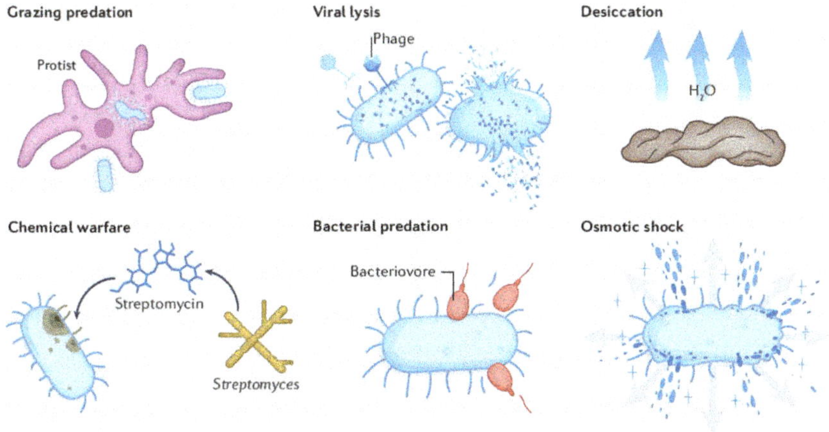

Figure 6 Various microbial death pathways in soil. Death pathways have direct consequences for the fate of microbial necromass, and thus for soil carbon cycling and storage. Source: Sokol et al. (2022), with permission.

(2022) describe how predation may result in the assimilation of necromass into a higher trophic level with resulting inefficiencies of trophic transfer. Conversely, death by cell lysis may result in the release of relatively labile organic matter derived from cell cytoplasms that may be subject to rapid mineralization to CO_2 (i.e. the so-called Birch effect) or may be subject to mineral association through molecular fractionation by soil minerals and therefore stabilized against further decomposition, thus resulting in the accumulation of necromass as described earlier.

3 Living soil microbiomes and organic carbon cycling

In addition to its contributions to soil organic matter formation, storage and stabilization through the formation of necromass, the soil microbiome plays an essential role in soil carbon cycling and loss. Rates of active soil carbon cycling (i.e. transformations in the chemical composition of soil organic matter) and soil CO_2 efflux (the ultimate product of heterotrophic carbon mineralization) are a function of variability in the amount and nature of the substrate, but also the size and metabolic activity of the microbiome. David Jenkinson (1977) described the soil microbial biomass as 'the eye of the needle through which all the organic materials must pass'. Soil microbiomes regulate carbon cycling through both catabolic and anabolic processes, modulated by metabolic properties and processes such as life history strategy, carbon use efficiency (CUE), and priming.

As demonstrated previously in Fig. 1, there is tremendous diversity in the soil microbiome participating in carbon mineralization. One organizing principle for describing the complex interactions between soil microbiomes and active soil carbon cycling has been adapted from the ecology of larger organisms – r–K selection theory. In response to selective pressures associated with the quantity and quality of substrates and nutrients, soil microbiomes can be categorized as copiotrophy (r-strategists) and oligotrophy (K-strategists). Copiotrophs are adapted to maximize their intrinsic growth rate when resources are abundant, and thus preferentially consume labile soil organic carbon pools and have high nutritional requirements. Conversely, oligotrophs are adapted to compete when resources are scarce, and thus exhibit slower growth rates and likely outcompete copiotrophs in low nutrient conditions (Fierer et al., 2007). Using a set of incubation experiments with added labile carbon (e.g. sucrose), Fierer et al. (2007) were able to make specific predictions about the ecological attributes and growth rates of specific bacterial taxa. In a similar set of experiments, Stone et al. (2023) concluded that a small proportion of the bacterial population demonstrated copiotrophy. Rather than a mutually exclusive dichotomy, these ecological categories are a continuum representing the trade-off between maintenance energy, growth efficiency, and growth rate (Stone et al., 2023). In soils, the conditions that favor copiotrophy versus oligotrophy (or vice versa) vary both in space (i.e. rhizosphere soil vs. non-rhizosphere soil) and time (i.e. through the span of a growing season).

Only a portion of the active soil carbon pool is taken up by the soil microbiome, retained as biomass, and thus potentially becomes necromass upon cell death and transformed into stabilized soil organic matter. The remainder is mineralized to CO_2 or exuded back into the soil environment as metabolites or other compounds. Microbial CUE is defined as the proportion of carbon substrates that the soil microbiome incorporates into biomass during microbial growth, or the fraction of carbon retained versus respired (Fig. 7). Values can range from 0 to 1 theoretically. A theorized and observed upper limit below 1 reflects that perfect conversion of substrate to biomass is impossible due to stoichiometric constraints, metabolic energy costs and thermodynamic inefficiencies (Sinsabaugh et al., 2013). CUE can be measured and expressed at the microbial population, community or ecosystem level (Geyer et al., 2016). High CUE values (i.e. efficient use of substrates) indicate anabolic, biomass-building processes, while low CUE values (i.e. inefficient substrate use) indicate catabolic, energy-generating processes. CUE estimates for soil vary widely depending on both the methods used to determine them (Geyer et al., 2019) and metabolic and environmental conditions. Community- and ecosystem-level estimates of CUE best represent the capacity of a diverse soil microbiomes to utilize substrates. Observed values range from 0.1 to 0.8 (Dijkstra et al., 2011; Frey et al., 2013; Geyer et al., 2016), with mean values around 0.5–0.55 (Manzoni et al., 2012; Qiao et al., 2019). Soil microbial CUE values vary in response to both substrate quantity and quality. Limited supply of substrate or

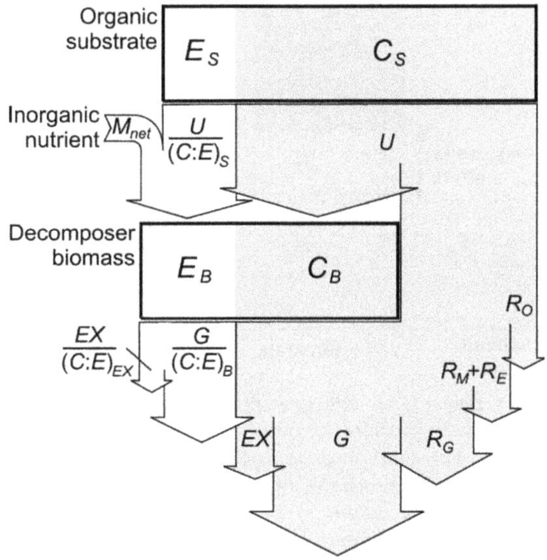

Figure 7 Mass balance of carbon (shaded portion) during substrate (C_s) uptake (U) and partitioning between microbial biomass (C_B), microbial exudation (EX), growth (G), and respiration (R_G, R_M, R_E, and R_O). Carbon utilization efficiency is defined as the proportion of substrate taken up and used for biomass, and is thus represent by U divided by G. Source: Manzoni et al. (2012), with permission.

low-quality substrate (i.e. relatively oxidized or low chemical energy content) will be mostly respired, resulting in low CUE. As substrate supply and energy content increase, CUE will increase. Qiao et al. (2019) observed differences in CUE estimates among ecosystems (0.41 in forests, 0.65 in grasslands, and 0.73 in shrublands) that reflected the complex interactions between substrate quantity, substrate quality, and metabolic diversity of the soil microbiomes. In terms of net soil carbon storage, it is hypothesized that high CUE values result in greater soil organic matter production via biomass and necromass accumulation, while low CUE values result in greater carbon mineralization and CO_2 production, and decreased soil organic matter formation or potentially net soil organic matter loss.

For over 100 years, experiments have shown an increase in soil-derived carbon mineralization (i.e. CO_2 efflux) in soils amended with fresh organic matter or nutrients compared to unamended soils (Fig. 8), a phenomenon called 'priming'. The priming effect is defined as a short-term change in the soil organic matter turnover rate caused by comparatively moderate changes in substrate or nutrient supply, such as input of organic or mineral fertilizer, root exudation, mechanical treatment of the soil (e.g. tillage), or wetting and drying. While priming was once suspected to be an artifact of experimental laboratory incubation conditions, pulse and continuous additions of fresh

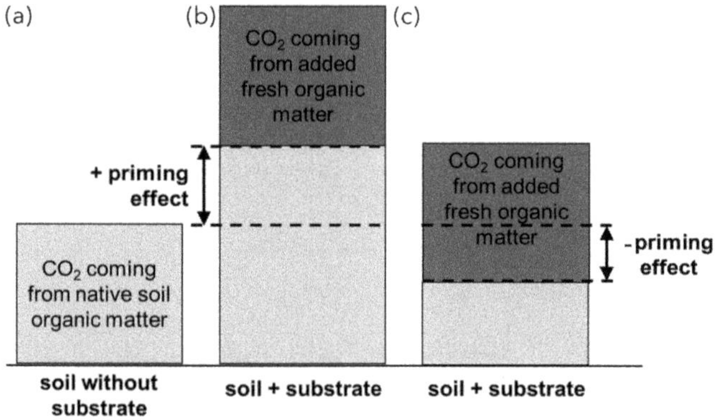

Figure 8 Soil mineralization (CO_2 efflux) in three experimental scenarios to illustrate the priming effect: (a) decomposition of native soil organic matter with no addition of fresh organic matter, (b) acceleration of soil organic matter decomposition in addition to decomposition of added fresh organic matter – positive priming, and (c) suppressed decomposition of soil organic matter in addition to decomposition of added fresh organic matter – negative priming. Source: Redrawn from Kuzyakov et al. (2000).

organic matter in natural soils are increasingly seen as important for how the soil microbiome regulates active carbon cycling, and it is now generally known that priming occurs immediately or very shortly after the addition/disturbance, priming effects are larger with increasing soil carbon concentrations and with increasing amount of added organic matter, and that priming is not observed under sterile conditions (Kuzyakov, 2010; Kuzyakov et al., 2000). The most frequently observed priming effect, termed positive priming, is an acceleration of native soil organic matter mineralization and CO_2 production in response to fresh organic matter addition, which may result in an overall net loss of soil carbon (Fig. 8b). However, priming can also be negative, where carbon mineralization is suppressed by immobilization of the added carbon and nutrients in the microbial biomass followed by necromass recycling (Fig. 8c). In addition, researchers make a distinction between apparent and real priming effects. Apparent priming effects are attributable to increases in CO_2 released from the increased turnover and mineralization of new microbial biomass, while real priming effects are the result of CO_2 released from the mineralization of native soil organic matter.

The complex patterns of decomposition and soil CO_2 efflux during litter decomposition or in response to pulses in carbon or nutrient availability represent multiple interactions between the soil microbiome and substrate quality. Multiple mechanisms have been proposed to explain the priming effect phenomenon (Bernard et al., 2022; Blagodatskaya and Kuzyakov, 2008; Fontaine et al., 2003; Kuzyakov et al., 2000), which may vary depending on the addition/disturbance in question, and bring together r–K selection theory and

CUE. Positive apparent priming may be caused by an acceleration of internal microbial metabolism or the acceleration of microbial turnover, triggered by added substrate. Fontaine et al. (2003) proposed two primary mechanisms for positive real priming. Mechanism 1 involves enhanced decomposition of soil organic matter by enzymes produced by r–strategists intended for the decomposition of the added organic matter, while Mechanism 2 suggests that K-strategists may benefits from polymerized substrates derived from added soil organic matter and thus grow and contribute to enhanced decomposition of native soil organic matter. A general explanation for priming effects invokes the energy supply of substrates for heterotrophic respiration via extracellular enzymes or catabolism. During negative real priming, the existing soil microbiome is presumed to switch from low-energy containing soil organic matter to more energy-rich fresh/added organic matter. Conversely, in positive real priming, the added organic matter provides a new energy source to the existing microbiome to allow it to undertake additional decomposition of the low-energy soil organic matter, which is commonly referred to as co-metabolism. Others have suggested that fresh inputs of organic matter, particularly organic acids exuded in the rhizosphere may enhance destabilization of soil organic matter from mineral surfaces, thus making this substrate newly available to decomposers (Keiluweit et al., 2015). While attributing priming effects to specific members of the soil microbiome suffers from the same challenges as the r-K selection theory, saprophytic fungi are believed to play a larger role in co-metabolism and priming than do bacteria, primarily through the stoichiometric balancing process of nitrogen mining (Fontaine et al., 2011).

Experimental evidence for the priming effect comes mostly from short-term laboratory incubations using additions of labile organic matter (e.g. sugars or plant residues) or less so from long-term field experiments manipulating litter addition rates such as the 'DIRT' experiments (Lajtha et al., 2018), but the large-scale and long-term implications of priming on soil carbon cycling are currently underexplored. Generally, climate change and rising atmospheric CO_2 concentration are projected to increase primary productivity and thus litter inputs to soil. Soil carbon cycling responses are believed to result in soil as a net sink of carbon due to a positive balance between increasing inputs versus increasing heterotrophic respiration due to temperature increase. However, accounting for priming challenges, this standard response because increases in fresh organic carbon inputs may generate positive real priming effects. For this reason, priming effects and the associated r–K strategist and CUE phenomena are increasingly being incorporated into models of soil carbon cycling to improve projections of soil carbon balances to future disturbances.

4 Modeling the role of soil microbiomes in soil carbon cycling

Models of soil carbon cycling are constructed, tested, and used to both represent our best conceptual understanding of the structure of the system (e.g. actors, agents, process, and rates), as well as to quantitatively predict the contributions of soil to global carbon cycling, which has implications for global climate change. At first glance, many mathematical and compartmental models used to describe and predict soil carbon dynamics do not account for soil microbiomes. While the first large-scale models of soil carbon cycling (RothC and CENTURY) included a pool identified as soil microbial biomass, this pool was largely operational and kinetically defined, such that the microbial biomass was lumped with actively cycling (i.e. labile) carbon. The soil microbiome is implicitly represented as a transfer function where microbial activity (i.e. decomposition or transformation of carbon from fast to slow cycling pools) is a linear function of the substrate (fast) pool size modified by environmental conditions (Fig. 9a). While these implicit first-order kinetics elide the soil microbiome, they are consistent with laboratory and field measurements. First-order kinetics are, in fact, a simplification under certain conditions of microbially explicit kinetics such as Michaelis–Menton or Monod kinetics (Parton et al., 2015).

However, new generations of soil organic matter dynamics models must do better than linear first-order models at representing the soil microbiome as the central actor in soil carbon cycling, and at being more responsive and dynamic (Schimel, 2023). The simplest microbially explicit models represent substrate and microbial biomass in separate pools linked by more complex, non-linear functions (Fig. 9b). More complex, microbially explicit models represent varying degrees of microbial diversity and their varying responses to environmental conditions. Some microbially explicit models now include microbial diversity by defining one or multiple distinct pools for microbial biomass based on functional traits. For example, Moorhead and Sinsabaugh (2006) represent the soil microbial biomass in three guilds (opportunists, decomposers, miners) each with their own distinct pools and kinetic parameters. The degree to which the vast genetic diversity of the soil microbiome can be meaningfully represented and adequately parameterized in models of soil carbon cycling remains unclear, and poses challenges to integrating modeling, conceptual, and measurable components (Blankinship et al., 2018). Alternatively, other microbially explicit models allow variation in functional traits representative of the physiology of the microbial biomass, such as stoichiometry (i.e. elemental ratios, C/N/P) or carbon use or growth efficiency, by modulating the fluxes among labile and more stable pools of organic matter. Parameter values for these functional traits can also vary in response to changes in environmental conditions such as warming temperatures. An example of new, microbially explicit models is the

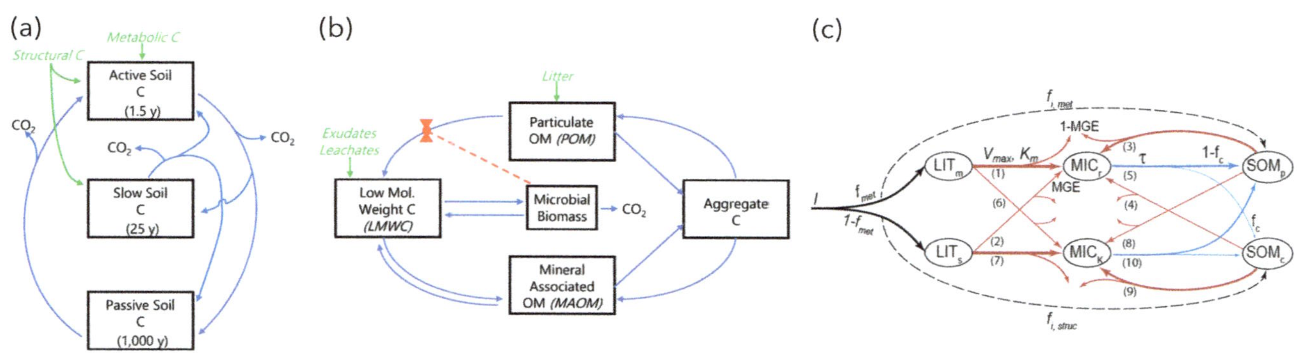

Figure 9 Soil carbon cycling model structures. (a) The CENTURY model is representative of conventional first-order kinetic models where microbial dynamics are implicit. (b) The Millennial model includes an explicit model pool for microbial biomass. (c) The MIMICS model represents the microbial biomass in two pools based on functional traits and their activity using Michaelis–Menten kinetics. Sources: (a, b) Schimel (2023), (c) Wieder et al. (2014), with permissions.

microbial-mineral carbon stabilization (MIMICS) model developed by Wieder et al. (2013, 2014), which implements both multiple pools and flux modulators to represent many of the soil microbiome properties and processes in soil carbon cycling mentioned in the previous sections (Fig. 9c). The microbial biomass is represented by two model pools distinguishing copiotrophic (r) and oligotrophic (K) functional growth strategies. The decomposition of litter and cycling of soil carbon are regulated by temperature-sensitive Michaelis–Menton kinetics. Microbial respiration is temperature sensitive and partitioned by microbial growth efficiency, which declines linearly with temperature. Microbial mortality is represented by a turnover rate and converts a portion of the dead microbial biomass into stabilized soil organic matter. When compared to conventional soil carbon cycling models, the MIMICS model better matched observations of soil carbon cycling and emissions, and better matched the responses to experimental warming, illustrating the importance of including characteristics of the soil microbiomes explicitly. The primary challenge in coupling models of the soil microbiome functions with the ecosystem function of soil C cycling is the two different time scales. While microbial functions operate on the scale of hours and days, global soil carbon cycling operates on the scale of years, decades, and centuries. It remains to be seen if greater incorporation of microbiomes improves prediction (Schimel, 2023).

5 Case study: Soil microbiomes as an indicator of biological soil health

The evolution in soil management perspectives from soil fertility, to soil quality, to soil health that has occurred over the past 50 years has involved a shift from a narrowly defined chemical view of soil capability for crop productivity to a biologically focused view of soils as providers of multiple ecosystem services (Lehmann et al., 2020). While air and water quality are defined mainly by the degree to which they are polluted and impact directly on human health, soil quality is defined more broadly as 'the capacity of a soil to function within ecosystem and land-use boundaries to sustain biological productivity, maintain environmental quality, and promote plant and animal health' (Doran and Parkin, 1997). While the term soil quality is still appropriate when the intended use is specified (typically production of individual plant species or communities such as row crops), the term and concept of soil health captures ecological attributes that have implications beyond quality. Soil health is defined by the US Department of Agriculture Natural Resources Conservation Service (USDA-NRCS) as 'the continued capacity of the soil to function as a vital living ecosystem that sustains plants, animals and humans' and invokes positive agricultural as well as environmental outcomes. While frequently used synonymously, soil health is generally preferred by farmers because of appreciation for the roles and functions of soil in the broader ecosystem, while scientists have generally preferred soil quality because it makes specific distinctions between inherent and dynamic properties. While some argue the centrality of microbes as the drivers of soil health is misplaced (Coyne et al., 2022), what conceptually distinguishes soil health is an emphasis on the integrated inclusion of soil biota and biotic processes. This places soil organic matter and soil microbiomes at the center of the evaluation of soil health. Soil microbiomes contribute directly to the maintenance of soil health by controlling the decomposition of organic matter and cycling of C, which is the central focus of this chapter, among other ecosystem functions such a nutrient and water regulation.

The overall assessment of soil health is typically through a set physical, chemical, and biological indicators. A large number of biological indicators of soil health have been proposed, including the biomass and diversity of soil macro- and microfauna, but those most germane to soil microbiomes and carbon cycling include soil microbial biomass, bacterial-to-fungal ratios, microbial community composition, microbial functional gene composition, microbial enzyme activities and ratios, carbon and nitrogen mineralization rates (Fierer et al., 2021). In spite of the rapid advances in molecular (DNA-based) methods to decipher microbiomes with respect to key functions such as carbon storage and cycling, finding specific keystone markers that could be used as indicators of microbial soil health remains challenging in the light of the huge functional redundancy in most soil microbiomes (Schloter et al., 2018). In general, microbial metrics of microbial soil health involve measuring the abundances of taxa or genes of interest (e.g. well-known pathogens), rates of microbial activities, or estimating pools of microbial biomass or enzymatic capacities (Fierer et al., 2021).

Developed in the northeast United States, the Cornell Framework for the Comprehensive Assessment of Soil Health (https://soilhealth.cals.cornell.edu/) is 'designed for farmers, gardeners, agricultural service providers, landscape managers and researchers who want to go beyond simply testing the nutrient levels of their soils'. According to the Cornell Soil Health Assessment Framework, a healthy soil supports a 'large population of beneficial organisms'. The core biological indicators of soil health adopted by the Cornell system are all directly or indirect relevant to carbon storage and cycling: total organic matter, soil protein, soil respiration, and active carbon. Total organic matter is measure by loss on ignition at 500°C, soil protein represents the fraction of organic matter that contains easily decomposable and mineralizable nitrogen, active carbon is a chemical measure of the easily decomposable and mineralizable carbon, and soil respiration is a measure of total heterotrophic activity of the soil microbiome. All of these indicators are scored in the Cornell framework using a 'more is better' scoring type, suggesting that any management strategy that increases organic matter content and microbial activity will improve soil health – a fundamental tenet of sustainable or regenerative agriculture.

More broadly, the connections between soil microbiomes, soil health, the health of other ecosystem components, and human health are increasingly recognized under the banner of the 'one health' concept (Fig. 10). Soils are a major source of microorganisms in terrestrial ecosystems, and thus represent the foundation of one health (Banerjee and van der Heijden, 2023). Soil microbiomes contribute directly to plant health in negative ways via soil-borne plant pathogens, but also promote plant health through synergistic rhizosphere interactions. Plants preferentially recruit soil microorganisms from the rhizosphere that are essential for their growth and development. The rhizosphere microbiome facilitates a range of processes: seed germination,

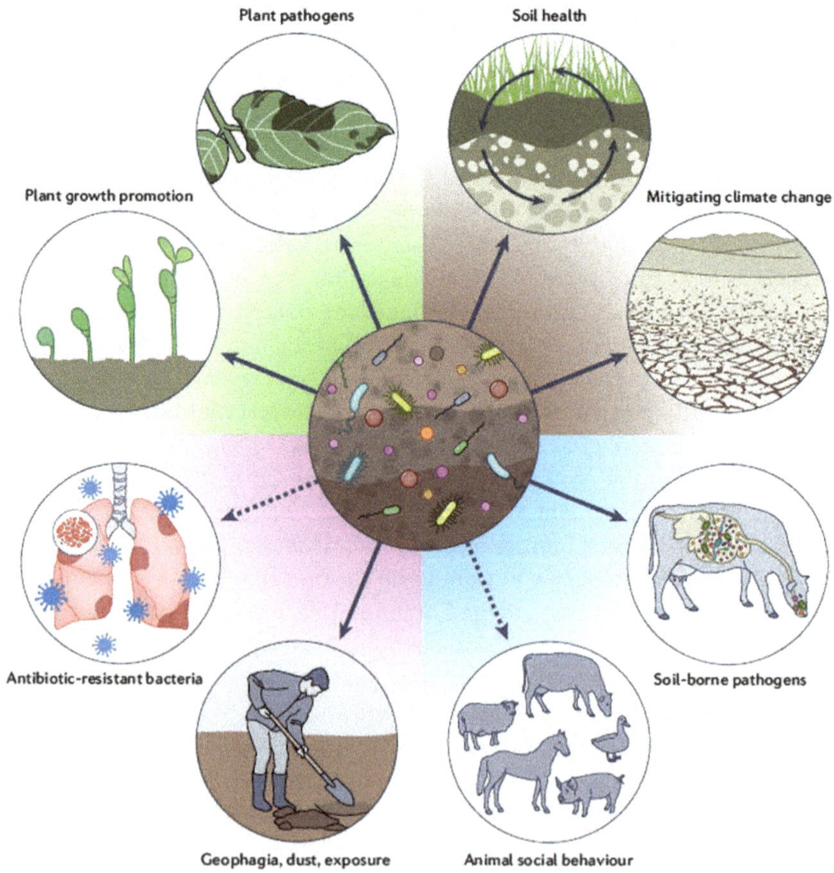

Figure 10 The centrality of the soil microbiome in the 'one health' concept that encompasses soil, plant, animal and human health. Source: Banarjee and van der Heijden (2023), with permission.

seedling establishment, nutrient acquisition, water regulation, growth promotion, pathogen and disease suppression, and stress tolerance. Soil microbiomes also affect animal and insect health. Soil-borne pathogens can cause deadly diseases in animals, including anthrax, but a diverse soil microbiome may contain important bacteria capable of resupplying the mammalian gut microbiota, with implications for gut health and mental health in ruminant herbivores. Lastly, humans are known to deliberately ingest soils as a nutrient supplement (i.e. geophagy), use soils as detoxifying agents for making some food products edible, and soil microbiomes are in important source of medicinal antibiotics. While there are emerging about the potential impacts of homogenization and reduction in diversity of soil microbiomes due to land use change, urbanization and agricultural intensification, this remains an area of active inquiry.

6 Conclusion and future trends

As illustrated in this chapter, soil microbiomes play central roles in soil carbon cycling. The importance of the contributions of soil microbiomes in these processes cannot be understated. Globally, soils contain 1700 Gt (1 Gt = 10^{15} g) of carbon, the largest actively cycling terrestrial pool, larger than the atmospheric (875 Gt C) and aboveground vegetation (450 Gt C) pools combined. In addition, soils exchange vast amounts of carbon with the atmosphere. Atmospheric inputs of carbon to soil through photosynthesis total 130 Gt C year^{-1}, which is nearly balanced by global outputs of carbon from soil through mineralization and respiration. The net balance of these processes determines whether soils represent a global sink or source for atmospheric carbon. While our understanding of the input component and its responses to disturbance are well advanced, the response of the output component is less well constrained. Disturbances such as land use change and management, as well as climate change, have direct and indirect impacts on how soil microbiomes influence land–atmosphere carbon exchanges that feedback on greenhouse gas emissions (Bardgett et al., 2008). A precise and quantitative understanding of how soil microbiomes both respond to and contribute to disturbance will improve our ability to predict soil carbon balances.

7 Where to look for further information

Several textbook chapters provide excellent summaries of the role of soil microbiomes in soil carbon cycling:

- Tate, K. R. (2017). *Microbial Biomass: A Paradigm Shift in Terrestrial Biogeochemistry*. London: World Scientific Publishing Europe.

- Paul, E. A. and S. Frey (2023). *Soil Microbiology, Ecology and Biochemistry*. 5th ed. New York: Academic Press.
- Gentry, T. J., Fuhrmann, J. J. and Zuberer, D. A. (2021) *Principles and Applications of Soil Microbiology*. 3rd ed. New York: Elsevier.

Key journals/conferences:

- *Soil Biology & Biochemistry;*
- The United States Department of Energy Environmental Microbiome Science project in the Genomic Science Program (https://genomicscience.energy.gov/environmental-microbiome-research/); and
- The Soil Health Institute (https://soilhealthinstitute.org/).

8 References

Angst, G., Mueller, K. E., Nierop, K. G. J. and Simpson, M. J. (2021). Plant- or microbial-derived? A review on the molecular composition of stabilized soil organic matter. *Soil Biology and Biochemistry* 156: 108189. DOI: 10.1016/j.soilbio.2021.108189.

Banerjee, S. and van der Heijden, M. G. A. (2023). Soil microbiomes and one health. *Nature Reviews: Microbiology*, Nature Publishing Group 21(1): 6–20. DOI: 10.1038/s41579-022-00779-w.

Bardgett, R. D., Freeman, C. and Ostle, N. J. (2008). Microbial contributions to climate change through carbon cycle feedbacks. *The ISME Journal*, Nature Publishing Group 2(8): 805–814. DOI: 10.1038/ismej.2008.58.

Bernard, L., Basile-Doelsch, I., Derrien, D., Fanin, N., Fontaine, S., Guenet, B., Karimi, B., Marsden, C. and Maron, P. (2022). Advancing the mechanistic understanding of the priming effect on soil organic matter mineralisation. *Functional Ecology* 36(6): 1355–1377. DOI: 10.1111/1365-2435.14038.

Blagodatskaya, E. and Kuzyakov, Y. (2008). Mechanisms of real and apparent priming effects and their dependence on soil microbial biomass and community structure: critical review. *Biology and Fertility of Soils* 45(2): 115–131. DOI: 10.1007/s00374-008-0334-y.

Blankinship, J. C., Berhe, A. A., Crow, S. E., Druhan, J. L., Heckman, K. A., Keiluweit, M., Lawrence, C. R., Marín-Spiotta, E., Plante, A. F., Rasmussen, C., Schädel, C., Schimel, J. P., Sierra, C. A., Thompson, A., Wagai, R. and Wieder, W. R. (2018). Improving understanding of soil organic matter dynamics by triangulating theories, measurements, and models. *Biogeochemistry* 140(1): 1–13. DOI: 10.1007/s10533-018-0478-2.

Camenzind, T., Mason-Jones, K., Mansour, I., Rillig, M. C. and Lehmann, J. (2023). Formation of necromass-derived soil organic carbon determined by microbial death pathways. *Nature Geoscience*, Nature Publishing Group 16(2): 115–122. DOI: 10.1038/s41561-022-01100-3.

Cleveland, C. C. and Liptzin, D. (2007). C:N:P stoichiometry in soil: is there a 'Redfield ratio' for the microbial biomass? *Biogeochemistry* 85(3): 235–252. DOI: 10.1007/s10533-007-9132-0.

Coyne, M. S., Pena-Yewtukhiw, E. M., Grove, J. H., Sant'Anna, A. C. and Mata-Padrino, D. (2022). Soil health: it's not all biology. *Soil Security* 6: 100051. DOI: 10.1016/j.soisec.2022.100051.

Crowther, T. W., van den Hoogen, J., Wan, J., Mayes, M. A., Keiser, A. D., Mo, L., Averill, C. and Maynard, D. S. (2019). The global soil community and its influence on biogeochemistry. *Science*, American Association for the Advancement of Science 365(6455): eaav0550. DOI: 10.1126/science.aav0550.

Dijkstra, P., Thomas, S. C., Heinrich, P. L., Koch, G. W., Schwartz, E. and Hungate, B. A. (2011). Effect of temperature on metabolic activity of intact microbial communities: evidence for altered metabolic pathway activity but not for increased maintenance respiration and reduced carbon use efficiency. *Soil Biology and Biochemistry* 43(10): 2023–2031. DOI: 10.1016/j.soilbio.2011.05.018.

Doran, J. W. and Parkin, T. B. (1997). Quantitative indicators of soil quality: a minimum data set. In: *Methods for Assessing Soil Quality*. John Wiley & Sons, Ltd, pp. 25–37. DOI: 10.2136/sssaspecpub49.c2.

Fierer, N., Bradford, M. A. and Jackson, R. B. (2007). Toward an ecological classification of soil bacteria. *Ecology* 88(6): 1354–1364. DOI: 10.1890/05-1839.

Fierer, N., Wood, S. A. and Bueno de Mesquita, C. P. (2021). How microbes can, and cannot, be used to assess soil health. *Soil Biology and Biochemistry* 153: 108111. DOI: 10.1016/j.soilbio.2020.108111.

Fontaine, S., Henault, C., Aamor, A., Bdioui, N., Bloor, J. M. G., Maire, V., Mary, B., Revaillot, S. and Maron, P. A. (2011). Fungi mediate long term sequestration of carbon and nitrogen in soil through their priming effect. *Soil Biology and Biochemistry* 43(1): 86–96. DOI: 10.1016/j.soilbio.2010.09.017.

Fontaine, S., Mariotti, A. and Abbadie, L. (2003). The priming effect of organic matter: a question of microbial competition? *Soil Biology and Biochemistry* 35(6): 837–843. DOI: 10.1016/S0038-0717(03)00123-8.

Frey, S. D., Lee, J., Melillo, J. M. and Six, J. (2013). The temperature response of soil microbial efficiency and its feedback to climate. *Nature Climate Change*, Nature Publishing Group 3(4): 395–398. DOI: 10.1038/nclimate1796.

Geyer, K. M., Dijkstra, P., Sinsabaugh, R. and Frey, S. D. (2019). Clarifying the interpretation of carbon use efficiency in soil through methods comparison. *Soil Biology and Biochemistry* 128: 79–88. DOI: 10.1016/j.soilbio.2018.09.036.

Geyer, K. M., Kyker-Snowman, E., Grandy, A. S. and Frey, S. D. (2016). Microbial carbon use efficiency: accounting for population, community, and ecosystem-scale controls over the fate of metabolized organic matter. *Biogeochemistry* 127(2): 173–188. DOI: 10.1007/s10533-016-0191-y.

Jenkinson, D. S. (1966). Studies on the decomposition of plant material in soil. *Journal of Soil Science* 17(2): 280–302. DOI: 10.1111/j.1365-2389.1966.tb01474.x.

Jenkinson, D. S. (1977). The soil biomass. *New Zealand Soil News* 25: 213–218.

Keiluweit, M., Bougoure, J. J., Nico, P. S., Pett-Ridge, J., Weber, P. K. and Kleber, M. (2015). Mineral protection of soil carbon counteracted by root exudates. *Nature Climate Change*, Nature Publishing Group 5(6): 588–595. DOI: 10.1038/nclimate2580.

Khan, K. S., Mack, R., Castillo, X., Kaiser, M. and Joergensen, R. G. (2016). Microbial biomass, fungal and bacterial residues, and their relationships to the soil organic matter C/N/P/S ratios. *Geoderma* 271: 115–123. DOI: 10.1016/j.geoderma.2016.02.019.

Kirkby, C. A., Kirkegaard, J. A., Richardson, A. E., Wade, L. J., Blanchard, C. and Batten, G. (2011). Stable soil organic matter: a comparison of C:N:P:S ratios in Australian and other world soils. *Geoderma* 163(3–4): 197–208. DOI: 10.1016/j.geoderma.2011.04.010.

Kuzyakov, Y. (2010). Priming effects: interactions between living and dead organic matter. *Soil Biology and Biochemistry* 42(9): 1363–1371. DOI: 10.1016/j.soilbio.2010.04.003.

Kuzyakov, Y., Friedel, J. K. and Stahr, K. (2000). Review of mechanisms and quantification of priming effects. *Soil Biology and Biochemistry* 32(11–12): 1485–1498. DOI: 10.1016/S0038-0717(00)00084-5.

Lajtha, K., Bowden, R. D., Crow, S., Fekete, I., Kotroczó, Z., Plante, A., Simpson, M. J. and Nadelhoffer, K. J. (2018). The detrital input and removal treatment (DIRT) network: insights into soil carbon stabilization. *Science of the Total Environment* 640–641: 1112–1120. DOI: 10.1016/j.scitotenv.2018.05.388.

Lehmann, J., Bossio, D. A., Kögel-Knabner, I. and Rillig, M. C. (2020). The concept and future prospects of soil health. *Nature Reviews. Earth and Environment*, Nature Publishing Group 1(10): 544–553. DOI: 10.1038/s43017-020-0080-8.

Liang, C., Amelung, W., Lehmann, J. and Kästner, M. (2019). Quantitative assessment of microbial necromass contribution to soil organic matter. *Global Change Biology* 25(11): 3578–3590. DOI: 10.1111/gcb.14781.

Liang, C., Kästner, M. and Joergensen, R. G. (2020). Microbial necromass on the rise: the growing focus on its role in soil organic matter development. *Soil Biology and Biochemistry* 150: 108000. DOI: 10.1016/j.soilbio.2020.108000.

Liang, C., Schimel, J. P. and Jastrow, J. D. (2017). The importance of anabolism in microbial control over soil carbon storage. *Nature Microbiology*, Nature Publishing Group 2(8): 17105. DOI: 10.1038/nmicrobiol.2017.105.

Manzoni, S., Taylor, P., Richter, A., Porporato, A. and Ågren, G. I. (2012). Environmental and stoichiometric controls on microbial carbon-use efficiency in soils. *New Phytologist* 196(1): 79–91. DOI: 10.1111/j.1469-8137.2012.04225.x.

Moorhead, D. L. and Sinsabaugh, R. L. (2006). A theoretical model of litter decay and microbial interaction. *Ecological Monographs* 76(2): 151–174. DOI: 10.1890/0012-9615(2006)076[0151:ATMOLD]2.0.CO;2.

Parton, W. J., Del Grosso, S. J., Plante, A. F., Adair, E. C. and Lutz, S. M. (2015). Modeling the dynamics of soil organic matter and nutrient cycling. In: Paul, E. A. (Ed.), *Soil Microbiology, Ecology and Biochemistry*. 4th ed. Boston: Academic Press, pp. 505–537. DOI: 10.1016/B978-0-12-415955-6.00017-7.

Qiao, Y., Wang, J., Liang, G., Du, Z., Zhou, J., Zhu, C., Huang, K., Zhou, X., Luo, Y., Yan, L. and Xia, J. (2019). Global variation of soil microbial carbon-use efficiency in relation to growth temperature and substrate supply. *Scientific Reports*, Nature Publishing Group 9(1): 5621. DOI: 10.1038/s41598-019-42145-6.

Schimel, J. (2023). Modeling ecosystem-scale carbon dynamics in soil: the microbial dimension. *Soil Biology and Biochemistry* 178: 108948. DOI: 10.1016/j.soilbio.2023.108948.

Schloter, M., Nannipieri, P., Sørensen, S. J. and van Elsas, J. D. (2018). Microbial indicators for soil quality. *Biology and Fertility of Soils* 54(1): 1–10. DOI: 10.1007/s00374-017-1248-3.

Schmidt, M. W. I., Torn, M. S., Abiven, S., Dittmar, T., Guggenberger, G., Janssens, I. A., Kleber, M., Kögel-Knabner, I., Lehmann, J., Manning, D. A., Nannipieri, P., Rasse, D. P., Weiner, S. and Trumbore, S. E. (2011). Persistence of soil organic matter as an ecosystem property. *Nature* 478(7367): 49–56. DOI: 10.1038/nature10386.

Simpson, A. J., Simpson, M. J., Smith, E. and Kelleher, B. P. (2007). Microbially derived inputs to soil organic matter: are current estimates too low? *Environmental Science and Technology*, American Chemical Society 41(23): 8070–8076. DOI: 10.1021/es071217x.

Sinsabaugh, R. L., Manzoni, S., Moorhead, D. L. and Richter, A. (2013). Carbon use efficiency of microbial communities: stoichiometry, methodology and modelling. *Ecology Letters* 16(7): 930–939. DOI: 10.1111/ele.12113.

Sokol, N. W., Slessarev, E., Marschmann, G. L., Nicolas, A., Blazewicz, S. J., Brodie, E. L., Firestone, M. K., Foley, M. M., Hestrin, R., Hungate, B. A., Koch, B. J., Stone, B. W., Sullivan, M. B., Zablocki, O., LLNL Soil Microbiome Consortium and Pett-Ridge, J. (2022). Life and death in the soil microbiome: how ecological processes influence biogeochemistry. *Nature Reviews. Microbiology*, Nature Publishing Group 20(7): 415–430. DOI: 10.1038/s41579-022-00695-z.

Stone, B. W. G., Dijkstra, P., Finley, B. K., Fitzpatrick, R., Foley, M. M., Hayer, M., Hofmockel, K. S., Koch, B. J., Li, J., Liu, X. J. A., Martinez, A., Mau, R. L., Marks, J., Monsaint-Queeney, V., Morrissey, E. M., Propster, J., Pett-Ridge, J., Purcell, A. M., Schwartz, E. and Hungate, B. A. (2023). Life history strategies among soil bacteria–dichotomy for few, continuum for many. *The ISME Journal*, Nature Publishing Group 17(4): 611–619. DOI: 10.1038/s41396-022-01354-0.

Tate, K. R. (2017). Microbial biomass: a paradigm shift in terrestrial biogeochemistry. *Europe*, World Scientific. DOI: 10.1142/q0038.

Tipping, E., Somerville, C. J. and Luster, J. (2016). The C:N:P:S stoichiometry of soil organic matter. *Biogeochemistry* 130(1): 117–131. DOI: 10.1007/s10533-016-0247-z.

Wang, B., An, S., Liang, C., Liu, Y. and Kuzyakov, Y. (2021). Microbial necromass as the source of soil organic carbon in global ecosystems. *Soil Biology and Biochemistry* 162: 108422. DOI: 10.1016/j.soilbio.2021.108422.

Whalen, E. D., Grandy, A. S., Sokol, N. W., Keiluweit, M., Ernakovich, J., Smith, R. G. and Frey, S. D. (2022). Clarifying the evidence for microbial- and plant-derived soil organic matter, and the path toward a more quantitative understanding. *Global Change Biology* 28(24): 7167–7185. DOI: 10.1111/gcb.16413.

Wieder, W. R., Bonan, G. B. and Allison, S. D. (2013). Global soil carbon projections are improved by modelling microbial processes. *Nature Climate Change*, Nature Publishing Group 3(10): 909–912. DOI: 10.1038/nclimate1951.

Wieder, W. R., Grandy, A. S., Kallenbach, C. M. and Bonan, G. B. (2014). Integrating microbial physiology and physio-chemical principles in soils with the MIcrobial-MIneral Carbon Stabilization (MIMICS) model. *Biogeosciences*, Copernicus GmbH 11(14): 3899–3917. DOI: 10.5194/bg-11-3899-2014.

Zhou, R., Liu, Y., Dungait, J. A. J., Kumar, A., Wang, J., Tiemann, L. K., Zhang, F., Kuzyakov, Y. and Tian, J. (2023). Microbial necromass in cropland soils: a global meta-analysis of management effects. *Global Change Biology* 29(7): 1998–2014. DOI: 10.1111/gcb.16613.

Chapter 10

Advances in understanding the role of the soil microbiome's effects on nitrogen cycling

Tim Clough, Lincoln University, New Zealand

1 Introduction

One of the Earth's largest, most diverse and complex microbiomes is found in the soil (Bardgett and van der Putten 2014; Banerjee and van der Heijden 2023). The soil microbiome may be defined as all microorganisms (archaea, bacteria, fungi, protists, phages and other microbial eukaryotes) contained in the soil (Fierer 2017). A single gram of soil contains as much as 0.5 mg of microbial biomass and over 50 000 species (Banerjee and van der Heijden 2023). This microbial biomass is dominated by fungi and bacteria, and global estimates of the microbial biomass in terrestrial soils indicate that there are 12, 7, 1.6 and 0.5 Gt of carbon (C) contained in fungi, bacteria, protists and archaea, respectively (Bar-On and Milo 2016). Bacteria and fungi typically have 10^2–10^4 times greater biomass than the other major components of the soil microbiome (protists, archaea and phages) (Fierer 2017). Humanity relies on the soil microbiome for the secure provision of food and fibre via ecosystem services that include but are not limited to, nutrient cycling, soil organic matter (SOM) development, carbon sequestration and soil structural stability.

http://dx.doi.org/10.19103/AS.2024.0136.22

In addition, the soil microbiome can contribute to unwanted outcomes with negative consequences (e.g. greenhouse gas emissions).

A soil's properties are determined by soil-forming factors (climate, organisms, topography, parent material and time). The resulting soil environment is not constant and varies spatially and temporally, resulting in spatial and temporal variation in the structure of the soil microbiome. Consequently, the soil's physico-chemical properties, along with climate and management, determine plant diversity. Plant type determines root architecture, and the ensuing biotic and abiotic processes in the root network's rhizosphere produce fine scale (mm) spatial changes in the soil environment (e.g. soil pH, oxygen and exudates), while changes in root physiology and plant growth status can result in temporal dynamics in the rhizosphere at this fine scale (Bardgett and van der Putten 2014). Such spatial and temporal variation results in differences in microbial abundances, microbial activity rates, and microbial community structure (Fierer 2017). At a greater scale (cm to m) spatial variability of the soil microbiome may result from variation in soil physico-chemical properties, determining water and/or nutrient availability, along with the quality and quantity of organic substrates returned to the soil as a result of soil and/or management conditions determining plant type (Bardgett and van der Putten 2014). Changes in the soil microbiome may also be driven at the ecosystem scale due to the interaction of climate, topography and soil abiotic factors (Bardgett and van der Putten 2014). In this chapter, the role of the soil microbiome in relation to nitrogen (N) cycling within the rhizosphere, detritusphere and bulk soil are examined.

2 The rhizosphere

Plants strongly influence both the soil microbiome structure and its function in the zone known as the rhizosphere: this is the volume of soil directly influenced by the living plant root and is measured in terms of mm, typically ≤2 mm (Nuccio et al. 2020; Sokol et al. 2022). Microbial diversity is lower in the rhizosphere due to the readily available root exudate supply (e.g. sugars, amino acids and organic acids) that creates a resource-rich environment, relative to bulk soil (Philippot et al. 2013; Sokol et al. 2022). However, microbial populations are higher in the rhizosphere (DeAngelis et al. 2008; Sokol et al. 2022). Root exudation rates and forms vary with root architecture and growth stage (Galindo-Castañeda et al. 2024). Plant nutrient uptake may also alter the rhizosphere soil's abiotic environment (e.g. soil pH): nitrate and ammonium uptake tend to raise and decrease the rhizosphere pH, respectively (Nye 1981). Root respiration, including the respiration of roots and microorganisms within the rhizosphere, varies with plant growth stage, root growth rates, root tissue age, root architecture, chemistry (e.g. N concentration), function and temperature (Ben-Noah and Friedman 2018). If soil oxygen supply cannot match the soil's oxygen respiration demand, hypoxic or anoxic conditions may be generated, leading to changes in microbial function within the rhizosphere.

Plant root exudation also plays a role in determining the rhizosphere microbial community through active recruitment to overcome biotic or abiotic stress. Initially, root colonisation results from contact with adjacent soil, but as the root develops, microbial distribution and function vary between the actively growing tip, the root elongation zone and mature zones of the root (Bonkowski et al. 2021). Microbial diversity loss can affect a plant's ability to recruit specific soil microorganisms, with changes in N cycling microbial composition (two- to four-fold increase in N_2O reducers and low abundance of bacterial ammonia oxidisers) observed to affect plant N status (Saghaï et al. 2022). Specific root exudate signals may also cause specific symbionts or pathogens to be attracted to the rhizosphere, affecting nutrient cycling. For example, isoflavones excreted from soybean roots were observed to attract both symbiotic N-fixing bacteria (*Bradyrhizobium japonicum*) and the pathogen *Phytophthora sojae*, which causes root rot and damping off of seedlings (Berg and Smalla 2009). Nutrient turnover and changes in microbial dynamics within the rhizosphere may also result from predatory activity. Protists, described as the puppet masters of the rhizosphere microbiome (Gao et al. 2018), make nutrients (e.g. N and phosphorus) available to other organisms and plants via predation of other microbes (Ceja-Navarro et al. 2021). For example, protists were observed to enhance shoot and root growth and increase N uptake by wheat plants due to the N embodied in the bacterial biomass being released (Henkes et al. 2018). However, this effect was not observed when arbuscular mycorrhizal fungi (AMF) were present, indicating that the AMF took up the N released by protists (Henkes et al. 2018). The protists and AMF were shown to alter the wheat plant's rhizosphere bacterial community (Henkes et al. 2018). Changes in the rhizosphere microbial community may result from top-down control by protists (Rosenberg et al. 2009) or via bottom-up change in resources induced by AMF (Henkes et al. 2018).

Previously, the uptake of organic N by AMF was also observed by Bukovská et al. (2018). Soil containing inputs of organic N, as opposed to inorganic N, contained higher colonisation rates of AMF hyphae, and mycorrhizal plants had greater biomass and higher accumulation of organic N sources compared to non-mycorrhizal sources of N (Bukovská et al. 2018). There was a suppression of other microbial abundances, such as bacteria (especially ammonia-oxidising bacteria) and fungi, in the presence of AMF hyphae indicating that AMF were strongly competing for ammonium ions and playing an important role in the recycling of organic N from soil to plants (Bukovská et al. 2018).

3 The detritusphere

The rhizosphere is a zone determined by the architecture and function of the living root. However, over time, roots die, and there ensues an abiotic change in the zone of soil around the dead root (the detritusphere), which consequently alters the microbial community and function. This microbial succession commences with increases in microbial taxa able to exploit compounds released early in decomposition, such as water-soluble compounds and simple carbohydrates, followed by microbes able to consume the litter's more complex compounds (Sokol et al. 2022). Using a time series of soil metatranscriptomes, Sieradzki et al. (2023) identified concurrent transformations of N in four discrete soil habitats relating to the annual grass *Avena fatua*: the rhizosphere (near live roots), detritusphere (near dead roots), rhizo-detritusphere (intermingled live and dead roots) and bulk soil (soil with no root influence). Over a period of days, N process rates were found to be affected by the interaction between soil habitat and plant development: in the rhizosphere, assimilatory nitrate reduction and ammonium assimilation through glutamate dehydrogenase and glutamine synthetase were upregulated relative to bulk soil. Such upregulation was considered indicative of the rhizosphere microbial community experiencing inorganic N limitation due to competition with the plant or due to higher microbial demand resulting from higher microbial density in the rhizosphere (Sieradzki et al. 2023). *A. fatua* is known to compete better than the rhizosphere community for nitrate, and this was thought to explain the observed upregulation (Sieradzki et al. 2023). With litter present in the detritusphere, the degradation of macromolecular organic N was enhanced along with the dissimilatory N transformation pathways, DNRA (dissimilatory nitrate reduction) and denitrification (Sieradzki et al. 2023). In the bulk soil nitrification, DNRA and the first step of denitrification were upregulated when compared to the rhizosphere (Sieradzki et al. 2023).

4 Bulk soil

Outside of the soil rhizosphere and detritus compartments exists the bulk soil mineral matrix, where microorganisms are sparser in number and distributed throughout the soil matrix depending on pore structure, water films and mineral type (Sokol et al. 2022). However, it is now recognised that most SOM, and of course the N associated with this, derives from microbial necromass (Liang et al. 2019). The turnover of microbial cells results in N-containing compounds, such as amino sugars and nucleic acids, being retained on mineral surfaces (Sokol et al. 2022).

This mineral-associated organic matter (MAOM) is a function of the soil's mineral properties, which determine the chemistry and stability of the organic–mineral interaction, the potential to accumulate MAOM and the regulation of MAOM turnover (Daly et al. 2021). Plant-derived N compounds may also contribute to MAOM. Sokol et al. (2018) state that MAOM formation in areas of high microbial density occurs through *in vivo* microbial turnover, while in areas of low microbial density, MAOM formation occurs through the sorption of plant-derived compounds to uncolonised mineral surfaces.

Most soil N occurs as MAOM, and being enriched in microbial residues and/or low molecular weight plant compounds, it has a relatively low C : N ratio that facilitates N mineralisation via microbial mining (Daly et al. 2021). The release of N from the MAOM complex may result from changes in soil chemistry (e.g. pH), soil chemistry and redox. The MAOM pool that resides in the soil matrix is not immune to disruption from rhizosphere processes, where localised changes in soil pH and chemistry, as well as root exudation of ligands and enzymes, can alter the sorption and mobilisation of N embodied in MAOM (Daly et al. 2021).

5 Nitrogen cycling and the rhizosphere

Plants influence microbial N turnover and transformations in order to both increase the availability of N and to retain N in the plant–soil system (Moreau et al. 2019). Plants may be adapted to nutrient-rich habitats, described as 'resource acquisitive species', and demonstrate fast growth, rapid uptake of N and high rhizodeposition rates. Conversely, plants adapted to nutrient-poor habitats described as 'resource conservative species' show low rates of nutrient turnover, respiration and growth (Moreau et al. 2019; Henneron et al. 2020). In the rhizosphere, plants have the ability to uptake available N through their roots, as reviewed elsewhere (Muratore et al. 2021). It is accepted that plants acquire soluble N as inorganic N or low-molecular-weight organic N, though the ecological significance of the later remains controversial (Moreau et al. 2019). When N is scarce, plants may acquire N by recruiting microbes. This may occur via symbiotic relationships; for example, AM fungi may take up organic or inorganic N and transfer N to plants, increasing both plant and fungal biomass (Thirkell et al. 2016). Fungal hyphae networks permit plant roots to acquire nutrients that are distant from the rhizosphere and without rhizosphere competition for the nutrient (Moreau et al. 2019). Legumes also have a well-recognised ability to form symbiotic relationships with N_2-fixing microbes that either form nodules on roots, reside in the rhizosphere, or occur as endophytes in plant tissues, generating an N supply for the plant (Moreau et al. 2019).

Rhizosphere priming has been defined as the short-term change in the mineralisation of SOM resulting from rhizodeposition of C (Moreau et al. 2019). This rhizo-deposited C is labile and results in increased microbial demand for N, which is sated by the microbial release of enzymes (proteases, deaminases) that release N from high-molecular-weight polymeric N contained in the SOM generating low molecular weight organic N in the rhizosphere soil (DeAngelis et al. 2008; Moreau et al. 2019). Priming of soil N has been correlated to root biomass, and there is evidence that microbial quorum sensing (density-dependent population behaviour) may regulate the soil N cycle and plant N uptake in the rhizosphere (DeAngelis et al. 2008; Gan et al. 2022). Dissolved organic N (DON) has been found to be two-fold higher in the rhizosphere compared to bulk soil, with low-molecular-weight DON comprising ~15% of the total DON in the rhizosphere, while being undetectable in bulk soil (DeAngelis et al. 2008). Thus, SOM decomposition can result in greater N uptake by acquisitive plants, which, when allocated to above-ground biomass, further promotes photosynthesis and growth (Henneron et al. 2020). Symbiotic relationships that capture N, such as legume-N_2 fixation, enhance photosynthesis and thus rhizodeposition, which in turn increases rhizosphere priming (Henneron et al. 2020).

The rhizosphere, especially, is a hotspot for denitrification (Graf et al. 2016), a process driven by heterotrophic organisms, where oxidised forms of N (e.g. nitrate and nitrite) act as electron acceptors under anaerobic conditions to form nitric oxide, nitrous oxide (N_2O) and dinitrogen (N_2). The abundance of genes involved in denitrification has been shown to increase in the rhizosphere following N addition, relative to bulk soil, due to the higher concentrations of C substrates in the rhizosphere resulting from exudation (Liao et al. 2024). The microbial reduction of N_2O to N_2 is catalysed by the N_2O reductase enzyme, which is encoded by the *nosZ* gene. The diversity of N_2O-reducing microbes is split between *nosZ1* and *nosZII* clades (Maheshwari et al. 2023). The type of rhizosphere C exuded can influence both the microbiome and the fate of N in the rhizosphere. For example, under anoxic conditions, niche differentiation of N_2O-reducing communities has been shown to depend on C substrate type: Betaproteobacterial N_2O reducers (clades I and II) were observed to become enriched in the presence of cellulose substrate, while Alphaproteobactetrial *nosZ* clade I lineages flourished under glucose and acetate substrates (Maheshwari et al. 2023). Such observed shifts in N_2O-reducing bacterial community compositions help to explain the observed effects of typical root exudate compounds on N_2O production and consumption rates in soil and the variation in the ratio of N_2O to N_2 produced (Morley et al. 2014; Li et al. 2021).

Nitrogen fertiliser effects on the rhizosphere priming release of N remain unclear (Gao et al. 2023). Microbes in the rhizosphere may respond to increased C availability following N fertilisation by either increasing extracellular enzyme production or increasing their biomass, which have contrasting effects on N cycling: the resource limitation for enzyme production and subsequent SOM mineralisation is removed if N fertiliser increases rhizodeposition (Gao et al. 2023). But if increased rhizodeposition, following N fertilisation is prioritised to biomass growth, the rate of decomposition of SOM declines (Zhu et al. 2014).

An example of the complex response to N fertilisation rate is demonstrated by Zhu et al. (2016) who showed that increasing urea-N fertiliser rate for maize increased both root exudation (sugars, sugar alcohols and phenolics) and the abundance of bacteria in the rhizosphere, which in turn affected microbial community composition. Notably, there was an increase in denitrifying and ammonia-oxidising bacteria with increasing N rate, which the authors speculated could explain the reduced N use efficiency observed with high N fertiliser rates (Zhu et al. 2016).

In addition to fertiliser, other agricultural management practices, such as plant selection and tillage, can also strongly influence the rhizosphere microbiome and associated N cycling. Schmidt et al. (2019) examined the interactions between management and plant type, finding that rhizosphere communities were impacted differently to bulk soil: plants recruit management-specific taxa, resulting in the rhizosphere N-cycling pathways being altered, differently from bulk soil (Schmidt et al. 2019). Differences between bulk and rhizosphere soil samples taken from under a maize crop were usually greater in conventional versus organic soils, which had been managed for 23 years. These differences may be a consequence of other management practices, such as fertilisation type and placement (Schmidt et al. 2019). Organic management increased the abundance of the *nosZ* gene in both bulk and rhizosphere soils, which may have been due to higher soil C resulting from long-term organic-matter applications (Schmidt et al. 2019). However, the plant management interaction on rhizosphere soil N cycling is complex and affected by factors such as plant species, soil type, climate, management factors and crop growth stage. Clearly, there remains a need to better understand this dynamic, as knowledge of the rhizosphere effects on N cycling driven by plant–soil interactions could assist in designing sustainable farming practices.

A rationale for improving our understanding of N cycling in soils is to improve N use efficiency in food production systems and to reduce adverse environmental effects of N fertiliser use, including greenhouse gas emissions and nitrate leaching. Potentially, the rhizosphere has a key role to play in achieving this aim. Plant root exudates may act as biological nitrification inhibitors (BNIs) with various secondary metabolites identified, from the roots of grasses and crops, that block ammonia-monooxygenase and hydroxylamine oxidoreductase during nitrification (Moreau et al. 2019; Lu et al. 2024). The efficacy of root exudate BNIs varies depending on soil type, BNI concentration and the N rate applied to an agricultural system, and this topic is reviewed elsewhere (Lu et al. 2024). The process is instigated when high ammonium ion concentrations occur in the root environment (Subbarao et al. 2007). Research has also identified that biological denitrification can be inhibited by up to 80% due to the exudation of procyanidins by roots (Bardon et al. 2014; Moreau et al. 2019; Lu et al. 2024).

6 Nitrogen cycling and the detritusphere

Under anaerobic conditions, denitrification has been shown to be the dominant mechanism responsible for N_2O emissions from the detritusphere (Rummel et al. 2021). The role of the detritusphere in promoting the microbiome responsible for denitrification has long been recognised, with 'hot-spots' of high denitrification activity linked to organic matter residues in soil (Parkin 1987).

In agricultural ecosystems, substantial litter inputs occur through various mechanisms, such as the incorporation of cover crop leaf tissue or root turnover in perennial crops. Several factors influence microbiome function and the resulting denitrification gas emissions from within the leaf-tissue detritusphere, including plant residue quality (Millar and Baggs 2005), soil moisture and the pore size distribution (Kravchemko et al. 2017; Kravchenko et al. 2018). Leaf residue acts as an anaerobic hotspot for denitrifiers: rapid mineralisation within the residue consumes oxygen, while high water content within the residue impedes oxygen supply (Kravchemko et al. 2017).

The relatively high water content of the leaf residue can exceed that of the surrounding soil due to its absorption of water from the surrounding soil, described as the 'sponge effect' enabling detritus to function as a hotspot for denitrifiers (Kravchemko et al. 2017). Studies have also observed the 'sponge effect,' occurring up to a distance of ~150 μm from young root tissues (Kim et al. 2020). Root growth alters the soil microbiome in the rhizosphere, and this microenvironment carries a legacy effect into the detritusphere (Kim et al. 2020, 2022).

There is a complex interaction between the detritusphere and the soil pore size matrix, which affects oxygen demand and supply, microbiome function and subsequent gas emission. Using *in situ* techniques Kim et al. (2022) found higher N_2O emissions from the detritusphere when large soil pores (>30 μm) were present relative to small soil pores (<10 μm). This greater N_2O production was not, as originally hypothesised, the consequence of reduced conversion of N_2O to N_2. Instead, the higher emissions of N_2O resulted from higher microbial activity due to faster decomposition of roots and microbial turnover in the former rhizosphere. While Kim et al. (2022) found that root-derived N dominated N_2O emissions, especially in the large pores, the *in situ* root decomposition also resulted in N_2O priming (Daly et al. 2024) in the former rhizosphere soil adjacent to the decaying root. Soil pore architecture is a key determining factor of microbial function with respect to the magnitude and temporal dynamics of N_2O emissions from detritus hotspots (Kim et al. 2022).

Detritus in the soil is also commonly described by the term 'particulate organic matter' (POM) (Cotrufo et al. 2019). As described above, POM promotes soil anaerobicity due to oxygen consumption exceeding supply. Attempts to model these hotspots have invoked a conceptual anaerobic soil volume (Arah and Smith 1989; Li et al. 1992). Studies have shown the anaerobic soil volume fraction and the distance of POM to air-filled pores are strongly correlated with denitrification products (Rohe et al. 2021; Lucas et al. 2023; Schlüter et al. 2024). Recent studies propose that dimensional denitrification models based on soil microstructural information should be used (Schlüter et al. 2024).

Other qualities of the detritus may also affect microbiome function in the detritusphere. For example, Chen et al. (2021) found that when leaf litter was infected with the endophyte *Epichloë,* there was a change in the absolute abundance of genes involved in denitrification (*nirS*, *nirK* and *nosZ*).

The interaction of microbes within the detritusphere has also been shown to affect N_2O emissions from the detritusphere: Li et al. (2023) found that AMF hyphae reduced N_2O emissions in the hyphosphere, which were derived from denitrification in C and N rich detritus. This reduction was due to AMF enhancing the abundance and expression of the *nosZ*I gene, while inconsistently increasing the *nirS* and *nirK* genes. This phenomenon was the result of hyphal exudation of carboxylates both recruiting denitrifiers (*Pseudomonas*) and upregulating *nosZ* expression in *P. fluorescens* (Li et al. 2023). In this study, *nosZ* gene copies per gram of soil were well correlated with hyphal length density in the soil, a finding the authors suggest may explain the observed uptake of N_2O in soil high in organic C but low in mineral N (Li et al. 2023). The activity of soil microbes on the detritusphere may also be sequential, for example, Rummel (2021) found that during maize litter decomposition, N_2O formation resulted from both fungal and bacterial denitrification, but a shift towards bacterial domination of this N_2O formation occurred after the onset of anoxic conditions.

Pertinent to a fuller understanding of N_2O sources and the interactive role of microbes in N_2O emissions from detritus is the study by Starr et al. (2019), who found that the structure of host and RNA viral communities depended on root litter: time-resolved analysis of hosts and viruses showed viral replication (Starr et al. 2019). While Starr et al. (2019) focused their conclusions on the implications for C cycling, the stoichiometry of microbial cell composition implies organic N will also be released and mineralised.

7 Conclusion and future trends

Understanding how N cycles in soils is fundamental to food security and to reducing adverse environmental impacts at regional and global levels (de Vries et al. 2023). Future research must focus on understanding how to improve N use efficiency in agricultural ecosystems while negating adverse environmental effects. Fundamental to this is increasing our understanding how N cycles in the soil–microbiome–plant continuum. Continued developments in methodologies, such as the use of metatranscriptomics (Sieradzki et al. 2023) and viral metagenomics, will facilitate this.

Primarily, we can expect to see research on MAOM to better understand its formation, persistence, and bioavailability (Daly et al. 2021; Sokol et al. 2022).

Future research trends will include an increased understanding of the role viruses play in cycling N, because of their effect on microbiome turnover and ecology. Viruses that infect bacteria (bacteriophages or phages) manipulate bacteria numbers through a process termed lytic infection, which releases nutrients embodied in the microbial biomass and influences biogeochemical cycling (Braga et al. 2020). Long recognised as having high relevance to nutrient cycling in marine environments, where ~20% of the microbial biomass in the ocean is turned over daily by phages (Suttle 2007), the understanding of how phages influence nutrient cycling in soils, where their abundance can be as high as 10^{10} per gram of soil, is still in its infancy (Williamson et al. 2017). Recent studies have begun to evaluate the potential for viruses to influence microbial mortality and the turnover of nutrients. For example, Santos-Medellín et al. (2023) rewetted Mediterranean grassland soils and found that the viral community was dependent on soil type, but in all soils, viral biomass and diversity flourished within 24 h of wetting prior to viral community turnover. It was found that the viruses responded to predicted host microbial activity upon rewetting with predation of the more abundant taxa (Santos-Medellín et al. 2023). Aligned with these viral–bacterial dynamics were observed decreases in relic DNA upon rewetting, the reaccumulation of this in the period after wetting was indicative of new microbial mortality (Santos-Medellín et al. 2023). Viral studies specific to soil N cycling are rare but include Braga et al. (2020), who examined varying community assembly scenarios by adding phages to soil, finding that changes in phage pressure impacted bacterial community composition and diversity, and that phage addition could increase soil ammonium concentrations: the latter was thought to result from the phages lysing their host cells and releasing organic N that was subsequently mineralised.

Current knowledge also needs to be continually exploited and improved. For example, secondary plant metabolites exuded by roots are recognised as altering nitrification and denitrification rates in the rhizosphere. The challenge is to implement such mechanisms within agronomically important crop species at a level where significant reductions in fertiliser inputs and adverse environmental effects are achieved.

Other trends in N cycling research will likely see further research on global anthropogenic pollutants and their effect on the soil N cycle, for example, the impact of micro-plastics (MP). Kim et al. (2023) examining bacterial abundance and diversity in soil exposed to varying rates of MP, found that MP increased plant N uptake due to effects on enzyme activities, the taxonomic profile of rhizosphere bacteria, and functional genes linked to nitrification and N fixation.

While Zhang et al. (2023) found that MP altered N cycling gene abundance to varying degrees depending on the plastic type and soil concentration.

8 Where to look for further information

Further reading

- A good review of how the soil microbiome influences the turnover of soil organic matter, the primary source of nutrients, is 'Life and Death in the soil microbiome: how ecological processes influence biogeochemistry' by NW Sokol et al. in Nature Reviews Microbiology, Volume 20, pages 415–430, 2022.
- To understand the potential of the soil microbiome to trigger denitrification there must be an understanding of the soil's anaerobic status, which is covered in the review 'The anaerobic soil volume as a controlling factor of denitrification: a review' by Schlüter et al. in Biology & Fertility of Soils, published on line, 2024.
- This paper provides the reader with the latest ideas and concepts associated with the bioavailability of nitrogen in the plant-microbe-soil mineral continuum 'A holistic framework integrating plant-microbial-mineral regulation of soil bioavailable nitrogen' by Daly et al. in Biogeochemistry, Volume 154, pages 211–229, 2021.

Key journals

- mSystems
- The ISME journal
- Soil Biology & Biochemistry
- Biogeochemistry
- FEMS Microbial Ecology

9 References

Arah, J. R. M. and Smith, K. A. 1989. Steady-state denitrification in aggregated soils: a mathematical model. *Journal of Soil Science* 40: 139–149.

Banerjee, S. and van der Heijden, M. G. A. 2023. Soil microbiomes and one health. *Nature Reviews Microbiology* 21: 6–20.

Bar-On, Y. M. and Milo, R. P. R. 2016. The biomass distribution on Earth. *Proceedings of the National Academy of Science* 1115(25): 6506–6511.

Bardgett, R. D. and van der Putten, W. H. 2014. Belowground biodiversity and ecosystem functioning. *Nature* 515: 505–511.

Bardon, C, Piola, F., Bellvert, F., Haichar, F., Comte, G., Meiffren, G., Pommier, T., Puijalon, S., Tsafack, N. and Poly, F. 2014. Evidence for biological denitrification in hibition (BDI) by plant secondary metabolites. *New Phytologist* 204: 620–630.

Ben-Noah, I. and Friedman, S. P. 2018. Review and evaluation of root respiration and of natural and agricultural processes of soil aeration. *Vadose Zone Journal* 17(1): 1–47.

Berg, G. and Smalla, K. 2009. Plant species and soil type cooperatively shape the structure and function of microbial communities in the rhizosphere. *FEMS Microbial Ecology* 68: 1–13.

Bonkowski, M., Tarkka, M., Razavi, B. S., Schmidt, H., Blagodatskaya, E., Koller, R., Yu, P., Knief, C., Hochholdinger, F. and Vetterlein, D. 2021. Spatiotemporal dynamics of Maize (Zea mays L.) Root growth and its potential consequences for the assembly of the rhizosphere microbiota. *Frontiers Microbiology* 12: 619499.

Braga, L. P. P., Spor, A., Kot, W., Breuil, M. C., Hansen, L., Setubal, J. C. and Philippot, L. 2020. Impact of phages on soil bacterial communities and nitrogen availability under different assembly scenarios. *Microbiome* 8: 52.

Bukovská, P., Bonkowski, M., Konvalinková, T., Beskid, O., Hujslová, M., Püschel, D., Řezáčová, V., Gutiérrez-Núñez, M. S., Gryndler, M. and Jansa, J. 2018. Utilization of organic nitrogen by arbuscular mycorrhizal fungi—is there a specific role for protists and ammonia oxidizers? *Mycorrhiza* 28: 269–283.

Ceja-Navarro, J. A., Wang, Y., Ning, D., Arellano, A., Ramanculova, L., Yuan, M. M., Byer, A., Craven, K. D., Saha, M. C., Brodie, E. L. and Others 2021. Protist diversity and community complexity in the rhizosphere of switchgrass are dynamic as plants develop. *Microbiome* 9: 96.

Chen, Z., Jin, Y., Yao, X., Li, X., Li, C., White, J. F. and Nan, Z. 2021. Gene analysis reveals that leaf litter from *Epichloë* endophyte-infected perennial ryegrass alters diversity and abundance of soil microbes involved in nitrification and denitrification. *Soil Biology & Biochemistry* 154: 108123.

Cotrufo, M. F., Ranalli, M. G., Haddix, M. L., Six, J. and Lugato, E. 2019. Soil carbon storage informed by particulate and mineral-associated organic matter. *Nature Geoscience* 12: 989–994.

Daly, A. B., Jilling, A., Bowles, T. M., Buchkowski, R. W., Frey, S. D., Kallenbach, C. M., Keiluweit, M., Mooshammer, M., Schimel, J. P. and Grandy, A. S. 2021. A holistic framework integrating plant-microbial-mineral regulation of soil bioavailable nitrogen. *Biogeochemistry* 154: 211–229.

Daly, E. J., Hernandez-Ramirez, G., Congreves, K. A., Clough, T., Voigt, C., Harris, E. and Ruser, R. 2024. Soil organic nitrogen priming to nitrous oxide: a synthesis. *Soil Biology & Biochemistry* 189: 109254.

de Vries, W., Kros, J., Kroeze, C. and Seitzinger, S. 2023. Assessing planetary and regional nitrogen boundaries related to food security and adverse environmental impacts. *Current Opinion in Environmental Sustainability* 5(3–4): 392–402.

DeAngelis, K., Lindow, S. E. and Firestone, M. K. 2008. Bacterial quorum sensing and nitrogen cycling in the rhizosphere soil. *FEMS Microbial Ecology* 66: 197–207.

Fierer, N. 2017. Embracing the unknown: disentangling the complexities of the soil microbiome. *Nature Reviews Microbiology* 15: 579–590.

Galindo-Castañeda1, T., Hartmann, M. and Lynch, J. P. 2024. Location: root architecture structures rhizosphere microbial associations. *Journal of Experimental Botany* 75(2): 594–604.

Gan, D., Zeng, H. and Zhu, B. 2022. The rhizosphere effect on soil gross nitrogen mineralization: a meta-analysis. *Soil Ecological Letters* 4(2): 144–154.

Gao, Y., Tan, Z., Wang, H. and Zhu, Y. 2023. Nitrogenfertilisation and the rhizosphere effect on nitrogen cycling: a meta-analysis. *Applied Soil Ecology* 186: 104788.

Gao, Z., Karlsson, I., Geisen, S., Kowalchuk, G. and Jousset, A. 2018. Protists: puppet masters of the rhizosphere microbiome. *Trends in Plant Science* 24(2): 165–176.

Graf, D. R. H., Zhao, M., Jones, C. M. and Hallin, S. 2016. Soil type overrides plant effect on genetic and enzymatic N2O production potential in arable soils. *Soil Biology & Biochemistry* 100: 125–128.

Henkes, G. J., Kandeler, E., Marhan, S., Scheu, S. and Bonkowski, M. 2018. Interactions of mycorrhiza and protists in the rhizosphere systemically alter microbial community composition, plant shoot-to-root ratio and within-root system nitrogen allocation. *Frontiers Environmental Science* 6. https://doi.org/10.3389/fenvs.2018.00117

Henneron, L., Kardol, P., Wardle, D. A., Cros, C. and Fontaine, S. 2020. Rhizosphere control of nitrogen cycling: a key component of plant economic strategies. *New Phytologist* 228: 1269–1282.

Kim, K., Guber, A., Rivers, M. and Kravchemko, A. 2020. Contribution of plant roots to N2O emissions by water absorption. *Geoderma* 375: 114506.

Kim, K., Song, I. G., Yoon, H. and Park, J. W. 2023. Sub-micron microplastics affect nitrogen cycling by altering microbial abundance and activities in a soil-legume system. *Journal of Hazardous Materials* 460: 132504.

Kim, K., Gil, J., Ostrom, N. E., Gandhi, H., Oerther, M. S., Kuzyakov, Y., Guber, A. K. and Kravchemko, A. N. 2022. Soil pore architecture and rhizosphere legacy define N_2O production in root detritusphere. *Soil Biology & Biochemistry* 166: 108565.

Kravchemko, A. N., Toosie, E. R., Guber, A. K., Ostrom, N. E., Yu, J., Azeem, K., Rivers, M. L. and Robertson, G. P. 2017. Hotspots of soil N_2O emission enhanced through water absorption by plant residue. *Nature Geoscience* 10: 496–500.

Kravchenko, A. N., Fry, J. E. and Guber, A. K. 2018. Water absorption capacity of soil-incorporated plant leaves can affect N_2O emissions and soil inorganic N concentrations. *Soil Biology & Biochemistry* 121: 113–119.

Li, C., Frolking, S. and Frolking, T. A. 1992. A model of nitrous oxide evolution from soil driven by rainfall events: 1 Model structure and sensitivity. *Journal of Geophysical Research: Atmospheres* 97: 99759–99776.

Li, X., Zhao, R., Li, D., Wang, G., Bei, S., Ju, X., An, R., Li, L., Kuyper, T. W., Christie, P. and Others 2023. Mycorrhiza-mediated recruitment of complete denitrifying Pseudomonas reduces N_2O emissions from soil. *Microbiome* 11(1): 45.

Li, Y., Clough, T. J., Moinet, G. Y. K. and Whitehead, D. 2021. Emissions of nitrous oxide, dinitrogen and carbon dioxide from three soils amended with carbon substrates under varying soil matric potentials. *European Journal of Soil Science* 72(5): 2261–2275.

Liang, C., Amelung, W., Lehmann, J. and Kästner, M. 2019. Quantitative assessment of microbial necromass contribution to soil organic matter. *Global Change Biology* 25: 3578–3590.

Liao, L., Wang, J., Dijkstra, F., Lei, S., Zhang, L., Wang, X., Liu, G. and Zhang, C. 2024. Nitrogen enrichment stimulates rhizosphere multi-element cycling genes via mediating plant biomass and root exudates. *Soil Biology and Biochemistry* 190: 109306.

Lu, Y., Kronzucker, H. J., Yu, M., Shabala, S. and Shi, W. 2024. Nitrogen-loss and carbon-footprint reduction by plant-rhizosphere exudates. *Trends in Plant Science* 29(4): 469–481.

Lucas, M., Gil, J., Robertson, G. P., Ostrom, N. E. and Kravchenko, A. 2023. Changes in soil pore structure generated by the root systems of maize, sorghum and switchgrass

affect in situ N_2O emissions and bacterial denitrification. *Biology & Fertility of Soils*. https://doi.org/10.1007/s00374-023-01761-1
Maheshwari, A., Jones, C., Tiemann, M. and Hallin, S. 2023. Carbon substrate selects for different lineages of N_2O reducing communities in soils under anoxic conditions. *Soil Biology & Biochemistry* 177: 108909.
Millar, N. and Baggs, E. M. 2005. Relationships between N_2O emissions and water-soluble C and N contents of agroforestry residues after their addition to soil. *Soil Biology & Biochemistry* 35: 605–608.
Moreau, D., Bardgett, R. D., Finlay, R. D., Jones, D. L. and Philippot, L. 2019. A plant perspective on nitrogen cycling in the rhizosphere. *Functional Ecology* 33: 540–552.
Morley, N. J., Richardson, D. J. and Baggs, E. M. 2014. Substrate induced denitrificaton over or under estimates shifts in soil N_2/N_2O ratios. *PLoS One* 9(9): e108144.
Muratore, C., Espen, L. and Prinsi, B. 2021. Nitrogen uptake in plants: the plasma membrane root transport systems from a physiological and proteomic perspective. *Plants* 10(4): 681.
Nuccio, E. E., Starr, E., Karaoz, U., Brodie, E. L., Zhou, J., Tringe, S. G., Malmstrom, R. R., Woyke, T., Banfield, J. F., Firestone, M. K. and Others 2020. Niche differentiation is spatially and temporally regulated in the rhizosphere. *The ISME Journal* 14(4): 999–1014.
Nye, P. H. 1981. Changes of pH across the rhizosphere induced by roots. *Plant and Soil* 61: 7–26.
Parkin, T. B. 1987. Soil microsites as a source of denitrification variability. *Soil Science Society of America Journal* 51: 1194–1199.
Philippot, L., Raaijmakers, J. M., Lemanceau, P. and van der Putten, W. H. 2013. Going back to the roots: the microbial ecology of the rhizosphere. *Nature Reviews Microbiology* 11: 789–799.
Rohe, L., Apelt, B., Vogel, H. J., Well, R., Wu, G. M. and Schluter, S. 2021. Denitrification in soil as a function of oxygen availability at the microscale. *Biogeosciences* 18: 1185–1201.
Rosenberg, K., Bertaux, J., Krome, K., Hartmann, A., Scheu, S. and Bonkowski, M. 2009. Soil amoebae rapidly change bacterial community composition in the rhizosphere of Arabidopsis thaliana. *The ISME Journal* 3: 675–684.
Rummel, P. S., Well, R., Pausch, J., Pfeiffer, B. and Dittert, K. 2021. Carbon availability and nitrogen mineralization control denitrification rates and product stoichiometry during initial maize litter decomposition. *Applied Sciences* 11: 5309.
Saghaï, A., Wittorf, L., Philippot, L. and Hallin, S. 2022. Loss in soil microbial diversity constrains microbiome selection and alters the abundance of N-cycling guilds in barley rhizosphere. *Applied Soil Ecology* 169: 104224.
Santos-Medellín, C., Blazewicz, S. J., Pett-Ridge, J., Firestone, M. K. and Emerson, J. B. 2023. Viral but not bacterial community successional patterns reflect extreme turnover shortly after rewetting dry soils. *Nature Ecology & Evolution* 7: 1809–1822.
Schlüter, S., Lucas, M., Grosz, B., Ippisch, O., Zawallich, J., He, H., Dechow, R., Kraus, D., Blagodatsky, S., Senbayram, M. and Others 2024. The anaerobic soil volume as a controlling factor of denitrification: a review. *Biology & Fertility of Soils*. Published online.
Schmidt, J. E., Kent, A. D., Brisson, V. L. and Gaudin, A. C. M. 2019. Agricultural management and plant selection interactively affect rhizosphere microbial community structure and nitrogen cycling. *Microbiome* 7: 146.

Sieradzki, E. T., Nuccio, E. E., Pett-Ridge, J. and Firestone, M. K. 2023. Rhizosphere and detritusphere habitats modulate expression of soil N-cycling genes during plant development. *mSystems* 8(5): e00315–23.

Sokol, N. W., Sanderman, J. and Bradford, M. A. 2018. Pathways of mineral-associated soil organic matter formation: integrating the role of plant carbon source, chemistry, and point of entry. *Global Change Biology* 25: 12–24.

Sokol, N. W., Slessarev, E., Marschmann, G. L., Nicolas, A., Blazewicz, S. J., Brodie, E. L., Firestone, M. K., Foley, M. M., Hestrin, R., Hungate, B. A. and Others 2022. Life and death in the soil microbiome: how ecological processes influence biogeochemistry. *Nature Reviews Microbiology* 20: 415–430.

Starr, E. P., Nuccio, E. E., Pett-Ridge, J., Banfield, J. F. and Firestone, M. K. 2019. Metatranscriptomic reconstruction reveals RNA viruses with the potential to shape carbon cycling in soil. *Proceedings of the National Academy of Sciences* 116(51): 25900–25908.

Subbarao, G., Wang, H., Ito, O., Ishikawa, T., Rao, I. M., Nakahara, K. and Berry, W. 2007. NH4+ triggers the synthesis and release of biological nitrification inhibition compounds in *Brachiaria humidicola* roots. *Plant and Soil* 290: 245–247.

Suttle, C. A. 2007. Marine viruses-major players in the global ecosystem. *Nature Reviews Microbiology* 5: 801–812.

Thirkell, T. J., Cameron, D. D. and Hodge, A. 2016. Resolving the 'nitrogen paradox' of *arbuscular mycorrhizzas*: fertilization with organic matter brings considerable benefits for plant nutrition and growth. *Plant, Cell & Environment* 39(8): 16383–16390.

Williamson, K. E., Fuhrmann, J. J., Wommack, F. E. and Radosevich, M. 2017. Viruses in soil ecosystems: an unknown quantity within an unexplored territory. *Annual Review of Virology* 4: 201–219.

Zhang, S., Pei, L., Zhao, Y., Shan, J., Zheng, X., Xu, G., Sun, Y. and Wang, F. 2023. Effects of microplastics and nitrogen deposition on soil multifunctionality, particularly C and N cycling. *Journal of Hazardous Materials* 451: 131152.

Zhu, B., Gutknecht, J. L. M., Herman, D. J., Keck, D. C., Firestone, M. K. and Cheng, W. 2014. Rhizosphere priming effects on soil carbon and nitrogen mineralization. *Soil Biology & Biochemistry* 76: 183–192.

Zhu, S., Vivanco, J. M. and Manter, D. K. 2016. Nitrogen fertilizer rate affects root exudation, the rhizosphere microbiome and nitrogen-use-efficiency of maize. *Applied Soil Ecology* 107: 324–333.

Chapter 11

Advances in understanding and exploiting the role of soil microbiomes in protecting crops against pathogens and pests

Rekha Kandaswamy, Bhagyam Agri-Enclave Pvt. Ltd, India; Himani Datta, SGS North America Inc., USA; Niladri Chaudhry, Food Corporation of India, India; Anuradha Gautam, Jawaharlal Nehru University, India; and Vasvi Chaudhry, University of Tübingen, Germany

1 Introduction

Estimated annual losses of US$40 billion have been attributed to the effects of biotic stress on crop losses and reduced yields (Rahman et al., 2018). The principal control measure in many agricultural systems has been to spray crops with synthetic pesticides. However, it is estimated that only 0.1% of sprayed chemicals end up contributing to crop protection (Fukamachi, 2019). The remainder is lost to the wider environment, where it poses potential risks to non-target species, human health, and broader agroecosystem function. This has resulted in increasing restrictions on pesticide use. In addition, because of continued exposure to chemical pesticides, predating insects and pathogens

http://dx.doi.org/10.19103/AS.2024.0136.23

have developed resistance to many synthetic pesticides, further limiting use and efficacy (Watson, 2018). Measures such as the development of genetically-modified (GM) crops have helped control some plant diseases (Bawa and Anilakumar, 2013). However, restrictions on GM crop production in many countries limit its role. This has led to the search for alternative, more sustainable solutions.

Soil microorganisms represent Earth's most abundant and diverse organisms (Locey and Lennon, 2016). It is estimated that 1 g of soil contains billions of microbes of thousands of different species consisting of prokaryotes (bacteria and archeae) and eukaryotes (including unicellular protozoa and multicellular nematodes, fungi, and viruses) (Torsvik and Øvreås, 2002). These microbes play a crucial role in the breakdown of organic matter from both plants and animals, contributing to plant growth, soil structure improvement, and fertility enhancement (Stavi and Lal, 2015). They play a crucial role in plant growth and protection against pathogens and pests (Lau and Lennon, 2012; Hubbard et al., 2019). Soil microbes provide indirect protection by promoting plant growth and health through enhanced nutrient availability (for essential minerals like phosphorus, iron, etc.) and enhanced production of plant growth hormones. Direct mechanisms involved in plant protection include microbial secretion of defense compounds (e.g. salicylic acid) and microbe-induced synthesis and exudation of phytochemicals with a protective function (in the form of root exudates, shoot exudates, and plant volatiles).

It is now well established that agricultural management practices, including soil tillage, fertilization, and pesticide applications, can have detrimental impacts on the structure, diversity, and dynamics of soil microbial communities (de Vries et al., 2012; Tsiafouli et al., 2015). Chaudhry et al. (2012), for example, have shown that long-term land management practices substantially influence the structure of microbial communities, their diversity, and the distribution of bacterial phyla within soils. This study showed differences in predominant bacterial phyla in different soil types, such as *Proteobacteria*, *Bacteroidetes*, *Gemmatimonadetes* (organically cultivated soil), *Acidobacteria* (fallow grassland), and *Actinobacteria* (chemically treated soil). This study also found that measures such as applying organic amendments can selectively promote the growth of specific bacterial phyla by optimizing various soil physicochemical properties, including pH, organic carbon, organic nitrogen, and microbial biomass carbon (Chaudhry et al., 2012).

However, due to the complex and dynamic nature of soil microbiomes, designing microbiome-focused crop protection strategies remains a challenge. This chapter reviews different interactions between microbiome components, factors affecting the soil microbiome, microbe-induced plant defense response mechanisms, and strategies for engineering the soil microbiome to protect crop plants against pests and pathogens.

2 Protective interactions between plant root exudates and the soil microbiome

It is reported that 5–20% of the photosynthetic yield of a plant is transferred to soil in the form of root exudates (REs), which, in turn, are used by the soil microbiome for growth and nutrition (Marschner, 1995). REs play a key role in promoting symbiotic relationships between plants and microbes in the rhizosphere, which benefit both in areas such as nutrient acquisition (Vives-Perez et al., 2020). This means that the rhizosphere microbiome possesses a larger diversity of microorganisms compared to the microbiome of plants and bulk soil (Smalla et al., 2001; Kowalchuk et al., 2002). REs are highly impacted by environmental factors such as pH, temperature, and water status; plant factors like species/cultivars and plant physiological stage (Chaparro et al., 2013a, 2013b); as well as microbial factors like pathogenic attack (De-la-Peña et al., 2012; Oburger et al., 2013).

Components of REs include high molecular weight compounds like intact root border cells, sloughed-off cells and tissue, proteins, and polysaccharides, as well as low molecular weight compounds including sugars, amino acids, amides, and organic acids, as well as aromatic and phenolic acids (Bais et al., 2006; Zhang et al., 2015). These RE components perform various functions in the soil. They can act as a source of nutrition, antimicrobials, and chemo-attractants or repellents for soil bacteria (Walker et al., 2003; Berendsen et al., 2012; Ellouze et al., 2012). The composition and diversity of rhizosphere microbiomes can be significantly affected by selective pressure exerted by antimicrobial compounds present in root exudates (Griffiths et al., 2006; Kielak et al., 2008; Bulgarelli et al., 2012; Peiffer et al., 2013).

RE components such as sugars, organic acids (malic acid), and secondary metabolites recruit beneficial bacteria, facilitating root colonization, while antimicrobial compounds limit the growth of pathogenic organisms. On being attacked by pathogens, plant roots exude specific metabolites to recruit beneficial bacteria and repel pathogenic organisms as a defense response. Certain RE components induce stress responses in microbes, synthesizing antibacterial and antifungal compounds (Cotton et al., 2019; Koprivova et al., 2019; Hart et al., 2018). RE also induces/interacts with the quorum-sensing signaling (QSS) molecules of bacteria. As an example, compounds like L-malic acid were shown to be exuded by *Arabidopsis* root under attack by the pathogenic *Pseudomonas syringae* to recruit beneficial organisms like *Bacillus subtilis,* which then compete with the pathogen (Bais et al., 2006). Another study in cucumber showed a significant increase in tryptophan and a decrease in raffinose levels in the REs upon infection by *Fusarium oxysporum*, along with alteration in 89 other primary metabolites, recruiting beneficial *Bacillus amyloliquefaciens* (Liu et al., 2017).

Root-exuding mechanisms used by plants for recruiting plant-beneficial organisms can also sometimes be exploited by pathogens. For instance, the root exudation of compounds like sesquiterpene (E)-b-caryophyllene (Ebc), benzoxazinoids (BXs) (Neal et al., 2012), and strigolactones has been reported to be used by soil-borne fungal pathogen *Colletotrichum graminicola* (Fantaye et al., 2015), fungi, and the western corn rootworm (Saunders et al., 2009; Robert et al., 2012), as well as pathogenic nematodes and parasitic weeds to locate their respective hosts (Lopez-Raez et al., 2017).

3 Biomolecules with a protective function against pathogen and pest attack

Biomolecules with a protective function include:

- Salicylic acid;
- Malic acid;
- Phenolics;
- Benzoxazinoids;
- Pyrrolizidine alkaloids; and
- Cardenolides.

These are discussed below.

Salicylic acid (SA) is a key defense signaling molecule and a secondary metabolite regulating the composition of the root microbiome. It has been reported that plant growth-promoting rhizobacteria (PGPR) colonization stimulates the synthesis of SA directly or indirectly. When attacked by a pathogen, SA enhances plants' systemic resistance (Malamy et al., 1992; Solano et al., 2008; Schilirò et al., 2012). Transcriptome and metabolite studies have shown that PGPR colonization, for example, gradually induces a defense response in rice plants systematically and externally, as observed through the enhanced accumulation of SA in roots and root exudates (Malamy et al., 1990; Rekha et al., 2019, 2020). In response to pests and pathogens, SA-dependent immune responses are triggered in plants (Song et al., 2015; Yang et al., 2011).

Malic acid (MA) is a tricarboxylic acid cycle metabolite. It is the major component of root exudates. Certain PGPR, such as *B. subtilis* and *Azospirillum brasilense*, favor MA as their secondary carbon source following glucose (Bashan and de-Bashan, 2002; Meyer et al., 2011). A transcriptome study performed on *B. subtilis* OKB105 grown in the presence of rice seedlings showed enhanced expression of the MA catabolic gene *ywkA* (Xie et al., 2015). Other studies have also demonstrated MA as a key chemoattractant for species such as *A. brasilense* and *Paenibacillus polymyxa* SQR-21, and *B. subtilis* RR4 (Zhulin and

Armitage, 1993; Ling et al., 2011; Rekha et al., 2018a, 2018b, 2020). Under pathogenic attack by *Pseudomonas syringae*, enhanced exudation of MA was observed in *Arabidopsis* roots, which facilitated recruitment and colonization by *B. subtilis (*Rudrappa et al., 2008).

Phenolics (e.g. coumarins) enhance plant growth by increasing the bioavailability of iron. Coumarins play a role in the selective recruitment of beneficial rhizobacteria and the suppression of soil-borne pathogens (Voges et al., 2019; Stringlis et al., 2019). In addition, they help deter herbivore attacks by imparting a bitter taste to the plants. Colonization by PGPR enhances the biosynthesis of phenolics and flavonoids, as observed from the enhanced levels of transcripts encoding corresponding biosynthetic genes, namely chalcone synthase C2 (flavonoid biosynthesis), (anthocyanin biosynthesis), and lysine decarboxylase-like protein (quinolizidine alkaloid biosynthesis) (Rekha et al., 2018).

BXs are pivotal in shaping fungal communities within the soil microbiome and providing plants with defense mechanisms against pest-related threats (Cotton et al., 2019).

Pyrrolizidine alkaloids are a group of plant defense compounds that act as toxins for various insect pests and soil organisms, including nematodes and microbes (Hol and Veen, 2002; Thoden and Boppre, 2010; Macel, 2011). Given their antifungal activity, PAs play a significant role in regulating the fungal population in the soil (Kowalchuk et al., 2006; Joosten and Van Veen, 2011). A study in *Jacobaea vulgaris* by Hol et al. (2004) demonstrated enhanced levels of PA in roots after mechanical damage, while a decreased level of PA was observed in roots on herbivore grazing. Reduced PA levels in roots increased the proliferation of soil fungal pathogens, subsequently leading to decreased root biomass (Hol et al., 2004).

Compounds like cardenolides also possess antimicrobial activity (Jacobsohn and Jacobsohn, 1985; Akhtar et al., 1992; Bertol et al., 2011; Agrawal et al., 2012). Plants like milkweeds synthesize this highly toxic steroidal secondary metabolite to prevent insect herbivores.

4 Microbiome-related defense mechanisms against pathogen and pest attack

It is well known that colonization of the rhizosphere by beneficial microbes reduces the incidence of pathogen infection due to competitive colonization (Lugtenberg and Kamilova, 2009). Like plants, microbes in soil secrete a variety of compounds such as phytohormones, organic acids, antimicrobials (siderophores, antibiotics, hydrogen cyanide), extracellular enzymes (cellulases, chitinases), signaling molecules (*N*-acyl-homoserine lactones), and cell surface factors that directly regulate other microbes. Species of Pseudomonads, *Bacillus,* and *Streptomyces* produce a variety of antibiotics, including phenazine, diacetyl-pluro-glucinol, hydrogen cyanide, oligomycin, etc.; bacteriocins like nisin; and antifungal compounds like pyoluteorin (Ping and Boland, 2004; Mabood et al., 2008; Paulin et al., 2009; Ryan et al., 2009).

Plants have the ability to identify bacterial cell surface elements such as flagellins, lipopolysaccharides, and microbe-associated molecular patterns (MAMPs) through specific high-affinity cell surface receptors known as pattern-recognition receptors (PRRs). Upon recognition of a pathogen, these receptors initiate immune responses in the plant (Ping and Boland, 2004; Dangl et al., 2013). The MAMP–PRR interaction is crucial for recruiting beneficial bacteria and deterring pathogenic organisms.

Colonization by beneficial microbes can promote plant immune responses through:

- systemic acquired resistance (SAR); and
- induced systemic resistance (ISR).

SAR is triggered after a localized attack by a pathogen or herbivore. The mechanisms involved in SAR include direct induction of defense responses through phytohormone biosynthesis and enhanced expression of pathogenesis-related genes (Durrant and Dong, 2004). In contrast, induced systemic resistance (ISR) entails the indirect stimulation of defense responses by modifying the plant's physiological condition, enabling enhanced and more effective reactions to biotic stress. This phenomenon is frequently described as the 'priming effect' (Conrath et al., 2006).

Beneficial microbes induce ISR, while pathogen attack stimulates SAR (Freeman and Beatie, 2008; Van der Ent et al., 2009). Signaling molecules like AHLs play a critical role in plant defense responses. For example, Ortiz-Castro et al. (2008) showed the bacterially synthesized AHL affected *Arabidopsis* root development and the induction of ISR. Compounds like jasmonic acid (JA) and ethylene (ET) play a critical role in ISR, while salicylic acid promotes SAR. Figure 1 illustrates the different defense signaling mechanisms mediated by JA, ET, and SA.

A study by Hu et al. (2018) showed increased root exudation of 6-methoxy-benzoxazolin-2-one after exposure to pathogens, enhancing JA-dependent defense responses in maize. Similar studies in *Arabidopsis* infected by *P. syringae* DC3000 (Pst) and downy mildew over multiple generations showed significant changes in root exudate composition, enhancing disease resistance in subsequent generations (Yuan et al., 2018; Berendsen et al., 2018). Long-chain organic acids were the major contributors to this resistance mechanism (Yuan et al., 2018).

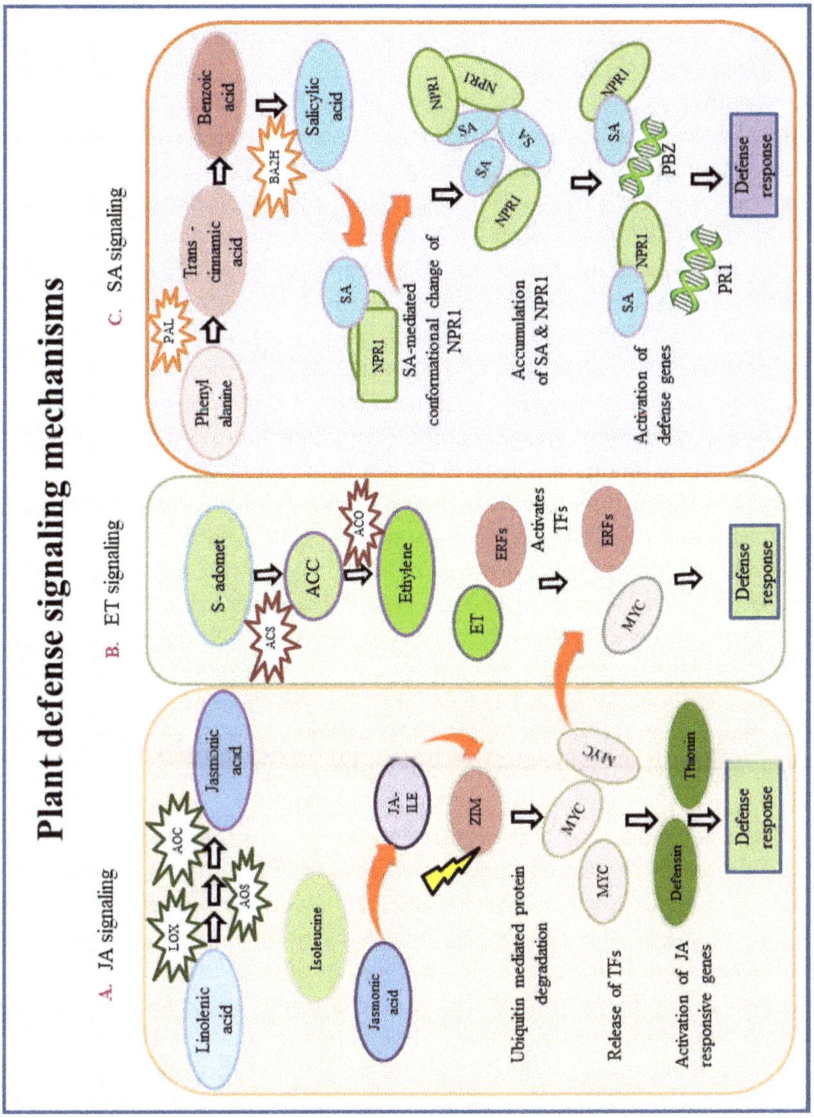

Figure 1 Illustration of distinct defense signaling mechanisms mediated by jasmonic acid (JA), ethylene (ET), and salicylic acid (SA) in plants.

A study of *Arabidopsis* showed that bacteria activated the ISR-regulatory transcription factor MYB72 and the downstream gene beta-glucosidase *BGLU42* (Stringlis et al., 2018). A similar study in *Arabidopsis* inoculated with a synthetic rhizobacterial community induced the expression of the *PHR1* gene, which regulates the ISR pathway through JA-mediated responses and synthesis of callose to stimulate production of glucosinolates (Castrillo et al., 2017; Khan et al., 2016; Pant et al., 2015).

Like pathogen infection, ISR and/or SAR mechanisms protect plants from pest attacks. Some critical metabolites involved in the defense mechanisms against insect pests include pyrrolizidine alkaloids, cardenolides, and glucosinolates. PGPRs like mycorrhizal fungi and Trichoderma have been reported to induce ISR against foliar herbivory through JA and ET-dependent signaling pathways (Pieterse et al., 2014; Pangesti et al., 2016). Studies in Arabidopsis have shown *LOX2*, *PDF1.2*, and *HEL* to be the essential genes involved in the PGPR-induced defense responses against insect pests (Pangesti et al., 2014, 2016; van Oosten et al., 2008). Other studies in cotton and Arabidopsis have demonstrated induction of genes of the JA-(*GhAOS*, *GhLOX1*, and *GhOPR3*) and ET-(*ORA59* and *PDF1.2*) signaling pathways, illustrating the priming effect of PGPRs against pest attack. Additionally, volatiles from PGPRs directly induce the synthesis of defense compounds such as glucosinolates in Arabidopsis and gossypol in cotton, conferring protection against insect herbivory (Pangesti et al., 2016; van de Mortel et al., 2012; Zebelo et al., 2016; Aziz et al., 2016). Figure 2 depicts the different modes of crop protection induced by PGPR. Some pests can avoid these defenses. For instance, phloem feeders like aphids and whiteflies feed on the phloem sap, avoiding damage to leaf tissues and the plant's immune system. This phloem-feeding characteristic also avoids the toxic effects of plant defense metabolites induced by PGPR colonization present in leaf tissue (Pineda et al., 2010).

It has been observed that disease incidence over time can be reduced due to an increase in disease-suppressive activity of the soil microbiome. For example, a study by Kwak and Weller (2013) showed that wheat crops grown in soil infested with the pathogenic fungus *Gaeumannomyces graminis* pv. *tritici* showed a gradual decrease in infection. Another study by Mendes et al. (2011) also showed a long-term shift toward a disease-suppressive soil after prolonged exposure to *Rhizoctonia solani*. Similar to the priming effect (ISR/SAR) observed in pathogen-affected plants, continued exposure to pathogenic microorganisms alters the composition of root exudates to deter pathogens, thereby modifying the soil and rhizosphere microbiome to control disease-causing pathogens. The mechanisms involved are also referred to as the soil-feedback responses (Gomez Exposito et al., 2017).

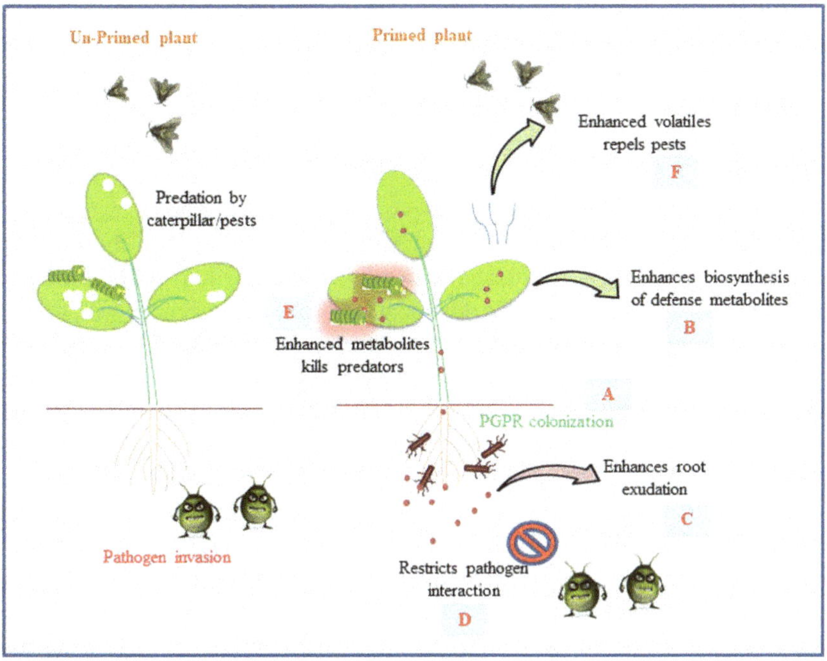

Figure 2 The figure depicts various modes of crop protection that are stimulated by plant growth-promoting rhizobacteria (PGPR). These mechanisms encompass a range of strategies employed by PGPR to enhance plant growth, bolster resistance against pests and pathogens, and promote overall crop health.

5 Engineering the soil microbiome for improved protection against pathogens and pests

The disease-suppressive activities of soil microbiomes have great potential in priming plants against pest and pathogen attacks. Several strategies have been adopted in the recent past to engineer the soil microbiome to exploit their disease-suppressive activity.

Techniques include:

- plant breeding;
- use of organic supplements;
- adding root exudate metabolites;
- adding artificial microbial consortia (AMC) as probiotics; and
- microbiome breeding and transplantation.

Among the above strategies, engineering microbial consortia and 'microbe-friendly' crops obtained from plant breeding/selection appear to be the most effective strategies in engineering microbiomes.

In the case of plant breeding, genetic variation can potentially enhance beneficial associations of plants with the rhizosphere microbiome. For example, Remans et al. (2007) found two quantitative trait loci (QTLs) involved in auxin sensing in common beans, which could play a significant role in associations with auxin-synthesizing bacteria, given that auxin is critical in plant defense responses.

The amendment of soil with compost and agricultural co-products such as biochar enriches the abundance of biocontrol organisms in the soil (Bell et al., 2011; De Corato, 2020). For instance, adding wheat bran or ethanol altered the soil microbiome and decreased disease incidence in tomatoes by *Pyrenochaeta lycopersici* (Testen and Miller, 2018). Carbon (C)-rich products like biochar (obtained from the partial or complete pyrolysis of organic matter like manure, bones, crop residues, greenhouse, and slaughterhouse wastes) have been shown to promote crop growth and resilience to disease (Whitman et al., 2016; Liu et al., 2017). Biochar application has been shown to possess a biocontrol effect against various pathogens like *Botrytis cinerea, F. oxysporum, Podosphaera aphanis, Pythium aphanidermatum, Phytophthora cactorum, Rhizoctonia solani,* and *Sclerotinia sclerotiorum* (Eizenberg et al., 2017). The incidence of disease from soilborne pathogens of crop plants (like Fusarium and *Ralstonia solanacearum*) and foliar fungal pathogens of pepper and tomato (like *B. cinerea*) was significantly reduced by the application of biochar to soil (Matsubara et al., 2002; Nerome et al., 2005; Elad et al., 2010). Volatile organic carbon from biochar also inhibits plant pathogens (Graber et al., 2010). Mechanisms underlying crop protection effects by biochar amendment involve the induction of priming of early- and late-acting defense genes of the ET and JA signaling pathways.

Incorporating biochar into soil systems also exerts suppressive effects on pest infestations. A study conducted by Elad et al. (2010) demonstrated a decrease in the damage inflicted by the broad mite pest (*Polyphagotarsonemus latus* (banks) on pepper plants due to the application of biochar amendments. Damage caused by nematodes like *Pratylenchus penetrans* (George et al., 2016) and parasitic weeds like *Phelipanche aegyptiaca* was significantly reduced by biochar amendment (Elad et al., 2012).

Greenhouse studies have shown that adding root exudate metabolites like MA and SA suppressed pathogen attack (*F. oxysporum*) and promoted the growth of rice plants in the face of abiotic stresses such as salinity (Rekha et al., 2019). A study by Liu et al. (2019) pretreated plants cultivated under saline conditions with root exudates and ethylene precursor 1-aminocyclopropane-1-carboxylic acid (ACC). The amendment of root exudates and ACC resulted in the recruitment of a beneficial microbiome that enhanced the saline tolerance capacity of plants and resistance to pathogen attack. ACC reduces the biosynthesis of the stress hormone ethylene, thereby shifting the plants' metabolism away from a stress response toward growth and greater potential resilience to pathogen attack. Later stress encountered by the plant stimulates the ACC-enhanced microbiome to increase root exudation levels of ACC, further enhancing resilience. Similar techniques can be used to enhance the soil microbiota for enhanced crop protection.

Metagenomics provides a useful tool for screening, identifying, and producing potentially beneficial bioactive compounds from soil (Mantri et al., 2021). Other approaches to identify bioactive compounds include sequencing the fosmid library of biosynthetic gene clusters (BGCs) and subsequent identification of BGCs (Santana-Pereira et al., 2020; Negri et al., 2022). A study by Rodrigo Mendes et al. (2011) of the rhizosphere microbiome identified key bacterial taxa and genes involved in suppressing a fungal root pathogen. Through PhyloChip-based metagenomics and culture-dependent functional analyses, Proteobacteria, Firmicutes, and Actinobacteria were consistently linked with disease suppression. Non-ribosomal peptide synthetases enzymes, specifically from γ-Proteobacteria, were found to be the key peptide secondary metabolites involved. The introduction of these bacteria or other bacteria overexpressing these genes in soil can help engineer microbiomes for enhanced crop protection (Weyens et al., 2010).

Developing AMC involves a number of steps, which include the selection of the microbes, culture of desired microorganisms, and testing the efficacy of the reconstructed synthetic consortia (Kong et al., 2018). A study by Baas et al. (2016) demonstrated enhanced availability of phosphate and a two-fold increase in crop production by using a consortium of microbes that included *Citrobacter freundii*, *Comamonas testosteroni*, *Enterobacter cloacae*, and *Pseudomonas putida*. A similar study in wheat under field conditions demonstrated enhanced nutrient uptake by using the bacteria *Azotobacter vinelandii* and the mycorrhizal fungus *Rhizophagus irregularis* as the diazotrophic N-fixers (Dal Cortivo et al., 2018). Another study in maize demonstrated enhanced drought tolerance using a consortium of *Acinetobacter* sp. EMM02, *Azospirillum brasilense* Sp7, *P. putida* KT2440, and *Sphingomonas* sp. OF178 (Molina-Romero et al., 2017).

Microbiome breeding and transplantation is a strategy that involves breeding beneficial microbes using host plants over multiple generations. Panke-Buisse et al. (2017) and Panke-Buisse et al. (2015) selected microbes bred for ten generations based on their influence on late or early flowering traits in *Arabidopsis thaliana*. Transplantation of the 'late-flowering stage microbiomes' enhanced inflorescence in *Arabidopsis* plants.

Efforts to introduce or remove specific microbial communities within soil ecosystems have encountered significant challenges and achieved only limited success (Thompson et al., 2005; Gerhardt et al., 2009; Qiu et al., 2014). Promising results obtained from microbiome engineering in laboratory studies are often difficult to translate into the field conditions due to differences in responses of soil inhabitants (culturable and non-culturable microbes) to biotic and abiotic factors, which differ with species (Rousk et al., 2009, 2010; Haichar et al., 2008; Berg and Smalla, 2009). Some examples include pre-treatment of tomato seedlings with cultures of *Bacillus amyloliquefaciens*, *Bacillus velezensis*, and *Acinetobacter* sp. UQ202 conferred protection against soil-borne *Phytophthora capsici* infection and promoted plant growth (Syed-Ab-Rahman et al., 2019). Another study in tomato seedlings showed that a consortia treatment of *Pseudomonas stutzeri* and *Stenotrophomonas maltophilia* emit compounds like dimethyl disulfide, conferring protection against the shoot pathogen *B. cinerea* (Rojas-Solís et al., 2018). Root microbiome and metagenome studies on *Ralstonia solanacearum*-resistant and tomato varieties identified flavobacteria in the rhizosphere of disease-resistant varieties (Kwak et al., 2018). Soil inoculation of this flavobacteria-rich microbiome conferred immunity against the pathogen.

Other approaches include the exogenous treatment of plants with defense hormones like SA or generating mutations in defense hormone synthesis pathways. For example, Tsolakidou et al. (2019) designed two synthetic bacterial communities with enhanced ACC deaminase activity. The introduction of these reconstructed microbial consortia demonstrated antimicrobial activity, reducing disease symptoms caused by *F. oxysporum* f. sp. *lycopersici* and improving the growth of tomatoes even in poor nutrient conditions.

Microbes can also play a significant role in enhancing plant defense against attack by insects through the priming effect. Research by Martínez-Medina et al. (2016) highlighted an elevated resistance in plants primed by soil microbes against leaf-chewing herbivores. Soil inoculation trials exhibited increased resistance to infestation by the thrips *Frankliniella occidentalis* compared to sterile soil conditions. Research by Badri et al. (2013) showed that soil microbiomes significantly enhanced the resistance in Arabidopsis plants against leaf-feeding herbivorous insects. Most research on plant–insect–microbe interactions has been performed with a limited number of microbial strains, the results of which are not feasible for field conditions (Gadhave et al., 2016; Timmusk et al., 2017). Studying the whole soil microbiome is critical in designing strategies for crop protection against insect pests. It is also important to take account of the fact that the structure of the soil microbiome is altered on the activity of root and shoot herbivores (Bardgett & Wardle 2010; Bennett 2010). Belowground herbivory (BG) changes root biomass, composition, and concentration of root exudates, which directly influence the soil microbiome, apart from directly interacting with the soil microbes through competition, facilitation, and/or predation (Anderson et al., 2011; Bardgett et al., 1999; van Dam, 2009). Aboveground herbivory (AG) alters the soil microbiome indirectly by modifying the levels and diversity of metabolites in root exudates and plant tissues (Bardgett et al., 1998; Mikola et al., 2001; Soler et al., 2007; Hamilton et al., 2008).

6 Conclusion

This chapter has discussed the soil microbiome, interactions between plants and microbes, the mechanisms involved in plant defense responses, and the critical biomolecules involved in plant–microbe–insect interactions. The engineering of microbiomes through different strategies and the factors associated with successful crop protection have been discussed. While microbiome engineering has the potential to revolutionize agricultural practices and mitigate the negative impacts of pests and pathogens, careful consideration of its potential effects (e.g. through appropriate risk analysis) is paramount to ensure that it aligns with sustainable, equitable, and environmentally responsible agricultural systems.

7 Acknowledgments

VC acknowledges support from Alexander von Humboldt-Stiftung, Cluster of Excellence 'Controlling Microbes to Fight Infections' (CMFI), and the University of Tübingen, Germany.

8 References

Agrawal A.A., Petschenka G., Bingham R.A., Weber M.G. and Rasmann S. (2012), Toxic cardenolides: chemical ecology and coevolution of specialized plant–herbivore interactions, *New Phytol.*, 194:28–45.

Akhtar N., Malik A., Ali S.N. and Kazmi S.U. (1992), Proceragenin, an antibacterial cardenolide from *Calotropis procera*, *Phytochem.*, 31:2821–2824.

Anderson P., Sadek M. and Wäckers F.L. (2011), Root herbivory affects oviposition and feeding behavior of a foliar herbivore, *Behav. Ecol.*, 22(6):1272–1277.

Aziz M. et al., (2016) Augmenting sulfur metabolism and herbivore defense in Arabidopsis by bacterial volatile signaling, *Front. Plant Sci.*, 7:458.

Baas P. et al., (2016), Phosphorus mobilizing consortium Mammoth P (TM) enhances plant growth, *Peer J.*, 4:e2121.

Badri D.V., Zolla G., Bakker M.G., Manter D.K. and Vivanco J.M., (2013), Potential impact of soil microbiomes on the leaf metabolome and on herbivore feeding behavior, *New Phytol.*, 198:264–273.

Bais H.P., Weir T.L., Perry L.G., Gilroy S. and Vivanco J.M., (2006), The role of root exudates in rhizosphere interactions with plants and other organisms, *Annu. Rev. Plant Biol.*, 57:233–266.

Bardgett R.D., Wardle D.A. and Yeates G.W., (1998), Linking above-ground and below-ground interactions: how plant responses to foliar herbivory influence soil organisms, *Soil Biol. Biochem.*, 30:1867–1878.

Bardgett R.D., Denton C.S. and Cook R., (1999), Belowground herbivory promotes soil nutrient transfer and root growth in grassland, *Ecol Lett.*, 2:357–360.

Bardgett R.D. and Wardle D.A., (2010), *Aboveground--belowground linkages: biotic interactions, ecosystem processes, and global change,* Oxford: Oxford University Press.

Bashan Y. and de-Bashan L.E., (2002), Protection of tomato seedlings against infection by *Pseudomonas syringae* pv. tomato by using the plant growth-promoting bacterium *Azospirillum brasilense*, *Appl. Environ. Microbiol.*, 68(6):2637–2643.

Bawa A.S., Anilakumar K.R., (2013), Genetically modified foods: safety, risks and public concerns-a review. *J. Food Sci. Technol.*, 50(6):1035–1046.

Bell T.H., Yergeau E., Martineau C., Juck D., Whyte L.G. and Greer C.W., (2011), Identification of nitrogen-incorporating bacteria in petroleum-contaminated Arctic soils by using [15N] DNA-based stable isotope probing and pyrosequencing, *Appl. Environ. Microbiol.*, 77:4163–4171.

Bennett A., (2010), The role of soil community biodiversity in insect biodiversity, *Insect Conserv. Diver.*, 3:157–171.

Berendsen R.L., Pieterse C.M. and Bakker P.A., (2012), The rhizosphere microbiome and health, *Trends Plant Sci.,* 17(8):478–486.

Berendsen R.L., Vismans G., Yu K., Song Y., de Jonge R., Burgman W.P., et al., (2018), Disease-induced assemblage of a plant-beneficial bacterial consortium, *ISME J.*, 12:1496–1507.

Berg G. and Smalla K., (2009), Plant species and soil type cooperatively shape the structure and function of microbial communities in the rhizosphere, *FEMS Microbiol. Ecol.,* 68:1–13.

Bertol J.W., Rigotto C., Pádua R.M., Kreis W., Barardi C.R., Braga F.C., et al., 2011. Antiherpes activity of glucoevatromonoside, a cardenolide isolated from a Brazilian cultivar of *Digitalis lanata*. *Antiviral Res.* 92:73–80.

Bulgarelli D., Rott M., Schlaeppi K., Ver Loren van Themaat E., Ahmadinejad N., Assenza F., et al., (2012), Revealing structure and assembly cues for Arabidopsis root-inhabiting bacterial microbiota. *Nature*, 488:91–95.

Castrillo G., Teixeira P.J., Paredes S.H., Law T.F., de Lorenzo L., Feltcher M.E., et al., (2017), Root microbiota drive direct integration of phosphate stress and immunity, *Nature*, 543:513–518.

Chaparro J.M., Badri D.V., Bakker M.G., Sugiyama A., Manter D.K., and Vivanco J.M., (2013a), Root exudation of phytochemicals in arabidopsis follows specific patterns that are developmentally programmed and correlate with soil microbial functions, *PLoS One,* 8:e55731.

Chaparro J.M., Badri D.V. and Vivanco J.M., (2013b), Rhizosphere microbiome assemblage is affected by plant development, *ISME J.*, 8:790–803.

Chaudhry V., Rehman A., Mishra A., Chauhan P.S., Nautiyal C.S. (2012), Changes in bacterial community structure of agricultural land due to long-term organic and chemical amendments, *Microb Ecol.*, 64:450–460.

Conrath U., Beckers G.J., Flors V., Garcia-Agustin P., Jakab G., Mauch F., et al., (2006), Priming: getting ready for battle, *Mol. Plant. Microbe Interact.*, 19:1062–1071.

Cotton T.E.A., et al., (2019), Metabolic regulation of the maize rhizobiome by benzoxazinoids, *ISME J.*, 13:1647–1658.

Dal Cortivo, C., et al., (2018), Effects of field inoculation with VAM and bacteria consortia on root growth and nutrients uptake in common wheat, *Sustainability,* 10:3286.

Dangl J.L., Horvath D.M. and Staskawicz B.J., (2013), Pivoting the plant immune system from dissection to deployment, *Science*, 341:746–751.

De Corato U., (2020), Disease-suppressive compost enhances natural soil suppressiveness against soil-borne plant pathogens: a critical review, *Rhizosphere*, 3:100192.

De-La-Peña C., Rangel-Cano A. and Alvarez-Venegas R., (2012), Regulation of disease-responsive genes mediated by epigenetic factors: interaction of Arabidopsis-Pseudomonas, *Mol. Plant Pathol.*, 13:388–398.

de Vries, F.T., Liiri, M.E., Bjørnlund, L., Bowker, M.A., Christensen, S., Setälä, H.M., Bardgett, R.D., (2012), Land use alters the resistance and resilience of soil food webs to drought, *Nat. Clim. Change*, 2:276.

Durrant W.E. and Dong X., (2004), Systemic acquired resistance. *Ann. Rev.Phytopathol.*, 42:185–209.

Eizenberg H., et al., (2017), Non-chemical control of root parasitic weeds with biochar, *Front. Plant Sci.*, 8:939.

Elad Y., Rav-David D., Harel Y.M., Borenshtein M., Ben Kalifa H., Silber A., et al., (2010), Induction of systemic resistance in plants by biochar, a soil-applied carbon sequestering agent, *Phytopathol.*, 100:913–921.

Elad Y., et al., (2012), The biochar effect: plant resistance to biotic stresses, *Phytopathol. Mediterr.*, 50:335–349.

Ellouze W., Hamel C., Cruz A.F., Ishii T., Gan Y., Bouzid S., et al., (2012), Phytochemicals and spore germination: at the root of AMF host preference?, *Appl. Soil Ecol.*, 60:98–104.

Fantaye C.A., Kopke D., Gershenzon J. and Degenhardt J., (2015), Restoring (E)-beta-caryophyllene production in a non-producing maize line compromises its resistance against the fungus *Colletotrichum graminicola*, *J. Chem. Ecol.*, 41:213-–223.

Freeman B.C. and Beattie G.A., (2008), An overview of plant defenses against pathogens and herbivores, *The Plant Health Instructor,* 8:94.

Fukamachi K., Konishi Y. and Nomura T., (2019), Disease control of phytophthora infestans using cyazofamid encapsulated in poly lactic-co-glycolic acid (PLGA) nanoparticles, *Colloids Surf. A Physicochem. Eng. Asp.*, 577:315–322.

Gadhave K.R., Finch P., Gibson T.M. and Gange A.C., (2016), Plant growth-promoting Bacillus suppress Brevicoryne brassicae field infestation and trigger density-dependent and density-independent natural enemy responses, *J. Pest Sci.*, 89(4):985–992.

George C., et al., (2016), Biochars reduce infection rates of the root-lesion nematode *Pratylenchus penetrans* and associated biomass loss in carrot, *Soil Biol. Biochem*, 95:11–18.

Gerhardt K.E., Huang X-D., Glick B.R. and Greenberg B.M., (2009), Phytoremediation and rhizoremediation of organic soil contaminants: potential and challenges, *Plant Sci.*, 176:20–30.

Gomez Exposito R., de Bruijn I., Postma J. and Raaijmakers J.M., (2017), Current insights into the role of rhizosphere bacteria in disease suppressive soils, *Front Microbiol.*, 8. doi:10.3389/fmicb.2017.02529.

Graber E.R., Meller-Harel Y., Kolton M., Cytryn E., Silber A., Rav-David D., et al., (2010), Biochar impact on development and productivity of pepper and tomato grown in fertigated soil-less media, *Plant Soil,* 337:481–496.

Griffiths R.I., Bailey M.J., McNamara N.P. and Whiteley A.S., (2006), The functions and components of the Sourhope soil microbiota, *Appl. Soil Ecol.*, 33:114–126.

Haichar F.Z., Marol C., Berge O., Rangel-Castro J.I., Prosser J.I., et al., (2008), Plant host habitat and root exudates shape soil bacterial community structure, *ISME J.*, 2(12):1221–1230.

Hamilton E.W., Frank D.A., Hinchey P.M. and Murray T.R., (2008), Defoliation induces root exudation and triggers positive rhizospheric feedbacks in a temperate grassland, *Soil Biol. Biochem.*, 40:2865–2873.

Hart M., et al., (2018), Fungal inoculants in the field: is the reward greater than the risk?, *Funct. Ecol.*, 32:126–135.

Hol W.H.G., Macel M., Van Veen J.A. and Van Der Meijden E., (2004), Root damage and aboveground herbivory change concentration and composition of pyrrolizidine alkaloids of *Senecio jacobaea*, *Bas. Appl. Ecol.*, 5:253–260.

Hol W.H.G. and van Veen J.A., (2002), Pyrrolizidine alkaloids from *Senecio jacobaea* affect fungal growth, *J. Chem. Ecol.*, 28:1763–1772.

Hu L., Robert C.A.M., Cadot S., Zhang X., Ye M., Li B., Manzo D., Chervet N., Steinger T., van der Heijden M.G.A., et al., (2018), Root exudate metabolites drive plant-soil feedbacks on growth and defense by shaping the rhizosphere microbiota, *Nature Commun.*, 9:2738.

Hubbard C.J., Li B., McMinn R., Brock M.T., Maignien L., Ewers B.E., et al., (2019), The effect of rhizosphere microbes outweighs host plant genetics in reducing insect herbivory, *Mol. Ecol.,* 28:1801–1811.

Jacobsohn M.K. and Jacobsohn G.M., (1985), Production of a fungistat and the role of fungi during germination of *Digitalis purpurea* L cv Gloxiniaflora seeds, *Ann. Bot.,* 56:543–552.

Joosten L. and Van Veen J.A., (2011), Defensive properties of pyrrolizidine alkaloids against microorganisms, *Phytochem. Rev.,* 10:127–136.

Khan G.A., Vogiatzaki E., Glauser G. and Poirier Y., (2016), Phosphate deficiency induces the jasmonate pathway and enhances resistance to insect herbivory, *Plant Physiol.*, 171:632–644.

Kielak A., Pijl A.S., van Veen J.A. and Kowalchuk G.A., (2008), Differences in vegetation composition and plant species identity lead to only minor changes in soil-borne microbial communities in a former arable field, *FEMS Microbiol. Ecol.*, 63:372–382, doi:10.1111/j.1574-6941.2007.00428.x.

Kong Z., et al., (2018), Paving the way from the lab to the field: using synthetic microbial consortia to produce high-quality crops, *Front. Plant Sci.*, 9:1467.

Koprivova A., Schuck S., Jacoby R.P., Klinkhammer I., Welter B., Leson L., et al., (2019), Root-specific camalexin biosynthesis controls the plant growth-promoting effects of multiple bacterial strains, *Proc. Natl. Acad. Sci. U.S.A.,* 116:15735–15744, doi:10.1073/pnas.1818604116.

Kowalchuk G.A., Buma D.S., de Boer W., Klinkhamer P.G.L. and van Veen J.A., (2002), Effects of above-ground plant species composition and diversity on the diversity of soil-borne microorganisms, *Antonie Van Leeuwenhoek,* 81:509–520.

Kowalchuk G.A., Hol W.H.G. and Van Veen J.A., (2006), Rhizosphere fungal communities are influenced by *Senecio jacobaea* pyrrolizidine alkaloid content and composition, *Soil Biol. Biochem.,* 38:2852–2859, doi:10.1016/j.soilbio.2006.04.043.

Kwak Y.S. and Weller D.M., (2013), Take-all of wheat and natural disease suppression: a review, *Plant Pathol. J.,* 29:125–135.

Kwak M.J., et al., (2018), Rhizosphere microbiome structure alters to enable wilt resistance in tomato, *Nat. Biotechnol.*, 36:1100–1109.

Lau J.A. and Lennon J.T., (2012), Rapid responses of soil microorganisms improve plant fitness in novel environments, *Proc. Natl. Acad. Sci. U.S.A.*, 109:14058–14062.

Ling N., Raza W., Ma J., Huang Q. and Shen Q., (2011), Identification and role of organic acids in watermelon root exudates for recruiting *Paenibacillus polymyxa* SQR-21 in the rhizosphere, *Eur. J. Soil Biol.*, 47(6):374–379, doi:10.1016/j.ejsobi.2011.08.009.

Liu Y., et al., (2017a), Identification of root-secreted compounds involved in the communication between cucumber, the beneficial Bacillus amyloliquefaciens, and the soil-borne pathogen Fusarium oxysporum, *Mol. Plant Microbe Interact.,* 30:53–62.

Liu L., et al., (2017b), Biochar amendments increase the yield advantage of legume-based intercropping systems over monoculture, *Agric. Ecosyst. Environ.*, 237:16–23.

Liu H., et al., (2019), Soil amendments with ethylene precursor alleviate negative impacts of salinity on soil microbial properties and productivity, *Sci. Rep.*, 9:6892.

Locey K.J. and Lennon, J.T., (2016), Scaling laws predict global microbial diversity, *Proc. Natl. Acad. Sci. U.S.A.*, 113:5970–5975.

Lopez-Raez J.A., Shirasu K. and Foo E., (2017), Strigolactones in plant interactions with beneficial and detrimental organisms: the Yin and Yang, *Trends Plant Sci.*, 22:527–537.

Lugtenberg B. and Kamilova F., (2009), Plant-growth-promoting rhizobacteria, *Annu Rev Microbiol.*, 63(1):541–556.

Mabood F., Jung W. and Smith D., (2008), Signals in the underground: microbial signaling and plant productivity. In *Molecular Mechanisms of Plant and Microbe Coexistence,* eds. Nautiyal C., Dion P. (Berlin: Springer), 291–318, doi:10.1007/978-3-540-75575-3_12.

Macel M., (2011), Attract and deter: a dual role for pyrrolizidine alkaloids in plant–insect interactions, *Phytochem. Rev.,* 1–8. doi:10.1007/s11101-010-9181-1.

Malamy J., Can J.P. and Klessig D.F., (1990), Salicylic acid: a likely endogenous signal in the resistance response of tobacco to viral infection, *Science*, 250:1001–1004.

Malamy J., Henning J. and Klessig D.F., (1992), Temperature-dependent induction of salicylic acid and its conjugates during the resistance response to tobacco mosaic virus infection, *Plant Cell,* 4:359–366.

Mantri S.S., Negri T., Sales-Ortells H., Angelov A., Peter S., Neidhardt H., et al., (2021) Metagenomic sequencing of multiple soil horizons and sites in close vicinity revealed novel secondary metabolite diversity, *mSystems.* 6(5):e0101821.

Marschner H., (1995), *Mineral nutrition of higher plants*, 2nd edn. Boston, MA, USA: Academic Press.

Martinez-Medina A., Flors V., Heil M., Mauch-Mani B., Pieterse C.M., Pozo M.J., et al., (2016), Recognizing plant defense priming, *Trends Plant Sci.,* 21:818–822, doi:10.1016/j.tplants.2016.07.009.

Matsubara Y.I., Hasegawa N. and Fukui H., (2002), Incidence of Fusarium root rot in asparagus seedlings infected with arbuscular mycorrhizal fungus as affected by several soil amendments, *J. Jpn. Soc. Hortic.*, 71:370–374.

Mendes R., Kruijt M., de Bruijn I., Dekkers E., van der Voort M., Schneider J.H., et al., (2011), Deciphering the rhizosphere microbiome for disease-suppressive bacteria, *Science*, 332:1097–1100.

Meyer F.M., Jules M., Mehne F.M.P., Le Coq D., Landmann J.J., Görke B., et al., (2011), Malate-mediated carbon catabolite repression in *Bacillus subtilis* involves the HPrK/

CcpA pathway, *J. Bacteriol.*, 193(24):6939–6949, doi:10.1128/JB.06197-11, PMID: 22001508.

Mikola J., Yeates G.W., Barker G.M., Wardle D.A. and Bonner K.I., (2001), Effects of defoliation intensity on soil food web properties in an experimental grassland community, *Oikos.*, 92:333–343.

Molina-Romero D., et al., (2017), Compatible bacterial mixture, tolerant to desiccation, improves maize plant growth, *PLoS One*, 12:1–21.

Neal A.L., Ahmad S., Gordon-Weeks R. and Ton J. (2012), Benzoxazinoids in root exudates of maize attract *Pseudomonas putida* to the rhizosphere, *PLoS One*, 7:e35498.

Negri T., Mantri S., Angelov A., Peter S., Muth G., Eustáquio A. S., et al., (2022), A rapid and efficient strategy to identify and recover biosynthetic gene clusters from soil metagenomes, *Appl. Microbiol. Biotechnol.,* 106:3293–3306.

Nerome M.K., Toyota T.M.D., Islam T., Nishijima T., Matsuoka K.S., Yamaguchi Y., et al., (2005), Suppression of bacterial wilt of tomato by incorporation of municipal biowaste charcoal into soil, *Soil Microorg.*, 59:9–14.

Oburger E., Dell'mour M., Hann S., Wieshammer G., Puschenreiter M. and Wenzel W.W. (2013), Evaluation of a novel tool for sampling root exudates from soil-grown plants compared to conventional techniques. *Environ. Exp. Bot.* 87:235–247.

Ortiz-Castro R., Martinez-Trujillo M. and Lopez-Bucio J., (2008), N-acyl-L-homoserine lactones: a class of bacterial quorum-sensing signals alter post-embryonic root development in Arabidopsis thaliana, *Plant Cell Environ.,* 31:1497–1509, doi:10.1111/j.1365-3040.2008.01863.x.

Pangesti N., et al., (2014), Variation in plant-mediated interactions between rhizobacteria and caterpillars: potential role of soil composition, *Plant Biol.*, 17:474–483.

Pangesti N., et al., (2016), Jasmonic acid and ethylene signaling pathways regulate glucosinolate levels in plants during rhizobacteria-induced systemic resistance against a leaf-chewing herbivore, *J. Chem. Ecol.,* 42:1212–1225.

Panke-Buisse K., et al., (2015), Selection on soil microbiomes reveals reproducible impacts on plant function, *ISME J.*, 9:980–989.

Panke-Buisse K., et al., (2017), Cultivated sub-populations of soil microbiomes retain early flowering plant trait, *Microb. Ecol.*, 73:394–403.

Pant B.D., Pant P., Erban A., Huhman D., Kopka J. and Scheible W.R., (2015), Identification of primary and secondary metabolites with phosphorus status-dependent abundance in Arabidopsis, and of the transcription factor PHR1 as a major regulator of metabolic changes during phosphorus limitation, *Plant Cell Environ.*, 38:172–187.

Paulin M.M., et al., (2009), Transcriptional activity of antifungal metabolite-encoding genes phlD and hcnBC in Pseudomonas spp. using qRT-PCR, *FEMS Microbiol. Ecol.,* 68:212–222.

Peiffer J.A., et al., (2013), Diversity and heritability of the maize rhizosphere microbiome under field conditions, *Proc. Nat. Acad. Sci. U.S.A.*, 110:6548–6553, doi:10.1073/pnas.1302837110.

Pieterse C.M.J., et al., (2014), Induced systemic resistance by beneficial microbes, *Annu. Rev. Phytopathol.*, 52:347–375.

Pineda A., et al., (2010), Helping plants to deal with insects: the role of beneficial soil-borne microbes, *Trends Plant Sci.*, 15:507–514.

Ping L. and Boland W., (2004), Signals from the underground: bacterial volatiles promote growth in Arabidopsis, *Trends Plant Sci.,* 9:263–266.

Qiu M., Li S., Zhou X., Cui X., Vivanco J., Zhang N., et al., (2014), De-coupling of root–microbiome associations followed by antagonist inoculation improves rhizosphere soil suppressiveness, *Biol. Fertil. Soils*, 50:217–224, doi:10.1007/s00374-013-0835-1.

Rekha K., Baskar B., Srinath S. and Usha B., (2018a), Plant growth promoting rhizobacteria *Bacillus subtilis* RR4 isolated from rice rhizosphere induces malic acid biosynthesis in rice roots, *Can J Microbiol.*, 64(1):20–27.

Rekha K., MohanKumar R., Ilango K., Rex A. and Usha B., (2018b), Transcriptome profiling of rice roots in early response to *Bacillus subtilis* (RR4) colonization, *Botany*, 96:749–765.

Rekha K., Mohan Kumar R., Rameshthangam P. and Usha B., (2019), Impact of *Pseudomonas putida* RRF3 on the root transcriptome of rice plants: Insights into defense response, secondary metabolism, and root exudation, *J. Biosci.*, 44:98.

Rekha K., Mohan Kumar R. and Usha B., (2020), Root exudation of organic acids as affected by plant growth-promoting rhizobacteria *Bacillus subtilis* RR4 in rice, *J. Crop. Improv.*, 34(4):571–586.

Robert C.A., et al., (2012), A specialist root herbivore exploits defensive metabolites to locate nutritious tissues, *Ecol. Lett.*, 15:55–64.

Remans R., Beebe S., Blair M., Manrique G., Tovar E. and Rao I., (2007), Physiological and genetic analysis of root responsiveness to auxin-producing plant growth-promoting bacteria in common bean (*Phaseolus vulgaris* L.), *Plant Soil*, 302:149–161.

Rojas-Solís D., et al., (2018), *Pseudomonas stutzeri* E25 and *Stenotrophomonas maltophilia* CR71 endophytes produce antifungal volatile organic compounds and exhibit additive plant growth-promoting effects, *Biocat. Agric. Biotech.*, 13:46–52.

Rousk J., Brookes P.C. and Bååth E., (2009), Contrasting soil pH effects on fungal and bacterial growth suggests functional redundancy in carbon mineralization, *Appl Environ Microbiol.*, 75:1589–1596.

Rousk J., Brookes P.C. and Bååth E., (2010), Investigating the mechanisms for the opposing pH-relationships of fungal and bacterial growth in soil, *Soil Biol-. Biochem.*, 42:926–934.

Rudrappa T., Czymmek K.J., Pare P.W. and Bais H.P., (2008), Root-secreted malic acid recruits beneficial soil bacteria, *Plant Physiol.*, 148:1547–1556.

Ryan P.G., Moore C.J., van Franeker J.A. and Moloney C.L., (2009), Monitoring the abundance of plastic debris in the marine environment, *Phil. Trans. R. Soc. B.*, 364:1999–2012.

Santana-Pereira A.L.R., Sandoval-Powers M., Monsma S., Zhou J., Santos S.R., Mead D.A., et al., (2020) Discovery of novel biosynthetic gene cluster diversity from a soil metagenomic library. *Front. Microbiol.*, 11:585398.

Saunders M. and Kohn L.M., (2009), Evidence for alteration of fungal endophyte community assembly by host defense compounds, *New Phytol.*, 182:229–238.

Schilirò E., Ferrara M., Nigro F. and Mercado-Blanco J., (2012), Genetic responses induced in olive roots upon colonization by the biocontrol endophytic bacterium *Pseudomonas fluorescens* PICF7, *PLoS O-ne*, 7:e48646.

Smalla K., Wieland G., Buchner A., Zock A., Parzy J., Kaiser S., et al., (2001), Bulk and rhizosphere soil bacterial communities studied by denaturing gradient gel electrophoresis: plant-dependent enrichment and seasonal shifts revealed, *Appl. Environ. Microbiol.*, 67:4742–4751.

Solano B.R., Maicas J.B., de la Iglesia M.T.P., Domenech J. and Mañero F.J.G., (2008), Systemic disease protection elicited by plant growth promoting rhizobacteria strains: relationship between metabolic responses, systemic disease protection, and biotic elicitors, *Phytopathol.*, 98(4):451–457.

Soler R., Bezemer T., Cortesero A., van der Putten W., Vet L. and Harvey J., (2007), Impact of foliar herbivory on the development of a root-feeding insect and its parasitoid, *Oecologia.*, 152:257–264, doi:10.1007/s00442-006-0649-z.

Song G.C., Lee S., Hong J., Choi H.K., Hong G.H., Bae D.W., et al., (2015), Aboveground insect infestation attenuates belowground Agrobacterium-mediated genetic transformation, *New Phytol.*, 207:148–158.

Stavi I., Lal R., (2015), Achieving zero net land degradation: challenges and opportunities, *J. Arid Environ.*, 112:44–51.

Stringlis I.A., Yu K., Feussner K., de Jonge R., Van Bentum S., Van Verk M.C., et al., (2018), MYB72-dependent coumarin exudation shapes root microbiome assembly to promote plant health, *Proc. Natl. Acad. Sci. U.S.A.,* 115:E5213–E5222.

Stringlis I.A., et al., (2019), The age of coumarins in plant–microbein teractions, *Plant Cell Physiol.*, 60:1405–1419.

Syed-Ab-Rahman S.F., et al., (2018), Emerging microbial biocontrol strategies for plant pathogens, *Plant Sci.*, 267:102–111.

Syed-Ab-Rahman S.F., et al., (2019), Suppression of *Phytophthora capsici* infection and promotion of tomato growth by soil bacteria, *Rhizosphere,* 9:72–75.

Testen A.L. and Miller S.A., (2018), Carbon source and soil origin shape soil microbiomes and tomato soilborne pathogen populations during anaerobic soil disinfestation, *Phytobiomes*, 2:138–150.

Thoden T.C. and Boppré M., (2010), Plants producing pyrrolizidine alkaloids: sustainable tools for nematode management?, *Nematology*, 12:1–24.

Thompson I. P., van der Gast C. J., Ciric L. and Singer A. C. (2005). Bioaugmentation for bioremediation: the challenge of strain selection. Environ. Microbiol. 7, 909–915.

Timmusk S., Behers L., Muthoni J., Muraya A. and Aronsson A.C., (2017), Perspectives and challenges of microbial application for crop improvement, *Front. Plant Sci.,* 8:49.

Torsvik V. and Øvreås L., (2002), Microbial diversity and function in soil: from genes to ecosystems, *Curr. Opin. Microbiol.*, 5:240, doi:10.1016/s1369-5274(02)00324-7.

Tsiafouli, M.A., et al., (2015), Intensive agriculture reduces soil biodiversity across Europe, *Glob. Change Biol.,* 21:973–985.

Tsolakidou M.D., et al., (2019), Rhizosphere-enriched microbes as a pool to design synthetic communities for reproducible beneficial outputs, *FEMS Microbiol. Ecol.*, 95:10.

van Dam N.M., (2009), Belowground herbivory and plant defenses, *Ann. Rev. Ecol. Evol. System.*, 40:373–391.

van de Mortel J.E., et al., (2012), Metabolic and transcriptomic changes induced in Arabidopsis by the rhizobacterium *Pseudomonas fluorescens* SS101, *Plant Physiol.*, 160:2173–2188.

Van der Ent S., Van Wees S.C.M., and Pieterse C.M.J., (2009), Jasmonate signaling in plant interactions with resistance-inducing beneficial microbes, *Phytochem.*, 70:1581–1588, doi:10.1016/j.phytochem.2009.06.009.

van Oosten V.R., et al., (2008), Differential effectiveness of microbially induced resistance against herbivorous insects in Arabidopsis, *Mol. Plant-Microbe Interact.,* 21:919–930.

Vives-Perez, V. et al. 2020. Root exudates: from plant to rhizosphere and beyond. *Plant Cell Rep.,* 39:3–17.

Voges M., et al., (2019), Plant-derived coumarins shape the composition of an Arabidopsis synthetic root microbiome, *Proc. Natl. Acad. Sci. U.S.A.,* 116:12558–12565.

Walker T.S., Bais H.P., Grotewold E. and Vivanco J.M., (2003), Root exudation and rhizosphere biology, *Plant Physiol.,* 1320:44.

Watson, D.H. 2018. *Pesticides and agriculture: profit, politics and policy*. Burleigh Dodds Science Publishing, Cambridge, UK.

Weyens N., Truyens S., Dupae J., Newman L., van der Lelie D., Carleer R., et al., (2010), Potential of *Pseudomonas putida* W619-TCE to reduce TCE phytotoxicity and evapotranspiration in poplar cuttings, *Environ Pollut.,* 158:2915–2919.

Whitman T., et al., (2016), Dynamics of microbial community composition and soil organic carbon mineralization in soil following addition of pyrogenic and fresh organic matter, *ISME J.*, 10:2918–2930.

Xie S., Wu H., Chen L., Zang H., Xie Y. and Gao X., (2015), Transcriptome profiling of *Bacillus subtilis* OKB105 in response to rice seedlings, *BMC Microbiol.*, 15:21, doi:10.1186/s12866-015-0353-4. PMID:25651892.

Yang J.W., Yi H-S., Kim H., Lee B., Lee S., Ghim S.-Y., et al., (2011), Whitefly infestation of pepper plants elicits defence responses against bacterial pathogens in leaves and roots and changes the below-ground microflora, *J. Ecol.*, 99:46–56.

Yuan J., et al., (2018), Root exudates drive the soilborne legacy of aboveground pathogen infection, *Microbiome*, 6:156.

Zebelo S., et al., (2016), Rhizobacteria activates (+)-d-cadinene synthase genes and induces systemic resistance in cotton against beet armyworm (*Spodoptera exigua*), *Plant Cell Environ.*, 39:935–943.

Zhang N., et al., (2015), Whole transcriptomic analysis of the plant-beneficial rhizobacterium *Bacillus amyloliquefaciens* SQR9 during enhanced biofilm formation regulated by maize root exudates, *BMC Genomics,* 16:685.

Zhulin I.B. and Armitage J.P., (1993), Motility, chemokinesis, and methylation-independent chemotaxis in *Azospirillum brasilense, J. Bacteriol.*, 175(4):952–958, doi:10.1128/jb.175.4.952-958.1993. PMID:8432718.

Chapter 12

Advances in understanding the role of soil microbiomes in promoting crop resilient to drought stress

Zakaria M. Solaiman and Kadambot H. M. Siddique, The UWA Institute of Agriculture, and UWA School of Agriculture and Environment, The University of Western Australia, Australia

1 Introduction

Abiotic stresses, including nutrient deficiencies, drought, heavy metals, and salinity, profoundly impact crop productivity and food security (Fig.1). Drought is a combined and complex stress where plants modify their physiology so that they can adapt to the water-scarce environment, including root architecture changes, stomatal closure, decrease of photosynthesis, reduction of germination rate, shortening the time for reproduction, and change in composition and amount of root exudates (Cantó et al., 2020; Farooq et al. 2012; Gupta et al. 2020; Rabbi et al. 2018). Drought also affects microbiome activity, physiology, and habitat, increasing osmotic compound buildup and sporulation rate, decreasing enzymatic activities, and causing nutrient cycling and habitat change. These can also lead to changes in microbial community

http://dx.doi.org/10.19103/AS.2024.0136.24

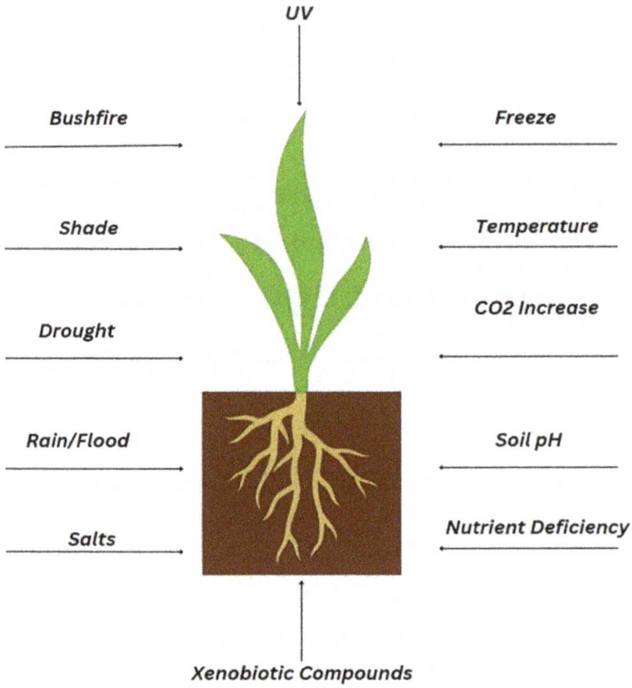

Figure 1 Abiotic stresses pose threats to crop plants.

compositions (Aamir et al., 2019; Cantó et al., 2020; Jansson and Hofmockel, 2020; Morcillo and Manzanera, 2021; Naylor and Coleman-Derr, 2018; Schimel, 2018).

Soil microbiomes consist of diverse communities of bacteria, archaea, fungi, and other microorganisms that interact with plants and influence their growth and stress responses. This review chapter investigates the latest advances in understanding how soil microbiomes promote crop resistance to drought stress, exploring mechanisms, applications, challenges, and future directions (Vurukonda et al., 2016).

We also discussed possible alternative strategies for more effective drought-stress management. Drylands are used to evaluate microbiome drought stress because of their limited agricultural productivity. Given the importance of water as a macromolecule for the physiological growth and development of plants, water scarcity and deprivation are phenomena that commonly cause drought stress (Siddique et al., 2000; Bartels and Phillips, 2010; Farooq et al., 2012; Askari-Khorasgani et al., 2021). Interventions with microbes can mitigate drought stress (Hossain et al., 2016; Li and Liu, 2016; Shinozaki et al., 2016). This review summarized drought stress mitigation using microbes, current control tactics, and unconventional control approaches.

We also discussed possible alternative strategies for more effective drought-stress management. Drought stress may be reduced through breeding, mass screening, and exogenous phytohormone production (Siddiqui et al., 2022). Plants can autoregulate to minimize stress by manufacturing plant hormones (e.g. abscisic acid, ABA, and gibberellins) and osmolytes (e.g. amino acids and polyols) by modifying tender leaves to reduce transpiration loss (Farooq et al., 2009; Rahdari and Hoseini, 2012).

Microbes can promote plant growth both directly and indirectly. The indirect activation of plant growth involves various actions by which microbes inhibit plant growth and development induced by pathogens (Ahmad et al., 2021; Khan and Irba, 2023). However, during direct activation, microbes biosynthesize compounds that increase the uptake of nutrients from the soil and stimulate plant growth and development (Choudhary et al., 2016). Microbes induce local or systemic stress mitigation response mechanisms that help plants survive under abiotic stress conditions, such as drought stress, and help plants sustain growth and development through the fixation and mobilization of nutrients, hormones, and organic phytostimulants (Martins et al., 2018; Yadav and Yadav, 2018; Kour and Yadav, 2020). We summarized the currently known implications using microbes for drought tolerance, including their fundamental mechanisms of action.

2 Microbes involved in the mitigation of drought stress

Microbes play a crucial role in drought stress mitigation by forming symbiotic relationships with plants, promoting soil health, and enhancing plant's ability to withstand water stress (Shaffique et al. 2022). Some key microbes involved in drought stress mitigation are described in the following sections.

2.1 Mycorrhizal fungi

Arbuscular mycorrhizal fungi (AMF) are an assembly of remarkable symbiotic organisms that play a vital role in terrestrial ecosystems. These fungi form mutualistic associations with the roots of most plants, including many crops and trees. The relationship between plants and AMF is ancient, dating back over 400 million years, and it is one of the most widespread and essential symbioses on Earth. AMF can form intricate structures known as arbuscules within the plant's root cells. AMF establish a mutualistic relationship with the roots of approximately 80% of terrestrial plant species, profoundly impacting plant health and productivity. Intriguingly, recent research underscores the pivotal role of AMF in enhancing plant nutrient uptake, particularly phosphorus (P) and nitrogen (N), by extending their hyphal networks into the soil (Bitterlich et al., 2018). Furthermore, AMF can improve plant stress tolerance, facilitating adaptation to challenging environmental conditions (Ruiz-Lozano et al., 2016).

Another innovative revelation is the contribution of AMF to soil structure stabilization and agroecosystem sustainability. Through the secretion of glomalin and the formation of a hyphal network, AMF can bind soil particles together, enhancing soil aggregation and reducing erosion (Leifheit et al., 2015). Moreover, AMF can act as biological bridges, connecting different plant individuals and enabling nutrient and signal exchange among plants, also known as the common mycorrhizal network (Figueiredo et al., 2021). Considering the mounting climate crisis, understanding and leveraging the potential of AMF in sustainable agriculture and ecosystem restoration are gaining momentum. Recent studies target manipulating AMF communities to improve crop yield, nutrient cycling, and plant resilience in changing environmental conditions such as drought stress (Thirkell et al., 2017). Other recent studies suggest that exploiting the symbiotic association of AMF might be pivotal for sustaining agricultural productivity in the face of escalating water scarcity and ensuring ecological stability (Chandrasekaran et al., 2019). Therefore, understanding the intricate interactions between AMF and plants is essential for developing innovative and sustainable strategies for plant production and ecosystem management in an era of changing climate.

2.2 Endophytes

Endophytes are microorganisms that exhibit symbiotic relationships with plants. These microorganisms, including fungi, bacteria, and actinomycetes, establish a symbiotic relationship with host plants, often enhancing their growth, stress tolerance, and resistance to pests (Hardoim et al., 2015). They have been comprehensively investigated because of their beneficial effects on environmental stress (Cheng et al., 2022; Meena et al., 2021; Verma et al., 2021). Endophytes can increase biomass accumulation in host plants under stressful conditions (Zhang et al., 2006; Barman et al., 2016). The drought stress mitigation effects of endophytes were comparable among different plant species. Under drought conditions in the presence of endophytes, dicots and C4 plant species showed increased biomass compared with monocots and C3 species.

One of the most well-known examples of endophytes is fungal endophytes in grasses, which enhance plant's resistance to herbivores and pathogens by producing secondary metabolites that inhibit these attackers (Davitt et al., 2011). In return, the plant provides a protected and nutrient-rich environment for the endophytes to thrive. Endophytes can also improve a plant's abiotic stress tolerance factors such as drought, salinity, and heavy metals, making them valuable allies in agriculture and environmental restoration efforts. Researchers are exploring the potential of endophytes to induce crop yields, reduce the need for chemical pesticides, and promote sustainable agriculture (Ren et al., 2011). Moreover, endophytes can produce bioactive compounds, which have been discovered for their antimicrobial and insecticidal properties (Strobel and Daisy, 2003). The vast metabolic abilities of endophytic microorganisms could be attributed to the excess secondary metabolites they synthesise (Dhakal et al., 2022), enhance host defence mechanisms and offer a promising reservoir of novel compounds for pharmaceutical and agricultural applications (Kusari et al., 2014).

Endophytes provide phytohormones and induce defence-related secondary metabolism within the plants (Barman et al., 2016; Cheng et al., 2022; Kozjek et al., 2021). Evidence of endophyte-mediated plant stress attenuation has been established in recent decades (Strobel, 2003; Rodriguez et al., 2009). Microbes are beneficial for reducing stress, which would otherwise limit the crop yield. When stressed, plants struggle to sustain their normal growth (Singh et al., 2020).

2.3 Phyllosphere microbiomes

The phyllosphere microbiomes are various microbial communities that inhabit the above-ground parts of plants, primarily the leaves. This ecosystem includes bacteria, fungi, archaea, viruses, and other organisms interacting with each other and the plant host. The phyllosphere microbiome is a dynamic and complex environment influenced by various factors, such as plant species, geography, climate, and agricultural practices (Jat et al., 2021). Microbes in the phyllosphere contribute to nutrient cycling, C sequestration, and the breakdown of plant debris. They can also affect volatile organic compounds (VOCs) emissions that influence atmospheric chemistry (Brilli et al., 2019). Understanding the phyllosphere microbiome has important implications for agriculture. Manipulating this microbiome through techniques such as biofertilization and biocontrol can help improve crop yield and reduce the use of pesticides and fertilizers. It plays an active role in plant health, agricultural practices, and ecosystem functioning, making it an essential area of microbiology, ecology, and agriculture research in drought stress (Arun et al., 2020). Phyllosphere microbiome improved growth attributes and crucial nutrients in plants under drought stress. It plays a role in C and N cycling in terrestrial ecosystems and contributes to the overall diversity of microbial life on Earth. The phyllosphere microbiome has broader ecological implications, as it can influence ecosystem stability and resilience.

2.4 Plant-growth-promoting rhizobacteria (PGPR)

Plant-growth-promoting rhizobacteria form mutually beneficial relationships with plants, promoting plant growth and health in various ways (Rolli et al.

2014). PGPR can enhance plant nutrient availability by solubilizing minerals like P, making them more accessible to plant roots. They can also fix atmospheric N_2 into a form that plants can use. Some PGPRs produce antibiotics or inhibit pathogenic microorganisms' growth, helping protect plants from soil-borne diseases. PGPRs can synthesize plant hormones like auxins, gibberellins, and cytokinins, influencing plant growth and development, including root and shoot growth. PGPR can help plants tolerate environmental stresses, such as drought, salinity, and heavy metal toxicity, by producing stress-related compounds or enhancing the plant's stress response mechanisms. Overall, PGPR is crucial in promoting plant growth and health, reducing the need for chemical fertilizers and pesticides, and contributing to sustainable agriculture practices. Their use in agriculture and horticulture can improve crop yields and reduce environmental impact, supporting sustainable agricultural practices (Bhattacharyya and Jha, 2012). A study by Vessey (2003) suggests that certain strains of PGPR can facilitate plant growth by fixing atmospheric N_2, solubilizing minerals, and producing plant growth-promoting hormones like indoleacetic acid (IAA). Moreover, PGPR has proven effective in reducing biotic and abiotic stress in plants, augmenting their resistance against harmful pathogens and unfavorable environmental conditions (Yang et al., 2009). Researchers such as Glick (2012) have indicated the potential role of PGPR in managing environmental stresses by triggering induced systemic resistance (ISR) and enhancing plant immune response. PGPR can ameliorate drought stress and improve agronomic sustainability (Kanwal et al., 2017; Vanamala et al., 2021; Aslam et al., 2022). PGPR enhances plant growth directly by assisting in resource acquisition (N, P, and essential minerals) or modulating plant hormone levels and indirectly by inhibiting pathogens by producing siderophores, lytic enzymes, and antibiotics.

2.5 Actinomycetes

Actinomycetes, a significant group of bacteria, are renowned for producing many secondary metabolites, including antibiotics such as streptomycin and nocardicin (Hopwood, 2007). They exhibit a complex life cycle and manifest a mycelial growth pattern similar to fungi. Often found in soil, actinomycetes are crucial in organic matter decomposition, contributing to ecosystems' C and N cycles (Davelos Baines et al., 2004). Moreover, actinomycetes are pivotal in biotechnology and medicine because they synthesize bioactive compounds and enzymes and display biocontrol properties against plant pathogens (Berdy, 2005). Remarkably, studying their diverse metabolic pathways has been pivotal in understanding microbial secondary metabolism and paved the way for biotechnological applications, including drug development (Watve et al.,

2001). Furthermore, ongoing genomic studies continue to unravel the intricate network of gene clusters associated with secondary metabolite biosynthesis in actinomycetes, offering promising avenues for discovering novel bioactive molecules (Bentley et al., 2002).

In a study involving Actinomycetes, *Streptomyces pactum* was inoculated into the wheat plant (Li et al., 2020), which significantly increased the overexpression of EXPA6, P5CS, EXPA2, and SnRK2 and increased the root length by 13.6%, shoot length by 10.3%, and fresh biomass by 21.3%. Chemical assays also revealed that the exposed seedlings showed stress reduction through an increase in the levels of sugars and antioxidant enzymes. In another experiment, foxtail millet crops were subjected to drought stress, followed by inoculation with *Acinetobacter calcoaceticus* and *Penicillium* sp. (Kour et al., 2020). Plant growth attributes showed drought stress mitigation and physiological growth due to inoculation. In this study, drought stress was mitigated by the accumulation of osmolytes, proline, and glycine betaine, as well as by increased chlorophyll a and b levels. An increased chlorophyll content resulted in plant growth and development, whereas increased proline and glycine betaine levels improved osmotic adjustment and membrane integrity, respectively. These results demonstrated that inoculation with microbes efficiently combats drought stress (Kour et al., 2020).

2.6 Methanotrophs

Methanotrophs are a remarkable group of bacteria capable of utilizing methane (CH_4) as their primary source of C and energy, thereby playing a critical role in mitigating the release of this potent greenhouse gas into the atmosphere. These microorganisms can be found in various environments, including soils, sediments, and aquatic ecosystems, where they act as a biological filter, oxidizing methane to CO_2 (Trotsenko and Murrell, 2008). Methanotrophs have garnered significant attention due to their potential applications in biotechnology and environmental remediation. For example, methanotrophs can be used for bioconversion of methane into valuable chemicals and fuels from landfills, wastewater treatment plants, and agricultural operations (Strong et al., 2015; Fei et al., 2014). Furthermore, studying methanotrophs provides insights into microbial ecology, evolution, and the global C cycle.

Methylotrophs increased the levels of auxin and cytokines, and the increased phytohormones promoted germination and growth. In addition to increasing the ACC deaminase activity, these bacteria enhance N_2 fixation and mineral solubilization. The novelty of this microbial community is that it is composed of type 1 and type 2 methanotrophs that have the potential to tolerate stress (Kumar et al., 2019). In another study, bacteria were isolated from the rhizosphere of *Zea mays* and reinoculated into plants (Danish et al.,

2020). Additionally, bioassays demonstrated the significant involvement of ACC deaminase. Due to the ACC deaminase activity, *E. cloacae* significantly increased grain yield by up to 73% among these microbes. Methanotrophs metabolize methane from C molecules, producing sulfate, nitrate, and energy via oxidation. They require only a single C atom for survival; therefore, their survival rate is higher, which mitigates drought stress (Shaffique et al. 2022). Methane is first catalyzed to methanol, then oxidated to formaldehyde by methanol dehydrogenase, and enters the main metabolic pathways of microorganisms.

3 Impact of drought stress on soil, plant, and microbes

3.1 Impact of drought stress on soil properties

Drought has become a significant limiting factor in developing terrestrial ecosystems due to global warming caused by anthropogenic disturbances. However, a comprehensive understanding of the impact of drought on soil physical, chemical, and biological properties at local and global scales is lacking. In a recent review, Omae and Tsuda (2022) reported that drought caused by decreased precipitation, heat, or fire changes soil features, including soil aggregation, soil moisture, the solubility and dispersion of nutrients in soil, and enzyme activities, as well as concentration of solutes (Cantó et al., 2020; Gupta et al. 2020; Morcillo and Manzanera 2021; Naylor and Coleman-Derr 2018; Schimel 2018). Qu et al. (2023) stated that drought significantly reduced soil microbial biomass and enzyme activity globally. Their results highlight the strong impact of drought on soil microbial biomass and C- and P-acquiring enzyme activity. Furthermore, excessive drought damages soil properties, such as structure, porosity, and pH (Chen et al., 2020; Lozano et al., 2021; Peng et al., 2023), creating unfavorable environmental conditions for the growth of microorganisms. Consequently, microbial metabolism, activity, and enzyme production are reduced (Manzoni et al., 2014; Fitzpatrick et al., 2017).

3.2 Impact of drought stress on plant growth and development

Drought stress affects plant growth and development, impacting agricultural productivity and ecosystem stability. The effects of drought stress on plants encompass a broad range of physiological, biochemical, and molecular responses (Wang et al., 2016). One of the first responses of plants to water-deficit conditions is the closure of stomata, the pores on the leaf surface, which reduces water loss through transpiration and restricts CO_2 intake, reducing photosynthesis (Dos Reis et al., 2012). Drought stress can also lead to a decline in photosynthesis due to potential damage to the photosynthetic apparatus. Reduced photosynthesis can limit the plant's energy and biomass production. Drought stress often reduces cell expansion and elongation, leading to stunted growth. Root growth may also be affected, though some plants might allocate more resources to root growth to access deeper water sources. Drought can lead to premature aging and death of plant tissues, especially leaves. This can result in early leaf drop, reducing the plant's photosynthetic capacity. Plants may accumulate solutes like proline, sugars, and other osmolytes to maintain cell turgor and osmotic balance under drought conditions. Drought stress can adversely affect flowering, pollen viability, fruit set, and seed production. This can result in reduced yields for many crops. Stress conditions can lead to the overproduction of reactive oxygen species (ROS) (Nair et al., 2008). In plants, greater levels of ROS cause membrane degeneration, lipid peroxidation, nucleic acid, protein, and lipid degradation (Nair et al., 2008). Drought stress causes an increase in the production of ethylene, which is synthesized in response to abiotic stress (Narayanasamy et al., 2020). However, because of the promotion of chlorosis, senescence, and leaf abscission, such an elevated ethylene concentration harms normal plant health and growth. To counteract this, plants may upregulate antioxidant systems, such as SOD and catalase.

Drought can affect the balance and concentration of plant hormones, including ABA, which plays a pivotal role in drought response by inducing stomatal closure. Some plants change their root architecture in response to water stress, either deepening their roots or increasing lateral roots to maximize water uptake. Drought can alter the expression of numerous genes. Plants have specific drought-responsive genes that are activated to help them cope with water-deficit conditions. Some plants, especially those from arid environments, have adaptations to survive extreme dehydration, essentially going dormant until water becomes available again. Drought can affect the uptake and distribution of nutrients within the plant, potentially leading to deficiencies or imbalances. For crop plants, drought stress typically results in significant reductions in yield, affecting food supply.

To mitigate the effects of drought stress, plants have evolved various adaptive strategies depending on their environment. For example, xerophytes (plants adapted to arid habitats) have specialized structures such as thick cuticles, sunken stomata, and succulent leaves to reduce water loss and store water. Given the increasing threat of climate change, understanding how plants respond to drought stress and developing approaches to improve drought tolerance in crop plants are critical research areas. As a result, leveraging the potential of plant and soil-related microorganisms to deal with the negative impacts of drought stress to boost crop output has gotten a lot of attention (Yang et al., 2009).

3.3 Impact of drought stress on microbial taxonomic and functional diversity

Drought stress can significantly impact microbial taxonomic and functional ecology in various ecosystems, including soils, aquatic environments, and even within host organisms (Bapiri et al., 2010; Barnard et al., 2013; Bouasria et al., 2012; Furze et al., 2017; Sun et al., 2017; Naylor et al., 2017). These impacts can disrupt microbial community structures, alter the composition of microbial populations, and change the functional roles that microbes play in their respective ecosystems. Some drought-tolerant species may thrive, while others struggle to survive or become dormant, which can result in reduced overall microbial diversity. Drought stress often changes the relative abundance of different microbial taxa. Some taxa may become more dominant, while others may decline. These shifts may have cascading effects on ecosystem functions. Microbes are essential for nutrient cycling in ecosystems. Drought stress can affect microbial nutrient cycling processes, such as N_2 fixation, mineralization, and decomposition. This can lead to imbalances in nutrient availability. Microbial communities typically exhibit functional redundancy, meaning multiple species perform similar ecological roles. Drought stress can reduce this redundancy as some species may be more sensitive to drought conditions than others, leading to potential functional shifts. Drought-stressed microbes may undergo metabolic changes to conserve energy and resources. Some may enter a state of dormancy or sporulation, reducing their metabolic activity and potentially slowing down ecosystem processes. Drought can impact C cycling in ecosystems by affecting microbial decomposition rates. Slower decomposition can accumulate organic matter, potentially altering C sequestration dynamics. Drought stress can disrupt microbial interactions, such as mutualistic relationships or competition for limited resources. This can further influence microbial community structure and function. Some microbial species have evolved mechanisms to withstand drought stress, such as producing osmoprotectants or developing drought-resistant biofilms. These adaptations can influence community dynamics. Drought stress can affect the prevalence and virulence of microbial pathogens in particular ecosystems, including in agricultural soils and human microbiomes. Microbial communities can exhibit resilience to drought stress, with the ability to recover once conditions become more favorable. However, the speed and extent of recovery can vary based on factors like ecosystem type and severity of the drought.

Changes in root microbial communities, such as enrichment of *Actinobacteria,* can occur locally or systemically (Simmons et al., 2020). Furthermore, the enrichment of Actinobacteria was not observed in dead roots under drought (Simmons et al., 2020). These results support the idea that plants actively regulate the enrichment of Actinobacteria. Although our understanding of how plants regulate the conserved enrichment of Actinobacteria genera such as Streptomyces during drought is limited, a mechanism was recently proposed (Xu et al., 2021). Time-series sorghum root transcriptome data revealed that drought affected host iron metabolism with reduced activity of the iron uptake system and increased iron storage in sorghum upon drought stress (Xu et al., 2021).

In rice plants experiencing severe drought, the bacterial community did not return to the state of well-watered control even after rewatering during the whole life cycle (Santos-Medellín et al., 2021). It would be interesting to investigate to what degree the difference between sorghum and rice in their ability to restructure microbiota is due to the host genetics, local soil, or microbial difference. Generally, the effects of drought on fungal community structure are smaller than those on bacterial community structure around roots or sometimes negligible (Bouasria et al., 2012; Furze et al., 2017; Ochoa-Hueso et al., 2018; Santos-Medellín et al., 2017). Furthermore, under drought, the cooccurrence of fungal networks was more stable than that of the bacterial community (de Vries et al., 2018). Nevertheless, some studies report drought-triggered fungal community change, particularly for AMF, which have essential roles in drought stress mitigation by providing nutrients and water to the host plant (Brunner et al., 2015; Talaat and Shawky, 2017).

3.4 Effects of drought on plant-microbe interaction

PGPRs are free-living in soil and are associated with the rhizosphere, rhizoplane, phyllosphere, and endosphere (Fadiji and Babalola, 2020a; Fadiji et al., 2020). The roles of endophytic bacteria are less reported as rhizospheric as PGPRs in promoting plant growth (Bashan et al., 2014; Fadiji et al., 2021; Sarkar et al., 2022). Plant development is helped by PGPR's acquisition of nutrients and secretion of phytohormones. In addition, priming of crop plants with PGPR boosts resistance to a variety of stresses, such as salt stress, heavy metals contamination, and drought (Sarkar and Rakshit, 2020; Fadiji et al., 2022; Sarkar et al., 2022). They also have several action mechanisms, such as siderophore production, ACC deaminase activity, excretion of cell wall degrading enzymes, quorum quenching, and antibiotics, which assures induced systemic tolerance and induced systemic resistance in plants to both biotic and abiotic stressors (Olanrewaju et al., 2017; Fadiji and Babalola, 2020b; Akanmu et al., 2021; Sarkar and Rakshit, 2021). *Agrobacterium, Bacillus, Bradyrhizobium, Burkholderia, Caulobacter, Erwinia, Pseudomonas,* and *Rhizobium* are among the well-studied PGPR (De Souza Vandenberghe et al., 2017).

Approximately 40% of C assimilated via photosynthesis is free into the soil from roots as root exudates (Bhattacharyya et al., 2021; Canarini et al., 2019; Cantó et al., 2020). Root exudates serve as signaling molecules, modifying the relationships between plants and microbes, between microbes and microbes, and between plants and plants; as nutrients for microbes; and as convertors of soil physical and chemical features. The amount and composition of root exudates changed due to drought stress were observed (Canarini et al., 2016; Karst et al., 2016; Preece et al., 2018; Song et al., 2012). The total amount of root exudates decreases, but the ratio of C allocation to root exudates increases with drought intensity; however, a different trend is observed under extreme drought (Preece and Peñnuelas, 2016).

Karlowsky et al. (2018) investigated assimilated C allocation to shoot, root, soil, and microbes during drought and rewatering. The C allocation to roots as sucrose and soil organic C increased during drought. Although C allocation to soil fungi and bacteria generally decreased during drought, allocation to Actinobacteria did not change, consistent with the observation that absolute and relative abundance of Actinobacteria increased during drought (Naylor et al., 2017). During the recovery phase, C allocation was back to normal. These suggest that plants keep releasing root exudates during drought for quick recovery of microbial activities upon rewatering, which may contribute to the recovery of plants from drought. Consistent with this idea, the soil respiration rate was higher in soil with added root exudates from drought-experienced plants than in drought-unexperienced plants (de Vries et al., 2019). Enzyme activities such as protease, urease, phosphatase, invertase, and catalase are altered during drought stress, which might affect the soil nutrient status (Song et al., 2012). As the osmotic pressure increased, exudation rates of organic acids increased, and the increase rate was higher in drought-tolerant than drought-sensitive maize plants (Song et al., 2012). Because organic acids, especially citrate, increase mobilization of P (Jones and Darrah 1994), increased organic acids such as citrate under drought stress might increase P mobilization, thereby mitigating nutrient stress under drought; this might be one of the mechanisms for drought tolerance of maize plants (Song et al. 2012).

AM fungi colonisation enhanced the osmoprotectants' proline and soluble sugar content, promoted the growth of macadamia plants (Yooyongwech et al., 2013), and enhanced water permeability in maize under drought (Quiroga et al., 2019). In a transcriptome study analyzing 22 diverse rice accessions with different drought tolerance, expression of genes involving in photosynthesis and symbiosis with AM fungi was higher in drought tolerant accessions, suggesting that drought tolerance related to AM fungi colonization shows within-species variation in rice (Groen et al., 2022). Moreover, in some cases, drought stress in tomato and lettuce induced root colonization by the AM fungus *Rhizophagus irregularis*, which alleviated drought stress (Ruiz-Lozano et al. 2016). AM fungi facilitate host plant survival by increasing water and nutrient absorption from the rhizosphere (Agapitos et al., 2008).

4 Mechanisms of microbes-mediated drought stress tolerance

The practicality of microbial inoculations for encouraging plant development exposed to drought stress has been investigated in several studies (Abdelaal et al., 2021; Santoyo et al., 2021). PGPRs are well known for boosting the development of various plants under harsh conditions, including vegetables, grains, and legumes (Sandhya et al., 2010; Abdelaal et al., 2021; Fadiji et al., 2022). PGPRs are used to lessen the adverse effects of environmental pressures on plant growth and production by enhancing nutrient absorption and increasing tolerance to environmental stress (Ngumbi and Kloepper, 2016). Numerous studies have shown that rhizobacteria can assist in mitigating the harmful effects of different stressors on crop development (Curá et al., 2017). Drought is one of the most detrimental conditions affecting plant productivity in semiarid and arid areas (Maheswari et al., 2013). PGPRs are useful and beneficial for enhancing plant development exposed to drought stress (Nadeem et al., 2010) because rhizobacteria may create exopolysaccharides (EPS) like cellulose and alginate, which are known to aid drought tolerance (Zahir et al., 2009; Glick, 2014). Exopolysaccharides may thus play a crucial role in stress reduction for microbial and plant populations under drought conditions. A few PGPRs create EPS, which may be used as a barrier to protect the roots and aid plant development in salinity-stressed conditions (Yang et al., 2016). For example, *B. subtilis* inaquosorum and *Marinobacter lypoliticus* SM19 were shown to mitigate the effect of drought stresses in soybeans (Miransari and Smith, 2009).

PGPR also releases lipo-chitooligosaccharide compounds produced by rhizobia and stimulated by flavonoids found in root exudates. *Bradyrhizobium japonicum* 532C inoculated soybean improved growth under drought stress (Gepstein and Glick, 2013). Trehalose is a nonreducing disaccharide found in plants, bacteria, fungi, and insects that helps plants tolerate abiotic conditions such as drought stress. Trehalose can ameliorate the effects of drought by avoiding protein aggregation and breakdown, which happens under various stress conditions (Jalili et al., 2009). The enhancement of plant growth under drought stress may result from the ACC-deaminase enzyme produced by certain strains of PGPR. This enzyme improves the absorption of essential nutrients such as K, N, and P, enhancing plant growth under different environmental conditions (Vaishnav et al., 2016). Applying PGPRs such as *A. brasilense* and *H. seropedicae* improved drought tolerance in maize (Curá et al., 2017) by incrementing water use efficiency and enhancing antioxidant enzyme activity exposed to drought stress.

Additionally, PGPR boosts growth characteristics by generating plant growth hormones such as cytokinins (CKs), gibberellins (GAs), and IAA, which increase N_2 fixation and promote nutrient absorption (Tiwari et al., 2020). Furthermore, due to the buildup of ABA, PGPR plays a vital role in reducing drought stress resistance. Conversely, PGPR accumulates osmoprotectants and antioxidants, which might help root development in response to stress (Sharma et al., 2019). The synthesis of IAA by Azospirillum species resulted in better root development and increased lateral root formation under drought (Zahir et al., 2009). Under drought stress, inoculating Lavandula dentate with *B. thuringiensis* improved plant metabolic activity and nutrient absorption (Vurukonda et al., 2016). Inoculation with Pseudomonas and Acinetobacter species also acclimated grapevine and Arabidopsis plants to drought stress (Armada et al., 2016). In another research, Pseudomonas inoculation of soybean plants increased the stem height and fresh weight under drought stress, with similar increases in ABA and salicylic acid (SA) concentrations than control (Abdelaal et al., 2021).

PGPR use is a promising way of boosting the yield and growth of drought-stressed crops (see Table 1). The role of microorganisms in plant growth, nutrient management, and biocontrol activity is well studied. In summary, these beneficial microorganisms colonize plants' rhizosphere and promote the development of plants through various direct and indirect mechanisms (Grover et al., 2011). Furthermore, microorganisms' role in managing drought stress is gaining importance. The possible explanation for these mechanisms related to plant drought tolerance induced by rhizobacteria includes (i) production of phytohormones, (ii) ACC-deaminase to reduce the level of ethylene in roots, (iii) induced systemic tolerance by bacterial compounds, and (iv) bacterial exopolysaccharides (Yang et al., 2009; Dimkpa et al., 2009; Timmusk and Nevo, 2011; Kim et al., 2013; Timmusk et al., 2014).

It is important to note that the specific mechanisms and effectiveness of microbe-mediated drought tolerance can vary depending on the plant species, the type of microbes involved, and environmental conditions. Several mechanisms have been proposed for microbiome-mediated drought stress tolerance in plants (Fig. 2). It includes phytohormonal activity, volatile compounds, alteration in root morphology, ACC deaminase activity, accumulation of osmolytes, EPS production, antioxidant defense and co-inoculations. Induced systemic tolerance has been coined to accommodate microbial-induced physicochemical changes in plants, which enhance tolerance to abiotic stresses (Yang et al., 2009).

4.1 Role of microbes in mineral uptake

Drought stress causes an inequity between the mineral and ion exchange. It disrupts the levels of nutrients that are responsible for plant growth. Plant microbial interactions can preserve the ionic balance by foraging the protein in

Table 1 Drought tolerance mechanisms of plants, their effects, and responses (reproduced from Ahluwalia et al. 2021, Resour. Eniron. Sustainab. 5:100032)

Secondary mechanism	Primary mechanism	Positive effects	Negative effects	References
Oxidative adjustment	Production of reactive oxygen species (ROS)	ROS act as secondary messengers during stresses like drought, salinity, pathogenic attack	• Decreased rate of photosynthesis • Lipid peroxidation • Protein mutations, peptide chain fragmentation • DNA oxidation, strand breakage, mismatch mutations • Cell death	Sharma et al. (2012); Chiappero et al. (2019)
Decreased transpiration and stomatal conductance	Stomatal closure	Prevention of water loss due to transpiration from stomata of stems and leaves	• Reduced CO_2 and nutrient uptake, decreased photosynthesis • Adaptations like leaf shedding, permanent wilting of leaves • Sclerophylly–hardening of leaves • Permanent changes in chlorophyll and mesophyll content of germinating seeds	Ngumbi and Kloepper (2016); Liu et al. (2019)
Modified root and shoot system	Increased uptake of water through roots, decreased transpiration rate through shoots	Thinning of roots, and short shoot growth for prevention of water loss through transpiration	• Stunted shoot growth • Cell damage, degradation of amyloplasts in columella cells of plant roots, hydrotropism • Decreased plant yield	Etesami et al. (2015); Barnawal et al. (2019)
Osmotic adjustment	Accumulation of inorganic cations and anions, organic acids, carbohydrates, amino acids, and sugars	• Protection of plants by cellular osmotic adjustment • Detoxification of ROS • Protection of cell membrane integrity • Enzymes or protein stabilization	• Ineffective osmotic adjustment under conditions of rapid desiccation • Water loss • Genotype-dependent solute accumulation	Etesami et al. (2015); Vurukonda et al. (2016)

(*Continued*)

Table 1 (*Continued*)

Secondary mechanism	Primary mechanism	Positive effects	Negative effects	References
Production of stress Hormone abscisic acid	Produced by the plant's stress signaling system	• Abscisic acid: Altered root architecture to prevent water loss through transpiration, increased water influx into roots • Increased root-to-shoot ratio • Leaf senescence • Stomatal closing to prevent water loss • Increased adaptivity of the plants	• Seed dormancy • Inhibition of stimulatory activity of some phytohormones • Inhibition of shoot growth by inhibition of cell division in the vascular cambium	Daszkowska-Golec (2016)
Production of stresshormone ethylene	Produced by the plant's stress signaling system	• Ethylene:Reduced water loss due to transpiration • Stimulation of fruit ripening, regulation of nodulation • Altered leaf senescence behavior	• Over accumulation of ethylene in leaves, increased leaf abscission, decreased stomatal conductance • Decreased growth of the root and shoot system • Increased epinasty • Inhibition of germination • Long-term hypoxia in the root system • Cell damage, cell death • Decreased plant yield	Iqbal et al. (2017); Ngumbi and Kloepper(2016); Danish et al. (2020)

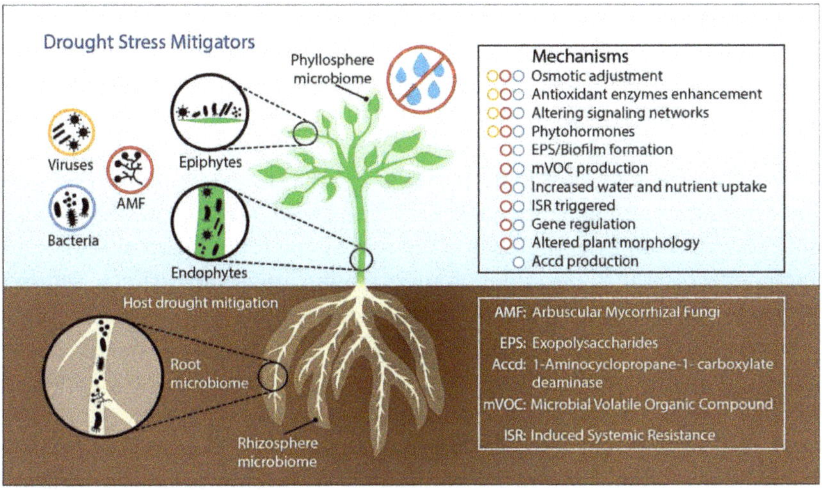

Figure 2 Drought stress alleviating mechanisms employed by plant-associated bacteria, fungi, and viruses. Bacteria, fungi, and viruses are members of the plant microbiome living in different plant compartments, on plant surfaces (epiphytes) and inside plant tissues (endophytes). (Source: Poudel et al. (2021). Front. Microbiol. 12: 743512).

the plant cytosol (Mukerji et al., 2006; Banfield and Nealson, 2018). Beneficial microbial populations produce many organic components that can be used as signaling molecules and nutritional elements for plants under drought stress. Various microbiota, such as fungi, cyanobacteria, and plant PGPR, play essential roles in mineral uptake and rehabilitation of the nutritional status (Zhu et al., 2016). They mediate the solubilization of phosphosulfate compounds, N_2 fixation, denitrification, and siderophore production. There are two pathways for mineral uptake, namely, direct and indirect (Zhu et al., 2016; Kapadia et al., 2021). Phosphate is primarily in the ecosystem as inositol phosphate; therefore, microbes solubilize P into the most soluble form by synthesizing low molecular weight components such as citric acid and gluconic acid. Indirectly, phosphorus is solubilized by lowering the external environment's pH and turning it into low molecular weight organic compounds (Sharma et al., 2020).

Therefore, a microbial population that interacts with plants in a beneficial environment can provide hemostasis for ionic exchange in the cytosol by maintaining the nutritional status of plants under drought stress. One of the fundamental roles of soil microbiomes is nutrient cycling, wherein microorganisms break down organic matter and release nutrients in forms that plants can assimilate. This process aids in alleviating nutrient deficiencies caused by stress. Additionally, some microbiota can solubilize minerals and fix N, thus enhancing nutrient availability to plants under stress conditions. Nutrient cycling is the biogeochemical process of transforming and recycling vital elements such as C, N, P, and other nutrients within ecosystems. Soil microbiomes, composed of bacteria, archaea, fungi, and other microorganisms, are the primary drivers of nutrient cycling. These microorganisms break down complex organic matter into simpler compounds that plants can take up. They also perform processes such as N_2 fixation, mineralization, nitrification, and denitrification, which are crucial for nutrient availability and balance in soil.

Soil microbiomes contribute to nutrient cycling by decomposing organic matter through enzymatic activities. This breakdown releases nutrients locked within complex compounds, making them available for plant uptake. Microbes secrete enzymes that degrade organic matter, releasing mineral nutrients that plants can assimilate. This enhanced mobilization can counteract nutrient limitations caused by abiotic stress. Certain soil microorganisms, particularly N2-fixing bacteria and archaea, are critical in converting atmospheric N_2 into plant-available forms. This process, known as N_2 fixation, contributes to soil fertility and can mitigate stress-induced N deficiencies. Additionally, ammonification, the conversion of organic N to ammonium, is facilitated by microbial communities, ensuring a steady supply of ammonium for plant uptake. Phosphorus availability is often limited due to its immobilization in soil minerals. Microbes that can solubilize phosphorus, such as phosphate-solubilizing bacteria and fungi, release this essential nutrient from mineral complexes, making it accessible to plants. Moreover, mycorrhizal fungi form symbiotic associations with plant roots, extending their reach and enhancing nutrient uptake, especially P, under stress conditions.

4.2 Role of phytohormones in imparting drought tolerance in plants

Phytohormones, such as GAs, ethylene, IAA, ABA, and SKs are chemical messengers that are organic, synchronize cellular events in plants, and play key roles in plant progression and drought stress (Khan et al., 2012; Jogawat et al., 2021; Shaffique et al. 2022.). Several studies have provided evidence to support the role of IAA in inducing drought-related signaling pathways when plants are exposed to osmotic stress (Fahad et al., 2015; Yu et al., 2020). The application of microbes also enhances the production of plant gibberellins and diterpenoids that cause hyperactive elongation of stems in response to stress. Gibberellins are associated with two bioactive components: carotenes and isoprene (Alagoz et al., 2018). Carotenes can stimulate photosynthesis and protect plant cells from harmful photodynamic reactions, whereas isoprene maintains turgor pressure and stabilizes the cell membrane structure (Mollah and Pratiwi, 2020). In response to abiotic stress, cytokinin biosynthesis increases in plants associated with auxin and regulation of developmental responses. In addition, cytokinin helps in the phosphorylation of sugar molecules, which leads to their accumulation in cells and prevents reverse diffusion (Mok and

Mok, 2001). Competitive phosphorylation helps plant cells combat stress, and cytokinin protects plant cells from oxidative stress (Werner and Schmülling, 2009; Wybouw and De Rybel, 2019).

Ethylene is a hormone that regulates various developmental processes, ranging from germination to the plant life cycle (Kanellis et al., 2012; Wen, 2014). Abscisic acid is a vital hormone secreted when plants are exposed to stress; thus, it is known as a universal stress hormone. ABA is critical in mitigating stress because it causes stomatal closure and potently inhibits stomatal opening (Bright et al., 2006). ABA also regulates genes responsible for desiccation tolerance (Sakata et al., 2014). The ability of these rhizobacteria to synthesize and produce phytohormones is the first key factor in mitigating stress through a complex signalling network in plants to respond to environmental conditions and maintain the plant's inner environment through regulation of these hormones, antioxidants, and osmolytes to enhance stress tolerance.

4.3 Role of volatiles in inducing drought tolerance

The use of beneficial microbes produces volatile compounds to increase growth, development, photosynthesis, iron uptake, and crop productivity while reducing the incidence of plant diseases and death. For example, stress-induced volatile compounds such as 3-hydroxy-2-butanone (acetoin), 2,3-butanediol, and 2-pentylfuran play a role in plant growth and development (Seco et al., 2015; Ali et al., 2017). Under stress conditions, these compounds cause stomatal closure and impart systemic stress resistance to plants, highlighting their role in plant growth and development under stress conditions (Liu and Zhang, 2015; Mohammadipanah and Zamanzadeh, 2019). Plant microbial interactions emit VOCs, which mitigate stress tolerance through the expression of genes and scavenge free radicals.

Rhizosphere microorganisms can assist the plant in dealing with drought stress through various methods (Fadiji et al., 2022). Recent research shows that insoluble metabolites produced by rhizosphere microbiomes improve plant drought tolerance. VOCs produced in bacteria are generally from metabolic end products by anaerobic fermentation and extracellular degradation of complex organic molecules (Choudhary et al., 2017; Poudel et al., 2021). According to Narayanasamy et al. (2020), VOCs are exciting candidates for stress decrease during plant development and growth. Wheat seedlings treated with *B. thuringiensis* AZP2 volatiles, for example, showed a five-fold improvement in survival under drought, owing to a large enhancement in plant biomass, which enhances photosynthetic rate (Narayanasamy et al., 2020). Some VOCs emitted by *B. thuringiensis*, such as geranyl, b-pinene, and benzaldehyde compounds, increase tolerance to drought in wheat. Some research has revealed that 2,3-butanediol stimulates the synthesis of nitric oxide (NO) in plants, which improves the plant's survival under drought, suggesting that NO signaling plays a vital role in improving tolerance to drought (Cho et al., 2013). Timmusk et al. (2014) presented that priming treatment with *Bacillus thuringiensis* AZP2 in wheat seedlings improved plant biomass where a five-fold higher plant survival rate under severe drought owing to emissions of stress-related volatiles.

4.4 PGPR alters root morphology under drought stress

The roots of plants are an essential tissue for the growth of the plant because they help with the absorption and transportation of nutrient water, anchoring, and symbiotic interaction with the beneficial microorganisms in soil which contribute to increased plant health and growth (Hodge et al., 2009; Abdelaal et al., 2021). In drought-stressed plants, the root architecture, such as secondary and primary roots, root number, and root diameter, is the most important characteristic. Root flexibility in the response of plants to water scarcity is a valuable tool for dealing with moisture stress (Ngumbi and Kloepper, 2016). PGPR-mediated alterations in root development and flexibility alter the architecture of the root. Some of the bacteria-associated changes in the structure of roots are projected to increase the area of the root surface and, as a result, nutrient and water uptake, which has a significant impact on drought resistance (Ngumbi and Kloepper, 2016; Abdelaal et al., 2021). Under drought stress, maize seedlings were primed with *A. faecalis* AF3. Ambreetha et al. (2018) quantified that *B. altitudinis* FD48 inoculation altered the architecture of the rice root system, and this was corroborated by the pattern of expression of defense-responsive genes encoding the main development of root and formation of lateral roots. These findings showed that inoculating rice with FD48 alters endogenous IAA levels and auxin-responsive genes, resulting in changes to the root system architecture. PGPR-mediated root architecture modification is important for increasing rice yield under extreme environmental circumstances (Paez-Garcia et al., 2015). Many studies have shown that modifying rice root system features can increase yield in stressful settings (Zou et al., 2015). The PGPR produces signaling molecules that naturally change the root system architecture of related crops (Ojuederie et al., 2019).

4.5 Role of aminocyclopropane-1-carboxylic acid (ACC) deaminase producing rhizobacteria in drought stress tolerance

Ethylene levels regulate plant activities, and ethylene biosynthesis is regulated by biotic and abiotic stresses (Vurukonda et al., 2016; Hardoim et al., 2008). Under drought stress, plants enhance their ethylene production, which inhibits plant growth by distressing seed germination and root enlargement. Higher ACC levels are induced to combat severe drought stress (Ustiatik et al., 2021).

Beneficial microbes can produce ACC deaminase. ACC regulates plant growth and development by sequestering and cleaving plant-produced ACC and lowering the level of ethylene in plants.

Under stress conditions, the plant hormone ethylene endogenously regulates plant homeostasis, reducing root and shoot growth. Plants' ACC is confiscated and degraded by ACC-deaminase-producing bacteria to supply N and energy (Vurukonda et al., 2016). Furthermore, by removing ACC, the bacteria reduce the deleterious effect of ethylene, ameliorating plant stress and promoting plant growth (Glick, 2005). ACC deaminase-producing PGPR *Achromobacter piechaudii* significantly increased tomato and pepper seedlings' biomass and reduced ethylene production under drought stress (Mayak et al., 2004).

Plants inoculated with ACC deaminase-producing bacterium gave more seed yield, seed number, seed N accumulation, and restoring nodulation, depressed in drought stress conditions (Dodd et al., 2005). ACC deaminase-producing bacteria eliminated the effects of drought stress on the growth, yield, and ripening of peas in both pot and field trials (Arshad et al., 2008). The successful deployment of bacteria in the rhizosphere increased plants' yield and nutritional value in dry soil via local and systemic hormone signaling. Such bacteria provide an easily realized economic means of sustaining crop yields in dryland agriculture (Belimov et al., 2009).

4.6 Osmotic adjustments in imparting drought tolerance in plants

When plants show abiotic stress, a sudden loss of osmolytes can shock the plant. However, microbial interactions can lead to osmotic adjustments. Secreted metabolites help plant–microbe interactions; plants upregulate the production of metabolites such as proline, sugars, glycine, organic acids, trehalose, betaine, potassium, calcium, and chloride ions as an adaptive response to drought stress (Shaffique et al., 2022). As one of the most critical osmolytes, proline accumulates in plants under osmotic stress and facilitates the maintenance and osmotic adjustment of cellular components in plant cells to overcome stress. Some rhizosphere bacteria produce proline (Mohammadkhani and Heidari, 2008; Dobra et al., 2010). Glycine betaine is a secondary metabolite that helps plants maintain growth and development. Trehalose is a safe, nonreducing sugar that contains two glucose molecules and stores energy for use under stress conditions. Microbes can accelerate trehalose biosynthesis through the trehalose-6-phosphate phosphatases pathway, maintain osmolytes concentrations, and stabilize turgor pressure in plant cells (Iordachescu and Imai, 2011). The microbes colonized by plant cells increase the exudation of organic acids (e.g. oxalic, citric, and malic acids) and minerals (e.g. chlorine, sodium, and potassium), which are essential for nutrient availability, metabolic reactions, and the maintenance of osmoregulation in plant cells (Chen and Jiang, 2010). The ability to regulate the osmotic amendment of PGPR in plant cells is due to generating a low water gradient potential in the cytosol. They maintain the turgor pressure and osmotic adjustment and improve stress tolerance.

Plant's adaptation to drought stress is allied with metabolic rules that lead to the accrual of compatible osmolytes such as sugars, proline, betaines, polyamines, quaternary ammonium compounds, polyhydric alcohols and some amino acids, and water stress proteins like dehydrins (Vurukonda et al., 2016). For example, PGPRs secrete osmolytes due to drought stress, which synergizes with plant-released osmolytes and motivates plant growth (Paul et al., 2008). *Pseudomonas putida* GAP-P45 inoculation improved plant biomass, relative water content, and leaf water potential by producing proline in maize plants that were exposed to drought stress (Sandhya et al., 2010). Inoculating plants with PGPR adds up to the existing proline concentrations; a sizeable quantity of proline increased when maize plants were inoculated with *P. fluorescens* under drought stress (Ansary et al., 2012). Increased proline content due to upregulation of the proline biosynthesis pathway keeps proline high to help maintain cell water status and protect membranes and proteins from stress (Sandhya et al., 2010). Higher proline accumulation in inoculated plants indicates higher plant tolerance to water stress (Gusain et al., 2015). Inoculation of maize plants with PGPR *P. putida* GAP-P45 (Sandhya et al., 2010) and *A. lipoferum* (Bano et al., 2013) improved plant growth by gathering free amino acids and soluble sugars than non-treated plants under drought stress. Trehalose acts as an osmoprotectant by stabilizing dehydrated enzymes and membranes those impart osmoprotection (Yang et al., 2010). Microarray analysis of 7200 sequence tags expressed from nodules inoculated with rhizobium strain that overexpressing trehalose-6-phosphate synthase gene exposed upregulation of genes involved in stress tolerance, C and N metabolism, suggesting a signaling mechanism for trehalose (Suarez et al., 2008).

4.7 Microbial production of exopolysaccharide (EPS) for drought stress mitigation

Drought stress affects soil's physicochemical and biological properties, making them unsuitable for microbial role to crop yield. The water availability controls the production and consumption of protein and polysaccharides by the bacteria (Roberson and Firestone, 1992) and thus indirectly influences soil structure. Exopolysaccharides are macromolecules composed of long-chain polymers of repeating sugar units such as glucose, galactose, and rhamnose in different ratios (Ilyas et al., 2020; Bhagat et al., 2021). Exopolysaccharides can be released into the soil as slime ingredients comprising cation hydrogen bridges, van der Waals linkages, and anion adsorption interactions, which act as defensive capsules around the soil and improve biological properties of the soil by increasing aggregation and macro-porosity (Ilyas et al., 2020). Exopolysaccharide production helps plants cope with abiotic stress (Naseem et al., 2018). The vital interaction of the plant and microbes facilitates the production of a biofilm that allows microbes to attach to plant roots, provides a defense to the roots, and imparts a solid root-adhering capability. Bacteria were successfully colonized in root-adhering rhizoplane and soil, increasing soil aggregate stability (Sandhya et al., 2009). More soil aggregation increases water and nutrient absorption from the rhizosphere, allowing plants to survive and flourish under induced drought stress (Sandhya et al., 2009).

4.8 Altering antioxidant defence system

Osmotic stress induces the creation of ROS in plants, including superoxide, radicals, singlet oxygen, hydrogen peroxide, superoxide anion radicals, and alkoxy radicals. ROS react with lipids, proteins, and DNA, causing oxidative damage and affecting redox regulation in plants (Ahmad et al., 2014). To protect against oxidative damage during osmotic stress, plants induced antioxidant defense systems involving enzymatic and non-enzymatic pathways (Chai et al., 2005). Enzymatic antioxidant pathways include enzymes such as the glutathione-reductase, superoxide dismutase (SOD), catalase, and ascorbate peroxidase, whereas non-enzymatic components include ABA, cysteine, and glutathione (Ahmad et al., 2014; Li et al., 2020). The presence and exogenous inoculation of microbes in soil can deliberate drought tolerance by encouraging antioxidant systems. Various microbes, such as actinomycetes, fungi, and algae, contain phenolic components secreted when plants are stressed (Kaushal and Wani, 2016; Ilyas et al., 2021). ROS scavenging ability of PGPR regulates antioxidant enzymes, which may provide a solid barrier against abiotic stress (Soni et al. 2023). The plant–microbe interaction ensures that there is always a homeostatic balance between ROS and removal machinery, which mitigates stress tolerance in plants.

Drought stress involves a cascade of events that cause oxidative damage, where the antioxidant defense system is activated by induction of gene expression to combat ROS. For example, inoculation of two AMF strains of *Gigaspora margarita* and *Glomus intraradices* in host plants resulted in the expression of GmarCuZnSOD, GintPDX1, GintMT1, and GintSOD. In addition, the mechanism for tolerating drought stress involves reducing cytoplasmic protein levels and regulating redox status via pyridoxamine synthesis (Zou et al., 2021). Drought stress negatively affects crop quality and quantity; however, drought stress can be mitigated through AMF inoculation (Al-Arjani et al., 2020). In another study, AMF and silicon were inoculated with 1-week-old strawberry seedlings to observe their synergistic effects against drought stress (Moradtalab et al., 2019). Following a randomized trial over 4 weeks, AMF and silicone were found to increase the biomass, water uptake, and mineral content and promote the antioxidant defense system. Thus, combined, the two are beneficial for combating drought stress (Moradtalab et al., 2019).

4.9 Stimulation of stress-response genes by plant-microbe interactions

Stress tolerance can be boosted by treating plants (Poudel et al. 2021) with microbes because this induces stress-response genes that regulate plant stress. The expression of such genes is modulated in plants under drought stress, which helps optimize plant growth and development. Stress-related genes and proteins involved in plant–microbe interactions include CaPR-10, sHSP, 11-pyrroline-5-carboxylate reductase, dehydrin-like protein, vacuolar ATPase, and pyrroline-5-carboxylate dehydrogenase (Shinozaki and Yamaguchi-Shinozaki, 2007). Microarray studies have suggested that gene expression can be categorized into functional or regulatory proteins depending on the role of the encoded protein. Regulatory proteins include transcription, protein phosphate kinases, and ABA biosynthetic factors. Functional proteins include detoxification enzymes, water channel transporters, proteases, osmolyte biosynthesis enzymes, and macromolecule protection factors. Genes and transcription factors can encode sustainable agricultural products (Shinozaki and Yamaguchi-Shinozaki, 2007; Kavar et al., 2008). The expression and upregulation of stress-related genes in response to drought stress can be a powerful tool for mitigating and enhancing drought tolerance in plants.

Studies involving the expression of a gene are a useful technique for determining and comparing plant responses to their surroundings. The transcriptome research examines the expression of a cell's whole collection of mRNA transcripts at a normal developmental stage or exposure to various environmental circumstances (Vurukonda et al., 2016). Plants colonized by microbes have distinct gene expression patterns than non-colonized plants, resulting in increased tolerance to drought. Comparison based on transcriptomic profiling aids in the identification of diverse groups of gene transcripts responsible for changes in two physiologically distinct expressions under different situations (Fadiji and Babalola, 2020c). Therefore, some of the most important tools in investigating plant–microbe interactions are transcriptome profiling from microarray and mRNA sequencing (Wang et al., 2016). The physiological function of PGPR was used to assess gene regulation impacted by moisture stress, according to Mohammadipanah and Zamanzadeh (2019). In Arabidopsis plants infected with *B. polymyxa* B2, the favorable impact of plant growth-promoting bacteria on improving plant stress tolerance at the transcriptomic level was investigated (Vurukonda et al., 2016; Abdelaal et al., 2021). These results show that inoculated plants produce the drought adaptive gene early response to ABA regulatory gene RAB18 (LEA) and dehydration 15 (ERD15), but uninoculated plants do not.

4.10 Co-inoculation of PGPR for alleviation of drought stress

In addition to single strains of PGPR, its combination with either AMF or *Rhizobium* elicits plant drought tolerance (Wang et al., 2012). For example, co-inoculation of common bean with *Rhizobium tropici*-CIAT 899, *P. polymyxa*-DSM36, and *P. polymyxa* Loutit strains increased growth compared to inoculation with *Rhizobium* alone. Furthermore, co-inoculation exhibited greater nodulation numbers, biomass, and N content than drought-stressed plants inoculated with only *Rhizobium* (Soni et al., 2023). Further, co-inoculation of wheat plants with *Azotobacter chrocoocum* (E1) and *Pseudomonas* sp. (E2) alleviated drought stress through increased anatomical changes such as the thickness of the epidermis, ground, mesophyll, and phloem tissues, xylem vessel diameter and vascular bundles dimensions of the root system. In contrast, water deficit levels (75%, 50%, and 25% field capacity) decreased the anatomical values of the un-inoculated wheat cultivars (El-Afry et al., 2012). Microbial consortium containing PGPR *Bacillus cereus* AR156, *B. subtilis* SM21, and *Serratia sp.* XY21 improved drought tolerance in cucumber plants (Vurukonda et al. 2016). After withholding watering for 13 days, PGPR-treated plants exhibited darker green leaves with lighter symptoms of wilt, decreased the leaf mono dehydroascorbate content and relative electrical conductivity, and increased the leaf proline and chlorophyll content and root recovery in cucumber plants under drought stress. In comparison with the control, the PGPR treatment enhanced the SOD activity and mitigated the drought-induced downregulation of genes cAPX, rbcL, and rbcS encoding cytosolic ascorbate peroxidase and ribulose-1,5-bisphosphate carboxy/oxygenase (Rubisco) large and small subunits, respectively, in cucumber leaves (Wang et al. 2012). Microbial consortia consisting of EPS-producing bacterial strains *P. penneri* (Pp1), *P. aeruginosa* (Pa2), and *A. faecalis* (AF3) showed greater potential for drought tolerance in maize than single PGPR strains (Naseem and Bano, 2014). PGPRs consortia eased drought stress in rice plants by means of reducing oxidative damage and buildup of proline, improving plant growth (Gusain et al., 2015; Vurukonda et al., 2016).

5 Conclusion

Plants have evolved sophisticatedly with microbiomes to overcome drought stress and its adverse effects within cells. Microbiomes strengthen plants' intrinsic stress tolerance mechanisms by producing exopolysaccharides, phytohormones, ACC-deaminase, osmolytes, and volatile compounds. In addition, cells synthesize diverse molecules that regulate redox damage and overcome antioxidant damage. This review highlights various adverse effects of drought stress on plant growth and reveals how microbiomes can mitigate such impacts. Previous studies have focused on drought mitigation by using a group of microbes. However, this review discusses a variety of microbes that reduce drought stress, their fundamental mechanisms of action, and the changes they produce at the molecular level. Microbial intervention has led to substantial progress in drought tolerance. Future investigations should evaluate whether critical and long-lasting effects benefit plant phenotypes. However, known microbial interventions are promising; their dose, frequency, and timing warrant further practical validation.

Drought stress is an awful environmental constraint to agricultural productivity. The microbiome plays a vital role in conferring resistance and adaptation of plants to this stress and potentially solving future food security issues. The interaction between plants and microbiomes under drought conditions affects the plant and changes the soil properties. The mechanisms elicited by the microbiome, prompting osmotic response due to the induction of novel genes, played a vital role in confirming plant survival under drought stress. Developing drought-tolerant crop varieties through genetic engineering and plant breeding is essential but is a long-drawn process. In contrast, microbiome inoculation to alleviate plant drought stresses opens a new era in their application in dryland agriculture. Presently, concerted future research is needed to identify the right kind of microbes, address the issue of delivery systems, and field evaluation of potential microbiomes.

The soil microbiome's vast and diverse community of microorganisms in the soil plays a crucial role in determining crops' health, productivity, and resilience to various abiotic stresses. Over the past few years, advances in understanding the role of the soil microbiome have shed light on its intricate interactions with plants and the potential benefits these interactions can offer in agriculture. Here are some critical conclusions drawn from these advances:

(i) Plants have co-evolved with their soil microbes to establish beneficial relationships (symbiosis). For instance, mycorrhizal fungi help plants absorb nutrients, especially P, while certain bacteria can fix N. These relationships often make crops more resilient to abiotic stresses like drought.

(ii) A diverse soil microbiome improves soil health. Diverse microbial communities can also enhance soil structure, improve nutrient cycling, and increase the resilience of crops to various stresses, including drought.

(iii) As our understanding grows, there's an increasing interest in developing biofertilizers and biostimulants that can introduce or boost beneficial microorganisms in the soil. These have shown potential in reducing the need for chemical fertilizers and enhancing crop resilience to drought.
(iv) Certain soil microorganisms have been found to produce organic compounds that help plants cope with drought stress. These compounds can stimulate root growth, improve water uptake, or even alter the plant's internal hormone balance to help it survive drought.
(v) While manipulating the soil microbiome to promote crop resilience has potential, it's a complex endeavor. Introducing a specific strain or community/consortium might have unforeseen effects on the native microbial community and soil health.
(vi) While there's clear evidence of the benefits of a healthy soil microbiome, there are challenges in harnessing this potential at a large scale. These challenges include variability in results across different soil types, the potential for unintended ecological consequences, and a lack of standardized methodologies.
(vii) Understanding the soil microbiome shouldn't be isolated from other soil health principles. Agronomic practices such as crop rotation, organic farming, and reduced tillage can also promote a healthy soil microbiome and should be integrated with microbiome-specific strategies.
(viii) Our current knowledge, although expanding rapidly, is still just the tip of the iceberg. More research is needed, especially long-term studies, to understand the dynamic nature of the soil microbiome and its interactions with crops under various conditions.
(ix) Abiotic stresses are a global challenge, and solutions can benefit from international cooperation. Sharing research, methodologies, and findings across regions can accelerate the application of microbiome-based solutions in agriculture.
(x) The soil microbiome holds great promise in promoting crop resilience to abiotic stresses.

6 Future trends

Recent development in the sector indicates that future study has enormous prospects to bring new insights into food production and sustainability. More unique formulations, such as nanoencapsulation, might be researched to ensure the efficiency of bioinoculants in field circumstances for better drought tolerance, colonization, and distribution of beneficial microbes on host plants. Furthermore, the microbiome affects not only the plants but also the soil qualities during drought. As a result, more molecular studies into plant–microbe interactions are needed to understand the routes employed by rhizosphere microorganisms in induced systemic tolerance and rhizosphere engineering under drought stress. To reap these benefits, future studies must focus on identifying the potential strain, adequate delivery mechanism, and field evaluation for long-term crop and food production. Some of these strategies are stated as follows:

(i) Investigating the entire plant microbiome and understanding the interactions among various microorganisms, including PGPR and fungal endophytes, will provide a more comprehensive view of how these interactions impact stress tolerance.
(ii) Advancements in technology can enable the targeted application of PGPB and fungal endophytes to specific stress-prone areas within fields, optimizing resource use.
(iii) Designing microbial consortia with specific traits can enhance stress resistance. Engineering microbial strains with desired characteristics might offer new avenues for stress mitigation.
(iv) Utilizing genomics, transcriptomics, and metabolomics delivers insights into genetic and molecular mechanisms underlying the interactions between plants and beneficial microorganisms.
(v) Tailoring research to specific crop species and local environmental conditions will provide more relevant insights into the practical applications of microbiomes.
(vi) Combining beneficial microorganisms with traditional breeding approaches can lead to developing climate-resilient crop varieties.
(vii) Educating farmers about the benefits and practicalities of using microbiomes in their farming practices will be crucial for successful adoption.

7 Challenges to research

While symbiogenics doesn't appear to be a widely recognized term in biological sciences, the use of microbiome in mitigating abiotic stresses is a well-explored area. However, like any scientific field, challenges and ongoing research directions exist. Here are some challenges related to the role of microbiome in addressing abiotic stresses:

(i) Not all strains of microbiomes are equally effective in enhancing plant tolerance to abiotic stresses. Identifying and selecting the most suitable strains for specific crops and stress conditions can be challenging.
(ii) The effectiveness of the microbiome can vary with changing environmental conditions, including soil type, climate, and agronomic practices. Ensuring consistent results across diverse environments is a challenge.

(iii) While the positive effects of microbiomes on plants under stress are apparent, the exact mechanisms behind these interactions are often complex and not fully understood.
(iv) Introducing beneficial microorganisms into agricultural systems might require regulatory approvals to ensure safety and prevent potential ecological disruptions.
(v) Plants often interact with a diverse microbial community. Understanding how microbiomes interact with other organisms in the rhizosphere and their combined effects on stress resistance is a complex endeavor.
(vi) While laboratory and greenhouse studies show promise, translating these findings to large-scale agricultural fields can be challenging due to logistical and economic constraints.

8 Author contributions

ZMS and KHMS contributed to the original draft writing, revision and publication preparation.

9 Where to look for further information

- Bashan, Y., and De-Bashan, L. (2005). Plant growth-promoting. In: *Encyclopedia of Soils in the Environment*, Vol. 1, D. Hillel, ed. Oxford, Elsevier, pp. 103–115. doi: 10.1016/B0-12-348530-4/00513-0.
- Farooq, M., et al. (2009). Plant drought stress: effects, mechanisms and management. In: *Sustainable Agriculture*, E. Lichtfouse, et al., eds. Dordrecht, Springer Netherlands, pp. 153–188.
- Figueiredo, M. D. V. B., et al. (2010). Plant growth promoting rhizobacteria: fundamentals and applications. In: *Plant Growth and Health Promoting Bacteria*, D. Maheshwari, ed. Berlin, Springer, pp. 21–43. doi: 10.1007/978-3-642-13612-2_2.
- Selvakumar, G., et al. (2012). Bacterial mediated alleviation of abiotic stress in crops. In: *Bacteria in Agrobiology: Stress Management*, D. Maheshwari, ed. Berlin, Heidelberg, Springer EBooks, pp. 205–224. https://doi.org/10.1007/978-3-642-23465-1_10.

10 Funding

There is no funding sorted for this work.

11 Acknowledgments

We thank the editor and anonymous reviewers for the critical reading of the manuscript for their constructive feedback.

12 References

Aamir, M., et al. (2019). Impact of climate change on soil carbon exchange, ecosystem dynamics, and plant-microbe interactions. In: *Climate Change and Agricultural Ecosystems,* K. K. Choudhary, A. Kuma, and A. K. Singh, eds. Duxford: Elsevier, Woodhead Publishing, pp. 379–413.

Abdelaal, K., et al. (2021). The role of plant growth-promoting bacteria in alleviating the adverse effects of drought on plants. *Biology* 10, 520.

Agapitos, A., et al. (2008). Generating diverse opponents with multiobjective evolution. In: *Proceedings of the IEEE Symposium on Computational Intelligence and Games, 2008. CIG'08*, Piscataway, NJ: IEEE, pp. 135–142.

Akanmu, A. O., et al. (2021). Plant disease management: leveraging on the plant microbe-soil interface in the biorational use of organic amendments. *Front. Plant Sci.* 12, 700507.

Alagoz, Y., et al. (2018). Cis-carotene biosynthesis, evolution and regulation in plants: the emergence of novel signaling metabolites. *Arch. Biochem. Biophys.* 654, 72–184.

Ali, F., et al. (2017). Recent methods of drought stress tolerance in plants. *Plant Growth Regul.* 82, 363–375. doi: 10.1007/s10725-017-0267-2.

Ahmad, A., et al. (2021). Exogenous salicylic acid-induced drought stress tolerance in wheat (*Triticum aestivum* L.) grown under hydroponic culture. *PLoS One* 16, e0260556.

Ahmad, P., et al. (2014). Drought stress induced oxidative damage and antioxidants in plants. In: *Oxidative Damage to Plants*, P. Ahmad, ed. New York: Elsevier, pp. 345–367.

Al-Arjani, A.-B. F., et al. (2020). Arbuscular mycorrhizal fungi modulates dynamics tolerance expression to mitigate drought stress in Ephedra foliata Boiss. *Saudi J. Biol. Sci.* 27, 380–394.

Ambreetha, S., et al. (2018). Plant-associated Bacillus modulates the expression of auxin-responsive genes of rice and modifies the root architecture. *Rhizosphere* 5, 57–66.

Ansary, M. H., et al. (2012). Effect of Pseudomonas fluorescens on proline and phytohormonal status of maize (*Zea mays* L.) under water deficit stress. *Annal. Biol. Res.* 3, 1054–1062.

Armada, E., et al. (2016). Native plant growth promoting bacteria Bacillus thuringiensis and mixed or individual mycorrhizal species improved drought tolerance and oxidative metabolism in Lavandula dentata plants. *J. Plant Physiol.* 192, 1–12.

Arshad, M., et al. (2008). Inoculation with Pseudomonas spp. containing ACC deaminase partially eliminate the effects of drought stress on growth, yield and ripening of pea (*P. sativum* L.). *Pedosphere* 18, 611–620.

Arun, K. D., et al. (2020). Mitigation of drought stress in rice rop with plant growth-promoting abiotic stress-tolerant rice phyllosphere bacteria. *J. Basic Microbiol.* 60, 768–786.

Askari-Khorasgani, O., et al. (2021). Drought stress sensing-signaling in plants. In: *Handbook of Plant and Crop Physiology*, M. Pessarakli, ed. Boca Raton, FL: CRC Press, pp. 465–481.

Aslam, M. M., et al. (2022). Rhizosphere microbiomes can regulate plant drought tolerance. *Pedosphere* 32, 61–74.

Banfield, J. F. and Nealson, K. H. (2018). *Geomicrobiology: Interactions between Microbes and Minerals*. Berlin: Walter de Gruyter GmbH & Co KG.

Bano, Q., et al. (2013). Effect of *Azospirillum* inoculation on maize (*Zea mays* L.) under drought stress. *Pak. J. Bot.* 45, 13–20.

Bapiri, A., et al. (2010). Drying–rewetting cycles affect fungal and bacterial growth differently in an arable soil. *Microb Ecol.* 60, 419–428.

Barman, J., et al. (2016). Mycorrhiza. *Resonance* 21, 1093–1104.

Barnard, R. L., et al. (2013). Responses of soil bacterial and fungal communities to extreme desiccation and rewetting. *ISME J.* 7, 2229–2241.

Barnawal, D., et al., (2019). Role of plant growth promoting rhizobacteria in drought tolerance. In: *PGPR Amelioration in Sustainable Agriculture*. pp. 107–128.

Bartels, D. and Phillips, J. (2010). Drought stress tolerance. In: *Genetic Modification of Plants*, F. Kempken and C. Jung, eds. Berlin: Springer, pp. 139–157.

Bashan, Y., et al. (2014). Advances in plant growth-promoting bacterial inoculant technology: formulations and practical perspectives (1998–2013). *Plant Soil* 378, 1–33.

Belimov, A. A., et al. (2009). Rhizosphere bacteria containing 1-aminocyclopropane-1-car boxylatedeaminase increase yield of plants grown in drying soil via both local and systemic hormone signaling. *New Phytol.* 181, 413–423.

Bentley, S. D., et al. (2002). Complete genome sequence of the model actinomycete Streptomyces coelicolor A3(2). *Nature* 417(6885), 141–147.

Berdy, J. (2005). Bioactive microbial metabolites. *J. Antibiotics* 58(1), 1–26.

Bhagat, N., et al. (2021). Bacterial exopolysaccharides: insight into their role in plant abiotic stress tolerance. *J. Microbiol. Biotechnol.* 31, 1045–1059.

Bhattacharyya, A., et al. (2021). Rhizosphere plant microbe interactions under water stress. In: *Advances in Applied Microbiology*, Volume 115, G. M. Gadd and S. Sariaslani, eds. Cambridge, MA: Elsevier, Academic Press, pp. 65–113.

Bhattacharyya, P. N. and Jha, D. K. (2012). Plant growth-promoting rhizobacteria (PGPR): emergence in agriculture. *World J. Microbiol. Biotechnol.* 28(4), 1327–1350.

Bitterlich, M., et al. (2018). Arbuscular mycorrhiza alleviates restrictions to substrate water flow and delays transpiration limitation to stronger drought in tomato. *Front. Plant Sci.* 9, 154.

Bouasria, A., et al. (2012). Changes in root-associated microbial communities are determined by species-specific plant growth responses to stress and disturbance. *Eur. J. Soil Biol.* 52, 59–66.

Bright, J., et al. (2006). ABA induced NO generation and stomatal closure in Arabidopsis are dependent on H2O2 synthesis. *Plant J.* 45, 113–122.

Brilli, F., et al. (2019). Exploiting plant volatile organic compounds (VOCs) in agriculture to improve sustainable defense strategies and productivity of crops. *Front. Plant Sci.* 10, 264.

Brunner, I., et al. (2015). How tree roots respond to drought? *Front. Plant Sci.* 6, 547.

Canarini, A., et al. (2016). Drought effects on Helianthus annuus and Glycine max metabolites: From phloem to root exudates. *Rhizosphere* 2, 85–97.

Canarini, A., et al. (2019). Root exudation of primary metabolites: mechanisms and their roles in plant responses to environmental stimuli. *Front. Plant Sci.* 10, 157.

Cantó, C. D., et al. (2020) An extended root phenotype: the rhizosphere, its formation and impacts on plant fitness. *Plant J.* 103(3), 951–964.

Chai, T.-T., et al. (2005). Water stress induced oxidative damage and antioxidant responses in micropropagated banana plantlets. *Biol. Plant.* 49, 153–156.

Chandrasekaran, M., et al. (2019). Endophytic and arbuscular mycorrhizal fungi as biopesticide: a review. *J. Fungi* 5(2), 41.

Chen, H. and Jiang, J.-G. (2010). Osmotic adjustment and plant adaptation to environmental changes related to drought and salinity. *Environ. Rev.* 18, 309–319.

Chen, W., et al. (2020). Effects of arbuscular mycorrhizal fungi on growth and physiological performance of catalpa bungei CAMey. under Drought Stress. *Forests* 11, 10.

Cheng, H.-Q., et al. (2022). Arbuscular mycorrhizal fungi mitigate drought stress in citrus by modulating root microenvironment. *Arch. Agron. Soil Sci.* 68(9), 1217-1228.

Cho, S.-M., et al. (2013). Nitric oxide and hydrogen peroxide production are involved in systemic drought tolerance induced by 2R, 3R-butanediol in *Arabidopsis thaliana*. *Plant Pathol. J.* 29, 427.

Chiappero, J., et al. (2019). Plant growth-promoting rhizobacteria improve the antioxidant status in Mentha piperita grown under drought stress leading to an enhancement of plant growth and total phenolic content. *Ind. Crops Prod.* 139, 111553.

Choudhary, D. K., et al. (2016). *Plant-Microbe Interaction: An Approach to Sustainable Agriculture*. Berlin: Springer.

Choudhary, D. K., et al. (2017). Characterization of bacterial volatiles and their impact on plant health under abiotic stress. In: *Volatiles and Food Security*, D. Choudhary et al., eds. Singapore: Springer, doi: 10.1007/978-981-10-5553-9.

Curá, J. A., et al. (2017). Inoculation with *Azospirillum* sp. and *Herbaspirillum* sp. bacteria increases the tolerance of maize to drought stress. *Microorganisms* 5, 41.

Danish, S., et al. (2020). ACC-deaminase producing plant growth promoting rhizobacteria and biochar mitigate adverse effects of drought stress on maize growth. *PLoS One* 15, e0230615.

Davelos Baines, A. L., et al. (2004). Ammonia production by fungal endophytes of tall fescue may enhance microbial antagonism toward a fungal pathogen. *Microbial Ecol.* 48(1), 47–54.

Davitt, A. J., et al. (2011) Understanding context-dependency in plant-microbe symbiosis: the influence of abiotic and biotic contexts on host fitness and the rate of symbiont transmission. *Environ. Exp. Botany* 71, 137–145.

De Souza Vandenberghe, L. P., et al. (2017). Potential applications of plant probiotic microorganisms in agriculture and forestry. *AIMS Microbiol.* 3, 629.

de Vries, F. T., et al. (2018). Soil bacterial networks are less stable under drought than fungal networks. *Nat. Commun.* 9, 3033.

de Vries, F. T., et al. (2019). Changes in root-exudate induced respiration reveal a novel mechanism through which drought affects ecosystem carbon cycling. *New Phytol.* 224, 132–145.

Dhakal, R., et al. (2022). Prophylactic and therapeutic potential of active phytoconstituents from *Amomum Subulatum* Roxb. *Food Front.* doi: 10.1002/fft2.184.

Dimkpa, C., et al. (2009). Plant-rhizobacteria interactions alleviate abiotic stress conditions. *Plant Cell Environ.* 32, 1682–1694.

Dobra, J., et al. (2010). Comparison of hormonal responses to heat, drought and combined stress in tobacco plants with elevated proline content. *J. Plant Physiol.* 167, 1360–1370.

Dodd, I. C., et al. (2005). Will modifying plant ethylene status improve plant productivity in water-limited environments? *4th International Crop Science Congress*, Brisbane, Australia.

Dos Reis, S. P., et al. (2012). Recent molecular advances on downstream plant responses to abiotic stress. *Int. J. Mol. Sci.* 13, 8628–8647.

El-Afry, M. M., et al. (2012). Anatomical studies on drought-stressed wheat plants (*Triticum aestivum* L.) treated with some bacterial strains. *Acta Biol. Szeged* 56, 165–174.

Etesami, H., et al., (2015). Indole-3-Acetic Acid and 1-Aminocyclopropane-1-Carboxylate Deaminase: Bacterial Traits Required in Rhizosphere, Rhizoplane and/or Endophytic Competence by Beneficial Bacteria. In: *Bacterial Metabolites in Sustainable Agroecosystem*, Maheshwari, D.K., ed., Cham: Springer International Publishing, pp. 183–258.

Fadiji, A. E. and Babalola, O. O. (2020a). Elucidating mechanisms of endophytes used in plant protection and other bioactivities with multifunctional prospects. *Front. Bioeng. Biotechnol.* 8, 467.

Fadiji, A. E. and Babalola, O. O. (2020b). Exploring the potentialities of beneficial endophytes for improved plant growth. *Saudi J. Biol. Sci.* 27, 3622–3633.

Fadiji, A. E., and Babalola, O. O. (2020c). Metagenomics methods for the study of plant-associated microbial communities: a review. *J. Microbiol. Methods* 170, 105860.

Fadiji, A. E., et al. (2020). Organic farming enhances the diversity and community structure of endophytic archaea and fungi in maize plant: a shotgun approach. *J. Soil Sci. Plant Nutr.* 20, 2587–2599.

Fadiji, A. E., et al. (2021). Metagenomic profiling of rhizosphere microbial community structure and diversity associated with maize plant as affected by cropping systems. *Int. Microbiol.* 24, 325–335.

Fadiji, A. E., et al. (2022) Efforts towards overcoming drought stress in crops: Revisiting the mechanisms employed by plant growth-promoting bacteria. *Front. Microbiol.* 13, 962427.

Fahad, S., et al. (2015). Potential role of phytohormones and plant growth-promoting rhizobacteria in abiotic stresses: consequences for changing environment. *Environ. Sci. Pollut. Res.* 22, 4907–4921.

Farooq, M., et al. (2012). Drought stress in plants: an overview. In: *Plant Responses to Drought Stress*, R. Aroca, ed. Berlin: Springer, pp. 1–33.

Farooq, M., et al. (2009). Plant drought stress: effects, mechanisms and management. *Sustain. Agric.* 29, 153–188.

Fei, Q., et al. (2014). Bioconversion of natural gas to liquid fuel: opportunities and challenges. *Biotechnol. Adv.* 32(3), 596–614.

Figueiredo, A. F., et al. (2021) Common mycorrhizae network: a review of the theories and mechanisms behind underground interactions. *Front. Fungal Biol.* 2, 735299.

Fitzpatrick, R. W., et al. (2017). Schwertmannite formation and properties in acidic drain environments following exposure and oxidation of acid sulfate soils in irrigation areas during extreme drought. *Geoderma* 308, 235–251.

Furze, J. R., et al. (2017). Resistance and resilience of root fungal communities to water limitation in a temperate agroecosystem. *Ecol. Evol.* 7, 3443–3454.

Gepstein, S. and Glick, B. R. (2013). Strategies to ameliorate abiotic stress induced plant senescence. *Plant Mol. Biol.* 82, 623–633. doi: 10.1007/s11103-013-0038-z.

Groen, S. C., et al. (2022). Evolutionary systems biology reveals patterns of rice adaptation to drought-prone agroecosystems. *Plant Cell* 34, 759–783.

Glick, B. R. (2005). Modulation of plant ethylene levels by the bacterial enzyme ACC deaminase. *FEMS Microbiol. Lett.* 251, 1–7.

Glick, B. R. (2012). Plant growth-promoting bacteria: mechanisms and applications. *Scientifica (Cairo)*, 2012, 963401. doi: 10.6064/2012/963401.

Glick, B. R. (2014). Bacteria with ACC deaminase can promote plant growth and help to feed the world. *Microbiol. Res.* 169, 30–39.

Grover, M., et al. (2011). Role of microorganisms in adaptation of agricultural crops to abiotic stresses. *World J. Microbiol. Biotechnol.* 27, 1231–1240.

Gupta, A., et al. (2020). The physiology of plant responses to drought. *Science* 368, 266–269.

Gusain, Y. S., et al. (2015). Bacterial mediated amelioration of drought stress in drought tolerant and susceptible cultivars of rice (*Oryza sativa* L.). *Afr. J. Biotechnol.* 14, 764–773.

Hardoim, P. R., et al. (2008). Properties of bacterialendophytes and their proposed role in plant growth. *Trends Microbiol.* 16, 463–471.

Hardoim, P. R., et al. (2015). The hidden world within plants: ecological and evolutionary considerations for defining functioning of microbial endophytes. *Microbiol. Mol. Biol. Rev.* 79(3), 293–320.

Hodge, A., et al. (2009). Plant root growth, architecture and function. *Plant Soil* 321, 153–187.

Hopwood, D. A. (2007). *Streptomyces in Nature and Medicine: The Antibiotic Makers*. Oxford: Oxford University Press.

Hossain, M. A., et al. (2016). *Drought Stress Tolerance in Plants, Vol 2: Molecular and Genetic Perspectives*. Berlin: Springer.

Ilyas, N., et al. (2020). Exopolysaccharides producing bacteria for the amelioration of drought stress in wheat. *Sustainability* 12, 8876.

Ilyas, M., et al. (2021). Drought tolerance strategies in plants: a mechanistic approach. *J. Plant Growth Regul.* 40, 926–944.

Iordachescu, M. and Imai, R. (2011). Trehalose and abiotic stress in biological systems. In: *Abiotic Stress in Plants-Mechanisms and Adaptations*, A. K. Shanker and B. Venkateswarlu, eds. Croatia: Intech, pp. 215–234.

Iqbal, N., et al., (2017). Ethylene role in plant growth, development and senescence: interaction with other phytohormones. *Front. Plant Sci.* 8, 475.

Jalili, F., et al. (2009). Isolation and characterization of ACC deaminase-producing fluorescent pseudomonads, to alleviate salinity stress on canola (*Brassica napus* L.) growth. *J. Plant Physiol.* 166, 667–674.

Jansson, J. K. and Hofmockel, K. S. (2020). Soil microbiomes and climate change. *Nat. Rev. Microbiol.* 18, 35–46.

Jat, S. L., et al. (2021) Microbiome for sustainable agriculture: a review with special reference to the corn production system. *Arch. Microbiol.* 203(6), 2771–2793.

Jogawat, A., et al. (2021). Crosstalk between phytohormones and secondary metabolites in the drought stress tolerance of crop plants: a review. *Physiol. Plant.* 172, 1106–1132.

Jones, D. L. and Darrah, P. R. (1994). Role of root derived organic acids in the mobilization of nutrients from the rhizosphere. *Plant Soil* 166, 247–257.

Kanellis, A., et al. (2012). *Biology and Biotechnology of the Plant Hormone Ethylene II*. Berlin: Springer Science & Business Media.

Kanwal, S., et al. (2017). Amelioration of drought stress in wheat by combined application of PGPR, compost, and mineral fertilizer. *J. Plant Nutr.* 40, 1250–1260.

Kapadia, C., et al. (2021). Halotolerant microbial consortia for sustainable mitigation of salinity stress, growth promotion, and mineral uptake in tomato plants and soil nutrient enrichment. *Sustainability* 13, 8369.

Karlowsky, S., et al. (2018). Drought-induced accumulation of root exudates supports post-drought recovery of microbes in mountain grassland. *Front. Plant Sci.* 9, 1593.
Karst, J., et al. (2016). Stress differentially causes roots of tree seedlings to exude carbon. *Tree Physiol.* 37, 154–164.
Kaushal, M. and Wani, S. P. (2016). Plant-growth-promoting rhizobacteria: drought stress alleviators to ameliorate crop production in drylands. *Ann. Microbiol.* 66, 35–42.
Kavar, T., et al. (2008). Identification of genes involved in the response of leaves of Phaseolus vulgaris to drought stress. *Mol. Breed.* 21, 159–172.
Khan, M. S. and Irba, Z. (2023). Microbial mitigation of drought stress in plants: adaptations to climate change. doi: 10.5772/intechopen.109669.
Khan, N. A., et al. (2012). *Phytohormones and Abiotic Stress Tolerance in Plants*. Berlin: Springer Science & Business Media.
Kim, Y. C., et al. (2013). Enhancement of plant drought tolerance by microbes. In: *Plant Responses to Drought Stress,* R. Aroca, ed. Berlin: Springer Verlag, pp. 383–413.
Kour, D., et al. (2020). Amelioration of drought stress in Foxtail millet (*Setaria italica* L.) by P-solubilizing drought-tolerant microbes with multifarious plant growth promoting attributes. *Environ. Sustainability* 3, 23–34.
Kour, D. and Yadav, A. N. (2020). Microbe mediated mitigation of drought stress in crops. *Agric. Lett.* 1, 79–82.
Kozjek, K., et al. (2021). Long-term agricultural management impacts arbuscular mycorrhizal fungi more than short-term experimental drought. *Appl. Soil Ecol.* 168, 104140.
Kumar, M., et al. (2019). Biodiversity of methylotrophic microbial communities and their potential role in mitigation of abiotic stresses in plants. *Biologia* 74, 287–308.
Kusari, S., et al. (2014). Chemical ecology of endophytic fungi: origins of secondary metabolites. *Chem. Biol.* 21(7), 855–863.
Leifheit, E. F., et al. (2015). Arbuscular mycorrhizal fungi reduce decomposition of woody plant litter while increasing soil aggregation. *Soil Biol. Biochem.* 81, 323–328.
Li, B., et al. (2020). Elevated CO2-induced changes in photosynthesis, antioxidant enzymes and signal transduction enzyme of soybean under drought stress. *Plant Physiol. Biochem.* 154, 105–114.
Lozano, Y. M., et al. (2021). Effects of microplastics and drought on soil ecosystem functions and multifunctionality. *J. Appl. Ecol.* 58, 988–996. doi: 10.1111/1365-2664.13839
Li, X. and Liu, F. (2016). Drought stress memory and drought stress tolerance in plants: biochemical and molecular basis. In: *Drought Stress Tolerance in Plants*, Vol. 1, I. M. A. Hossain, S. H. Wani, S. Bhattacharjee, D. J. Burritt, and L.-S. P. Tran, eds. Berlin: Springer, pp. 17–44. doi: 10.1007/978-3-319-28899-4_2.
Liu, F., et al., (2019). Effect of the inoculation of plant growth-promoting rhizobacteria on the photosynthetic characteristics of Sambucus williamsii Hance container seedlings under drought stress. *AMB Expr.* 9, 169.
Liu, X.-M. and Zhang, H. (2015). The effects of bacterial volatile emissions on plant abiotic stress tolerance. *Front. Plant Sci.* 6, 774.
Maheswari, T. U., et al. (2013). Studies on phytohormone producing ability of indigenous endophytic bacteria isolated from tropical legume crops. *Int. J. Curr. Microbiol. Appl. Sci.* 2, 127–136.
Manzoni, S., et al. (2014). A theoretical analysis of microbial eco-physiological and diffusion limitations to carbon cycling in drying soils. *Soil Biol. Biochem.* 73, 69–83.
Martins, S. J., et al. (2018). Plant-associated bacteria mitigate drought stress in soybean. Environ. *Sci. Pollut. Res.* 25, 13676–13686.

Mayak, S., et al. (2004). Plant growth-promoting bacteria that confer resistance to water stress in tomatoes and peppers. *Plant Sci.* 166, 525–530.

Meena, K. K., et al. (2021). Root-endophytes and their contribution to plant abiotic stress tolerance. In: *Microbial Management of Plant Stresses*, A. Kumar and S. Droby, eds. Elsevier, pp. 119–129. doi: 10.1016/b978-0-323-85193-0.00002-4.

Miransari, M. and Smith, D. (2009). Alleviating salt stress on soybean (Glycine max (L.) Merr.)–Bradyrhizobium japonicum symbiosis, using signal molecule genistein. *Eur. J. Soil Biol.* 45, 146–152.

Mohammadipanah, F. and Zamanzadeh, M. (2019). Bacterial mechanisms promoting the tolerance to drought stress in plants. In: *Secondary Metabolites of Plant Growth Promoting Rhizomicroorganisms*, H. B. Singh et al., eds. Singapore: Springer, pp. 185–224. doi: 10.1007/978-981-13-5862-3_10.

Mohammadkhani, N. and Heidari, R. (2008). Drought-induced accumulation of soluble sugars and proline in two maize varieties. *World Appl. Sci. J.* 3, 448–453.

Mok, D. W. and Mok, M. C. (2001). Cytokinin metabolism and action. *Annu. Rev. Plant Biol.* 52, 89–118.

Mollah, A. and Pratiwi, E. (2020). Application of various types of local microorganisms to the growth and production of two rice varieties. *IOP Conf. Ser. Earth Environ. Sci.* 575, 012148.

Moradtalab, N., et al. (2019). Silicon and the association with an arbuscular-mycorrhizal fungus (*Rhizophagus clarus*) mitigate the adverse effects of drought stress on strawberry. *Agronomy* 9, 41.

Morcillo, R. J. L. and Manzanera, M. (2021). The effects of plant associated bacterial exopolysaccharides on plant abiotic stress tolerance. *Metabolites* 11, 337.

Mukerji, K. G., et al. (2006). *Microbial Activity in the Rhizosphere*. Berlin: Springer Science & Business Media.

Nadeem, S. M., et al. (2010). Rhizobacteria capable of producing ACC-deaminase may mitigate salt stress in wheat. *Soil Sci. Soc. Amer. J.* 74, 533–542.

Nair, A., et al. (2008). Studies on the changes in lipid peroxidation and antioxidants in drought stress induced Cowpea (*Vigna unguiculata* L.) varieties. *J. Environ. Biol.* 29, 689–691.

Narayanasamy, S., et al. (2020). Plant growth promoting Bacillus sp. cahoots moisture stress alleviation in rice genotypes by triggering antioxidant defense system. *Microbiol. Res.* 239, 126518.

Naseem, H., et al. (2018). Exopolysaccharides producing rhizobacteria and their role in plant growth and drought tolerance. *J. Basic Microbiol.* 58, 1009–1022.

Naseem, H. and Bano, A. (2014). Role of plant growth-promoting rhizobacteria and their exopolysaccharide in drought tolerance of maize. *J. Plant Interact.* 9, 689–701.

Naylor, D., et al. (2018). Microbial community composition reveals spatial variation and distinct sub-community structure in the rhizosphere of agricultural soil. *Soil Biol. Biochem.* 125, 251–260.

Naylor, D., et al. (2017). Drought and host selection influence bacterial community dynamics in the grass root microbiome. *ISME J.* 11, 2691–2704.

Ngumbi, E. and Kloepper, J. (2016). Bacterial-mediated drought tolerance: current and future prospects. *Appl. Soil Ecol.* 105, 109–125.

Ochoa-Hueso, R. et al. (2018). Drought consistently alters the composition of soil fungal and bacterial communities in grasslands from two continents. *Glob. Change Biol.* 24, 2818–2827.

Omae, N., and Tsuda, K. (2022). Plant-microbiota interactions in abiotic stress environments. *MPMI* 35, 511–526.

Ojuederie, O. B., et al. (2019). Plant growth promoting rhizobacterial mitigation of drought stress in crop plants: implications for sustainable agriculture. *Agronomy* 9, 712.

Olanrewaju, O. S., et al. (2017). Mechanisms of action of plant growth promoting bacteria. *World J. Microbiol. Biotechnol.* 33, 191–116.

Paul, M. J., et al. (2008). Trehalose metabolism and signaling. *Annu. Rev. Plant Biol.* 59, 417–441.

Paez-Garcia, A., et al. (2015). Root traits and phenotyping strategies for plant improvement. *Plants* 4, 334–355.

Peng, Y. M., et al. (2023). Effects of intercropping and drought on soil aggregation and associated organic carbon and nitrogen. *Soil Use Manage.* 39, 316–328.

Poudel, M., et al. (2021) The role of plant-associated bacteria, fungi, and viruses in drought stress mitigation. *Front. Microbiol.* 12, 743512.

Preece, C. and Peñnuelas, J. (2016). Rhizodeposition under drought and consequences for soil communities and ecosystem resilience. *Plant Soil* 409, 1–17.

Preece, C., et al. (2018). Thirsty tree roots exude more carbon. *Tree Physiol.* 38, 690–695.

Qu, Q., et al. (2023) Impact of drought on soil microbial biomass and extracellular enzyme activity. *Front. Plant Sci.* 14, 1221288.

Quiroga, G., et al. (2019). The arbuscular mycorrhizal symbiosis regulates aquaporins activity and improves root cell water permeability in maize plants subjected to water stress. *Plant Cell Environ.* 42, 2274–2290.

Rabbi, S. M. F., et al. (2018). Plant roots redesign the rhizosphere to alter the three-dimensional physical architecture and water dynamics. *New Phytol.* 219, 542–550.

Rahdari, P. and Hoseini, S. M. (2012). Drought stress, a review. *Int. J. Agron. Plant Prod.* 3, 443–446.

Ren A. Z., et al. (2011) Benefits of a symbiotic association with endophytic fungi are subject to water and nutrient availability in *Achnatherum sibiricum*. *Plant Soil* 346, 363–373.

Roberson, E. B. and Firestone, M. K. (1992). Relationship between desiccation and exopolysaccharide production in soil Pseudomonas sp. *Appl. Environ. Microbiol.* 58, 1284–1291.

Rodriguez, S. J., et al. (2009). Trehalose accumulation in Azospirillum brasilense improves drought tolerance and biomass in maize plants. *FEMS Microbiol. Lett.* 296, 52–59.

Rolli, E., et al. (2014). Improved plant resistance to drought is promoted by the root-associated microbiome as a water stress-dependent trait. *Environ. Microbiol.* 17, 316–331.

Ruiz-Lozano, J. M., et al. (2016). Arbuscular mycorrhizal symbiosis induces strigolactone biosynthesis under drought and improves drought tolerance in lettuce and tomato. *Plant Cell Environ*. 39, 441–452.

Sandhya, V., et al. (2009). Alleviation of drought stress effects in sunflower seedlings by the exopolysaccharides producing *Pseudomonas putida* strain GAP-P45. *Biol. Fertil. Soils* 46, 17–26.

Sandhya, V., et al. (2010). Effect of plant growth promoting Pseudomonas spp. on compatible solutes antioxidant status and plant growth of maize under drought stress. *Plant Growth Regul.* 62, 21–30.

Schlauch, K.A., et al. (2010). Transcriptomics Analysis Methods: Microarray Data Processing, Analysis and Visualization Using the Affymetrix Genechip® Vitis Vinifera

Genome Array. In: *Methodologies and Results in Grapevine Research*. Delrot, S., eds. Dordrecht: Springer. https://doi.org/10.1007/978-90-481-9283-0_22.

Sharma, A., et al. (2019). Phytohormones regulate accumulation of osmolytes under abiotic stress. *Biomolecules* 9, 285.

Siddique, M., et al. (2000). Drought stress effects on water relations of wheat. *Bot. Bull. Acad. Sin.* 41, 35–39.

Sakata, Y., et al. (2014). ABA as a universal plant hormone. In: *Progress in Botany*, U. Lüttge, W. Beyschlag, and J. Cushman, eds. Berlin: Springer, pp. 57–96.

Santos-Medellín, C., et al. (2017). Drought stress results in a compartment-specific restructuring of the rice root-associated microbiomes. *mBio* 8, e00764-17.

Santos-Medellín, C., et al. (2021). Prolonged drought imparts lasting compositional changes to the rice root microbiome. *Nat. Plants* 7, 1065–1077.

Santoyo, G., et al. (2021). Plant growth stimulation by microbial consortia. *Agronomy* 11, 219.

Sarkar, D. and Rakshit, A. (2020). Safeguarding the fragile rice–wheat ecosystem of the indo-gangetic plains through bio-priming and bioaugmentation interventions. *FEMS Microbiol. Ecol.* 96, fiaa221.

Sarkar, D. and Rakshit, A. (2021). Bio-priming in combination with mineral fertilizer improves nutritional quality and yield of red cabbage under Middle Gangetic Plains, India. *Sci. Hortic.* 283, 110075.

Sarkar, D., et al. (2022). Bio-priming with compatible rhizospheric microbes enhances growth and micronutrient uptake of red cabbage. *Land* 11, 536.

Schimel, J. P. (2018). Life in dry soils: Effects of drought on soil microbial communities and processes. *Annu. Rev. Ecol. Evol. Syst.* 49, 409–432.

Seco, R., et al. (2015). Ecosystem-scale volatile organic compound fluxes during an extreme drought in a broadleaf temperate forest of the Missouri Ozarks (central USA). *Glob. Change Biol.* 21, 3657–3674.

Selvakumar, G., et al. (2012). Bacterial mediated alleviation of abiotic stress in crops. In: *Bacteria in Agrobiology: Stress Management,* D. K. Maheshwari, ed. Berlin and Heidelberg: Springer Verlag, pp. 205–224.

Shaffique, S., et al. (2022) Research progress in the field of microbial mitigation of drought stress in plants. *Front. Plant Sci.* 13, 870626. doi: 10.3389/fpls.2022.870626.

Sharma, P., et al., (2012). Reactive oxygen species, oxidative damage, and antioxidative defence mechanism in plants under stressful conditions. J. Bot., 217037.

Sharma, V., et al. (2020). *Molecular Aspects of Plant Beneficial Microbes in Agriculture*. Cambridge, MA: Academic Press.

Shinozaki, K. and Yamaguchi-Shinozaki, K. (2007). Gene networks involved in drought stress esponse and tolerance. *J. Exp. Bot.* 58, 221–227.

Shinozaki, K., et al. (2016). Drought stress. *Plant Physiol. Dev.* 1, 8–15.

Siddiqui, M. H., et al. (2022). *Hot and Dry: How Plants can Thrive in Future Climates*. Berlin: Springer.

Simmons, T., et al. (2020). Drought drives spatial variation in the millet root microbiome. *Front. Plant Sci.* 11, 599.

Singh, A., et al. (2020). Tiny microbes, big yields: microorganisms for enhancing food crop production sustainable development. In: *Trends of Microbial Biotechnology for Sustainable Agriculture and Biomedicine Systems: Diversity and Functional Perspectives*, A. A. Rastegari, A. N. Yadav, and N. Yadav, eds. Amsterdam: Elsevier, pp. 1–15. doi: 10.1016/B978-0-12-820526-6.00001-4

Song, F., et al. (2012). Response to water stress of soil enzymes and root exudates from drought and nondrought tolerant corn hybrids at different growth stages. *Can. J. Soil Sci.* 92, 501–507.

Soni, R., et al. (2023) Role of microbes in alleviating abiotic stress in plants. *Plant Science Today* 10(3), 160–187.

Strobel, G. and Daisy, B. (2003). Bioprospecting for microbial endophytes and their natural products. *Microbiol. Mol. Biol. Rev.* 67(4), 491–502.

Strong, P. J., et al. (2015). Methane as a resource: can the methanotrophs add value? *Environ. Sci. Technol.* 49(7), 4001–4018.

Sun, S., et al. (2017). Soil bacterial and fungal communities show distinct recovery patterns during forest ecosystem restoration. *Appl. Environ. Microbiol.* 83, e00966-17.

Suarez, R., et al., (2008). Improvement of drought tolerance and grain yield in common bean byover expressing trehalose-6-phosphate synthase in rhizobia. *Mol. Plant Microb. Interact.* 21, 958–966.

Talaat, N. B. and Shawky, B. T. (2017). Microbe-mediated induced abiotic stress tolerance responses in plants. In: *Plant-Microbe Interactions in Agro-Ecological Perspectives,* D. P. Singh, H. B. Singh, and R. Prabha, eds. Singapore: Springer, pp. 101–133.

Thirkell, T. J., et al. (2017). Are mycorrhizal fungi our sustainable saviours? Considerations for achieving food security. *J. Ecol.* 105, 921–929.

Timmusk, S., et al., (2014). Drought-tolerance of wheat improved by rhizosphere bacteria from harsh environments: Enhanced biomass production and reduced emissions of stress volatiles. *PLoS One* 9, 1–13.

Timmusk, S. and Nevo, E. (2011). Plant root associated biofilms. In: *Bacteria in Agrobiology. Plant Nutrient Management, 3,* D. K. Maheshwari, ed. Berlin: Springer Verlag, pp. 285–300.

Tiwari, P., et al. (2020). Phytohormones producing fungal communities: metabolic engineering for abiotic stress tolerance in crops. In: *Agriculturally Important Fungi for Sustainable Agriculture.* Cham: Springer, pp. 56–86. doi: 10.1007/978-3-030-45971-0_8.

Trotsenko, Y. A. and Murrell, J. C. (2008). Metabolic aspects of aerobic obligate methanotrophy. *Adv. Appl. Microbiol.* 63, 183–229.

Ustiatik, R., et al. (2021). Mercury resistance and plant growth promoting traits of endophytic bacteria isolated from mercury-contaminated soil. *Bioremediat. J.* 26, 1–20.

Vaishnav, A., et al. (2016). PGPR mediated amelioration of crops under salt stress. In: *Plant Microbe Interaction: An Approach to Sustainable Agriculture*, D. K. Choudhary et al., eds. Singapore: Springer, pp. 205–226.

Vanamala, P., et al. (2021). "Plant Growth-Promoting Rhizobacteria (PGPR): a unique strategy for sustainable agriculture. In: *Handbook of Research on Microbial Remediation and Microbial Biotechnology for Sustainable Soil*, J. Ahmad Malik, ed. Hershey, PA: IGI Global, pp. 332–357.

Verma, H., et al. (2021). The potential application of endophytes in management of stress from drought and salinity in crop plants. *Microorganisms* 9, 1729.

Vessey, J. K. (2003). Plant growth promoting rhizobacteria as biofertilizers. *Plant Soil* 255(2), 571–586.

Vurukonda S. S. K. P., et al. (2016) Enhancement of drought stress tolerance in crops by plant growth promoting rhizobacteria. *Microbiol. Res.* 184, 13–24.

Wang, C. J., et al. (2012). Induction of drought tolerance in cucumber plants by a consortium of three plant growth-promoting rhizobacterium strains. *PLoS One* 7, e52565.

Wang, X., et al. (2016). Drought-responsive mechanisms in plant leaves revealed by proteomics. *Int. J. Mol. Sci.* 17, 1706.

Watve, M. G., et al. (2001). How many antibiotics are produced by the genus Streptomyces? *Arch. Microbiol.* 176(5), 386–390.

Wen, C.-K. (2014). *Ethylene in Plants*. Berlin: Springer.

Werner, T. and Schmülling, T. (2009). Cytokinin action in plant development. *Curr. Opin. Plant Biol.* 12, 527–538.

Wybouw, B. and De Rybel, B. (2019). Cytokinin–a developing story. *Trends Plant Sci.* 24, 177–185.

Xu, L., et al. (2021). Genome-resolved metagenomics reveals role of iron metabolism in drought-induced rhizosphere microbiome dynamics. *Nat. Commun.* 12, 3209.

Yadav, A. N. and Yadav, N. (2018). Stress-adaptive microbes for plant growth promotion and alleviation of drought stress in plants. *Acta Sci. Agric.* 2, 85–88.

Yang, A., et al. (2016). Enhancing salt tolerance in quinoa by halotolerant bacterial inoculation. *Funct. Plant Biol.* 43, 632–642.

Yang, J., et al. (2009). Rhizosphere bacteria help plants tolerate abiotic stress. *Trends Plant Sci.*, 14(1), 1–4.

Yang, S., et al. (2010). Narrowing down the targets, towards successful genetic engineering of drought-tolerant crops. *Mol. Plant* 3, 469–490.

Yooyongwech, S., et al. (2013). Arbuscular mycorrhiza improved growth performance in Macadamia tetraphylla L. grown under water deficit stress involves soluble sugar and proline accumulation. *Plant Growth Regul.* 69, 285–293.

Yu, Z., et al. (2020). How plant hormones mediate salt stress responses. *Trends Plant Sci.* 25, 1117–1130.

Zahir, Z. A., et al. (2009). Comparative effectiveness of Pseudomonas and Serratia sp. containing ACC deaminase for improving growth and yield of wheat (*Triticum aestivum* L.) under salt-stressed conditions. *Arch. Microbiol.* 191, 415–424.

Zhang, H. W., et al. (2006). Biology and chemistry of endophytes. *Nat. Prod. Rep.* 23, 753–771.

Zhu, Q., et al. (2016). Multiple soil nutrient competition between plants, microbes, and mineral surfaces: model development, parameterization, and example applications in several tropical forests. *Biogeosciences* 13, 341–363.

Zou, H., et al. (2015). OsEXPB2, a b-expansin gene, is involved in rice root system architecture. *Mol. Breed.* 35, 1–14.

Zou, Y. N., et al. (2021). Unravelling the role of arbuscular mycorrhizal fungi in mitigating the oxidative burst of plants under drought stress. *Plant Biol.* 23, 50–57.

Part 5

Promoting soil microbiomes and their contribution to soil ecosystem services

Chapter 13

Understanding and optimizing the physical properties of soil to promote soil microbiomes

Wei Shi, North Carolina State University, USA

1 Introduction

An extraordinary abundance and diversity of microorganisms inhabit soil and contribute to many ecosystem services that soil provides to humans, such as those related to agricultural production and environmental quality. Both theoretical analysis and empirical data generally support the view that the microbial diversity of soil is critical to sustain plant health and ecosystem productivity (Bell et al., 2009; Saleem et al., 2019; Wagg et al., 2021). The positive relationship between the microbial diversity of soil and ecosystem functioning has been documented by most studies using controlled experiments, although such a clear relationship cannot be easily revealed by direct examination of natural ecosystems (Bell et al., 2009).

The coexistence of microbes differing in phylogeny, morphology and physiology is the basis of soil multifunctionality, as different microbes exhibit particular and unique functions. For example, some microbes are able to degrade complex plant material, such as lignin, whereas others can oxidize ammonium to nitrate, a process important to plant nutrition and the environmental fate of nitrogen (Janusz et al., 2017; Lehtovirta-Morlety, 2018). Given the divergence of growth requirements among microbes, an increased microbial diversity may help improve resource use efficiency by the microbial

http://dx.doi.org/10.19103/AS.2024.0136.26

community, i.e. the microbial community exploits the environment to an extent that individual species alone cannot achieve. The coexistence of diverse microbes is also crucial to stabilize ecosystem services against perturbation, as microbes that perform similar functions may differ in sensitivity to environmental stresses.

It is well known that natural processes, anthropogenic activities and the associated changes in the environment can all affect the diversity, composition and function of the soil microbial community. On a geographical scale, soil pH is the primary driver dictating microbial diversity and distribution in soil (Fierer and Jackson, 2006; Lauber et al., 2009; Dasgupta and Brahmaprakash, 2021). However, on a field scale, management appears to be more influential in modulating the soil microbial community (Trivedi et al., 2016; Gupta et al., 2022). Many practices directly impact the soil physical environment, which controls the distribution of growth essentials for microbes, such as organic carbon, nitrogen, mineral elements, water, oxygen and other growth factors. The allocation of these growth essentials within the soil matrix is usually non-uniform, even on the nanometre scale (Lehmann et al., 2008). The fine-scale heterogeneity in spatial distribution of growth essentials underlies the vast microbial diversity in soil (Zhou et al., 2002; Vos et al., 2013; Portell et al., 2018).

Microbes are spread unevenly and sparsely within the soil matrix. While a gram of productive soil encompasses billions of individual microbes and thousands of different types, they cover no more than 1% of the soil matrix surface and may be hundreds of microns apart (Young et al., 2008; Tecon and Or, 2017). Spatial separation limits microbial competition for growth essentials and overrides competitive exclusion, allowing competitive microbes to coexist in soil (Zhou et al., 2002; Dechesne et al., 2008). Spatial separation also prevents the trophic interaction between microbes and predators, such as protozoa and nematodes (Heynen et al., 1988; Wright et al., 1995). However, this doesn't imply that microbes are not 'locally overcrowded' in the soil matrix, given that their distribution is highly non-uniform. Microbes may be completely absent in some microsites yet abundant in others, depending on stochastic dispersal within the soil matrix as well as growth rate and yield.

Microbes in natural systems, including soil often conglomerate in a social structure, e.g. microcolonies and biofilms where inter- and intra-species interactions are probably strong (van Elsas et al., 2019). They may affect each other antagonistically or synergistically, but synergistic interactions appear to be highly prevalent (Ren et al., 2015). Social behaviour in biofilms may help microbes to improve environmental fitness and the efficiency of resource use, thereby boosting their distribution in soil (Lennon and Lehmkuhl, 2016; Cai et al., 2019). Understanding microbial interactions and responses to the environment at the fine spatial scale will undoubtedly help elucidate the connection between the soil microbiome and physical properties and thus improve reliability in predicting microbiome functions and dynamics, which benefits management strategies.

Agricultural management has been continuously reformed towards practices that not only improve agricultural productivity but also minimize undesirable environmental consequences. Widely adopted and used practices include, but are not limited to, crop rotation, conservation tillage and soil amendments, which may substantially affect the soil physical environment. For example, 30 long-term experiments conducted in Europe and China demonstrated that both soil organic amendments and no-tillage promoted soil aggregate stability, although the magnitude depended on soil texture (Bai et al., 2018). Cover crops were found to reduce soil bulk density and increase soil macroporosity as well as water infiltration (Haruna et al., 2020). Conservation Agriculture, with minimal or no tillage, also enhanced soil organic matter content and in turn improved soil structure, water filtration and soil water storage (Page et al., 2020). Nonetheless, the links between physical properties and the soil microbiome need to be better understood in order to develop informed management decisions.

This chapter focuses on the predictive understanding of how the physical environment of the soil modulates the soil microbiome. The goal is to provide a gateway to enable informed management decisions and practical applications but not a complete review of the literature. The mechanics underlying how the soil physical environment acts on the microbiome are briefly delineated. Trends or patterns between the soil microbiome and soil physical properties are highlighted. Scalable fine spatial-scale microbial interactions and functions will be discussed from the perspective of ecosystem services provided by the soil microbiome. A few management practices will be further exemplified to illustrate the importance of understanding and optimizing soil physical properties in promoting the soil microbiome.

2 The soil physical environment and impact on soil microbiota

The physical environment of soil is defined by a number of macro- and microscale physical properties (Table 1). It controls the movement of water and air through soil, directs the allocation of chemicals within the soil matrix and mediates microbial responses to the surroundings. The heterogenous distribution of water, air and resources that are essential to microbial survival and growth occurs on any spatial scale. Yet, among the most important of these scales are soil particles, aggregates and pores, because these are not only within the immediate range of microbial activities or influences but also connect to macroscale metrics for assessing soil health and sustainability (Fig. 1).

Table 1 Soil physical environment defined by a number of macro- and microscale physical properties

Physical properties	Definition or description
Macroscale	
Texture	Relative proportion of sand, silt and clay in a given amount of soil; 12 textural classes per USDA system
Structure	Ways of single particles (sand, silt and clay) to be assembled into large and compound particles (aggregates); well-structured soil having a continuous pore network that allows water and air movement and root development
Bulk density	Mass or dry weight per unit of soil volume; the higher the bulk density, the more compacted the soil
Porosity	The fraction of total pores per unit of soil volume; the lower the porosity, the more compacted the soil
Permeability	The ability of soil to allow fluid to pass through
Consistency	The resistance of soil to deformation and rupture
Colour	Different shades, such as black, brown, red and grey; used to define soil horizons or layers
Microscale	
Particles	Sand, silt and clay; often characterized in size, surface area and roughness
Aggregates	Clumps of sand, silt and clay held together; often characterized in size, organic matter contained, and interior pores and pore space
Pores	Voids formed between soil solids; often characterized in size, geometry and entrance size

Often-used descriptive properties in soil microbial ecology are soil texture, structure, bulk density, porosity, particles, aggregates and pores.

2.1 How the heterogeneity of soil particles regulates the soil microbiota

The soil mineral fraction is composed of sand- (2.00–0.05 mm), silt- (0.05–0.002) and clay (<0.002 mm)-sized particles that differ in surface area, roughness, charge density, hydrophobicity and shape. Particle characteristics can considerably influence how microbes anchor to the particle surface; for example, bacterial attachment on clay minerals was found to increase with the valency of cations present (Santoro and Stozky, 1968). Such dependency on the nature of particles was highlighted with a bacterial strain of *Azospirillum brasilense*, showing strong and weak sorption on clays and quartz sands, respectively (Bashan and Levanony, 1988). Microbial sorption on soil particles also varies with microbial species or strains (Chen et al., 2003). However, more strikingly, only a small fraction of the particle surface can be colonized by microbes. It was estimated that between 4% and 30% of individual sand

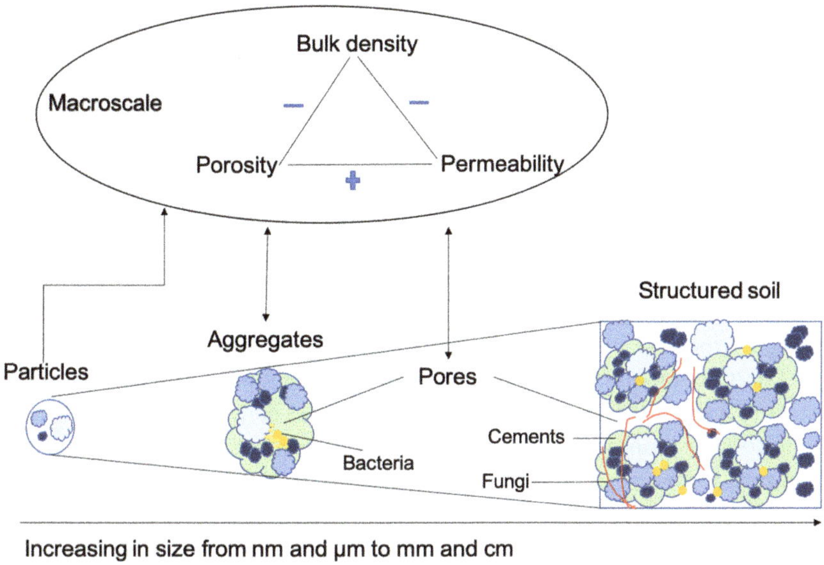

Figure 1 The schematic diagram showing arrangement of soil particles (sand, silt and clay) into aggregates and pores among soil particles and aggregates. Macroscale properties (e.g. bulk density, porosity and permeability) and microscale properties (i.e. particles, aggregates and pores) are related in some way. Symbols '+' and '−' denote positive and negative relationships.

particles of marine sediment were colonized by microbes as a result of surface roughness or topography (Meadow and Anderson, 1968; Probandt et al., 2018). The particle surface contains both convex/smooth areas and concave/protected areas. More microbes have been found in concave areas because they were well protected against predators, as well as mechanical abrasion associated with water advection and strong and frequent sediment transport processes (Probandt et al., 2018). While mechanical abrasion in unsaturated soil is unlikely to be as strong as the sediment, other processes affect physical interactions between microbes and soil particles. By affecting the sorption of organic matter and thickness of surface water films (Doerr et al., 2000; Or et al., 2007), particle surface characteristics might indirectly impact microbial colonization. An increase in organic matter content favours microbial colonization, whereas soil drying could reduce microbial sorption on soil particles (Bashan and Levanony, 1988).

Fine-scale variations in particle surface characteristics may contribute to great microbial diversity. Even a single sand grain was found to harbour several thousand different microbial species, defined as sequences at 97% similarity (Probandt et al., 2018). The authors showed that microbial community compositions varied significantly among sand grains and that individual sand grains accounted for 27–42% of total microbial species present in the bulk sediment. Nonetheless, a combination of a few sand grains contributed to 50% of the total microbial species, suggesting some factors are common to all sand grains that are key to microbial diversity. This raises an important question concerning whether microbial community diversity and composition diverge significantly among sand, silt and clay particles, given that variations in particle surface characteristics between particle groups are probably greater than those within a particle group. Here, it is assumed that within-group variations will be smaller for sand particles than for clay particles, considering that soil clay particles are more diverse in minerology, such as kaolin, smectite and illite. In addition, physical interactions between microbes and soil particles are likely governed by chemical interactions, specifically for clay particles. Vieira et al. (2020) compared bacterial communities assembled on illite and goethite that were exposed to natural grassland soils and showed that clay minerals overrode soil properties and the source of organic substances to select bacterial colonizers, forming specific mineral-dependent bacterial communities.

On the coarse level, soil heterogeneity in terms of particle distribution is described by the proportion of sand, silt and clay grains and classified into 12 textural classes. At the finer level, each sand or silt group can be further divided into very coarse, coarse, medium, very fine and fine subgroups. Together with different shapes and surface characteristics, the variability in soil particles may be substantial. Such complexity or heterogeneity has been assessed using a combination of laser granulometry and fractal geometry, contributing to a predictive understanding of how soil texture regulates microbial diversity and composition (Seaton et al., 2020).

2.2 How the heterogeneity of soil aggregates controls the soil microbiota

Special attention has been given to the heterogenous distribution of organic carbon among aggregates of different sizes. Investigators have focused on the quantity, quality, origin and dynamics of the organic carbon. In general, the amount of organic carbon is greater in macroaggregates than in microaggregates (Tisdall and Oades, 1980; Elliott, 1986; Gupta and Germida, 1988; Puget et al., 1998; Mendes et al., 1999; Six et al., 2000). Particulate organic matter originating mainly from plant residues is constrained to macroaggregates (Beare et al., 1994), whereas mineral-associated organic carbon is highly concentrated in microaggregates. As a consequence, the turnover rate of organic carbon is faster in macroaggregates than in microaggregates (Jastrow et al., 1996), suggesting that organic carbon associated with microaggregates is more recalcitrant and older.

The consideration of microbial contributions to soil aggregate formation and stability has led to an extensive examination of aggregate-associated microbial metabolic products. Microbial products can act as cementing agents to bind primary mineral particles into microaggregates and then tie microaggregates with plant residues into macroaggregates. Extracellular microbial products are usually water insoluble and heat resistant, and thus may account partially for the resistance of soil aggregates to breakdown upon wetting. Polysaccharides and glycoproteins of microbial origin have been well documented as the cementing agents of soil aggregates. The concentration of microbial metabolic products is usually greater in macroaggregates than in microaggregates (Hu et al., 1995; Roberson et al., 1995; Wright and Upadhyaya, 1996, 1998; Haynes and Beare, 1997; Puget et al., 1998).

An early hierarchical model of aggregate formation emphasizes microaggregates as the building blocks of macroaggregate formation (Tisdall and Oades, 1982). Subsequent studies have provided evidence that microaggregates are formed and stabilized in macroaggregates (Jastrow, 1996; Six et al., 2000). The top-down hierarchical model proposed by Six et al. (2000) underscores that macroaggregates are generated around fresh plant residues and these residues form the coarse intra-aggregate particular organic matter (iPOM). The decomposition and subsequent fragmentation of coarse iPOM within macroaggregates produce fine iPOM, which is then coated with microbial products and fine mineral particles to generate microaggregates. When cementing agents decompose, macroaggregates become unstable, leading to the release of microaggregates. Noticeably, the dynamic turnover of aggregates is associated with the alteration of organic carbon quality and quality. Macroaggregate-associated organic carbon is relatively abundant and bioavailable, whereas organic carbon associated with microaggregates is more recalcitrant to decomposition.

A deeper insight into the aggregate-associated soil heterogeneity has been gained through examining the heterogenous distribution of the microbial population. It has been suggested that the structural difference between micro- and macroaggregates is responsible for variations in the microbial community among aggregates, as nutrient restriction and protection from predatory grazing only occur in certain aggregate size classes (Mendes et al., 1999). The aggregate-associated microbial community is of interest due to its role in soil structure formation and stability. Fungi play an important role in the formation of macroaggregates due to physical enmeshment of fungal hyphae (Tisdall and Oades, 1982; Gupta and Germida, 1988). Fungal metabolic products have also been highlighted for macroaggregate stability (Caesar-TonThat and Cochran, 2000). In addition, direct counts have demonstrated that fungal hyphae are primarily within macroaggregates but to a lesser extent in microaggregates (Guggenberger et al., 1999). These studies have shown that fungi are tightly coupled with soil aggregates of large size. In general, the ratio of fungi to bacteria is greater in macroaggregates than in microaggregates

(Gupta and Germida, 1988; Singh and Singh, 1995). Indirect measurements also shed light on the heterogenous microbial distribution across aggregate size classes. The microbial metabolic quotient, i.e. qCO_2, has been used to signify the qualitative change in microbial community composition. A negative correlation between qCO_2 and aggregate size was reported (Miller and Dick, 1995), and authors hypothesized that the microbial community differed between macro- and microaggregates. The ratio of microbial biomass carbon to nitrogen can also be used to differentiate community composition. It has been reported that the ratio of microbial biomass carbon to nitrogen is lower in macroaggregates than in microaggregates (Singh and Singh, 1995), which suggests different community compositions reside in soil aggregates of different sizes.

Soil aggregates play a crucial role in controlling microbial distribution and diversity in soil. Aggregates have been considered as the base unit for understanding microbial structure–function relationships in soil, as aggregates help stabilize the soil microbial community, enhance microbial species interactions and further modulate microbial traits and functions via spatial confinement (Wilpiszeski et al., 2019). Aggregates have also been viewed as 'massively concurrent evolutionary incubators' to cultivate the soil microbiome into a fine-scale organization, considering that aggregate interiors are relatively isolated from the surrounding matrix or that physical and chemical conditions inside aggregates are generally different from conditions in the bulk soil (Rillig et al., 2017). Aggregates can also modulate the soil microbial community by serving as shelters protecting microbes against predators.

Soil aggregates need to be classified and separated in order to explore the aggregate-scale soil heterogeneity and its impact on the soil microbiome. Over time, assorted aggregate size classifications have been arbitrarily chosen to achieve specific objectives and accommodate particular soils; however, a consensus has been reached regarding the size classification of micro- and macroaggregates. In general, soil aggregates $\leq$0.25 mm and >0.25 mm are referred to as microaggregates and macroaggregates, respectively. Often explored aggregate sizes include classes of >2000 μm (large macroaggregates), 2000–250 μm (small macroaggregates), 250–53 μm (microaggregates) and <53 μm (silt and clay fraction). Separating aggregates of different sizes is often conducted using the wet-sieving method (Kemper and Rosenau, 1986). Wet-sieving of air-dried soil leads to slaking-resistant aggregates, as rapid re-wetting produces internal air pressures that can disrupt unstable aggregates (Kemper et al., 1985). Several modifications, which include the alteration of re-wetting velocity and use of soil at different moisture regimes, have been used to create water-stable aggregates of different sizes (Bossuyt et al., 2001). However, the wet-sieving method may be problematic for analysing the aggregate-associated microbial community, as microbes on the surface of soil aggregates may be lost through washing. Instead, a dry-sieving technique has been employed to gain insight into the microbiota associated with aggregates of different sizes (Mendes et al., 1999; Schutter and Dick, 2002).

2.3 How the heterogeneity of soil pores influences the soil microbiota

Soil is comprised of solids and voids, the so-called pores. Based on the size, soil pores are classified into cryptopores (<0.1μm), ultramicropores (0.1–5 μm), micropores (5–30 μm), mesopores (30–75 μm) and macropores (>75 μm). In addition, pores vary in the size of neck, geometry and accessibility, thereby generating a complex transportation and habitat network for microbes within the soil matrix. Pores of different sizes exhibit great heterogeneity in hydration status, chemical environment and roles involved in soil processes (Or et al., 2007; Quigley et al., 2018). Macropores are the major path of preferential flow in soil, whereas meso- and micropores play a greater role in holding water in soil. Even within a pore, water does not uniformly cover the whole pore space, with thick water films are often present in the corner (Tecon and Or, 2017). Pore hydration status and hydraulic connectivity among pores influence not only microbial mobility and dispersal but also substrate and oxygen diffusion with the soil matrix. By feeding microbial communities with carbon-13-labelled fructose and then examining phospholipid fatty acid profiles, Ruamps et al. (2011) demonstrated a high degree of variation in microbial communities within pores of different size classes. The pore-scale heterogeneity of essential resources for microbial growth is the key driver for the coexistence of diverse microbes in soil; put simply, the pore network affects microbial dispersal in soil (Yang and van Elsas, 2018). Bacteria dominate the microbial community in small pores due to their smaller size and lack of migration between these pores, particularly under drier conditions that could promote separation of bacterial communities and maintain great diversity. In contrast, fungi are capable of moving through dry pore space, less likely to become restricted to the specific soil environment and can transfer resources through the hyphal network to compensate for changes in the local environment (Tecon and Or, 2017; Whiteside et al., 2019).

Pore-scale heterogeneity in size and geometry also plays a key role in allowing diverse microbes to coexist in soil. Microbes cannot reside in all soil pores because some pores or the pore neck are too small to allow access. Due to size differentiation, soil organisms at different trophic levels appear to have a favourable pore size to inhabit. For example, bacteria biomass was found to be positively associated with soil pore volumes with a diameter of 0.2–1.2 μm, whereas nematode biomass had a greater association with soil volume comprising pores of 30–90 μm diameter (Hassink et al., 1993). This physical isolation helps protect microbes against predators. By comparing nematode biomass in fine- versus coarse-textured soils (i.e. fewer versus a greater number of >30 μm habitable pores), Elliott et al. (1980) demonstrated that nematode biomass increased to a greater degree in the fine-textured soil than in coarse-textured soil in the presence of amoebae, a type of protozoa. Their data suggest that protozoa enter soil pores inaccessible to nematodes to feed on bacteria and then emerge as food unavailable to nematodes. This experiment elegantly demonstrates the pore-scale habitat influences trophic interactions. Similarly, by introducing bacterial cells into small (<6 μm) and intermediate (6–30 μm) pores but protozoa into large pores (30–60 μm), Wright et al. (1995) showed that the bacterial population decreased to a greater extent in intermediate pores than in small pores as a result of protozoa feeding. It further highlights that soil pores can provide significant protection from predation. The physical structure of soil affects trophic interactions via several modes of action, including constraining accessibility, confinement of resource/prey in small pores and limited mobility reducing opportunities of consumer–resource interactions (Erktan et al., 2020).

Pore-size heterogeneity can be measured by both water retention and particle size distribution data. Briefly, water retention data collected using a desorption method are modelled to estimate the soil pore size distribution (Klute, 1986; Banwart et al., 2017). Particle size distribution, obtained via the hydrometer method can also be modelled to estimate the soil pore size distribution (Arya et al., 1999). The approach of particle size distribution provides the advantage of being linked more directly to soil texture, which is relatively easier information to obtain and also to generalize (via sand and clay content) for prediction purposes.

2.4 Heterogeneity in the soil physical environment

While soil structural heterogeneity can be quantified by the heterogenous distribution of particles, aggregates or pores of different sizes, its strength and complexity may far surpass measures that individual metrics offer. Advances in scanning and imaging techniques have now enabled visualization and examination of the soil physical environment over two- or three-dimensions at high resolutions. X-ray micro-computed tomography has been particularly effective for soil biology-relevant research, as this tool causes little disturbance to the soil microbiome (Bouckaert et al., 2013). The confined heterogenous distribution of microbes in the physical environment of soil can also be visualized via scanning or transmission electron microscopy (Watteau and Vilemin, 2018).

As the distribution of microbes is heterogeneous, the community analysis in a spatial resolution down to the level of soil particles, aggregates or pores may help elucidate microbial responses to the surroundings, in addition to microbial interactions. The often-used soil sampling size (0.25–1 g) to assess the soil microbiome is in a lower spatial resolution if we consider how many particles, aggregates or pores of different sizes a gram of soil may contain. Several attempts have been made to examine the environmental microbiome at the fine scale and successfully demonstrate the heterogeneity of prokaryotic abundance and diversity at the single sand or macroaggregate dimension (Probandt et al., 2018; Szoboszlay and Tebbe, 2021). For example, a sufficient quantity and quality of metagenomic DNA was extracted from 1 mg of soil, equivalent to a single macroaggregate and then subjected to 16S marker gene library preparation and high-throughput sequencing (Szoboszlay and Tebbe, 2021). Bypassing metagenomic DNA extraction, Probandt et al. (2018) proved that direct PCR amplification of a single grain of sand was feasible for downstream sequencing and analysis of the microbial community. With the enhanced capability of probing the microstructure-associated soil physical and chemical environment and the soil microbiota, we will likely discover emergent traits of the soil microbiome, in particular, pertaining to microbial competition and co-operation for mediating the biogeochemical cycling of elements as well as soil carbon dynamics.

3 Microbial distribution related to the soil physical environment

3.1 Specific aggregate size-preferred microbial taxa

Optimizing the soil environment to promote microbiomes hinges on whether we are able to extract trends or patterns from the current data of microbial distribution in relation to the physical properties of soil. Despite a number of investigations on the soil aggregate-associated microbiome, it is not easy to correlate microbial distribution with aggregate size classes at the taxon level. This is perhaps because aggregates are the organization of soil particles and pores and, thus, associated physical and chemical environments may differ largely even among aggregates of the same size. Microbiota of individual macroaggregates from the same soil have been shown to diverge substantially from each other (Szoboszlay and Tebbe, 2021). Only when many macroaggregates (i.e. >25) are pooled together can abundant amplicon sequence variants represent the bulk soil. Similar conclusions have been drawn by Bailey et al. (2013a). They employed synchrotron radiation-based transmission X-ray tomography to gauge aggregate structure and found that pore distribution within macroaggregates was highly heterogenous. As a result, microbial community composition varied considerably among individual macroaggregates. No apparent trend for microbial distribution across aggregate size classes is further substantiated by the observation that differences in the microbial community between aggregates of differing sizes are mainly attributed to a small fraction of rare phylotypes (Davinic et al., 2012). Kravchenko et al. (2014) also documented that microbial communities from different agricultural systems diverged more widely between individual aggregates than between bulk soils.

Aggregate dynamics augment the difficulty of discerning patterns of microbial distribution with aggregate size classes. Miller et al. (2009) examined the distribution of components of the bacterial denitrifier community, i.e. *Pseudomonas mandelii* and related species and also denitrifiers carrying *cnorBp* (i.e. cytochrome bc-type nitric oxide reductase gene) and *nosZ* (i.e. nitrous oxide reductase gene) between aggregates of different sizes under permanent grassland and an arable copping system. The two systems showed stark contrasts, with denitrifier populations being similar among aggregate size classes in the arable cropping system, yet being more abundant in aggregates of smaller size in the permanent grassland system. This was presumably due to mechanical shearing (e.g. intensive tillage) in the arable cropping system breaking up macroaggregates and therefore increasing the frequency of soil mixing.

Various methods used by different studies to examine the microbiome also increases the difficulty of finding trends or patterns between the microbiota and aggregate size classes. Lupwayi et al. (2001) performed a catabolic potential study assessing Biolog substrate utilization and did not observe any effects of aggregate size classes on microbial community composition and diversity. Using the same approach, Schutter and Dick (2002) also showed that the microbial community was not distinguishable among aggregate size classes. In contrast, an examination with fingerprints of extractable fatty acid methyl esters identified some differences in microbial community composition among aggregate size classes (Schutter and Dick, 2002). In particular, the biomarker of fungi, such as 18:2ω6 increased significantly with increasing size of soil aggregates. Smith et al. (2014) also showed that the fungal-to-bacterial ratio increased with aggregate size classes. Despite complexity in trending relationships between the microbiota and aggregate size classes, there is a consensus that microbial diversity, composition and responses to the environment generally differ between micro- and macroaggregates, and between the exterior and interior of aggregates (Seech and Beauchamp, 1988; Ranjard et al., 2000; Davinic et al., 2012; Constancias et al., 2014; Ivanova et al., 2015; Fox et al., 2018). Fungi prefer macroaggregates and are more abundant on the outside of aggregates, whereas bacteria are associated to a greater degree with aggregates of smaller size and are richer within aggregates.

3.2 Specific particle size-preferred microbial taxa

Although aggregates constantly change via natural processes and anthropogenic activities, soil particles and thus soil textural classes are relatively stable. Particle size distribution-based microbial communities tend to be consistent among investigations. Texture effects on microbial diversity are often indirect and of secondary importance. When no other parameters were put into the equation, e.g., there was no apparent relationship between microbial alpha diversity metrics and texture heterogeneity (Seaton et al., 2020). Only after controlling land use, soil carbon and pH, texture heterogeneity showed slightly positive effects on bacterial diversity, but not on fungi. Sessitsch et al. (2001) directly examined microbial diversity associated with fine sand, silt and clay fractions and demonstrated that microbial diversity was greater in particles of smaller size than those of large size. Low nutrient availability, weak protozoan grazing and less competition with fungi were thought to promote bacterial diversity in smaller particles. By manipulating a quartz-based sandy soil with 10% addition of silt- and clay-sized particles of quartz, Carson et al. (2010) demonstrated that texture had little effect on bacterial richness and diversity, but instead drier soils promoted bacterial diversity. They suggested that low pore connectivity was the driver of microbial diversity in soil. Several other studies also attributed the 'apparent' relationship between soil texture classes and microbial diversity to low hydraulic connection through soil pores (Chau et al., 2011; Naveed et al., 2016). For example, Naveed et al. (2016) showed that the soil bacterial Shannon index increased with the percentage of clay fraction as well as the volume of micropores, i.e. water retention capacity. Although Chau et al. (2011) reported that soil bacterial richness increased with the percentage of sand fraction, they still attributed it to the increased number of disconnected water films in larger pores of coarse-textured soils. By designing a complex experiment that used 72 intact soil cores across seven textural classes and three soil moisture regimes, equivalent to dry soil, soil at field moisture and soil at saturation-field capacity, Xia et al. (2020) revealed that microbial diversity was moderately associated with soil textural classes, particularly for bacteria with no more than 6% of the coefficient variation in Shannon index and Pielou's evenness. Again, their data suggested that texture effects on microbial diversity were through control of pore hydraulic connectivity. Their results were also in agreement with those of Naveed et al. (2016), revealing that fungal diversity was enhanced by the percentage of soil sand fraction, presumably due to greater macroporosity. Overall, bacteria are sensitive to particle size distribution and increasing clay content promotes bacterial diversity. The increase in bacterial diversity is due to the physical protection of bacteria in small pores from their predators and the restriction of microbial competition due to low pore connectivity.

Both bacteria and fungi also express a specific soil particle size preference. One approach to examine how soil constituents structure the microbial community is to separate particles of different sizes for survey. Direct examination of the bacterial community on fine sands, silts and clays revealed that Alphaproteobacteria dominated particles of large size, but members of Acidobacteria preferred particles of smaller size, regardless of different types of organic amendments (Sessitsch et al., 2001). Their results suggested that specific microbe–particle associations were minimally affected by external factors. Hemkemeyer et al. (2018) also showed that Alphaproteobacteria and Bacteroidetes exhibited greater abundance in the sand fraction. Given the filamentous shape and extensive hyphal growth, fungi are not expected to have soil particle size preference. However, a taxon-level fungal preference for distinct soil particle size fractions has been found (Hemkemeyer et al., 2019); several members of the class Dothideomycetes prefer the sand fraction, whereas members of the class Sordariomycetes are relatively more abundant in the fine silt fraction. Mineral particle-associated carbon and nutrients as well as taxon-specific interactions with particle surfaces are considered crucial for structuring the microbial community (Sessitsch et al., 2001; Hemkemeyer et al., 2018, 2019).

Fractionation procedures for separating soil particles of different sizes are not free from concern, such as how to avoid cross contamination in probing biological relevance of a given particle size fraction. Therefore, an implicit approach, e.g. correlation analysis is also used to identify particle size-preferred bacterial and fungal taxa. A nationwide survey of soil bacteria from more than 2000 French soil samples showed that among 35 taxonomically assigned bacterial phyla or subphyla, Actinobacteria, Chloroflexi and Verrucomicrobia were most positively associated with the soil clay fraction, whereas Betaproteobacteria and Gemmatimonadetes were most negatively associated with the clay fraction (Karimi et al., 2018). Similarly, Xia et al. (2020) examined both bacterial and fungal communities of 72 soil samples and derived particle size-preferred bacterial and fungal taxa via correlation analysis. Their data showed that filamentous bacteria (i.e. Actinobacteria and Chloroflexi), together with the fungi Basidiomycota and Eurotiomycetes, were highly associated with silt and/or clay fractions. Overall, direct and indirect approaches provide consistent views on the particle size preferred by bacterial taxa. Although both approaches agree that some fungal taxa show particle size preference, no trends or patterns can be drawn at present, due to limited studies investigating the subject.

3.3 Specific pore size-preferred microbial taxa

The soil pore network determines how soil microbes access growth essentials, including organic substrate, water and oxygen and, therefore, affects the soil microbial community. Pore size distribution can even override other soil factors, such as substrate quality and pH to structure the soil microbial community (Sleutel et al., 2012). In unsaturated soil, both fungi and filamentous bacteria may have the advantage over non-filamentous bacteria to grow in macropores as filamentous growth can bridge air-filled gaps (Wolf et al., 2013). Metagenomic analysis for soils of different pore size distributions demonstrated that specific microbes preferred certain pore sizes at all taxonomic levels (Xia et al., 2022). Soil with a greater proportion of macro- and mesopores harboured more Betaproteobacteria, Bacteroidetes and Eurotiales, but soils with a greater proportion of micropores favoured Alphaproteobacteria, Sordariomycetes and Chaetothyriales. It seems that microbial distribution trends in terms of pore size and particle size distribution are, to some degree, congruent, perhaps because particle size distribution can significantly affect soil pore size distribution (He et al., 2021). The authors showed that an increase in particle size can cause a remarkable increase in the maximum pore size. However, more surveys are needed, given that particle size and pore size distributions are not always correlated (Wu et al., 1990). Nonetheless, the pore size-preferred distribution of specific microbial taxa appears to be caused by pore-associated organic carbon distribution and minimally by water distribution. As demonstrated by Xia et al. (2022), soil moisture only slightly affected microbial distribution in soils with a great proportion of micropores.

The chemistry of organic matter in pores may also regulate specific pore size-preferred microbial taxa. In a study comparing pore-associated bacterial communities between cropping systems with or without legume cover crops, Kravchenko et al. (2014) demonstrated the preference of Actinobacteria and Firmicutes for pores of >110 μm in the cropping system with legume cover crops but not without legume crops. Perhaps pores of >110 μm with legume crops originated mainly from legume root activities that provided readily available carbon for microbial survival and growth. Bacteria that previously remained inactive in this pore size were therefore out of their concealed state and became well populated. An interesting finding of the study by Kravchenko et al. was that among pores of different sizes, medium-sized pores (e.g. 32–84 μm) were significantly correlated with a high fraction of the top 100 abundant bacterial species, defined as operational taxonomic units, such as those belonging to Acidobacteria and Proteobacteria.

4 Fine spatial-scale microbial interactions and functions

The single most important question is how relevant an understanding of the soil microbiome at the fine spatial scale is to ecosystem services that the microbiome can offer. To answer this question, we first need to prove that the community composition and function are able to connect at the fine spatial scale. We also need to know what information obtained at the microscale can mirror the soil microbiome at the macroscale. Further, we need to identify which microscale properties are more important from the perspective of management. Recent studies on the soil microbiome at the fine spatial scale have shed light on these key questions.

Bailey et al. (2013b) collected 450 individual macroaggregates of similar size and examined beta-glucosidase activity of each macroaggregate that presumably harboured a unique microbial community. These macroaggregates were separated into two groups of low and high beta-glucosidase activity after discarding aggregates for which the glucosidase activities showed no substantial differences (i.e. activities are within one standard deviation from the mean activity value). 10 to 11 aggregates were pooled into one sample for DNA extraction and pyrosequencing, leading to a total of five and seven soil samples for groups of high and low beta-glucosidase activity, respectively. They demonstrated that bacterial communities diverged significantly between aggregates of high and low beta-glucosidase activity. The aggregates of high activity were more abundant in species (i.e. operational taxonomic units) belonging to the phylum Bacteroidetes. Their results indicated that at the fine spatial scale, microbial community function could be highly linked with the community structure.

By examining the prokaryotic community of individual macroaggregates, followed by co-occurrence analysis, Szoboszlay and Tebbe (2021) extracted ecophysiology-based microbial species interactions that could not be realized by interrogating the bulk soil microbiome. As stated by Cordero and Datta (2016), co-occurrence patterns found at the fine spatial scale are most likely authentic to microbial interactions. Nonetheless, Szoboszlay and Tebbe (2021) demonstrated how microbes were organized in the community and formed two clusters to perform carbon degradation and nitrogen transformations, respectively. The cluster dominated by the members of Thaumarchaeota, Verrucomicrobia and Actinobacteria was probably involved in ammonia oxidation. By contrast, the cluster dominated by the members of Acidobacterial subgroup 6 contributed to the decomposition of relatively recalcitrant organic matter. Thus, fine-scale sampling and analysis unveiled microbial interactions and provided new ways of linking microbial community structure to function.

The link between microbial community structure and function is also apparent at the pore scale. Negassa et al. (2015) showed that 75–80% of added plant residue was decomposed in soil with a greater volume of air-filled and well-connected large pores (>30 μm), whereas a lesser amount of added plant residue (~60%) was decomposed in soil with a greater volume of water-filled and smaller pores. The microbial community in soil with a greater volume of large pores contained more known cellulose decomposers, such as Bacteroidetes, Proteobacteria, Actinobacteria and Firmicutes. On the contrary, a number of oligotrophic Acidobacteria were more abundant in soil with a greater volume of small pores. Their work provided experimental evidence that specific pore size-preferred microbes were highly related to pore-scale processes.

Fine spatial-scale understanding of the microbiome has provided insight into soil carbon stabilization (Kravchenko et al., 2019). Using a unique combination of X-ray micro-computed tomography and microscale enzyme mapping, the authors clearly showed that pore characteristics were the driver dictating the direction and magnitude of the fate of newly added carbon in soil. When a greater number of 30–150 μm pores were formed, partially decomposed and labile organic carbon generated within these pores might rapidly migrate into the adjacent soil matrix and be stored to prevent further microbial degradation. In contrast, when a low number of pores of 30–150 μm were developed, less volume of the soil matrix would be within the influence of these pores, resulting in less carbon translocated into the surrounding areas and protected from microbial degradation. It seemed that pores of 30–150 μm were the main location of new carbon input as well as active microbial communities. Pores of this size demonstrated faster organic matter decomposition, greater presence of labile organic compounds and higher microbial abundance. The greater number of these pores in the soil, the greater amount of microbial products can be transported into the surroundings and protected against further microbial access. Therefore, a better understanding of microbial interactions and microbial responses to the pore environment may allow the effective management of microbial resource use efficiency in porous soil.

5 Management, macroscale physical properties and the soil microbiome

A few studies have examined relationships between the soil microbiome and macroscale physical properties, such as bulk density and porosity. Macroscale physical properties represent the average state of microscale properties and, therefore, cannot usually provide high-resolution measures or forecasts regarding abundance, diversity, composition and function of the soil microbiome. A salient example is the relationship between the percentage of soil clay particles and soil microbial population size. The more clay a soil contains, the more carbon and nutrients it can carry and the more surface area it can provide to support microbial inhabitants. Accordingly, a positive relationship between clay content and microbial biomass is expected. However, microbial biomass has been found to increase with soil clay content and plateau at ~50% (Dalal, 1998). A possible explanation was that a greater number of pores were too small to harbour microbes at the high clay content, in addition to the explanation that the extraction efficiency of microbial biomass was low due to micropore-constrained diffusion. Furthermore, fungi are often considered to preferably remain on the exterior of aggregates or in large pores, but fungi can be present at smaller pores, perhaps in the form of spores, and compete successfully with bacteria for labile carbon compounds (Ruamps et al., 2011).

Knowledge regarding the microbiome of soil pores may help determine why similar management practices may result in contrasting consequences. This is particularly necessary to mitigate the emission of nitrous oxide, a potent greenhouse gas and ozone deletion substance from arable land. Nitrous oxide production via prokaryotic and fungal processes are promoted by labile carbon and microaerobic conditions, which are highly dependent on how organic carbon, water and oxygen transport through the soil pore network. For the same macroscale properties, e.g. bulk density and total porosity, pore-scale variations in organic carbon, water and oxygen availability can be substantial due to heterogeneity in pore size, geometry and connectivity. It is necessary to understand microscale physical properties in order to improve ecosystem services concerning soil compaction and conservation tillage.

5.1 Effects of soil compaction on the microbiome

Soil compaction accompanies an increase of soil bulk density, a measure of soil dry weight per unit of soil volume. As soil bulk density increases, soil porosity will decrease if soil particle density remains the same. This means that soil compaction will lower the total pore volume for microbes to inhabit and also limit the diffusion of carbon, nutrients, water and oxygen required for microbial growth; therefore, soil compaction will reduce microbial population size or microbial biomass. Indeed, when the soil bulk density was increased from 1.00 g cm^{-3} to 1.60 g cm^{-3}, the population density of bacteria, fungi or actinomycetes decreased by 26–39%; the optimal bulk density for soil enzyme activities was in the range of 1.0–1.3 g cm^{-3} (Li et al., 2002); however, exceptions exist. Harris et al. (2003) demonstrated that soil fungal distribution was irregular initially and became widespread when the soil bulk density increased from 1.2 g cm^{-3} to 1.4 g cm^{-3}. The authors attributed these counterintuitive observations to substantial changes in pore size and geometry. Soil compaction reduced the average pore size, broke up the continuity of macro- and mesopore space, and accordingly limited water connectivity through large pores. However, fungi explored new territories via filamentous growth in this hydraulic disconnection situation. Their data corroborated those of Toyota et al. (1996), substantiating that by influencing pore size distribution, soil compaction regulated how fungi moved through the soil matrix and reorganized across the pore network. So, the inconsistency regarding how compaction affected the fungal density appeared to be reliably determined when microscale physical properties were considered.

By altering total porosity, soil compaction can also affect gas diffusion and, in turn, soil aeration. A small change in soil bulk density can even cause large variations in air permeability and gas diffusion. Air permeability was found to be reduced by 60–94%, and gas diffusion by 48–66% when the soil bulk density was increased by 10–15%. As a consequence, the microbial community structure changed, specifically for the fungal community. The microbial community also showed little resilience to soil compaction, which modulated microbial diversity and abundance, but mixed results have been observed (Hartmann et al., 2014). Methods of different resolutions used for examining the microbial community may help explain the discrepancy. Still, the direction and magnitude of changes in the microscale physical environment by soil compaction should not be excluded. Pore-scale processes, such as predation and fungal spreading behaviour (Shestak and Busse, 2005; Mishra et al., 2023), can increase the complexity of connecting the soil microbiome with macroscale physical properties.

Given that soil compaction can disrupt large pores and pore continuity, and hence water transport and distribution through the pore network, anaerobic microbes, e.g. nitrous oxide-producing prokaryotes and fungi, are likely more favoured by soil compaction. This supposition is supported by empirical data obtained at both organismal and process levels. As soil was compacted, the abundance of denitrifiers estimated by functional gene copy numbers was increased and soil nitrous oxide emission was enhanced (Hartmann et al., 2014). Soil relative gas diffusivity, a macroscale measure that links soil moisture content with soil structure (i.e. pore connectivity and continuity), proved to be reliable in predicting when nitrous oxide emissions peaked, but it could not explain large variations in the magnitude of nitrous oxide emissions (Rousset et al., 2020). This suggests that microbial processes, such as those involved in nitrous oxide production and consumption need to be better understood at the pore scale. The soil microbiome has also been found to correlate with bulk density and soil macroporosity, although the association with soil pH is much stronger (Hermans et al., 2020).

5.2 Effects of conservation tillage on the soil microbiome

Minimum and no-tillage practices improve soil structure by increasing the number, dimension and stability of soil aggregates. A comprehensive review has shown that mean weight diameter of water-stable aggregates was larger by approximately three-fold after long-term no-tillage compared with conventional tillage (Paustian et al., 2000). Frequent exposure to wet-dry and freeze-thaw cycles under the conventional tillage system also contributed to the decrease in number, dimension and stability of soil aggregates (Paustian et al., 2000). These alterations in soil aggregate structure may have profound impacts on the soil microbiome and associated carbon and nitrogen processes.

Six et al. (2000) reported that the ratio of soil micro- to macroaggregates was 1.7-fold lower and intra-aggregate particulate organic matter was approximately three-fold greater in no-tillage compared with the conventional tillage system. The difference in intra-aggregate particulate organic matter explained 21% of the variation in soil organic carbon between conventional and no-tillage systems. Soil carbon accumulation under no-tillage was attributed to the slow turnover of macroaggregates (Six et al., 1999). The alteration in soil aggregate structure and associated changes in physico-chemical and microbiological features may regulate aggregate-associated microbial activity and processes.

Frey et al. (1999) reported that the ratio of fungi to bacteria was higher in reduced tillage compared with the conventional tillage system. According to their data, fungi were more abundant in soil of no-tillage than conventional tillage. Similarly, fungi were found to be significantly greater in no-tillage compared with the tilled system (Drijber et al., 2000). Using microbial metabolic products, i.e. glucosamine and muramic acid as the indicator of fungal and bacterial abundance, respectively, Guggenberger et al. (1999) also came to a similar conclusion that fungi were greatly enhanced under no-tillage. In recent years, metagenomic analysis has provided detailed and comprehensive information regarding aggregate-associated microbiomes. However, differences in the microbiome among aggregates of different sizes are often management practice specific (Pellegrino et al., 2022), suggesting that microhabitat heterogeneities are more complex than expected from the simple dichotomy between no-till versus tillage practices.

Mixed results have been documented regarding whether no-tillage increases soil nitrous oxide emissions. Generally, no-tillage is believed to promote soil denitrification by enhancing the activity and abundance of the denitrifier community, based on hundreds of observations and data compilation; however, nitrous oxide efflux rates vary largely with other management practices as well as soil texture under no-tillage (Wang and Zhou, 2020). A highly controversial question focuses on how no-tillage versus tillage affects soil nitrogen processes, in particular, denitrification and soil nitrous oxide emissions. A recent study has shown that functional genes encoding enzymes that catalyse nitrite and nitrous oxide reductions were no difference between no-till and tillage systems or even lower under no-tillage (Cloutier et al., 2023). Taken together, these data suggest that we need to understand the distribution of denitrifiers at the finer spatial scale that contribute to microbial processes within aggregates and pores.

6 Conclusion and future trends

Continuing advances in molecular biology, imaging and modelling will enable further investigation of soil microbes and their microhabitats. One topic requiring elucidation is how microbes interact to mediate soil organic carbon degradation and stability. Microbes reside in soil pores of different sizes and mostly remain anchored to the solid–pore interface in a complex structure, i.e. microcolonies and biofilms. They interact with each other and likely evolve microhabitat-specific strategies to partition resources between growth and energy needs. Carbon use efficiency is an important trait of the microbiome as it links to the rate of carbon dioxide lost from an ecosystem via microbial respiration. Diverse microbes differ in carbon use efficiency, which varies with resource quality, quantity and environmental conditions such as temperature, moisture and oxygen availability. Yet it is challenging to understand how a spatially heterogeneous environment, like soil, selects a microbiome of high carbon use efficiency (Roller and Schmidt, 2015). Understanding microbiota assembly and function at the pore scale may be a good starting point for enhancing our ability to effectively manage microbial communities.

Microscale examination of the microbiota requires robust imaging of soil pore structure and collecting organisms that inhabit the pores, which is extremely challenging as not all pores can be accessed. Microaggregates, for example, are relatively stable and cannot be easily disrupted; therefore, it may not be feasible to obtain bio-samples from pores within microaggregates. It is also difficult to prevent cross-contamination of samples of small size, and the sample size may be too small to reliably analyse the microbial community. Still, there are ways to partially overcome these obstacles. Recently, the microbiota of a macroaggregate was successfully analysed using metagenomic tools (Szoboszlay and Tebbe, 2021). Aggregates of similar size may vary in pore size and geometry, but sorting individual aggregates based on pore characteristics will aid pore size-based understanding of the soil microbiome. Visualization methods to locate organic matter within aggregates or spectroscopic methods to characterize organic matter chemistry will further help decipher microbial interactions and microbiome functional traits.

To scale up the microbiota and associated processes from microsites to the bulk soil level, mathematical models need to be developed. Data fitting can help build mechanisms emphasizing different roles of soil pores of varying size in organic carbon degradation, translocation and stabilization (Kravchenko et al., 2019). Modelling is also necessary to improve understanding of how the soil pore network regulates nutrient transformation processes, such as nitrification and denitrification. The connection of microbial metabolism between the microscale and bulk soil level is essential to resolve why the same macroscale and bulk soil properties can lead to different ecological consequences. Fine spatial-scale knowledge of the soil microbiome is expected to provide profound insight into soil health and offer new ways to optimize the soil environment to improve ecosystem services.

7 References

Arya, L. M., Leij, F. J., van Genuchten, M. T. and Shouse, P. J. 1999. Scaling parameter to predict the soil water characteristic from particle-size distribution data. *Soil Science Society of America Journal* 63(3), 510–519.

Bai, Z., Caspari, T., Gonzalez, M. R., Batjes, N. H., Mäder, P., Bünemann, E. K., de Goede, R., Brussaard, L., Xu, M., Ferreira, C. S. S., Reintam, E., Fan, H., Mihelič, R., Glavan, M. and Tóth, Z. 2018. Effects of agricultural management practices on soil quality: A review of long-term experiments for Europe and China. *Agriculture, Ecosystems and Environment* 265, 1–7.

Bailey, V. L., McCue, L. A., Fansler, S. J., Boyanov, M. I., DeCarlo, F., Kemner, K. M. and Konopka, A. 2013a. Micrometer-scale physical structure and microbial composition of soil aggregates. *Soil Biology and Biochemistry* 65, 60–68.

Bailey, V. L., Fansler, S. J., Stegen, J. C. and McCue, L. A. 2013b. Linking microbial community structure to beta-glucosidic function in soil aggregates. *ISME Journal* 7(10), 2044–2053.

Banwart, S. A., Bernasconi, S. M., Blum, W. E. H., de Souza, D. M., Chabaux, F., et al. 2017. Quantifying and managing soil functions in Earth's critical zone-combining experimentation and mathematical modeling. Chapter one-soil functions in Earth's critical zone: Key results and conclusions. *Advances in Agronomy* 142, 1–27.

Bashan, Y. and Levanony, H. 1988. Adsorption of the rhizosphere bacterium *Azospirillum brasilense* Cd to soil, sand, and peat particles. *Microbiology* 134(7), 1811–1820. doi:10.1099/00221287-134-7-1811.

Beare, M. H., Hendrix, P. F. and Coleman, D. C. 1994. Water-stable aggregates and organic matter fractions in conventional- and no-tillage soils. *Soil Science Society of America Journal* 58(3), 777–786.

Bell, T., Gessner, M. O., Griffiths, R. I., McLaren, J. R., Morin, P. J., van der Heijden, M. and van der Putten, W. H. 2009. Microbial biodiversity and ecosystem functioning under controlled conditions and in the wild. In: Naeem, S. and others (Eds.), *Biodiversity, Ecosystem Functioning, and Human Wellbeing: An Ecological and Economic Perspective*. Oxford: Academic Press, pp. 121–133.

Bossuyt, H., Denef, K., Six, J., Frey, S. D., Merckx, R. and Paustian, K. 2001. Influence of microbial populations and residue quality on aggregate stability. *Applied Soil Ecology* 16(3), 195–208.

Bouckaert, L., van Loo, D., Ameloot, N., Buchan, D., van Hoorebeke, L. and Sleutel, S. 2013. Compatibility of X-ray micro-computed tomography with soil biological experiments. *Soil Biology and Biochemistry* 56, 10–12.

Caesar-TonThat, T.-C. and Cochran, V. L. 2000. Soil aggregate stabilization by a saprophytic lignin-decomposing basidiomycete fungus I. Microbiological aspects. *Biology and Fertility of Soils* 32(5), 374–380.

Cai, P., Sun, X., Wu, Y., Gao, C., Mortimer, M., Holden, P. A., Redmile-Gordon, M. and Huang, Q. 2019. Soil biofilms: Microbial interactions, challenges, and advanced techniques for ex-situ characterization. *Soil Ecology Letters* 1(3–4), 85–93.

Carson, J. K., Gonzalez-Quiñones, V., Murphy, D. V., Hinz, C., Shaw, J. A. and Gleeson, D. B. 2010. Low pore connectivity increases bacterial diversity in soil. *Applied and Environmental Microbiology* 76(12), 3936–3942.

Chau, J. F., Bagtzoglou, A. C. and Willig, M. R. 2011. The effect of soil texture on richness and diversity of bacterial communities. *Environmental Forensics* 12(4), 333–341.

Chen, G., Rockhold, M. and Strevett, K. A. 2003. Equilibrium and kinetic adsorption of bacteria on alluvial sand and surface thermodynamic interpretation. *Research in Microbiology* 154(3), 175–181.

Cloutier, M., Alcaide, T., Duiker, S. and Bruns, M. A. 2023. Tillage intensity and plant rhizosphere selection shape bacterial-archaeal assemblage diversity and nitrogen cycling genes. *Soil and Tillage Research* 225, 105525.

Constancias, F., Prévost-Bouré, N. C., Terrat, S., Aussems, S., Nowak, V., Guillemin, J., Bonnotte, A., Biju-Duval, L., Navel, A., Martins, J. M., Maron, P. and Ranjard, L. 2014. Microscale evidence for a high decrease of soil bacterial density and diversity by cropping. *Agronomy for Sustainable Development* 34(4), 831–840.

Cordero, O. X. and Datta, M. S. 2016. Microbial interactions and community assembly at microscales. *Current Opinion in Microbiology* 31, 227–234.

Dalal, R. C. 1998. Soil microbial biomass – what do the numbers really mean? *Australian Journal of Experimental Agriculture* 38(7), 649–665.

Dasgupta, D. and Brahmaprakash, G. P. 2021. Soil microbes are shaped by soil physico-chemical properties: A brief review of existing literature. *International Journal of Plant and Soil Science* 33, 59–71.

Davinic, M., Fulta, L. M., Acosta-Martinez, V., Calderón, F. J., Cox, S. B., Dowd, S. E., Allen, V. G., Zak, J. C. and Moore-Kucera, J. 2012. Pyrosequencing and mid-infrared spectroscopy reveal distinct aggregate stratification of soil bacterial communities and organic matter composition. *Soil Biology and Biochemistry* 46, 63–72.

Dechesne, A., Or, D. and Smets, B. F. 2008. Limited diffusive fluxes of substrate facilitate coexistence of two competing bacterial strains. *FEMS Microbiology Ecology* 64(1), 1–8.

Doerr, S. H., Shakesby, R. A. and Walsh, R. P. D. 2000. Soil water repellency: its causes, characteristics and hydro-geomorphological significance. *Earth-Science Reviews* 51(1-4), 33-65.

Drijber, R. A., Doran, J. W., Parkhurst, A. M. and Lyon, D. J. 2000. Changes in soil microbial community structure with tillage under long-term wheat-fallow management. *Soil Biology and Biochemistry* 32(10), 1419–1430.

Elliott, E. T. 1986. Aggregate structure and carbon, nitrogen, and phosphorus in native and cultivated soils. *Soil Science Society of America Journal* 50(3), 627–633.

Elliott, E. T., Anderson, R. V., Coleman, D. C. and Cole, C. V. 1980. Habitable pore space and microbial trophic interactions. *Oikos* 35(3), 327–335.

Erktan, A., Or, D. and Scheu, S. 2020. The physical structure of soil: Determinant and consequence of trophic interactions. *Soil Biology and Biochemistry* 148, 107876.

Fierer, N. and Jackson, R. B. 2006. The diversity and biogeography of soil bacterial communities. *Proceedings of the National Academy of Sciences of the United States of America* 103(3), 626–631.

Fox, A., Ikoyi, I., Torres-Sallan, G., Lanigan, G., Schmalenberger, A., Wakelin, S. and Creamer, R. 2018. The influence of aggregate size fraction and horizon position on microbial community composition. *Applied Soil Ecology* 127, 19-29.

Frey, S. D., Elliott, E. T. and Paustian, K. 1999. Bacterial and fungal abundance and biomass in conventional and no-tillage agroecosystems along two climatic gradients. *Soil Biology and Biochemistry* 31(4), 573–585.

Guggenberger, G., Frey, S. D., Six, J., Paustian, K. and Elliot, E. T. 1999. Bacterial and fungal cell-wall residues in conventional and no-tillage agroecosystems. *Soil Science Society of America Journal* 63(5), 1188–1198.

Gupta, V. V. S. R. and Germida, J. J. 1988. Distribution of microbial biomass and its activity in different soil aggregate size classes as affected by cultivation. *Soil Biology and Biochemistry* 20(6), 777–786.

Gupta, A., Singh, U. B., Sahu, P. K., Paul, S., Kumar, A., Malviya, D., Singh, S., Kuppusamy, P., Singh, P., Paul, D., Rai, J. P., Singh, H. V., Manna, M. C., Crusberg, T. C., Kumar, A. and Saxena, A. K. 2022. Linking soil microbial diversity to modern agricultural practices: A review. *International Journal of Environmental Research and Public Health* 19(5), 3141.

Harris, K., Young, I. M., Gilligan, A., Otten, W. and Ritz, K. 2003. Effect of bulk density of the spatial organization of the fungus *Rhizoctonia solani* in soil. *FEMS Microbial Ecology* 44, 45–56.

Hartmann, M., Nikalus, P. A., Zimmermann, S., Schumutz, S., Kremer, J., Abarenkov, K., Lüscher, P., Widmer, F. and Frey, B. 2014. Resistance and resilience of the forest soil microbiome to logging-associated compaction. *ISME Journal* 8(1), 226–244.

Haruna, S. I., Anderson, S. H., Udawatta, R. P., Gantzer, C. J., Phillips, N. C., Cui, S. and Gao, Y. 2020. Improving soil physical properties through the use of cover crops: A review. *Agrosystems, Geosciences and Environment* 3(1), e20105.

Hassink, J., Bouwman, L. A., Zwart, K. B. and Brussaard, L. 1993. Relationships between habitable pore space, soil biota and mineralization rates in grassland soils. *Soil Biology and Biochemistry* 25(1), 47–55.

Haynes, R. J. and Beare, M. H. 1997. Influence of six crop species on aggregate stability and some labile organic matter fractions. *Soil Biology and Biochemistry* 29(11–12), 1647–1653.

40He, S.-H., Ding, Z., Hu, H.-B. and Gao, M. 2021. Effect of Grain Size on Microscopic Pore Structure and Fractal Characteristics of Carbonate-Based Sand and Silicate-Based Sand. *Fractal and Fractional* 5(4) 152. https://doi.org/10.3390/fractalfract5040152.

Hemkemeyer, M., Dohrmann, A. B., Christensen, B. T. and Tebbe, C. C. 2018. Bacterial preferences for specific soil particle size fractions revealed by community analyses. *Frontiers in Microbiology* 9, 149.

Hemkemeyer, M., Christensen, B. T., Tebbe, C. C. and Hartmann, M. 2019. Taxon-specific fungal preference for distant soil particle fractions. *European Journal of Soil Biology* 94, 103103.

Hermans, S. M., Buckley, H. L., Case, B. S., Curran-Cournane, F., Taylor, M. and Lear, G. 2020. Using soil bacterial communities to predict physico-chemical variables and soil quality. *Microbiome* 8(1), 79.

Heynen, C. E., Vanelsas, J. D., Kuikman, P. J. and Vanveen, J. A. 1988. Dynamics of *Rhizobium-leguminosarum biovar trifolii* introduced into soil – the effect of bentonite clay on predation by protozoa. *Soil Biology and Biochemistry* 20(4), 483–488.

Hu, S., Coleman, D. C., Beare, M. H. and Hendrix, P. F. 1995. Soil carbohydrates in aggrading and degrading agroecosystems: Influences of fungi and aggregates. *Agriculture, Ecosystems and Environment* 54(1–2), 77–88.

Ivanova, E. A., Kutovaya, O. V., Tkhakakhova, A. K., Chernov, T. I., Pershina, E. V., Markina, L. G., Andronov, E. E. and Kogut, B. M. 2015. The structure of microbial community in aggregates of a typical chernozem aggregates under contrasting variants of its agricultural use. *Eurasian Soil Science* 48(11), 1242–1256.

Janusz, G., Pawlik, A., Sulej, J., Swideska-Burek, U., Jarosz-Wilkotazka, A. and Paszczynski, A. 2017. Lignin degradation: Microorganisms, enzymes involved, genomes analysis and evolution. *FEMS Microbiology Reviews* 41(6), 941–962.

Jastrow, J. D. 1996. Soil aggregate formation and the accrual of particulate and mineral-associated organic matter. *Soil Biology and Biochemistry* 28(4–5), 665–676.

Jastrow, J. D., Miller, R. M. and Boutton, T. W. 1996. Carbon dynamics of aggregate-associated organic matter estimated by carbon-13 natural abundance. *Soil Science Society of America Journal* 60(3), 801–807.

Karimi, B., Terrat, S., Dequiedt, S., Saby, N. P. A., Horrigue, W., Lelièvre, M., Nowak, V., Jolivet, C., Arrouays, D., Wincker, P., Cruaud, C., Bispo, A., Maron, P. A., Bouré, N. C. P. and Ranjard, L. 2018. Biogeography of soil bacteria and archaea across France. *Science Advances* 4(7), eaat1808.

Kemper, W. D. and Rosenau, R. C. 1986. Aggregate stability and size distribution. In: Klute, A. (Ed.), *Methods of Soil Analysis, Part I. Physical and Mineralogical Methods*. Madison, WI: Soil Science Society of America, pp. 383–409.

Kemper, W. D., Rosenau, R. and Nelson, S. 1985. Gas displacement and aggregate stability of soils. *Soil Science Society of America Journal* 49(1), 25–28.

Klute, A. 1986. Water retention: Laboratory methods. In: Gambell, G. S. (Eds.), *Methods of Soil Analysis Part I. Physical and Mineralogical Methods* (2nd edn.). Madison, WI: ASA-CSSA-SSSA, pp. 635–662.

Kravchenko, A. N., Guber, A. K., Razavi, B. S., Koestel, J., Quigley, M. Y., Roberston, G. P. and Kuzyakov, Y. 2019. Microbial spatial footprint as driver of soil carbon stabilization. *Nature Communications* 10(1), 3121.

Kravchenko, A. N., Negassa, W. C., Guber, A. K., Hildebrandt, B., Marsh, T. L. and Rivers, M. L. 2014. Intra-aggregate pore structure influences phylogenetic composition of bacterial community in macroaggregates. *Soil Science Society of America Journal* 78(6), 1924–1939.

Lauber, C. L., Hamady, M., Knight, R. and Fierer, N. 2009. Pyrosequencing-based assessment of soil pH as a predictor of soil bacterial community structure at the continental scale. *Applied and Environmental Microbiology* 75(15), 5111–5120.

Lehmann, J., Solomon, D., Kinyangi, J., Dathe, L., Wirick, S. and Jacobsen, C. 2008. Spatial complexity of soil organic matter forms at nanometer scales. *Nature Geoscience* 1(4), 238–242.

Lehtovirta-Morlety, L. E. 2018. Ammonia oxidation: Ecology, physiology, biochemistry and why they must all come together. *FEMS Microbiology Letters* 365(9), fny058.

Lennon, J. T. and Lehmkuhl, B. K. 2016. A trait-based approach to bacterial biofilms in soil. *Environmental Microbiology* 18(8), 2732–2742.

Li, C. H., Ma, B. L. and Zhang, T. Q. 2002. Soil bulk density effects on soil microbial populations and enzyme activities during the growth of maize (*Zea Mays* L) planted in large pots under filed exposure. *Canadian Journal of Soil Science* 82(2), 147–154.

Lupwayi, N. Z., Arshad, M. A., Rice, W. A. and Glyton, G. W. 2001. Bacterial diversity in water-stable aggregates of soils under conventional and zero-tillage management. *Applied Soil Ecology* 16(3), 251–261.

Meadow, P. S. and Anderson, J. G. 1968. Micro-organisms attached to marine sand grains. *Journal of the Marine Biological Association of the United Kingdom* 48(1), 161–175.

Mendes, I. C., Bandick, A. K., Dick, R. P. and Bottomley, P. J. 1999. Microbial biomass and activities in soil aggregates affected by winter cover crops. *Soil Science Society of America Journal* 63(4), 873–881.

Miller, M. and Dick, R. P. 1995. Dynamics of soil C and microbial biomass in whole soil and aggregates in two cropping systems. *Applied Soil Ecology* 2(4), 253–261.

Miller, M. N., Zebarth, B. J., Dandie, C. E., Burton, D. L., Goyer, C. and Trevors, J. T. 2009. Denitrifier community dynamics in soil aggregates under permanent grassland and Arable Cropping Systems. *Soil Science Society of America Journal* 73(6), 1843–1851.

Mishra, A., Singh, L. and Singh, D. 2023. Unboxing the black box – one step forward to understand the soil microbiome: A systematic review. *Microbial Ecology* 85(2), 669–683. doi:10.1007/s00248-022-01962-5.

Naveed, M., Herath, L., Moldrup, P., Authur, E., Nicolaisen, M., Norgaard, T., Ferré, T. P. A. and de Jonge, L. W. 2016. Spatial variability of microbial richness and diversity and relationships with soil organic carbon, texture and structure across an agricultural field. *Applied Soil Ecology* 103, 44–55.

Negassa, W. C., Guber, A. K., Kravchenko, A. N., Marsh, T. L., Hildebrandt, B. and Rivers, M. L. 2015. Properties of soil pore space regulate pathways of plant residue decomposition and community structure of associated bacteria. *PLoS ONE* 10(4), e0123999.

Or, D., Smets, B. F., Wraith, J. M., Dechesne, A. and Friedman, S. P. 2007. Physical constrains affect bacterial habitats and activity in unsaturated porous media – a review. *Advances in Water Resources* 30(6–7), 1505–1527.

Page, K. L., Dang, Y. P. and Dalal, R. C. 2020. The ability of conservation agriculture to conserve soil organic carbon and the subsequent impact on soil physical, chemical and biological properties and yield. *Frontiers in Sustainable Food Systems* 4, 31.

Paustian, K., Six, J., Elliott, E. T. and Hunt, H. W. 2000. Management options for reducing CO_2 emissions from agricultural soils. *Biogeochemistry* 48(1), 147–163.

Pellegrino, E., Piazza, G., Helgason, T. and Ercoli, L. 2022. Microbiome structure and interconnection in soil aggregates across conservation and conventional agricultural practices allow to identify main prokaryotic and fungal taxa related to soil functioning. *Soil Biology and Biochemistry* 175, 108833.

Portell, X., Pot, V., Garnier, P., Otten, W. and Baveye, P. C. 2018. Microscale heterogeneity of the spatial distribution of organic matter can promote bacterial diversity in soils: Insights from computer simulations. *Frontiers in Microbiology* 9, 1583.

Probandt, D., Eickhorst, T., Eilrott, A., Amann, R. and Knittel, K. 2018. Microbial life on a sand grain: From bulk sediment to single grains. *ISME Journal* 12(2), 623–633.

Puget, P., Angers, D. A. and Chenu, C. 1998. Nature of carbohydrates associated with water-stable aggregates of two cultivated soils. *Soil Biology and Biochemistry* 31(1), 55–63.

Quigley, M. Y., Negassa, W. C., Guber, A. K., Rivers, M. L. and Kravchenko, A. N. 2018. Influence of pore characteristics on the fate and distribution of newly added carbon. *Frontiers in Environmental Science* 6, 51.

Ranjard, L., Poly, F., Combrisson, J., Richaume, A., Gourbière, F., Thioulouse, J. and Nazaret, S. 2000. Heterogeneous cell density and genetic structure of bacterial pools associated with various soil microenvironments as determined by enumeration and DNA fingerprinting approach (RISA). *Microbial Ecology* 39(4), 263–272.

Ren, D., Madsen, J. S., Sørensen, S. and Burmølle, M. 2015. High prevalence of biofilm synergy among bacterial soil isolates in cocultures indicates bacterial interspecific cooperation. *ISME J* 9, 81–89.

Rillig, M. C., Muller, L. A. H. and Lehmann, A. 2017. Soil aggregates as massively concurrent evolutionary incubators. *ISME Journal* 11(9), 1943–1948.

Roberson, E. B., Shennan, C., Firestone, M. K. and Sarig, S. 1995. Nutritional management of microbial polysaccharide production and aggregation in an agricultural soil. *Soil Science Society of America Journal* 59(6), 1587–1594.

Roller, B. R. K. and Schmidt, T. M. 2015. The physiology and ecological implications of efficient growth. *The ISME Journal* 9(7), 1481–1487.

Rousset, C., Clough, T. J., Grace, P. R., Rowlings, D. W. and Scheer, C. 2020. Soil type, bulk density, and drainage effects on relative gas diffusivity and N_2O emission. *Soil Research* 58(8), 726–736.

Ruamps, L. S., Nunan, N. and Chenu, C. 2011. Microbial biogeography at the soil pore scale. *Soil Biology and Biochemistry* 43(2), 280–286.

Saleem, M., Hu, J. and Jousset, A. 2019. More than the sum of its parts: Microbiome biodiversity as a driver of plant growth and soil health. *Annual Review of Ecology, Evolution, and Systematics* 50(1), 145–168.

Santoro, T. and Stozky, G. 1968. Sorption between microorganisms and clay minerals as determined by the electrical sensing zone particle analyzer. *Canadian Journal of Microbiology* 14(4), 299–307.

Schutter, M. E. and Dick, R. P. 2002. Microbial community profiles and activities among aggregates of winter fallow and cover-cropped soil. *Soil Science Society of America Journal* 66(1), 142–153.

Seaton, F., George, P. B. L., Lebron, I. and Jones, D. L. 2020. Soil texture heterogeneity impacts bacterial but not fungal diversity. *Soil Biology and Biochemistry* 144, 10776.

Seech, A. G. and Beauchamp, E. G. 1988. Denitrification in soil aggregates of different sizes. *Soil Science Society of America Journal* 52(6), 1616–1621.

Sessitsch, A., Weillharter, A., Gerzabek, M. H., Kirchmann, H. and Kandeler, E. 2001. Microbial population structures in soil particle size fractions of a long-term fertilizers field experiment. *Applied and Environmental Microbiology* 67(9), 4215–4224.

Shestak, C. J. and Busse, M. D. 2005. Compaction alters physical but not biological indices of soil health. *Soil Science Society of America Journal* 69(1), 236–246.

Singh, S. and Singh, J. S. 1995. Microbial biomass associated with water-stable aggregates in forest, savanna and cropland soils of a seasonally dry tropic region, India. *Soil Biology and Biochemistry* 27, 1027–1033.

70Six, J., Elliott, E.T. and Paustian, K. 1999. Aggregate and Soil Organic Matter Dynamics under Conventional and No-Tillage Systems. *Soil Science Society of America Journal* 63, 1350-1358. https://doi.org/10.2136/sssaj1999.6351350x.

Six, J., Paustian, K., Elliott, E. T. and Combrink, C. 2000. Soil structure and organic matter. I. Distribution of aggregate-size classes and aggregate-associated carbon. *Soil Science Society of America Journal* 64(2), 681–689.

Sleutel, S., Bouckaert, L., Buchan, D., Van Loo, D., Cornelis, W. M. and Sanga, H. G. 2012. Manipulation of the soil pore and microbial community structure in soil mesocosm incubation studies. *Soil Biology and Biochemistry* 45, 40–48.

Smith, A. P., Marín-Spiotta, E., de Graaff, M. A. and Balser, T. C. 2014. Microbial community structure varies across soil organic matter aggregate pools during tropical land cover change. *Soil Biology and Biochemistry* 77, 292–303.

Szoboszlay, M. and Tebbe, C. C. 2021. Hidden heterogeneity and co-occurrence networks of soil prokaryotic communities revealed at the scale of individual soil aggregates. *MicrobiologyOpen* 10(1), e1144.

Tecon, R. and Or, D. 2017. Biophysical processes supporting the diversity of microbial life in soil. *FEMS Microbiology Reviews* 41(5), 599–623.

Tisdall, J. M. and Oades, J. M. 1980. The effect of crop rotation on aggregation in red-brown earth. *Soil Research* 18(4), 423–433.

Tisdall, J. M. and Oades, J. M. 1982. Organic matter and water-stable aggregates in soils. *Journal of Soil Science* 33(2), 141–163.

Toyota, K., Young, I. M. and Ritz, K. 1996. Effects of soil matric potential and bulk density on the growth of *Fusarium oxysporum* f. sp. *raphani*. *Soil Biology and Biochemistry* 28(9), 1139–1145.

Trivedi, P., Delgado-Baquerizo, M., Anderson, I. C. and Singh, B. K. 2016. Response of soil properties and microbial communities to agriculture: Implications for primary productivity and soil health indictors. *Frontiers in Plant Science* 7, 990.

Van Elsas, J. D., Pratama, A. A. and de Araújo, T. J. T. 2019. Microbial interactions in soil. In: van Elsas, J. D., Trevors, J. T., Rosado, A. S. and Nannipieri, P. (Eds.), *Modern Soil Microbiology* (3rd edn.). CRC Press, Boca Raton, Florida, USA.

Vieira, S., Sikorski, J., Gebala, A., Boeddinghaus, R. S., Marhan, S., Rennert, T., Kandeler, E. and Overmann, J. 2020. Bacterial colonization of minerals in grassland soils is selective and highly dynamic. *Environmental Microbiology* 22(3), 917–933.

Vos, M., Wolf, A. B., Jennings, S. J. and Kowalchuk, G. A. 2013. Micro-scale determinants of bacterial diversity in soil. *FEMS Microbiology Reviews* 37(6), 936–954.

Wagg, C., Hautier, Y., Pellkofer, S., Banerjee, S., Schmid, B. and van der Heijden, M. G. A. 2021. Diversity and asynchrony in soil microbial communities stabilizes ecosystem functioning. *eLife* 10, e62813.

Wang, J. and Zhou, J. 2020. No-till increases soil denitrification via its positive effects on the activity and abundance of the denitrifying community. *Soil Biology and Biochemistry* 142, 10776.

Watteau, F. and Vilemin, G. 2018. Soil microstructures examined through transmission electron microscopy reveal soil-microorganisms interactions. *Frontiers in Environmental Science* 6, 106.

Whiteside, M. D., Werner, G. D. A., Caldas, V. E. A., Padje, A. V., Dupin, S. E., Elbers, B., Bakker, M., Wyatt, G. A. K., Klein, M., Hink, M. A., Postma, M., Vaitla, B., Noë, R., Shimizu, T. S., West, S. A. and Kiers, E. T. 2019. Mycorrhizal fungi respond to resource inequality by moving phosphorus from rich to poor patches across networks. *Current Biology* 29(12), 2043–2050.e8.

Wilpiszeski, R. L., Aufrecht, J. A., Retterer, S. T., Sullivan, M. B., Graham, D. E., Pierce, E. M., Zablocki, O. D., Palumbo, A. V. and Elias, D. A. 2019. Soil aggregate microbial communities: Towards understanding microbiome interactions at biologically relevant scales. *Applied and Environmental Microbiology* 85(14), e00324-19.

Wolf, A. B., Vos, M., de Boer, W. and Kowalchuk, G. A. 2013. Impact of matric potential and pore size distribution on growth dynamics of filamentous and non-filamentous soil bacteria. *PLoS ONE* 8(12), e83661.

Wright, D. A., Killhan, K., Glover, L. A. and Prosser, J. I. 1995. Role of pore size location in determining bacterial activity during predation by protozoa in soil. *Applied and Environmental Microbiology* 61(10), 3537–3543.

Wright, S. F. and Upadhyaya, A. 1996. Extraction of an abundant and unusual protein from soil and comparison with hyphal protein of arbuscular mycorrhizal fungi. *Soil Science* 161(9), 575–586.

Wright, S. F. and Upadhyaya, A. 1998. A survey of soils for aggregate stability and glomalin, a glycoprotein produced by hyphae of arbuscular mycorrhizal fungi. *Plant and Soil* 198(1), 97–107.

Wu, L., Vomocil, J. A. and Childs, S. W. 1990. Pore size, particle size, aggregate size, and water retention. *Soil Science Society of America Journal* 54(4), 952–956.

Xia, Q., Rufty, T. and Shi, W. 2020. Soil microbial diversity and composition: Links to soil texture and associated properties. *Soil Biology and Biochemistry* 149, 107953.

Xia, Q., Zheng, N., Heitman, J. L. and Shi, W. 2022. Soil pore size distribution shaped not only compositions but also networks of the soil microbial community. *Applied Soil Ecology* 170, 104273.

Yang, P. and van Elsas, J. D. 2018. Mechanisms and ecological implications of the movement of bacterial in soil. *Applied Soil Ecology* 129, 112–120.

Young, I. M., Crawford, J. W., Nunan, N., Otten, W. and Spiers, A. 2008. Microbial distribution in soils: Physics and scaling. *Advances in Agronomy* 100, 81–121.

Zhou, J., Xia, B., Treves, D. S., Wu, L. Y., Marsh, T. L., O'Neill, R. V., Palumbo, A. V. and Tiedje, J. M. 2002. Spatial and resource factors influencing high soil microbial diversity. *Applied and Environmental Microbiology* 68(1), 326–334.

Chapter 14

Assessing the impact of inorganic fertilizers on soil microbiomes

Joann K. Whalen, McGill University, Canada and Center for Sustainable Soil Sciences (C3S), Mohammed VI Polytechnic University, Morocco; and Shamim Gul, McGill University, Canada and University of Balochistan, Pakistan

1 Introduction

Grains are a staple of human diets. The global abundance of these nutritious foodstuffs depends on agronomic sciences, which have the know-how to develop high-yielding, stress-tolerant varieties and the methods to create a suitable environment for crop growth by managing fields and agricultural inputs sustainably. Once the agronomist has optimized plant access to sunlight and water in a non-compacted soil and controlled the weeds, diseases, and pests that could limit crop growth, their attention turns to a responsible nutrient management program. Soil testing and analysis allows the agronomist to estimate how much of the 14 essential plant nutrients – nitrogen (N), phosphorus (P), potassium (K), sulfur, calcium, magnesium, iron, boron, manganese, copper, zinc, molybdenum, nickel, and chlorine – will be supplied by natural soil fertility, and which nutrients may need to be added. Agronomists may advise the use of inorganic fertilizer if it is a convenient and cost-effective choice for the cropping system. Inorganic fertilizers are typically water-soluble salts formulated as solid granules, liquid solution (e.g. urea ammonium nitrate), or high-pressure gas (e.g. anhydrous ammonia). The most common are the solid fertilizer granules that supply a single nutrient or a mixture of nutrients. Inorganic fertilizers are

http://dx.doi.org/10.19103/AS.2024.0136.28

nutrient-dense with a higher guaranteed analysis than organic fertilizers, which are typically rated around 1–1–1, meaning 1% by dry weight of N, 1% of P in the form of P_2O_5 and 1% K on the basis of K_2O. The physical and chemical properties of inorganic fertilizers makes it easy to apply them at the right depth and distance from the seed or growing plant, at a suitable time in the growing season. Consequently, inorganic fertilizers are applied to supplement the natural soil fertility and increase the growth of grains and other agricultural plants, including vegetables, perennial hayfield, vine, and tree crops.

Solid granules of inorganic fertilizer dissolve in water and the salts dissociate into nutrient ions, which are bioavailable, meaning the ions absorb through cell membranes by passive and active transport. Fertilizer application rates are calibrated to meet the nutrient requirements of plants, but soluble nutrient ions are also absorbed and metabolized by soil microorganisms. The soil microbiome – specifically the bacteria, archaea, and fungi that inhabit soil for part or all of their lifespan – requires nutrients for growth and reproduction. The cellular biomass of the soil microbiome is composed predominantly of carbon, hydrogen, and oxygen that form carbohydrates and lipids. Nitrogen and sulfur are incorporated into proteins, enzymes, and genetic materials like DNA and RNA. Phosphorus is part of the phospholipids in cell membranes and is used for energetic reactions. Potassium, calcium, and magnesium maintain the water balance and cytoplasmic functions, and together with other cations and anions are instrumental in enzyme activation, electron transport chain, and biosynthesis reactions. In field soils, nutrients contained in the soil microbiome depend upon access to metabolizable carbon and nutrient ions, resulting in a stoichiometry in the molar ratios for C:N of about 5–8, and a molar ratio in the C:P of 18–79 (Heuck et al., 2018). When microorganisms die, the nutrients in their biomass are released into the environment. If microbial cells lyse around plant roots, nutrient ions may be absorbed by the plant. Alternatively, nutrients released from microbial cells coud become bound to soil solids or transported in soil pore water to the bulk soil.

The application of inorganic fertilizers is expected to impact the abundance, diversity, and functions of the soil microbiome. These effects may be direct, due to the provision of water-soluble nutrients that create a nutrient-rich environment for fast-growing microbial populations and alter the community structure of the soil microbiome. Furthermore, nutrient ions change the ionic strength, affecting the osmolarity and pH of water in soil microhabitats. Indirectly, inorganic fertilizers affect soil microbiomes by stimulating plant growth, resulting in more extensive root development and altering plant–microbe interactions. Plant roots distinguish nutrient-limited and nutrient-rich environments, and this affects the signal exchange between plants and associated soil microorganisms. Plants provide substrates for the soil microbiome, during their growth and after harvest. For example, sloughed root cells, exudates and secretions attract and support the growth of diverse microorganisms, some of which are facultative root symbionts, such as rhizobia and mycorrhizal fungi. Non-edible residues left in the field after crop harvest are colonized by saprophytic bacteria and fungi, and the quantity and chemical composition of these residues determines the speed at which residues decompose and replenish the soil nutrient supply. How inorganic fertilizers can directly and indirectly impact the soil microbiome is discussed in the following sections. Examples focus mainly on inorganic fertilizers containing N, P, and K since these are the most applied for agriculture worldwide.

2 Direct effects of inorganic fertilizer on the soil microbiome

The application of inorganic fertilizers in agricultural fields involves the distribution of nutrient-rich granules and solutions. Conventional fertilizer spreading equipment is designed to broadcast the fertilizing material uniformly across the field, where it is left on the soil surface or mixed in the topsoil with tillage implements. In contrast, modern precision fertilizer applicators distribute fertilizer granules and solutions in a specific location of the agricultural field, matching the nutrient inputs with the nutrient demands of the growing crop (IPNI, 2012). For example, fertilizer can be coated on the seed, applied with the seed as a pop-up fertilizer, or placed in a band within several centimeters of the planted seed. Fertilizer banding is popular in grains and other row crops because the machinery can place the seeds and fertilizer several centimeters apart. Seed is physically separated from fertilizer granules to avoid salt stress during germination. After the seedling emerges, its root system will reach the fertilized area within the first weeks of growth. Fertigation is another option. After mixing fertilizer with water, the dissolved nutrients are delivered to the plant through a water-saving irrigation system, such as drip-line irrigation in the crop row. Finally, dilute nutrient solutions can be sprayed onto the foliage, where soluble nutrients are absorbed through pores in leaf surfaces. Foliar solutions also percolate through the canopy and into the soil, where nutrients are absorbed by the roots.

Since the first terrestrial surfaces emerged from the primordial ocean more than a billion years ago, soil microbiomes evolved to survive, grow, and reproduce in natural environments with scarce nutrients. When they encounter a nutrient-rich fertilizer band or crop row, opportunistic bacteria likely outcompete other microorganisms for nutrient ions, provided there is sufficient carbon substrate to avoid growth limitation. This may occur soon after nutrient ions appear in the microhabitat. The microhabitat refers, in part, to the water-filled pore space where planktonic cells, whether single or aggregated, circulate passively or in motile forms. The microhabitat is also the biofilm-coated soil surface in contact with water films, since flowing water delivers ions for cellular growth and biofilm expansion on the solid surface and may be a trigger for the dynamic disaggregation and detachment of cells that occurs continuously in natural biofilms (Sauer et al., 2022). Still, the dissolution of inorganic fertilizers is expected to change the chemistry of soil water, increasing its electrolyte concentration and altering the pH to more acidic or alkaline, depending on the fertilizer source. Therefore, the direct impact of inorganic fertilizers on the soil microbiome will be observed in the nutrient-rich zone around dissolving fertilizer granules.

3 Soil microbiome in the fertosphere

We define the fertosphere as the microenvironment around the water-soluble salt that is an inorganic fertilizer granule. The granule dissolves when infused with water, a polar molecule that associates with cations and anions in the salt. Since water flows osmotically from low salt concentration to higher salt concentration, this means that the fertilizer granule attracts water. This is supported by the observation that water-soluble salts of N dissolve completely within 2.5–21 h of exposure to humid air (Sigtryggsson et al., 2020; Fig. 1). Furthermore, it takes less than 7 min for 2 mm of simulated rainfall to dissolve about 50% of the mass in granular ammonium nitrate and calcium ammonium nitrate fertilizers, and the granules are virtually liquified in <25 min when exposed to 18 mm rainfall h^{-1} (Sigtryggsson et al., 2020; Fig. 2). Consequently, most of the salts in inorganic fertilizer granules will dissolve within hours to days of fertilizer application in a moist agricultural soil.

3.1 Osmotic stress in the fertosphere

Nutrient ions from the granule move into soil water according to the molecular and ionic diffusion rates of the electrolyte. Strong electrolytes that dissociate completely in water, such as potassium chloride (KCl), diffuse further than weak electrolytes like diammonium phosphate ($(NH_4)_2HPO_4$) and non-electrolytes like urea ($(NH_2)_2CO$). This creates a gradient of ionic concentration radiating from the center of the fertilizer granule, extending from about 10–100 mm or more, depending on the nutrient ion (Fig. 3). Greater diffusion of K, followed by N and P, is observed experimentally and in mathematical models. For example, the estimated effective diffusion coefficients are $D_{eff} = 7.2 \times 10^{-10}$ m^2 s^{-1} for K salts, with a D_{eff} from 1.5×10^{-10} m^2 s^{-1} to 1.3×10^{-11} m^2 s^{-1} for N salts, and $D_{eff} = 5.8 \times 10^{-12}$ m^2 s^{-1} for P salts (Martínez-Hilario et al., 2023; Basu et al., 2010). Still, many factors affect the movement of ions from the granule into the fertosphere. The size of the fertilizer granule and its physicochemical properties (e.g. sphericity, coatings), soil moisture, temperature, pH, and porosity are included in models of nutrient diffusion from fertilizer (Basu et al., 2010). Since nutrient

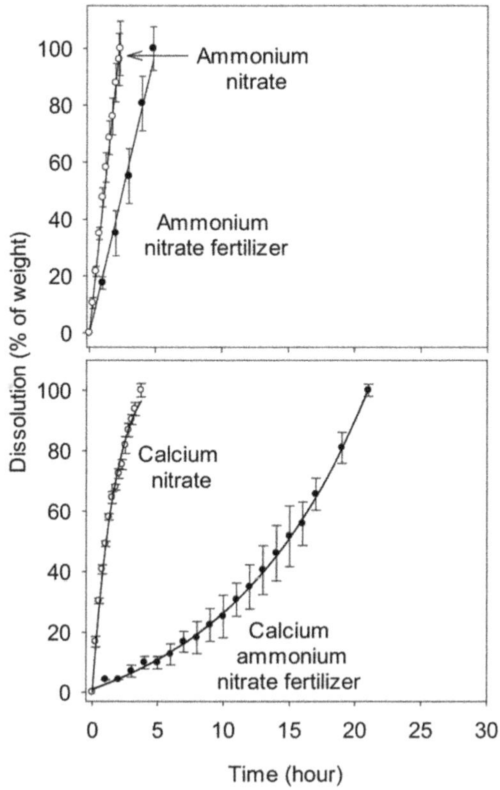

Figure 1 Dissolution of salt crystals and granules of calcium and ammonium nitrate, ammonium nitrate fertilizer and calcium ammonium nitrate fertilizer when exposed to air with 90–99% relative humidity at 25°C, ($n = 4$, except $n = 3$ for calcium ammonium nitrate; datapoints are the mean and standard deviation). Source: Adapted from Sigtryggsson et al. (2020).

ions are electrolytes, their movement can be detected by electrical conductivity probes, allowing for the measurement and modeling of nutrient transport in the fertosphere (Fig. 3; Phong et al., 2021).

Fertilizer salts are hygroscopic, meaning they attract and retain water due to their high salt concentration. This creates an osmotic gradient that has consequences for the survival and growth of the soil microbiome. Since microbial cells attract water osmotically by having a higher salt concentration than the surrounding soil water, and the fertosphere has an even higher salt concentration than microbial cells, this has the tendency to pull water across microbial cell membranes toward the fertosphere. This effect, termed osmotic shock, is also observed in seeds and seedlings that come into contact with granular fertilizer. To avoid the physiological dysfunctions that accompany osmotic shock, fertilizer applicators are calibrated to place granules several

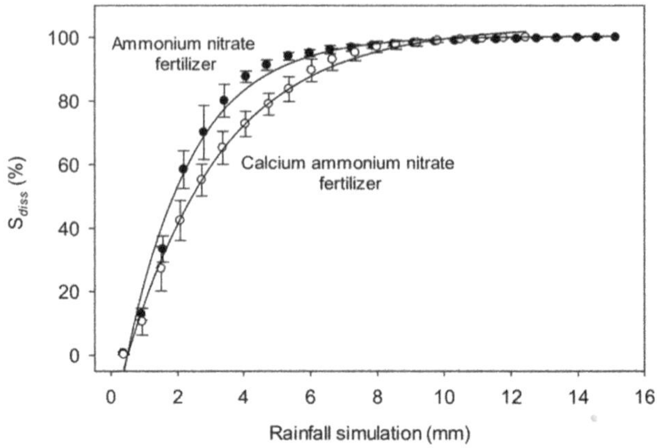

Figure 2 Dissolution (S_{diss}) rate of ammonium nitrate fertilizer or calcium ammonium nitrate fertilizer wetted by simulated rainfall of 18 mm h^{-1} (each symbol represents four replicates; error bars are the standard deviation of the mean for each datapoint). Source: Adapted from Sigtryggsson et al. (2020).

centimeters away from the seed (IPNI, 2012). However, soil microorganisms cannot avoid the osmotic gradient of the fertosphere, particularly around granules with a high salt index like KCl and ammonium-based fertilizers such as NH_4NO_3 (Table 1).

The soil microbiome is adept at sensing osmotic gradients. Because rapid, uncontrolled swelling or shrinkage of cells would be fatal, microorganisms have strategies to maintain cellular homeostasis. When the osmolarity inside the cell exceeds that of the water flowing along the cell membrane (hypertonic stress), the transmembrane Na^+/H^+ exchangers are activated within ~0.5 s to control the intracellular concentrations of these two ions via mechanosensitive channels, thereby maintaining a constant cell volume (Călinescu et al., 2014). The high salt concentration of the fertosphere produces hyperosmotic stress at the cell membrane. Soil bacteria respond to the high external salt concentration by accumulating compatible solutes, such as betaine, ectoine, proline, and trehalose in their cytoplasmic fluid (Jentsch, 2016). Compatible solutes elevate the internal osmolarity by transporting salt ions into the cell. For instance, the osmoregulated secondary carrier BetP of the betaine/carnitine/choline transporter family co-transports betaine along with two Na^+ ions into cells of the gram-positive bacterium *Corynebacterium glutamicum*. The maximum steady-state accumulation of glycine betaine is 4×10^6 times inside bacterial cells than in the external electrolyte solution (Morbach and Krämer, 2006). Cell turgor is maintained when compatible solutes transport Na^+ inside cells, and these organic solutes protect enzymes, membranes, and other intracellular components from reactive oxidative stress due to high Na^+ concentration. The gram-negative bacterium *Gluconacetobacter diazotrophicus* is an endophyte

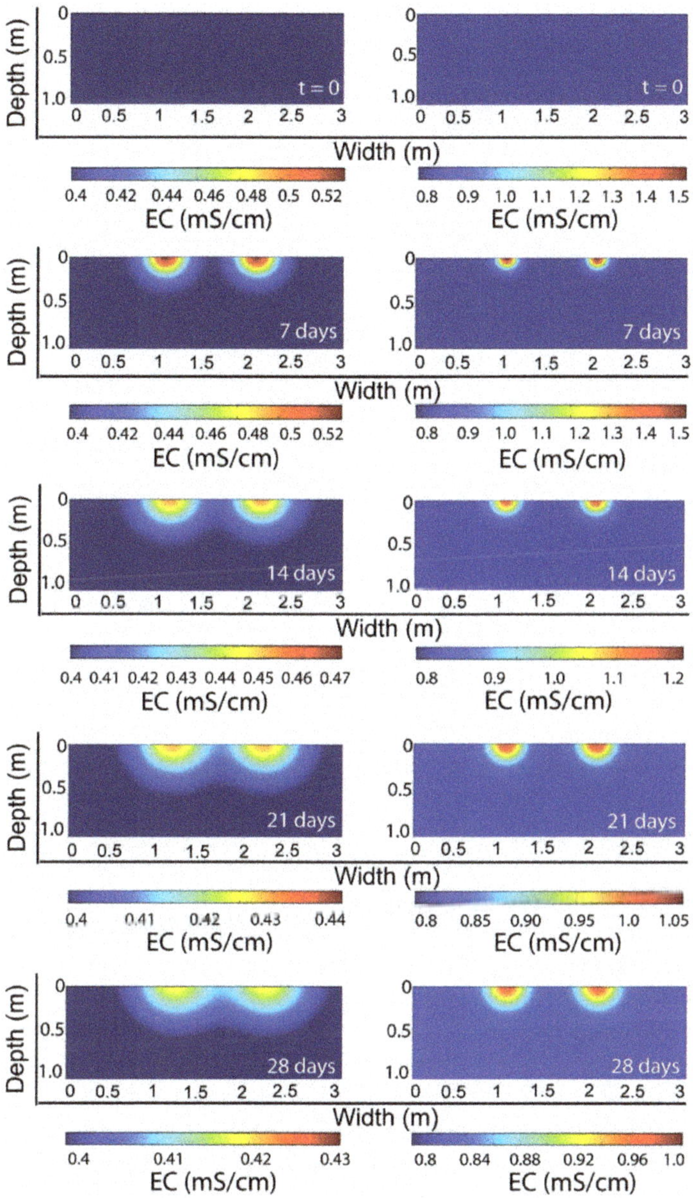

Figure 3 Spatial distribution of the soil electrical conductivity (EC) around two granules of NPK fertilizer containing 10% total N, 30% total P_2O_5, and 20% total K_2O measured at 7-day intervals during a 28-day study. Changes in soil electrical conductivity are an indication of the rate of electrolyte movement from the granule into the bulk soil, which was faster in the sandy-loam soil than the clay-loam soil. Agricultural soils were S1, a sandy-loam with 484 kg sand kg^{-1} and 103 g clay kg^{-1} and 8.3 cmol(+) kg^{-1}; and S2, a clay-loam soil with 248 kg sand kg^{-1} and 361 g clay kg^{-1} and 15.2 cmol(+) kg^{-1}. Source: Adapted from Phong et al. (2021).

Table 1 Salt content and acidity index of selected inorganic fertilizers. Fertilizer grade is the percentage of total nitrogen (N) - total phosphate (P_2O_5) and total potash (K_2O). The salt index is expressed relative to sodium nitrate, which has a salt index of 100. The acidity index is the calcium carbonate equivalent, meaning the kilograms of agricultural lime required to neutralize acidity in 100 kg of fertilizer. A negative acidity index is the kilograms of H^+ consumed following the application of 100 kg of fertilizer

Fertilizer	Grade	Salt index	Acidity index	Density(kg L^{-1})
Urea	46-0-0	74	166	0.80
Calcium ammonium nitrate	27-0-0	93	120	0.90
Ammonium sulfate	21-0-0	68	151	1.10
Calcium nitrate	15-0-0	65	-(20)	1.20
Potassium nitrate	14-0-41	80	-(26)	1.20
Sodium nitrate	17	100	-(29)	1.25
Superphosphate	0-20-0	8	Neutral	1.10
Triple superphosphate	0-46-0	10	Neutral	1.10
Monoammonium phosphate	11-52-0	27	95	1.00
Diammonium phosphate	18-46-0	29	102	1.00
Muriate of potash	0-0-60	115	Neutral	1.10
Potassium sulfate	0-0-50	43	Neutral	1.20
Sulfate of potash-magnesium	0-0-22	43	Neutral	1.50

Source: Adapted from Jones (2003).

of several economically important crops like sugarcane, cocoa, and rice. It tolerates osmotic stress by regulating uptake proteins for iron, sugar, and other nutrients, by synthesizing compatible solutes (trehalose, mannitol) for osmoregulation, and by adjusting its cell envelope structure (Leandro et al., 2021). Osmotolerance is widely reported in the soil microbiome, with functional metagenomics revealing osmotolerant phylotypes affiliated to Proteobacteria, Actinobacteria, Gemmatimonadetes, Bacteroidetes, Firmicutes, and Acidobacteria (Ahmed et al., 2018).

Soil fungi also have strategies to avoid dehydration in environments with high osmolarity due to the presence of salts. Like bacteria, fungi use active transport to maintain ionic homeostasis in the cytoplasm. They synthesize melanin, which adheres to and strengthens the cell wall by cross-linking with polysaccharides to prevent solutes from leaking through the cell wall (Kogej et al., 2007). Also, they synthesize compatible solutes for osmotic adjustment, including glycerol, erythritol, arabitol, mannitol, and trehalose, to maintain the intracellular ionic balance without diminishing enzyme functions and growth under osmotic stress. Many fungal species are mild to strongly osmotolerant. More than half of the isolates from 40 fungal species were mild to highly osmotolerant when directly exposed to 0.1–4.0 M KCl, including many saprophytic fungi (e.g. *Mucor* and *Penicillium* spp.), plant-pathogens such as *Colletotrichum* and *Fusarium*, and mycoparasites like *Clonostachys* spp. (Araújo et al., 2020). Highly osmotolerant *Aspergillus* spp. can be coated on granules or co-applied with inorganic fertilizer to improve the solubility of P, copper, zinc, and manganese and to contribute to elemental sulfur oxidation (Majaron et al., 2022). Although some fungi are osmo-sensitive, notably *Gliocladium* and *Trichoderma* (Araújo et al., 2020), many fungi in the soil microbiome have mechanisms to tolerate salts in the fertosphere.

The osmotic stress experienced by the soil microbiome depends on the water content of the porous soil matrix. Soil pores fill with water after rainfall and irrigation events. Soil water moves preferentially through macropores and diffuses between connected micropores via matricial flow. Nutrient-rich water around the fertilizer granule is diluted by the natural soil water, which contains fewer electrolytes. When soil is drier, micropores close to the fertosphere become saturated with salt ions that can be cytotoxic to microorganisms. Using a combination of mathematically modeling and experimental micro X-ray computed tomography scanning, Ruiz et al. (2020) mapped the NH_4^+ and NO_3^- released from NH_4Cl, NH_4NO_3, and KNO_3 granules into the fertosphere. The fertosphere develops more rapidly due to faster salt dissolution in wetter soil (30–70% water saturation) than in drier soil (15% water saturation; Ruiz et al., 2020). Furthermore, NO_3^- diffuses further than NH_4^+ due to greater mobility of K^+ and NO_3^- electrolytes (Fig. 4). In micropores of dry soil, the soil microbiome is exposed to higher salt concentration, which can be cytotoxic. Predictions from a three-dimensional mathematical model suggest that a higher NH_4^+ and K^+ concentration in the soil pore water will increase the risk of microbial inhibition or lysis, which occurs in <10 min in the soil 1–4 mm from the center of the granule (Fig. 5; Ruiz et al., 2020).

Soil bacteria and fungi may experience short-term osmotic stress when N and K fertilizer granules begin to dissolve. However, this is probably a transient stress on the soil microbiome. Natural rainfall and irrigation events add to water that dilutes the ionic concentration of the fertosphere, while increasing the electrolyte flow into pores of the surrounding soil matrix. Ions exchange between the soil pore water and the surface of organo-minerals that comprise the solid soil matrix. Net removal of NH_4^+ and K^+ occurs through inner-sphere and outer-sphere reactions on reactive surfaces and by ion fixation between the lattices of expanding and non-expanding clay minerals. In soil slurries, the sorption equilibrium is reached in <1 h for K^+ and in <12 h for NH_4^+. Silty soil sorbs 14–18 cmol(+) K^+ kg^{-1} and 1.8–2.6 cmol(+) NH_4^+ kg^{-1}, removing nearly 20% of the K^+ and NH_4^+ from a saturated salt–soil mixture in less than a day (Abasiyan and Towfighi, 2018). Also, the soil microbiome absorbs NH_4^+, NO_3^- , and K^+ because these ions are essential for vital functions like energy production, protein and nucleic acid biosynthesis, and intracellular

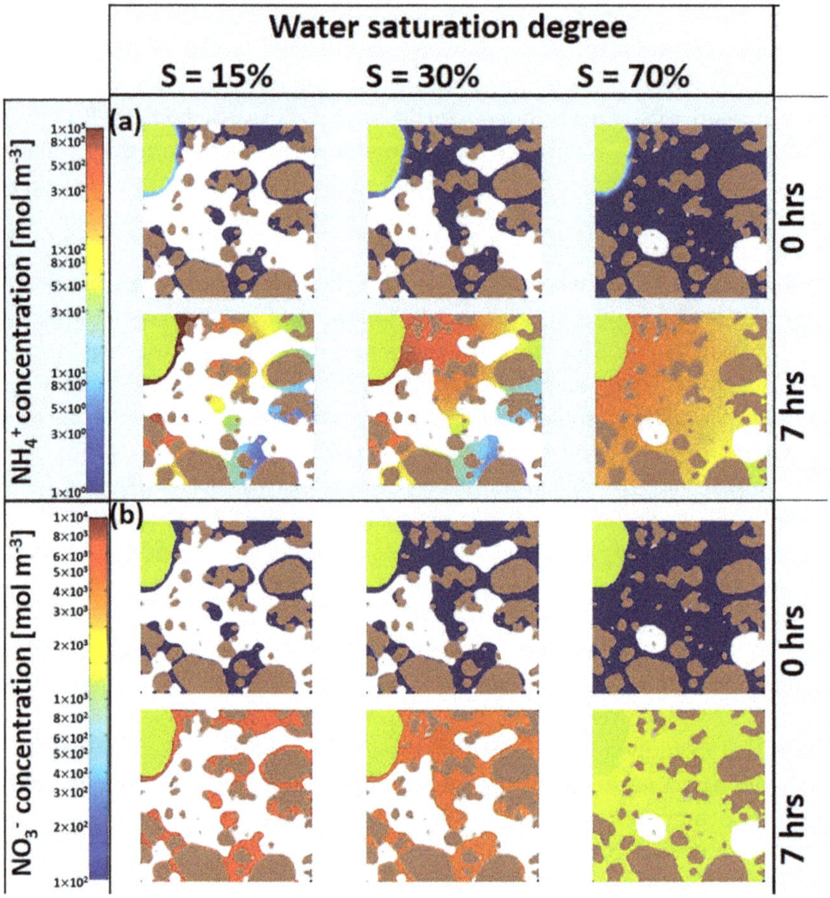

Figure 4 Diffusion of NH_4^+ from ammonium-based fertilizer (NH_4Cl) and NO_3^- from nitrate-based fertilizer (KNO_3) in the fertosphere from 0 h to 7 h. Electrolyte movement from the granule into soil pores containing 15%, 30%, and 70% water saturation is mapped for an area of 3.8 mm × 3.8 mm. Source: Adapted from Ruiz et al. (2020).

homeostasis. The soil microbiome consumes 7.4 ± 0.7 mg NH_4-N kg^{-1} day^{-1} and 2.0 ± 0.3 mg NO_3-N kg^{-1} day^{-1}, based on a global analysis of 274 studies that used ^{15}N stable isotopes to track soil N immobilization (Elrys et al., 2022). Microbial cells maintain a stoichiometry of about 47 times more carbon than K in their biomass, regardless of the fertilizer source and soil conditions (Schwalb et al., 2023). If microbial growth is stimulated in the fertosphere, some nutrient ions will be immobilized in microbial biomass. Together, the combined ion removal from the soil pore water due to surface exchange reactions and immobilization should buffer the osmotic stress in the fertosphere of N and K fertilizer granules.

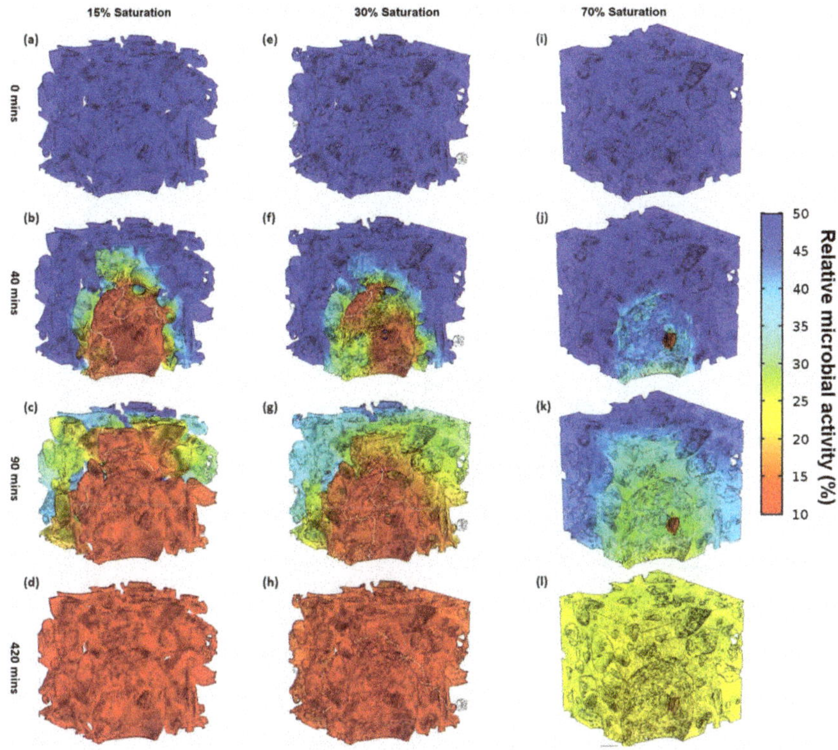

Figure 5 Microbial-inhibited zones, based on the relative microbial activity, in the soil pore space of the fertosphere develops with time (from 0 to 420 min), due to the diffusion of electrolytes from a dissolving NH_4NO_3 granule. The inhibited zone, in red, depends upon the soil water content. Soil pores contained 15%, 30%, and 70% water saturation. Source: Adapted from Ruiz et al. (2020).

The fertosphere around a P fertilizer granule is unlikely to cause salt stress in the soil microbiome for two reasons. First, the salt index of P salts is considerably lower than that of ammonium-based or K-based salts (Table 1). Second, phosphate ions react with other soil components, which lowers the ionic strength in the fertosphere. When monoammonium phosphate ($NH_4H_2PO_4$) and diammonium phosphate ($[NH_4]_2HPO_4$) are applied to soil, the phosphate ions ($H_2PO_4^-$, HPO_4^{2-}) tend to precipitate with reactive cations like Mg^{2+}, Ca^{2+}, Al^{3+}, and Fe^{3+} in soil water and adsorb to charged surfaces on clays, minerals, and organic matter (Meyer et al., 2023; Fig. 6). Adding monoammonium phosphate to moist agricultural soil ($\theta = 0.35$) resulted in $H_2PO_4^-$ diffusion of 10 to >45 mm from the center of the fertilizer granule after 7 days (Degryse and McLaughlin, 2014). Greater $H_2PO_4^-$ diffusion occurred in soils with low phosphate adsorption capacity, based on Freundlich *k* values, and those soils also had lower $CaCO_3$ concentration. The $H_2PO_4^-$ and HPO_4^{2-} concentration in

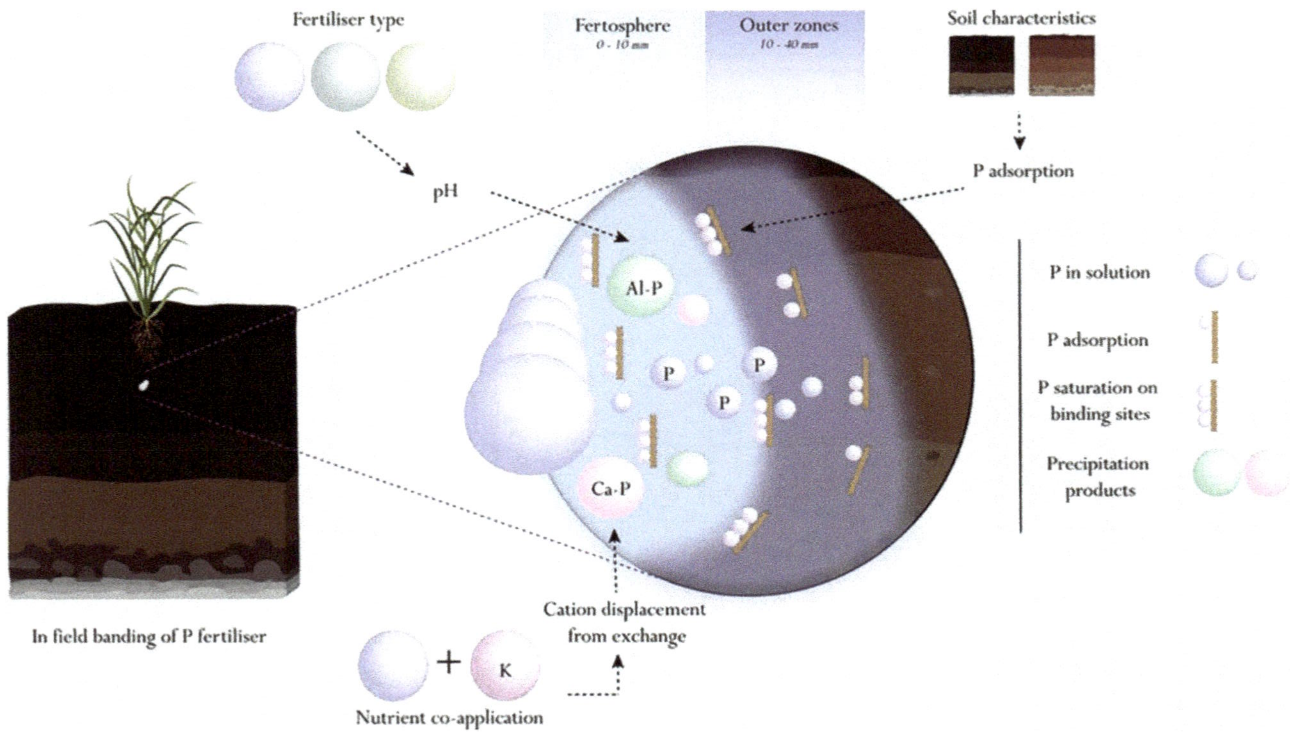

Figure 6 Inorganic fertilizers containing water-soluble salts of phosphorus (P) may be applied in a band beneath the crop row. In-field banding is a strategy to place P fertilizer where it is likely to be accessed by roots and their associated microbiome. However, dissolved phosphate ions ($H_2PO_4^-$, HPO_4^{2-}) released from the fertilizer granule react by adsorbing, binding, and precipitating with other soil components, which limits their diffusion beyond about 25 mm from the center of the fertosphere. Source: Adapted from Meyer et al. (2023).

the fertosphere declines with time, and only 41–73% remained in a labile P form in the fertosphere after 28 days (Degryse and McLaughlin, 2014).

In the fertosphere of a P fertilizer granule, the soil microbiome controls its uptake of phosphate ions. Absorption of $H_2PO_4^-$ and HPO_4^{2-} may occur passively through outer membrane proteins such as classic porins, which allow nonspecific diffusion of solutes with a molecular mass of ≤600 Da (Vergalli et al., 2020). However, it is more likely that bacteria, archaea and fungi uptake $H_2PO_4^-$ and HPO_4^{2-} (collectively referred to as inorganic phosphate, P_i) through active membrane transporters that use H^+/P_i symporters powered by H^+-ATPases and Na^+/P_i symporters driven by Na^+-ATPases. High P_i conditions activate the low-affinity PitH transporter system in bacteria, archaea, and fungi. A P_i-depleted microenvironment induces the high-affinity phosphate transporters in the phosphate-specific transport (Pst) family for bacteria and archaea, and the PHT1 (PHOSPHATE TRANSPORTER 1) family in fungi (Ezawa and Saito, 2018; Park et al., 2022). Since P fertilizer does not have an elevated salt index, and phosphate ion uptake through cell membranes is actively controlled, there is no physiological or metabolic reason for the soil microbiome to experience osmotic stress in the fertosphere of a P fertilizer granule.

3.2 Soil pH fluctuations in the fertosphere

The fertosphere is often more acidic than the surrounding soil, which might affect the activity and diversity of the soil microbiome. The affinity of NH_4^+, K^+, and other cations for negatively charged surfaces on organo-minerals, clays, and within clay interlattices displaces H+, resulting in H^+ accumulation in the soil pore water as a result of chemical exchange in the soil. Active transport of cations like NH_4^+ and K^+ into cells via transporter proteins results in H^+ release through electron pumps, which maintain the electrical neutrality of the cell membrane. Biosynthesis of amino acids, peptides, and nucleic acids from NH_4^+ releases 1 mole of H^+ for every mole of NH_4^+ consumed, shown in equation (1):

$$NH_4^+ + R - OH \rightarrow R - NH_2 + H^+ + H_2O \tag{1}$$

In addition, biological processing of NH_4^+ through the sequential ammonia oxidation and nitrification reactions releases 2 moles of H^+ for every mole of NH_4^+ consumed (equation (2)):

$$NH_4^+ + 2O_2 \rightarrow NO_3^- + H_2O + 2H^+ \tag{2}$$

If the nitrification product, NO_3^-, leaches out of the crop root zone, it is accompanied by base cations, such as Na^+, K^+, Mg^{2+}, and Ca^{2+}, leaving behind H^+ ions that add to the soil acidity.

The H^+ generated by soil chemical and biological reactions following the application of ammonium-based fertilizer is expressed as the acidity index or the calcium carbonate ($CaCO_3$) equivalent. This indicates the amount of agricultural lime required to neutralize acidity in the fertosphere, with 55–166 kg of $CaCO_3$ needed per 100 kg of fertilizer containing NH_4^+ (Table 1). Nitrate fertilizers like calcium nitrate, $Ca(NO_3)_2$, potassium nitrate, KNO_3 and sodium nitrate, and $NaNO_3$ are acid-neutralizing and consume 20–29 kg H^+ per 100 kg of fertilizer. Nitrate fertilizers add base cations that replace H^+ in the soil pore water and on exchange sites. Furthermore, NO_3^- uptake by microorganisms and crops and other biota consumes H^+ and produces OH^- (equation (3)):

$$NO_3^- + 8H^+ + 8e^- \rightarrow NH_2 + 2H_2O + OH^- \quad (3)$$

Inorganic fertilizers of P and K generally have a neutral effect on soil pH. However, the dissolution of ammonium phosphates has a variable effect on the pH around the granule. The localized zone of acidity (pH of 4.0–4.5) around a dissolving granule of monoammonium phosphate contrasts with the alkaline conditions (on average, pH 7.5–8.0) in the fertosphere of diammonium phosphate (IPNI, 2012). The temporary alkalinity around diammonium phosphate granules is due to H^+ consumption by the phosphate ion HPO_4^{2-}, but this dissipates within a few days as dissolved NH_4^+ undergoes biological ammonia oxidation and nitrification.

The activity and diversity of the soil microbiome may be controlled by pH more than any other environmental variable. A neutral pH of 7 is associated with the maximum metabolic activity and diversity of soil bacteria, the dominant microorganisms with the fastest growth rate in the soil microbiome. Bacterial survival and growth in acidic or alkaline conditions involves cellular adaptation (e.g. in the structure and function of bacterial proteins at the cell membrane) that affecta transport of molecules and energy transfer reactions, which are directly related to metabolic activity. Experimentally, the model bacterium *Pseudomonas fluorescens* produces the maximum membrane protein content at pH 7, and its membrane protein content correlates significantly to its metabolic rate from pH 5 to 9 (Luan et al., 2023). This is consistent with the pH-dependent response of microbial respiration, which increased linearly from pH 4.5 to 7.5 in soils collected from a semi-arid grassland (Chen et al., 2016). Therefore, acidification of the fertosphere could inhibit the functions of a diverse soil microbiome.

Soil acidification is commonly reported in agricultural soils that received repeated or long-term applications of ammonium-based inorganic fertilizer. Acidic soils have lower alpha diversity and functionality (i.e. biogeochemical processes) although it does not consistently change the beta diversity of the soil microbiome (Zhou et al., 2020). Still, soil microbial responses appear to be context-specific. For example, the negative impact of inorganic fertilizers on soil bacterial diversity in acidic soils from humid regions was not generalizable to alkaline soils from semi-arid and arid regions of China (Feng et al., 2022). Furthermore, specialized N-cycling bacteria like *Nitrosospira* and *Nitrososphaera* are more abundant in soil that receives inorganic fertilizers. In a global meta-analysis, the taxonomic diversity of prokaryotes (bacteria, archaea) and fungi was the same in NPK-fertilized and unfertilized control soils, while the functional diversity (i.e. metabolic capacity) of the soil microbiome was 2.8% greater, on average, in the NPK-fertilized than unfertilized control soils (Bebber and Richards, 2022).

These field-based results show an inconsistent response of the soil microbiome to pH changes that are attributed to the use of inorganic fertilizers. One possibility is that the soil microbiome is more resilient to the effects of soil acidification from inorganic fertilizers than predicted by the metabolic theory of microbial ecology of Luan et al. (2023). Fluctuations in soil pH due to rainfall, chemical buffering, and biological processes are seldom reported in relation to the soil microbiome. In addition, sampling methods may not be inadequate to distinguish the microbiome within the fertosphere from microorganisms in the surrounding bulk soil, based on the relatively small size of the fertosphere (e.g. radius of ~40 mm from the dissolving fertilizer granule) and the fact that water-soluble nutrients are chemically or biologically transformed in the soil in a matter of days after the fertilizer granule dissolves. To understand how the soil microbiome is responding to soil acidification from inorganic fertilizers, future studies should focus on the acidifying ammonium-based fertilizer sources, after developing techniques to study the microbial dynamics at a fine spatiotemporal scale (<100 mm of the fertilizer granule, within a few days of fertilizer application).

4 Indirect effects of inorganic fertilizer on the soil microbiome

Agronomists apply inorganic fertilizer to make crops grow better and produce higher yields of nutritious grain, tubers, oilseed, and other edible foods. Fertilized crops are taller, heavier, and have more non-edible biomass in their fibrous stems and leaves, with deeper, laterally branching roots. Banding inorganic fertilizer in the crop row, below the seed or several centimeters away from the seedlng, creates a nutrient-rich zone that stimulates root elongation in the soil profile. Roots sense water and soluble nutrients, and they use that information to initiate lateral root growth. Hydropatterning, the process whereby axial roots bend and grow toward water, produces more lateral roots in moist soil than in dry soil (Orosa-Puente et al., 2018). Lateral root development is controlled by auxin, a plant hormone that is transported through the vascular systems as a water and nutrient sensor. Basipetal auxin transport (i.e. from root tips to the shoot) initiates lateral root primordia, while acropetal auxin transported from the shoot to root tips controls the elongation of lateral roots (Casimiro et al., 2001). Practically, this allowed maize (*Zea mays* L.) roots to produce denser and longer lateral roots in a location with higher NO_3^- concentration (Yu et al., 2016). The long-distance signaling of other plant hormones likes cytokinins also modulates plant responses to N and P concentrations in the soil pore water, including negative feedback and regulation of the NH_4^+ and NO_3^- transporters in roots of N-fertilized *Arabidopsis* plants, and proliferation of lateral roots and expression of Pi-deprivation-inducible cells in P-limited *Arabidopsis* roots (Rubio et al., 2009). Since grains and other crops that receive inorganic fertilizers often have a more extensive root system and greater biomass than their unfertilized counterparts, this will have an indirect effect on the activity, diversity and structure of the soil microbiome.

4.1 Root-microbial interactions of crops receiving inorganic fertilizer

The soil microbiome has interacted with roots since the appearance of terrestrial plants on Earth, about 450 million years ago. The common symbiotic signaling pathway, a conserved molecular signaling pathway, is found in all terrestrial plants that support rhizobial, actinorhizal, and mycorrhizal fungal symbioses (Oldroyd, 2013). These symbiotic interactions are well documented. The rhizobial symbiosis occurs in nitrogen-fixing root nodules that provide about 30–70% of the nitrogen required by annual legumes like common beans (*Phaseolus vulgaris*), chickpea (*Cicer arietinum*), lentil (*Lens culinaris*), soybean (*Glycine max*), groundnut (*Arachis hypogaea*), and faba bean (*Vicia faba*; Herridge et al., 2008). Biological N_2 fixation by rhizobia supplies 23–176 kg N ha^{-1} $year^{-1}$ to these annual crops, with about 33 kg N ha^{-1} $year^{-1}$ in rice paddies with N_2-fixing cyanobacteria, approximately 25 kg N ha^{-1} $year^{-1}$ in sugarcane associated with endophytic diazotrophic bacteria, and <5 kg N ha^{-1} $year^{-1}$ in other annual crops that associate with free-living, non-nodulating N_2-fixing bacteria (Herridge et al., 2008). Symbiotic rhizobia are from the genera *Rhizobium* and *Bradyrhizobium*. Endophytic and free-living bacteria with the genetic and physiological capacity for N_2 fixation are distributed in many phyla, including the Actinobacteria, Bacteroidetes, Cyanobacteria, Chlorobi, Chloroflexi, Firmicutes, and Proteobacteria. Of these, Proteobacteria has the largest number of confirmed diazotrophs, which are members of the genera *Azohydromonas*, *Azospirillum*, *Azospira*, *Azoarcus*, *Azotobacter*, *Burkholderia*, *Herbaspirillum*, *Pelomonas*, *Pseudacidovorax*, and *Sphingomonas* (Solanki et al., 2020).

Application of inorganic N fertilizer can reduce the immediate growth benefits of rhizobial nodulation for legumes. From an energetic perspective, it is less costly for plants to absorb NH_4^+ and NO_3^- from soil pore water than for the legume host to supply its symbiotic rhizobia with a carbon source to power the nitrogenase-catalyzed reaction of N_2 to NH_3 (Regus et al., 2017). In the short term, some legumes might form fewer nodules when exposed to high NO_3^- concentrations, but this is not universally true. The native annual legume *Acmispon strigosus* commonly known as hairy locus is always highly nodulated, regardless of the amount of atmospheric N deposition or the soil NH_4^+ and NO_3^- concentration in its growing environment (Regus et al., 2017). This suggests that nodulation has other critical functions for legume survival and development. Evidence for this function comes from the nodules of *Lotus japonicus*, which are responsible for whole-plant S-partitioning and metabolism, including S-uptake, assimilation, and thiol biosynthesis (Kalloniati et al., 2015).

The abundance and diversity of non-nodulating diazotrophs could be impacted by inorganic fertilizers. For example, 8 years of inorganic N fertilizer applications in a wheat (*Triticum aestivum*) and sweet potato (*Ipomoea batatas*) rotation significantly decreased the abundance of the nifH gene that encodes for the nitrogenase enzyme, but this was related to a decline in the soil bioavailable P concentration rather than to a direct effect of the N fertilizer (Wang et al., 2022). The abundance and diversity of diazotrophs was the same with no fertilizer and 60 kg N ha^{-1}, but higher N fertilizer rates of 120 kg N ha^{-1} and 180 kg N ha^{-1} caused soil acidification, which was associated with fewer Proteobacteria and Cyanobacteria, and more Bacteroidetes and Firmicutes (Wang et al., 2022). While N fertilizers do not change the fundamental nature of the plant–diazotroph interaction, they can change the proportion of N_2 that is fixed by the bacterial partner. In flooded rice systems, associated diazotrophs captured about 33 kg N ha^{-1} when no inorganic fertilizer was added, and this declined by more than half, to 13 kg N ha^{-1} when 140 kg N ha^{-1} of ammonium sulfate fertilizer was applied (Carreres et al., 1996). Thus, crops benefit from symbiotic and associative diazotrophs for biological N_2 fixation and other nutritional benefits, which can reduce reliance on inorganic N fertilizer in cropping systems, as part of a balanced and rational fertilizer plan.

Mycorrhizal fungi are symbiotic endophytes that grow on the extraradial or interradial surfaces of most vascular plants. The smaller size of fungal hyphae (4–6 μm on average), relative to plant roots (i.e. the smallest root hairs are 15–17 μm), permits the exploration of a greater volume of micropores. As a result, mycorrhizal fungi are responsible for water and nutrient transfer from the tiniest micropores toward the root surface. These transfers may not only occur through active processes such as cell-to-cell symplastic transport but also occurs passively through streaming (i.e. water and dissolved electrolytes flow along the surface of the fungal hyphae, outside the hyphal cell membrane, toward the root because of cohesion and transpiration; Kakouridis et al., 2022). There is evidence that high bioavailable P concentration in soil associated with inorganic P fertilizer use could reduce the colonization and functions of mycorrhizal fungi, based on greenhouse studies. However, this result is not borne out by field experiments that show no change, a decline or an increase in crop root colonization by arbuscular mycorrhizal fungi (Hagh et al., 2016; de Figueiredo et al., 2019; Li et al., 2019). Still, it is difficult to identify a genetic or physiological reason for the limited growth of mycorrhizal fungi in the presence of inorganic P fertilizer. Agricultural soils have a large, diverse population of indigenous arbuscular mycorrhizal fungi that respond to signals from crops and colonize their roots. In fact, mycorrhizal symbiosis should be advantageous to field-grown maize that is constrained by a low soil N supply and no P limitation, as demonstrated by larger maize plants having more mycorrhizal colonization in soil with elevated bioavailable P concentration (Sobat and Whalen, 2022). It still remains to be determined how much of the plant nutrition is acquired actively or passively by the mycorrhiza in inorganically fertilized soil.

4.2 Microbiome and substrates in the rhizosphere

Within the rhizosphere, the soil surrounding and adhering to the roots, are numerous microenvironments that are attractive to the diverse, free-living microorganisms of the soil microbiome. These microenvironments include the rhizosheath, comprising the polysaccharide-rich mucilage secreted from border cells in the plant root tip, and the rhizoplane. The rhizoplane is a root-impacted zone that accumulates secretions, rhizodeposits, and cellular debris, including fine root hairs. These microenvironments make the rhizosphere attractive to soil microorganisms, which then associate symbiotically, become endophytes, or colonize the rhizosphere. These root–microbial associations can be referred to as a rhizobiome, although it contains microbial species that also appear in the soil microbiome, some of which are more abundant around the roots than in the bulk soil.

Roots are a favorable growing environment for the soil microbiome because they provide substrates for microbial metabolism. Photosynthates sent to the root system are used to build plant structural components, for cellular respiration, and released into the rhizosphere as secretions and rhizodeposits. The readily metabolizable substrates stimulate the growth of the soil microbiome, which together with roots respired about 27% of the photosynthates allocated to the roots of spring rye (*Secale cereale*) during early vegetative growth, which gradually declines to about 5–8% of photosynthates respired as the plant matures (i.e. during stem elongation and ear emergence stages; Remus et al., 2016). Well-fertilized crops typically have higher photosynthesis rates, meaning more energy and carbon to produce tissues and byproducts, including rhizodeposits. While some secretions are bioactive secondary compounds, actively exuded from roots through energy-dependent active transporters, the vast majority are exudates – mainly sugars, amino acids, and organic acids – passively lost through the meristematic root cells (Sasse et al., 2018). These compounds serve to attract microorganisms to the rhizosphere.

Our emerging understanding of root–microbial interactions indicates an intricate feedback, whereby plants and microbes are both modulating the concentration gradients that exist in the soil pore water around the root. The exudation and diffusion of plant-derived compounds serve as a signal that modifies root growth, by changing the source/sink processes as well as regulating of efflux carrier (Canarini et al., 2019). On the other hand, the soil microbiome is a strong sink for the plant carbon, which increases the concentration gradient of metabolites, as illustrated for N and P in Fig. 7. How the soil microbiome mediates the N concentration and transfer of N from soil to roots still needs to be understood mechanistically.

Adding inorganic N fertilizer creates more complexity in root–microbial interactions, in part due to the local acidification associated with ammonium-based fertilizers. In the root zone of maize, the soil bacterial communities were affected significantly by the addition of NH_4^+ or NO_3^- in the root zone (<50 mm from the root). Maize roots had a larger population of acidophilic bacteria (e.g. Acidibacter, Burkholderia, and Catenulispora) when NH_4^+ fertilizer was added, while the NO_3^- fertilizer was favorable to bacteria that prefer higher pH (e.g. Sphingomonas, Sphingobium, Azospirillum, and Novosphingobium; Zhang et al., 2021). Roots together with microorganisms are expected to acquire a larger share of the P needed for their growth, thereby overcoming P limitation. For example, wheat seedlings grew within 10 mm of granules of water-soluble triple superphosphate, $Ca(H_2PO_4)_2 \cdot H_2O$ and water-insoluble struvite, $MgNH_4PO_4 \cdot 6\ H_2O$, as illustrated with micro X-ray computed tomography (Ahmed et al., 2016). The attraction of roots to the fertilizer granules was not limited by the low solubility of struvite; although struvite has a solubility of 0.018 g 100 mL^{-1} in water, it readily dissolves in dilute acid solutions at a rate of 0.033 g 100 mL^{-1} in 0.001 M HCl and 0.18 g 100 mL^{-1} in 0.01 M HCl at 25°C (Le Corre et al., 2009). Solubilization of struvite could be achieved by root-produced tricarboxylic acids (citric and tartaric acids), dicarboxylic acids (maleic and fumaric acids), or monocarboxylic acids (oxalic and acetic acids), or by organic acids and protons released by root-compatible microorganisms. Phosphate-solubilizing bacteria in the genera *Burkholderia*, *Cronobacter*, *Pseudomonas*, and others release glycolic, oxalic, malonic, and succinic acids, with additional inputs of oxalic, maleic, succinic, citric, and fumaric acids from rhizosphere-compatible fungi such as *Alternaria*, *Aspergillus*, *Fusarium*, and others (Whalen and Gul, 2023). Consequently, the presence of inorganic N and P fertilizers is expected to have a stimulatory effect on the soil microbiome in the rhizosphere, for the mutual benefit of the plant and their rhizobiome.

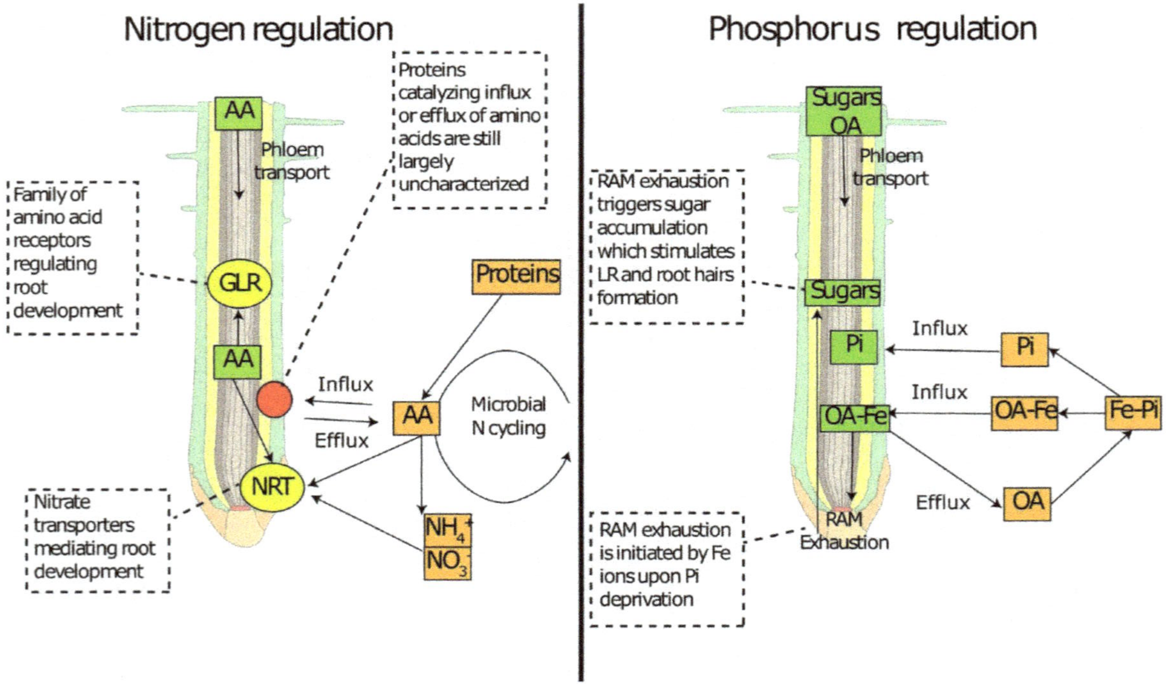

Figure 7 Involvement of common metabolites in nutrient sensing at the root tip. Schematic representation of how common primary metabolites allows nutrient sensing at the root tip for nitrogen (left panel) and phosphorus (right panel). Amino acid concentration can be sensed by specific receptors (GLR) and can interfere with N transporter expression (NRT). Amino acid concentration inside the root tip depends on the delivery from the phloem and on root exudation, the latter of which depends on concentration gradients mediated by soil micro-organisms. Organic acids can deliver Fe to the root tip, which activates the RAM exhaustion and temporarily increases sugar concentrations. Sugars can be sensed at the root tip and activate changes in root architecture. AA, amino acids; OA, organic acids; Pi, inorganic phosphorus; RAM,root apical meristem; NRT, nitrate transporters; GLR, glutamate-like receptors. Source: Adapted from Canarini et al. (2019).

5 Conclusion and future trends

The soil microbiome is abundant, diverse, and resilient to the effects of inorganic fertilizers used in agriculture. The direct effects of fertilizer salts on the soil microbiome are largely confined to the fertosphere. The osmolarity of the nutrient-rich zone generated around a dissolving fertilizer granule, which may also be more acidic or alkaline than the surrounding soil pore water, could be stressful for the soil microbiome. However, prokaryotes and fungi have adapted strategies to tolerate osmotic and pH stresses, which they activate rapidly. The stressful effects of inorganic fertilizer are more pronounced within a few millimeters of the granule and dissipate with increasing distance and time as the salts and protons are diluted by the soil water and react with the soil matrix. The immobilization of fertilizer nutrients in microbial cells and nutrient absorption by root cells also controls the electrolyte concentration of soil pore water. Agronomic practices of co-applying, mixing, and coating soil bacteria on fertilizer granules demonstrate the essential compatibility of the soil microbiome with inorganic fertilizer. For example, the plant growth-promoting bacteria *Bacillus* sp. strain KAP6 combined with compost and kaolin clay formed a sticky mixture that adhered to the outer surface of diammonium phosphate and urea granules (Ahmad et al., 2017). When the bacterial-coated fertilizer was applied to a calcareous soil, wheat plants were taller, had a greater photosynthesis rate and produced more grain, with 20–25% more nitrogen and up to 17% more phosphorus in the grain of plants exposed to bacterial-coated fertilizer than unmodified fertilizer (Ahmad et al., 2017). The ability of diverse bacteria and fungi to survive in close contact with inorganic fertilizer, in the bulk soil and particularly in the rhizosphere, suggests that the soil microbiome is well-adapted to thrive in the fertosphere. Further work is needed to understand how the soil microbiome mediates the transfer of nutrients from the fertosphere to crops through regulated pathways that control root exudation, the activity of N, P and K transporters, and root system architecture.

6 Where to look for further information

Further reading:

- Four new reports highlight the importance of the microbiome for food safety, soils and nutrition. Food and Agriculture Organization of the United Nations. Available from: https://www.fao.org/newsroom/detail/four-new-reports-highlight-importance-of-the-microbiome-for-food-safety--soils-and-nutrition/en.
- Management impacts on the soil microbiome. Soil Health Institute. Available from: https://soilhealthinstitute.org/our-work/initiatives/management-impacts-on-the-soil-microbiome/?highlight=microbiome.

- Soil biodiversity. Global Soil Partnership. Available from: https://www.fao.org/global-soil-partnership/areas-of-work/soil-biodiversity/en/.

Key journals/conferences:

- Soil Biology and Biochemistry.
- The ISME Journal: *Multidisciplinary Journal of Microbial Ecology*.
- Microbiome Virtual International Forum (MVIF). Available at: https://www.microbiome-vif.org/en-US/-/info/the-event.

7 References

Abasiyan, S. M. A. and Towfighi, H. (2018), Kinetics of competitive fixation of potassium and ammonium ions by silt component of soils from different agro-climatic regions, *Comm. Soil Sci. Plant Anal.* 49: 675–688. doi: 10.1080/00103624.2018.1432635

Ahmed, S., et al. (2016), Imaging the interaction of roots and phosphate fertiliser granules using 4D X-ray tomography, *Plant Soil* 401: 125–134. doi: 10.1007/s11104-015-2425-5

Ahmad, S., et al. (2017), Bacterial impregnation of mineral fertilizers improves yield and nutrient use efficiency of wheat, *J. Sci. Food Agric.* 97: 3685–3690. doi: 10.1002/jsfa.8228

Ahmed, V., et al. (2018), Metagenomic profiling of soil microbes to mine salt stress tolerance genes, *Front. Microbiol.* 9: 159. doi: 10.3389/fmicb.2018.00159

Araújo, C. A. S., et al. (2020), Osmotolerance as a determinant of microbial ecology: a study of phylogenetically diverse fungi, *Fungal Biol.* 124: 273-288. doi: 10.1016/j.funbio.2019.09.001

Basu, S. K., et al. (2010), Modeling NPK release from spherically coated fertilizer granules, *Simul. Model. Pract. Theory.* 18: 820–835. doi: 10.1016/j.simpat.2010.01.018

Bebber, D. P. and Richards, V. R. (2022), A meta-analysis of the effect of organic and mineral fertilizers on soil microbial diversity, *Appl. Soil Ecol.* 175: 104450. doi: 10.1016/j.apsoil.2022.104450

Călinescu, O., et al. (2014), Species differences in bacterial NhaA Na^+/H^+ exchangers, *FEBS Lett.* 588, 3111–3116. doi: 10.1016/j.febslet.2014.05.066

Canarini, A., et al. (2019), Root exudation of primary metabolites: Mechanisms and their roles in plant responses to environmental stimuli, *Front. Plant Sci.* 10: 157. doi: 10.3389/fpls.2019.00157

Carreres, R., et al. (1996), Effect of nitrogen rates on rice growth and biological nitrogen fixation, *J. Agric. Sci.* 127: 295–302. doi:10.1017/S002185960007845X

Casimiro, I., et al. (2001), Auxin transport promotes Arabidopsis lateral root initiation, *Plant Cell* 13: 843–852. doi: 10.1105/tpc.13.4.843

Chen, D., et al. (2016), Soil acidification exerts a greater control on soil respiration than soil nitrogen availability in grasslands subjected to long-term nitrogen enrichment, *Functional Ecol.* 30: 658–669. doi: 10.1111/1365-2435.1252

de Figueiredo, C. C., et al. (2019), Sewage sludge biochar alters root colonization of mycorrhizal fungi in a soil cultivated with corn, *Eur. J. Soil Biol.* 93: 103092. doi: 10.1016/j.ejsobi.2019.103092

Degryse, F. and McLaughlin, M. J. (2014), Phosphorus diffusion from fertilizer: Visualization, chemical measurements, and modeling, *Soil Sci. Soc. Am. J.* 78: 832–842. doi: 10.2136/sssaj2013.07.0293

Elrys, A. S., et al. (2022), Global patterns of soil gross immobilization of ammonium and nitrate in terrestrial ecosystems, *Global Change Biol.* 28, 4472–4488. https://doi.org/10.1111/gcb.16202

Ezawa, T. and Saito, K. (2018), How do arbuscular mycorrhizal fungi handle phosphate? New insight into fine-tuning of phosphate metabolism, *New Phytol.* 220: 1116–1121. doi: 10.1111/nph.15187

Feng, Y. Z., et al. (2022), Responses of soil bacterial diversity to fertilization are driven by local environmental context across China, *Engineering* 12: 164–170. doi: 10.1016/j.eng.2021.09.012

Hagh, E. D., et al. (2016), Optimizing phosphorus use in sustainable maize cropping via mycorrhizal inoculation, *J. Plant Nutr.* 39: 1348–1356. doi: 10.1080/01904167.2015.1086797

Herridge, D. F., et al. (2008), Global inputs of biological nitrogen fixation in agricultural systems, *Plant Soil* 311: 1–18. doi: 10.1007/s11104-008-9668-3

Heuck, C., et al. (2018), Corrigendum to Heuck et al. (2015), Soil microbial biomass C:N:P stoichiometry and microbial use of organic phosphorus, *Soil Biol. Biochem.* 127: 329–330. doi: 10.1016/j.soilbio.2018.09.011

International Plant Nutrition Institute. (2012), *4R Plant Nutrition: A Manual for Improving the Management of Plant Nutrition*, International Plant Nutrition Institute, Norcross, GA. p. 130.

Jentsch, T. J. (2016), VRACs and other ion channels and transporters in the regulation of cell volume and beyond, *Nat. Rev. Mol. Cell Biol.* 17: 293–307. doi: 10.1038/nrm.2016.29

Jones, J. B., Jr. (2003), *Agronomic Handbook: Management of Crops, Soils and Their Fertility*, CRC Press, Boca Raton, FL.

Kakouridis, A., et al. (2022), Routes to roots: direct evidence of water transport by arbuscular mycorrhizal fungi to host plants. *New Phytol.* 236: 210–221. doi: 10.1111/nph.18281

Kalloniati, C., et al. (2015), Nitrogen-fixing nodules are an important source of reduced sulfur, which triggers global changes in sulfur metabolism in *Lotus japonicus*. *Plant Cell* 27: 2384–2400. doi: 10.1105/tpc.15.00108

Kogej, T., et al. (2007), Osmotic adaptation of the halophilic fungus *Hortaea werneckii*: role of osmolytes and melanization, *Microbiology* 153, 4261–4273. doi: 10.1099/mic.0.2007/010751-0

Leandro, M., et al. (2021), Comparative proteomics reveals essential mechanisms for osmotolerance in *Gluconacetobacter diazotrophicus*, *Res. Microbiol.* 172: 103785. doi: 10.1016/j.resmic.2020.09.005

Le Corre, K. S., et al. (2009), Phosphorus recovery from wastewater by struvite crystallization: a review, *Crit. Rev. Environ. Sci. Technol.* 39: 433–477. doi: 10.1080/10643380701640573

Li, Z., et al. (2019), Effects of nitrogen feeding for extraradical mycelium of *Rhizophagus irregularis* maize symbiosis incorporated with phosphorus availability, *J. Plant Nutr. Soil Sci.* 182: 647–655. doi: 10.1002/jpln.201800674

Luan, L., et al. (2023), Integrating pH into the metabolic theory of ecology to predict bacterial diversity in soil, *Proc. Nat. Acad. Sci. USA* 120: e2207832120. doi: 10.1073/pnas.2207832120

Majaron, V. F., et al. (2022), Bioactive material with microorganisms can enhance the micronutrients solubilization and sulfate availability from low reactive sources: insight for application as coating fertilizer granules, *J. Polym. Environ.* 30: 2602–2613. doi: 10.1007/s10924-022-02380-x

Martínez-Hilario, D. G., et al. (2023), Modeling NPK nutrients release via nanoparticle/hydrogel system CONTENIDO, *Rev. Mex. Ing. Quim.* 22: Sim2399. doi: 10.24275/rmiq/Sim2399

Meyer, G., et al. (2023), Mobility and lability of phosphorus from highly concentrated fertiliser bands, *Geoderma* 429: 116248. doi: 10.1016/j.geoderma.2022.116248

Morbach, S. and Krämer, R. (2006), Structure and function of the betaine uptake system BetP of *Corynebacterium glutamicum*: Strategies to sense osmotic and chill stress, *J. Mol. Microbiol. Biotechnol.* 10: 143–153. doi: 10.1159/000091561

Oldroyd, G. E. (2013), Speak friend and enter: signaling systems that promote beneficial symbiotic associations in plants, *Nat. Rev. Microbiol.* 11: 252–263. doi: 10.1038/nrmicro2990

Orosa-Puente, B., et al. (2018), Root branching toward water involves posttranslational modification of transcription factor ARF7, *Science* 362: 1407–1410. doi: 10.1126/science.aau3956

Park, Y., et al. (2022), Strategies of organic phosphorus recycling by soil bacteria: acquisition, metabolism, and regulation, *Environ. Microbiol. Rep.* 14: 3–24. doi: 10.1111/1758-2229.13040

Phong, P. H., et al. (2021), Simulating and monitoring the temporal and spatial transfer of NPK fertilizer in agricultural soils using a mathematical model and multi-channel electrical conductivity measurement, *J. Soil Sci. Plant Nutr.* 21: 374–388. doi: 10.1007/s42729-020-00367-y

Regus, J. U., et al. (2017), Nitrogen deposition decreases the benefits of symbiosis in a native legume, *Plant Soil* 414: 159–170. doi: 10.1007/s11104-016-3114-8

Remus, R., et al. (2016), Determining the timepoint when ^{14}C tracer accurately reflect photosynthate use in the plant-soil system, *Plant Soil* 408: 457–474. doi: 10.1007/s11104-016-3002-2

Rubio, V., et al. (2009), Plant hormones and nutrient signalling, *Plant Mol. Biol.* 69: 361–373. doi: 10.1007/s11103-008-9380-y

Ruiz, S. A., et al. (2020), Image-based quantification of soil microbial dead zones induced by nitrogen fertilization, *Sci. Total Environ.* 727: 138197. doi: 10.1016/j.scitotenv.2020.138197

Sasse, J., et al. (2018), Feed your friends: do plant exudates shape the root microbiome? *Trends Plant Sci.* 23: 25–41. doi: 10.1016/j.tplants.2017.09.003

Sauer, K., et al. (2022), The biofilm life cycle: expanding the conceptual model of biofilm formation. *Nat. Rev. Microbiol* 20: 608–620. doi: 10.1038/s41579-022-00767-0

Schwalb, S. A., et al. (2023), Long-term differences in fertilisation type change the bacteria:archaea:fungi ratios and reveal a heterogeneous response of the soil microbial ionome in a Haplic Luvisol, *Soil Biol. Biochem.* 177: 108892. doi: 10.1016/j.soilbio.2022.108892

Sigtryggsson, C., et al. (2020), Laboratory studies on dissolution of nitrogen fertilizers by humidity and precipitation, *Agric. Environ. Lett.* 5: e20016. doi: 10.1002/ael2.20016

Sobat, E. and Whalen, J. K. (2022), Mycorrhizal colonization associated with roots of field-grown maize does not decline with increasing plant-available phosphorus, *Soil Use Manage.* 38: 1370–1379. doi: 10.1111/sum.12786

Solanki, M. K., et al. (2020), Assessment of diazotrophic Proteobacteria in sugarcane rhizosphere when intercropped with legumes (peanut and soybean) in the field, *Front. Microbiol.* 11: 1814. doi: 10.3389/fmicb.2020.01814

Vergalli, J., et al. (2020), Porins and small-molecule translocation across the outer membrane of Gram-negative bacteria, *Nat. Rev. Microbiol.* 18: 164–176. doi: 10.1038/s41579-019-0294-2

Wang, L., et al. (2022), Long-term fertilization with high nitrogen rates decreased diversity and stability of diazotroph communities in soils of sweet potato, *Appl. Soil Ecol.* 170: 104266. doi: 10.1016/j.apsoil.2021.104266

Whalen, J. K. and Gul, S. (2023), Root interactions with the microbiome from the rhizoplane to the bulk soil: an overview, in: Goss, M. and Oliver, M. (eds) *Encyclopedia of Soils in the Environment*. 2nd edn. Vol. 1, pp. 357–368. Elsevier. doi: 10.1016/B978-0-12-822974-3.00243-3

Yu, P., et al. (2016), Root type-specific reprogramming of maize pericycle transcriptomes by local high nitrate results in disparate lateral root branching patterns, *Plant Physiol.* 170: 1783–1798. doi: 10.1104/pp.15.01885

Zhang, H. Q., et al. (2021), Changes in soil bacterial communities with increasing distance from maize roots affected by ammonium and nitrate additions, *Geoderma* 398: 115102. doi: 10.1016/j.geoderma.2021.115102

Zhou, Z. H., et al. (2020), Meta-analysis of the impacts of global change factors on soil microbial diversity and functionality, *Nat. Commun.* 11: 3072. doi: 10.1038/s41467-020-16881-7

Chapter 15

Understanding how land-use management affects soil microbiomes

Lucas William Mendes, Thierry Alexandre Pellegrinetti and Alexandre Pedrinho, Center for Nuclear Energy in Agriculture, University of São Paulo, Brazil; and Dennis Goss-Souza, Federal Institute of Paraná, Brazil

1 Introduction

Soil is a crucial biological ecosystem, serving as a habitat for billions of microorganisms and supporting rich biodiversity (Bach et al., 2020). The soil microbiome, a complex network of microorganisms inhabiting the soil, plays a pivotal role in sustaining the health and functionality of terrestrial ecosystems. This microbiome encompasses bacteria, fungi, archaea, protozoa, and viruses, collectively contributing to crucial ecosystem services. They participate in nutrient cycling, decomposition of organic matter, soil formation, and even the regulation of greenhouse gas (GHG) emissions (Mendes et al., 2017). The diversity and functions of these microbial communities are intricately linked to soil health and fertility. Land-use change, such as deforestation, urbanization, agricultural intensification, and other human activities, significantly impacts the soil microbiome, leading to alterations in its composition and functionality (Averill et al., 2022).

For instance, deforestation disrupts the delicate balance of microbial communities by altering environmental conditions and reducing organic matter inputs. Likewise, intensive agricultural practices, which involve pesticides, fertilizers, and tillage methods, can significantly impact the diversity and functions

http://dx.doi.org/10.19103/AS.2024.0136.33

of soil microbes. These alterations in the soil microbiome trigger cascading effects on ecosystem functions, agricultural productivity, and even global carbon cycling. Furthermore, disruptions in the soil microbiome resulting from land-use changes can lead to unforeseen consequences, such as increased soil erosion, reduced fertility, and decreased resilience to environmental stressors like droughts or extreme weather events. Understanding these intricate relationships is crucial as they directly influence soil fertility, plant health, and ultimately, the sustainability of ecosystems and human livelihoods.

The Global Land Outlook report (United Nations, 2017) highlights that approximately 25% of the world's land is currently degraded. This degradation often leads to biodiversity loss, soil erosion, and diminished agricultural productivity (Cardinale et al., 2012), affecting the soil's natural productivity and its capacity to support both human and natural systems (Lal, 2020). Hence, studying the soil microbiome in the context of land-use changes becomes imperative for developing sustainable land management practices. This approach aims to preserve soil health, enhance agricultural productivity, mitigate climate change impacts, and safeguard biodiversity in terrestrial ecosystems. This chapter delves into examples illustrating how land-use changes affect soil microbiome composition and functions, focusing on various biomes in Brazil.

2 Effects of land-use change on soil microbiome composition

Land-use changes have profound implications for soil microbial communities, altering their composition, diversity, and functionality. When natural habitats are converted for agricultural, urban, or industrial purposes, these alterations can disrupt the delicate balance within soil ecosystems. Intensive land-use practices like deforestation or agricultural expansion often lead to decreased microbial diversity and altered community structures. For instance, the application of pesticides and chemical fertilizers in agriculture can result in the depletion of certain microbial species while favoring the proliferation of others, thus impacting the overall microbial balance. Additionally, the removal of vegetation cover through land conversion diminishes organic matter inputs to the soil, affecting microbial food sources and disrupting their ecological functions, such as nutrient cycling and decomposition processes.

Furthermore, land-use changes can induce shifts in soil microbial activity and functionality. For example, deforestation along with agricultural practices can lead to increased soil compaction, altering microbial habitats and reducing their ability to carry out essential functions. Changes in land use may also modify soil pH, moisture levels, and nutrient availability, directly influencing the metabolic activities and survival of various microbial communities. Such alterations can affect the breakdown of organic matter and nutrient cycling rates, ultimately impacting the overall soil fertility. Understanding these intricate relationships between land-use changes and soil microbial communities is crucial for devising sustainable land management strategies that aim to preserve soil health and ecosystem functionality. Figure 1 illustrates the results discussed in this section.

2.1 Effects of land-use change on soil microbiome in the Amazon region

The Amazon rainforest holds 40% of the remaining tropical rainforests, serving as a crucial force in conserving biodiversity, regulating biogeochemical cycles, and managing the climate. Considered the planet's largest repository of plant and animal diversity, the Amazon also shelters a myriad of microorganisms crucial to its ecological balance (Rodrigues et al., 2013). However, despite its immense global significance, the Amazon faces an ongoing threat due to agricultural expansion, marking it as the most active frontier of deforestation worldwide. Deforestation in the Brazilian Amazon involves the clearing and transformation of native forests into alternative land uses, predominantly for agriculture and cattle ranching. The rapid expansion of agricultural activities has become the primary force disrupting the Amazon region, exerting profound consequences on its intricate network of soil microorganisms (Mendes et al., 2015a). Consequently, there has been a recent surge in attention directed toward exploring the belowground microbial communities thriving in Amazonian soils.

The pioneering exploration of microbial diversity in Amazon soils traces back to Borneman and Triplett (1997), who utilized the Sanger sequencing technique to examine sequences extracted from both native forest and pasture areas. Their initial findings unveiled an expansive microbial panorama, underscoring stark disparities in composition between forest and pasture ecosystems. *Clostridia*-affiliated sequences were prevalent in forest samples, while *Bacillus* dominated the pasture environments. Subsequent studies in the Brazilian Amazon have significantly enriched our understanding of how alterations in land use affect soil microbial communities (Jesus et al., 2009; Grossman et al., 2010; Pazinato et al., 2010; Navarrete et al., 2011, 2015; Germano et al., 2012; Rodrigues et al., 2013; Taketani et al., 2013; Brossi et al., 2014; Mendes et al., 2014). These investigations underscore the remarkably diverse microbial communities thriving in Amazonian soils and illuminate how land-use changes profoundly alter the structure, diversity, and composition of these microbial assemblages.

Recent studies aimed at understanding the impact of land-use changes on Amazon soils utilized a soil DNA shotgun metagenomics approach to examine

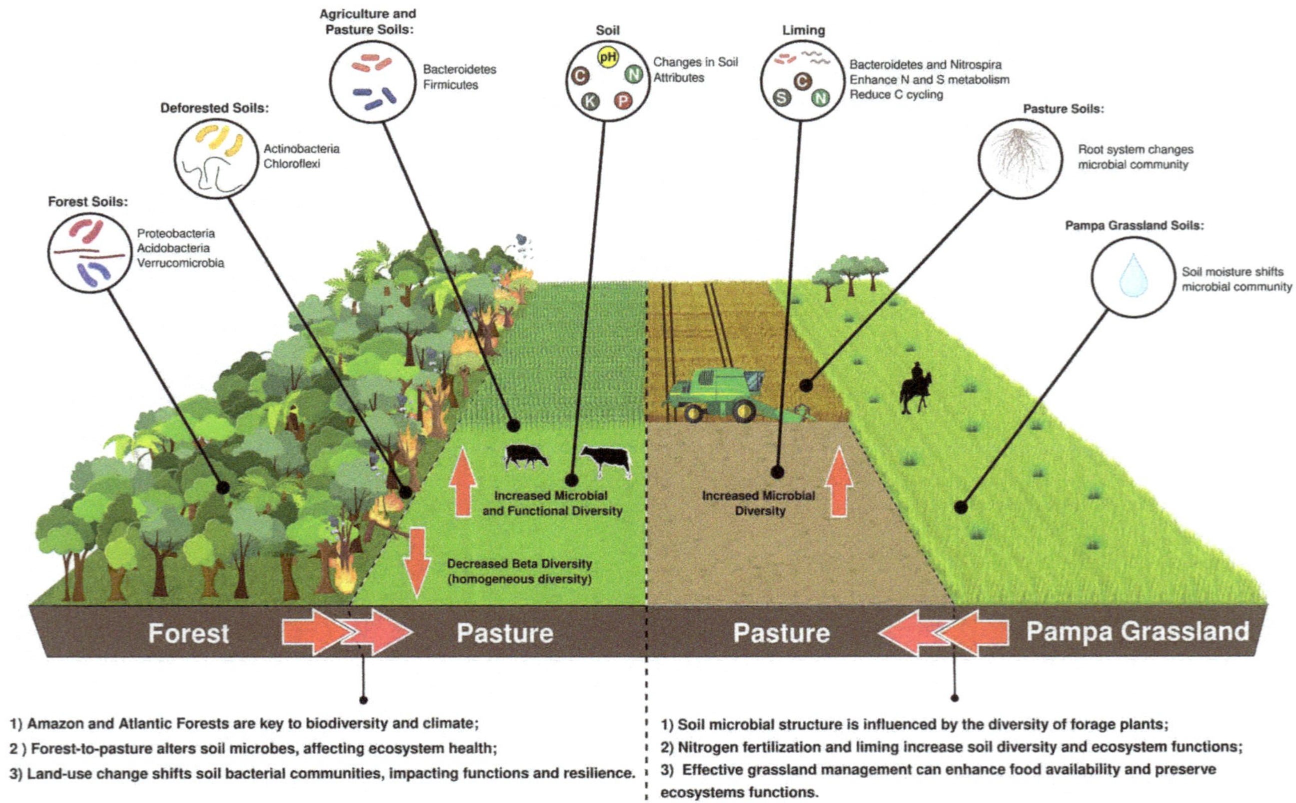

Figure 1 Effects of land-use change on the soil microbiome. The figure illustrates how land-use change affects soil chemical attributes and subsequently influences the soil microbiome.

microbes in various zones: native forest, deforested land converted into soybean fields, and pasture (Mendes et al., 2015a; Pedrinho et al., 2019). These investigations revealed that different land-use systems significantly influenced the types of bacteria present in the soil. Factors such as pH levels and the presence of specific elements, such as carbon, nitrogen, nitrate, and potassium, were closely associated with the variations in bacterial communities. For instance, Proteobacteria, Acidobacteria, and Verrucomicrobia were dominant in forest soils. Conversely, deforested regions showed a higher abundance of Chloroflexi and Actinobacteria, while agriculture and pasture displayed an increased abundance of Bacteroidetes and Firmicutes, respectively. These distinct bacterial populations were linked to their specialized roles within the environment. Proteobacteria, abundant in forest soils, play a crucial role in the global cycling of carbon, nitrogen, and sulfur. Acidobacteria influence the acidity and nutrient properties of Amazon soils, while Verrucomicrobia thrive in forests due to their ability to break down robust carbon compounds found in moist forest soils. Changes in bacterial composition post-deforestation were attributed to heightened soil temperatures and carbon deposition from burning. Consequently, Chloroflexi, thriving in higher temperatures, and Actinobacteria, known for decomposing organic matter, increased in these areas. Bacteroidetes, associated with aiding plant growth and cellulose decomposition, were more abundant in soybean fields. Firmicutes, known for their resilience in harsh conditions, dominated pasture soils, which are typically more challenging environments.

In summary, diverse bacterial communities thrive in different environments based on their unique abilities to adapt to specific conditions and fulfill distinct ecological functions. Thus, the transformation of land use plays a pivotal role in shaping soil microbial communities. Alterations in land use, whether transitioning from native forests to agricultural fields or converting landscapes for different purposes like pasture or cultivation, exert a significant influence on the composition and diversity of soil microbes. The shift from one land use to another triggers a cascade of modifications in soil conditions, leading to specific adaptations among microbial communities to survive and thrive in these altered environments. Consequently, the way land is utilized stands as a crucial determinant in structuring and molding the intricate and diverse microbial communities found within soil ecosystems.

2.2 Effects of land-use change on soil microbiome in the Atlantic Forest

The Brazilian Atlantic Forest ranks as one of the world's richest biodiversity hotspots, hosting 2.7% and 2.1% of global endemic plant and vertebrate species, respectively (Myers et al., 2000). Regrettably, in recent decades, we have witnessed extensive fragmentation and deforestation, resulting in the depletion of the original vegetation to a mere 11.7% (Ribeiro et al., 2009). The conversion of these forests into croplands and pasturelands, constituting 20% and 42% of total human net primary production appropriation, significantly contributes to this loss (Weinzettel et al., 2018). Projections for 2100 suggest a potential 58% decrease in natural vegetative cover across 34 global biodiversity hotspots, with up to 1/3 of habitat loss and 16% of species loss attributed to forest conversion alone (Jantz et al., 2015).

Soil microorganisms, like plants and animals, exhibit significant responses to changes in land use (Lauber et al., 2013; Kaiser et al., 2016; Goss-Souza et al., 2017; Ceola et al., 2021). Studies in Brazilian tropics indicate that converting forests into pasturelands and croplands often results in the loss of bacterial diversity (Jesus et al., 2009; Rodrigues et al., 2013; Mendes et al., 2015b; Goss-Souza et al., 2020) and affects ecosystem services linked to microbial activity (Paula et al., 2014; Meyer et al., 2017; Goss-Souza et al., 2019; Pedrinho et al., 2019). Most research highlights taxa trade-offs, diversity turnover, and shifts in microbial functions due to land-use changes, driven by local abiotic environmental filters such as soil pH and fertility, suggesting a dominance of homogeneous selection (Goss-Souza et al., 2020).

In the Atlantic Forest, limited research, especially in the subtropical region (Southern Brazil), has explored the diversity of soil bacterial communities and the consequences of forest-to-agriculture conversion on both bacterial and arbuscular mycorrhizal (AM) fungi diversities and ecological processes shaping their distribution (Faoro et al., 2010; Goss-Souza et al., 2017; Ceola et al., 2021; Goss-Souza et al., 2022). From a biogeographic perspective, some authors propose that soil type, a historical contingency along with dispersal (an evolutionary contingency), may locally filter taxa distributions more strongly than the influence of land-use change (Ceola et al., 2021; Goss-Souza et al., 2022). These works indicate that soil microbial diversity and niche occupancy are influenced by spatial distance and long-term historical contingencies related to soil origin (soil type and climate). This leads to significant patterns of dispersal limitation and spatial correlations, with stochastic processes outweighing the effect of deterministic selection processes caused by soil historical contingencies and the formation of small geographic islands shaped by soil type and climate. These patterns become particularly evident when evaluating microbial niche specialists, especially at a local scale, with implications for biotic interactions among members of local microbial communities. This emphasizes the importance of expanding our understanding of the microbial component in ecological studies and conservation efforts within the context of the Brazilian Atlantic Forest.

2.3 Effects of land-use change on soil microbiome in grasslands

Over the last century, the escalating demand for primary products such as wood, food, and fiber has precipitated extensive conversion of natural grasslands into cultivated pastures (Sühs et al., 2020). This transition frequently leads to a decrease in the diversity of forage species and consequent impacts on soil microbial diversity (Delory et al., 2019; Yin et al., 2019). Natural and improved natural grasslands often contend with challenges such as overgrazing or exclusion. The heightened demand for forage has instigated the transformation of natural grasslands into cultivated pastures (Zanella et al., 2021). The intensification of soil management has been observed to stimulate the deterministic (niche-based) assembly process of soil microbial communities (Wang et al., 2021). Indeed, soil management enhances microbial functions, potentially owing to niche specialization (Wu et al., 2019). Nevertheless, the impacts of replacing natural grasslands with cultivated pastures on habitat specialization and the resultant assembly of microbial communities remain unknown.

In a recent study (Tomazelli et al., 2023a), the authors demonstrated that the structure of soil microbial communities was influenced by the diversity of forage plants following the conversion of natural grasslands to cultivated pastures. Soil management practices such as nitrogen fertilization and liming increased soil microbial alpha diversity and altered potential ecosystem functions. The study revealed that soil liming was the primary driver of changes in microbial communities by reducing available aluminum levels in the soil. This favored microbial groups related to nitrogen and sulfur metabolism, including 'nitrification,' 'nitrate reduction,' 'nitrogen respiration,' and 'respiration of sulfur compounds,' represented by Bacteroidetes and Nitrospirae. This came with a decrease in potential functions associated with C cycling, such as 'cellulose,' as represented by Acidobacteria.

Another recent study underscored the importance of grassland management and plant diversity in shaping microbial communities in grasslands (Tomazelli et al., 2023b). The authors found that generalists in the grasslands were shaped by niche-based models, while specialists in natural grasslands exhibited more stochastic behavior. In improved pastures, lognormal and preemption models were predominant, suggesting that homogenizing dispersal outweighed variable selection. The study demonstrated that improved pastures are environmentally beneficial, being more deterministic and having a higher proportion of niche-specialist microbes.

In the cultivated pasture soils of southern Brazil, the architecture of root systems, providing high interconnectivity, often results in stochastic behavior in microbial communities, leading to community homogenization in the face of environmental changes (Goss-Souza et al., 2017). In the same region, natural grasslands of the pampa biome exhibited both stochastic and deterministic processes, with soil moisture being the primary deterministic driver of soil microbial community assembly patterns (Lupatini et al., 2019). These results have highlighted that improved management of natural grasslands could serve as an alternative to increase cattle food availability. This approach maintains the endemic diversity of forage plants and soil biological quality while preserving essential ecosystem functions performed by soil microbes.

3 Effects of land-use change on soil microbiome functions

Forest-to-pasture conversion stands as the most prevalent and high-impact type of land-use change in the Brazilian Amazon region (Nascimento et al., 2019; Nunes et al., 2022). Over the years, various studies conducted in this region consistently demonstrated the negative effects of such conversions on vegetation cover (Nunes et al., 2022; Vieira, 2019), soil physicochemical properties (Navarrete et al., 2015; Melo et al., 2017), and soil microorganisms (Goss-Souza et al., 2017; Mendes et al., 2015a; Rodrigues et al., 2013). Recent research has further revealed that forest-to-pasture conversion influences soil microbial functions (Mendes et al., 2015b; Navarrete et al., 2015; Paula et al., 2014; Pedrinho et al., 2019). In this context, we present studies illustrating how land-use change, primarily forest-to-pasture conversion in the Brazilian Amazon region, impacts soil microbial functions, focusing particularly on nitrogen cycling, phosphorus transformation, antibiotic resistance genes (ARGs), and GHG emissions. Figure 2 illustrates the results discussed in this section.

3.1 Effects of land-use change on nitrogen cycling in the Amazon region

The Amazon rainforest plays a crucial role in nitrogen cycling (Pajares and Bohannan, 2016; Pedrinho et al., 2020). Typically, Amazonian soils, along with many other tropical soils, are highly weathered and nutrient-depleted (Fujii et al., 2018). Their fertility heavily relies on the cycling of a thin layer of organic matter associated with a substantial amount of plant litter material (Pajares and Bohannan, 2016). In this context, soil microorganisms play a fundamental role, responsible for both (i) mineralization, which involves the decomposition and/or oxidation of organic matter into forms easily available to plants and/or other soil microorganisms (Elrys et al., 2023), and (ii) assimilative/dissimilative processes, such as nitrogen fixation, nitrification, denitrification, dissimilatory nitrate reduction to ammonium (DNRA), and anammox (Pajares and Bohannan,

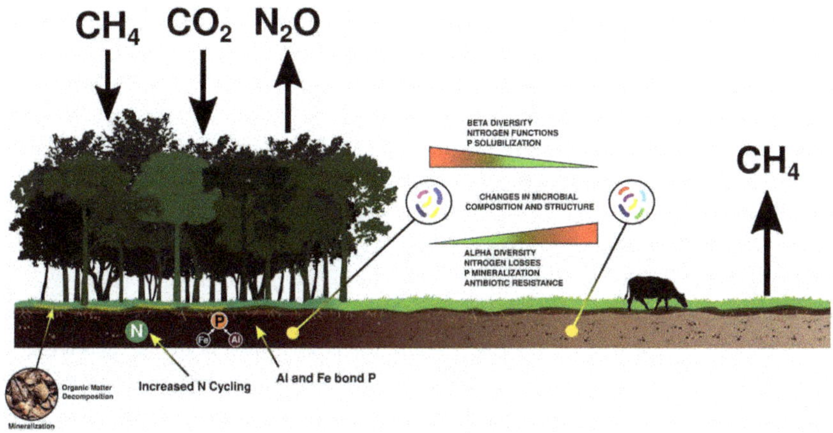

Figure 2 Effects of land-use change on the soil microbiome functions. Native forests act as a vital sink for greenhouse gases, whereas land conversion to agriculture and pasture transforms the land into a source of greenhouse gases. Additionally, land-use change significantly impacts the soil's nutrient metabolism.

2016). These microbial processes significantly contribute to N availability (Camenzind et al., 2018).

However, land-use change in the Amazon region notably disrupts the N cycle (Merloti et al., 2019; Pedrinho et al., 2020). This is primarily due to the high sensitivity of soil microorganisms to environmental disturbances (Mendes et al., 2015b). Therefore, when tropical forests are converted into pasture or agricultural fields, the processes mediated by soil microorganisms undergo significant alterations, thereby affecting N dynamics and availability. Different studies performed using eDNA sequencing have consistently demonstrated that forest-to-pasture or forest-to-agriculture conversion in the Amazon region can alter the structure, diversity, and composition of the soil microbial communities, subsequently affecting their functionality (Goss-Souza et al., 2017; Mendes et al., 2015b; Merloti et al., 2019; Pedrinho et al., 2019, 2020).

For example, biological nitrogen fixation (BNF) transforms atmospheric dinitrogen gas (N_2) into ammonia (NH_3), a vital process supplying new nitrogen to terrestrial ecosystems. BNF relies on free-living or symbiotic diazotrophic microorganisms that contain the *nif*H gene, encoding the reductase subunit of nitrogenase, the enzyme responsible for this conversion (Pajares and Bohannan, 2016). Previous studies in the Amazon region have highlighted that forest-to-pasture or forest-to-agriculture conversions induce changes in the structure, diversity, composition, and functioning of these soil microorganisms (Merloti et al., 2019; Mirza et al., 2014; Pedrinho et al., 2020). Mirza et al. (2014) and Pedrinho et al. (2020) observed a significant increase in the abundance of the *nif*H gene in pasture soils post-forest conversion compared to forest soils. Similarly, Merloti et al. (2019) noted a substantial rise in the abundance of the *nif*H gene in agricultural soils following forest-to-agriculture conversion. Such changes in *nif*H gene abundance and nitrogen fixation generally reflect the response of diazotrophic microorganisms to alterations in land use, plant community composition, and management practices (Danielson and Rodrigues, 2022). These alterations may contribute to an increase (Navarrete et al., 2015), decrease (Merloti et al., 2019; Pedrinho et al., 2020), or stability (Durigan et al., 2017) of soil carbon and nitrogen levels over time.

Nitrification is a multistep reaction in which ammonium (NH_4^+) is converted to nitrate (NO_3^-). This process plays a crucial role in the N cycle of terrestrial ecosystems as it regulates nitrogen availability and loss. Nitrification is facilitated by various microorganisms, including ammonia-oxidizing archaea (AOA) and bacteria (AOB), nitrite-oxidizing bacteria (NOB), and heterotrophic bacteria and fungi capable of utilizing organic nitrogen as a substrate (Pajares and Bohannan, 2016). These ammonia-oxidizing microorganisms (AOMs) carry the *amo*A gene, encoding the α-subunit of ammonia monooxygenase, the enzyme responsible for catalyzing this reaction. According to Zhang et al. (2016), changes in soil physicochemical properties, such as NH_4^+ concentration, soil pH, temperature, moisture, etc., significantly impact AOMs (both AOA and AOB). Consequently, the conversion of tropical forests to pasture or agricultural fields can lead to dramatic alterations in ammonia-oxidizer structure, composition, and activity (Hamaoui Jr et al., 2016; Merloti et al., 2019). Previous studies conducted in the Amazon region have indeed indicated that forest-to-pasture or forest-to-agriculture conversions can modify the abundance of AOMs and increase net mineralization and nitrification rates (Merloti et al., 2019; Neill et al., 1997).

Denitrification, the final step in the nitrogen cycle, is regulated by several microbial groups through an anaerobic respiration pathway, converting NO_3^- and/or nitrite (NO_2^-) to nitrous oxide (N_2O) and/or N_2. The denitrification process comprises four reactions catalyzed by the metalloenzymes nitrate reductase (*nar*G and *nap*A marker-genes), nitrite reductase (*nir*K and *nir*S), nitric oxide reductase (*nor*B), and nitrous oxide reductase (*nos*Z). Like the microorganisms involved in BNF and nitrification, denitrifiers are sensitive to changes in soil physicochemical properties, such as soil moisture and NO_3^- concentration, induced by forest-to-pasture and/or forest-to-agriculture conversions (Merloti et al., 2019; Pedrinho et al., 2020; Pajares and Bohannan, 2016). Previous studies conducted in the Brazilian Amazon region have observed an increase in the abundance of *nir*K and *nos*Z genes, alongside denitrification rates, following forest-to-pasture and/or forest-to-agriculture conversions (Merloti et al., 2019; Pedrinho et al., 2020). This suggests that such conversions lead to a system more susceptible to nitrogen losses.

3.2 *Effects of land-use change on phosphorus dynamics in the Amazon region*

Amazonian soils, along with many other tropical soils, contain high levels of iron (Fe) and aluminum (Al) oxides, a result of the intense weathering processes prevalent in this region, characterized by heavy rainfall and high temperatures (Pedrinho et al., 2023; Soltangheisi et al., 2019). These Fe and Al oxides possess geochemical properties enabling them to bind to phosphorus (P), rendering them unavailable for plant uptake or other soil organisms (Gama-Rodrigues et al., 2014). Consequently, many plants and soil organisms rely on the recycling of P from soil and litterfall into forms readily usable. It is important to note that this phenomenon is particularly prominent in tropical forests, where the process of P transformation is essentially closed, exhibiting minimal losses or gains (Solomon et al., 2002).

In this context, soil microorganisms, including bacteria and fungi, play a pivotal role as they participate in three principal processes that influence P availability in soils: mineralization, solubilization, and immobilization. Microorganisms engaged in P mineralization possess the capability to produce enzymes that release P from recalcitrant organic forms (Po), thereby converting them into accessible inorganic forms (Pi) for plants and other soil organisms (Arenberg and Arai, 2019). Those involved in P solubilization can generate and discharge organic acids like citric acid, gluconic acid, and malic acid, which effectively solubilize recalcitrant P forms in soils (Alori et al., 2017). Lastly, microorganisms engaged in P immobilization can assimilate inorganic P into their biomass, competing with plants and other soil microbes for the available P (Richardson and Simpson, 2011)

Recent studies have highlighted the impact of land-use changes on the microbial community involved in P transformation and their functionality. Pedrinho et al. (2023) found that forest-to-pasture conversion results in an increase in microbial groups associated with P mineralization in pasture soil. These microorganisms typically harbor genes that encode enzymes like C–P lyases (*phn*) and alkaline phosphatase (*pho*D and *pho*A), known for their ability to release P from recalcitrant organic forms in soils. Consequently, such land-use changes can modify the total amount of P and its fractions in soils (Pedrinho et al., 2023; Soltangheisi et al., 2019), ultimately exerting negative effects on soil P dynamics in the Amazon region (Chavarro-Bermeo et al., 2022).

3.3 *Antibiotic resistance genes in land-use-altered environments*

Soil microorganisms constitute the primary source of antibiotics in natural environments (Xie et al., 2018). Typically, antibiotics produced by soil microorganisms function as signaling molecules at minimal concentrations. At higher concentrations, these antibiotics serve as a mechanism for microbial 'arming,' allowing them to competitively outmatch neighboring soil organisms for resources and space (Martínez et al., 2015). However, anthropogenic activities, including land-use changes and antibiotic usage in agriculture, progressively alter these microbial communities and the types and quantities of antibiotics they produce (Lemos et al., 2021b; Xie et al., 2018). These human-induced activities also influence the emergence and dissemination of antibiotic-resistant microbial groups (Wright, 2007). The resistance capability in these microbial groups is conferred by ARGs, often transferable within and among microbial communities through horizontal gene transfer (Xie et al., 2018). Hence, understanding how anthropogenic activities, particularly those in the Amazon region, impact the spread of ARGs is crucial. Such insights are pivotal in devising practices to mitigate this global health and food security threat (Lemos et al., 2021b).

While numerous studies have assessed the effects of land-use change on soil microbial communities and functions in the Amazon region, its influence on the soil resistome, encompassing all ARGs (Wright, 2007), remains largely unexplored. To date, only one study has investigated this aspect, demonstrating that land-use change significantly impacts the structure, diversity, and abundance of ARGs (Lemos et al., 2021b). Following forest-to-pasture conversion, a notable increase in various functional mechanisms and ARGs was observed in pasture soils. Particularly concerning was the significant rise in genes associated with resistance to tetracycline, a widely used antibiotic in livestock husbandry for treating animal diseases (He et al., 2020). This poses a substantial concern, as anthropogenic activities in the Amazon region have the potential to markedly alter the diversity of ARGs in the soil reservoir, potentially entering the food chain through the consumption of meat, vegetables, and other food sources (Cheng et al., 2019; Lemos et al., 2021b).

3.4 Effects of land-use change on greenhouse gas emissions in the Amazon region

The Amazon rainforest functions primarily as a carbon sink, sequestering carbon dioxide (CO_2) from the atmosphere through photosynthesis and storing it in both aboveground (trees and vegetation) and belowground compartments, including dead wood, litter, and soil (Mitchard, 2018; Singh et al., 2020). However, anthropogenic activities like deforestation and fires lead to the release of stored carbon, resulting in increased emissions of CO_2 and methane (CH_4) into the atmosphere (Baccini et al., 2017). These disturbances disrupt the carbon balance, altering the ratio between photosynthesis and respiration and consequently impacting soil microorganisms, such as archaea, bacteria, and fungi, which play crucial roles in carbon cycling (Gougoulias et al., 2014; Singh et al., 2020).

These soil microorganisms play a pivotal role in both producing and consuming GHGs, including carbon dioxide (CO_2), methane (CH_4), and nitrous oxide (N_2O), within terrestrial ecosystems (Stein, 2020). As mentioned earlier, these microorganisms break down organic matter into forms accessible to plants and other soil organisms (Elrys et al., 2023). In this process, they release CO_2 and methane CH_4 into the atmosphere as byproducts (Stein, 2020). However, land-use changes, especially forest-to-pasture or forest-to-agriculture conversions, can disrupt the activity, composition, and functionality of these soil microbial communities. Such disturbances negatively impact carbon decomposition rates, increase GHG emissions, and reduce the Amazonian soils' capacity to act as a carbon sink, potentially transforming them into a carbon source, thereby intensifying climate change (Danielson and Rodrigues, 2022; Obregon Alvarez et al., 2023; Steudler et al., 1996).

Several studies conducted in the Amazon region have reported divergent effects on soil carbon levels after forest-to-pasture conversion: an increase (Durrer et al., 2021), a decrease (Maia et al., 2010), or no significant change (Durigan et al., 2017). This variability suggests that the influence of forest-to-pasture or forest-to-agriculture conversions on soil carbon is contingent upon pre-existing conditions. As highlighted by Danielson and Rodrigues (2022), these conditions might be linked to factors such as initial carbon stocks, soil characteristics (e.g. texture, nutrient content), and land management practices (e.g. fires, grazing intensity, fertilization). The impact of land-use change on CH_4 flux in the Amazon has received greater attention compared to CO_2, mainly due to the involvement of a smaller number of genes and microorganisms in CH_4 cycling, and the significantly higher global warming potential of CH_4, which is 27–30 times greater than CO_2 (IPCC, 2014). The flux of CH_4 depends on how much is made versus how much is used by microorganisms. Under anaerobic conditions, methanogenic archaea produce CH_4 using the *mcr*A gene, converting CO_2 and methanol (CH_3OH) (Conrad, 2007). In contrast, under aerobic conditions, methanotrophic bacteria, possessing the *pmo*A gene, turn CH_4 into methanol using O_2 as an electron acceptor (Conrad, 2007).

Studies conducted in the Amazon region highlight the significant impact of forest-to-pasture conversion on both methanogenic archaea and methanotrophic bacteria. After the transition from forest to pasture, there is a notable increase in the abundance and diversity of methanogenic archaea in pasture soils, as noted by Obregon Alvarez et al. (2023). Additionally, observations by Meyer et al. (2017) and Paula et al. (2014) indicate that forest-to-pasture conversion leads to a decline in the abundance and alters the composition of methanotrophic bacteria in pasture soils. Consequently, this change in land use consistently affects CH_4 flux across the Amazon basin. Research suggests that tropical forests usually function as a sink for CH_4 (−470 mg CH_4 m^2), while pastures serve as a source (+270 mg CH_4 m^2) (Danielson and Rodrigues, 2022; Steudler et al., 1996).

It is also crucial to highlight N_2O as another significant GHG, with a global warming potential 300 times higher than CO_2 (Forster et al., 2007; UNEP, 2013). As discussed previously, N_2O is generated during the denitrification process by a group of microorganisms known as denitrifiers. A meta-analysis of N_2O flux studies across the Amazon basin reveals that tropical forests generally exhibit higher annual emissions (2.42 kg N ha^{-1}) compared to old pasture soils (0.9 kg N ha^{-1}), which, in certain cases, may even function as a slight sink (Meurer et al., 2016). However, the same analysis by Meurer et al. (2016) noted that recently converted pasture areas (less than 10 years) present elevated annual emissions (2.52 kg N ha^{-1}), akin to levels observed in tropical forests. Moreover, it is worth noting that appropriate pasture management can also stimulate methane capture by the soil. A recent study conducted in the Amazon has shown that better management of grasses in pastures can enhance the soil's capacity for carbon sequestration, mainly due to an increase in the abundance of methanotrophic microorganisms in the rhizosphere of the plants (Souza et al., 2022).

4 Effects of land recovery on soil microbiome properties

Extensive land use for agriculture and pasture often leads to soil degradation, a process detrimental to the health and fertility of the land. Continuous cultivation of crops or grazing in pastures without adequate rest or sustainable management practices can strip the soil of essential nutrients, decrease organic matter, and erode its structure. This overexploitation diminishes soil quality, reducing its ability to retain water, support plant growth, and maintain biodiversity. Land degradation poses a significant obstacle to sustainable development, with far-reaching consequences for food security, biodiversity conservation, and climate change mitigation and adaptation. Therefore, urgent action is required to devise strategies aimed at reducing and restoring degraded ecosystems. There are several strategies to reduce and restore degraded land and ecosystems, including conservation, reforestation, and regenerative agriculture. Figure 3 illustrates the results discussed in this section.

4.1 Land recovery through secondary forest

Typically, the conversion of forest to pasture in the Amazon region follows a process involving selective logging of valuable timber, subsequent burning of

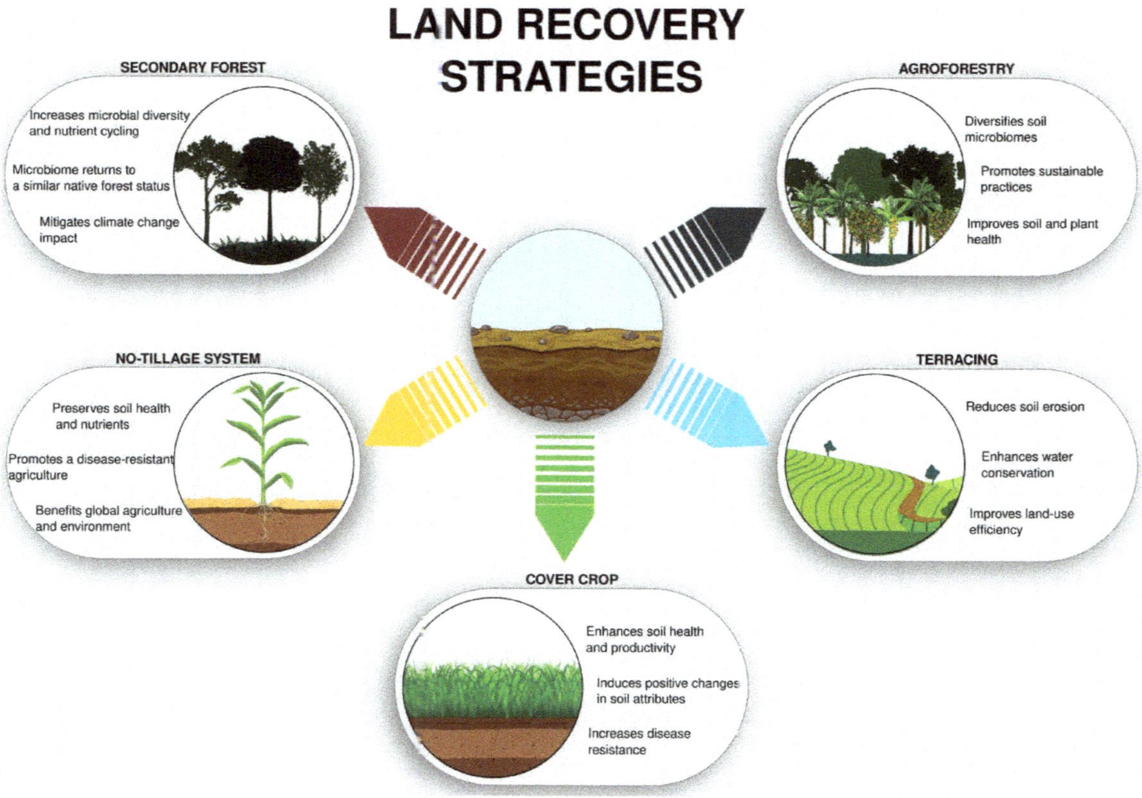

Figure 3 Strategies for enhancing ecosystem recovery. The figure illustrates diverse approaches utilized to mitigate land degradation and promote restoration, with particular emphasis on their effects on the microbiome and the environment.

the remaining vegetation, and the seeding of forage grasses (*Urochloa* genera) (Meyer et al., 2017; Navarrete et al., 2015; Pedrinho et al., 2019). However, within 5–15 years of exploitation and poor management practices, pastures become degraded and unproductive and, consequently, are abandoned (Souza Braz et al., 2013; Pedrinho et al., 2019; Pedrinho et al., 2020). As a result, the secondary forest vegetation starts to build up a new forest stand (Cenciani et al., 2009; Pedrinho et al., 2020). Several studies performed in the Amazon region have compared microbial communities in primary forests and pastures with those in secondary forests (Durrer et al., 2021; Paula et al., 2014; Pedrinho et al., 2019; Pedrinho et al., 2020; Pedrinho et al., 2023). Many of these studies concluded that microbial diversity indices in secondary forests are more similar to primary forests and that specific microbial groups and functions can be recovered over time (especially those linked to nutrient cycling) (Paula et al., 2014; Pedrinho et al., 2020; Pedrinho et al., 2023). For example, when focusing on the nitrogen cycle, different studies suggested that, during secondary forest succession, the abundance of diazotrophs decreases in comparison to pastures, converging toward levels observed in primary forests (Mirza et al., 2014; Pedrinho et al., 2020). The same pattern is observed for the abundance of the *amo*A gene (maker gene of AOMs) in secondary forest soil, which presented values similar to those observed in primary forests (Hamaoui et al., 2016).

For instance, numerous studies have shown that the conversion of natural forest areas into pasturelands disrupts the methane gas flow in Amazonian soils, further contributing to the problem of climate change (Meyer et al., 2017; Kroeger et al., 2021; Venturini et al., 2022; Alvarez et al., 2023). However, the recovery of degraded areas through the growth of secondary forests has shown that microbial functions have started to return to the state of native vegetation (Pedrinho et al., 2019, 2020, 2023). Moreover, it is worth noting that appropriate pasture management can also stimulate methane capture by the soil. A recent study conducted in the Amazon has shown that better management of grasses in pastures can enhance the soil's capacity for carbon sequestration, mainly due to an increase in the abundance of methanotrophic microorganisms in the rhizosphere of the plants (Souza et al., 2022). These examples illustrate how the recovery of vegetation and the proper management of agricultural areas can effectively balance GHG emissions by altering the abundance of specific microorganisms involved in the carbon cycle. Considering that microbial diversity is essential in determining the global biogeochemical balance, improved management practices along with microbiome manipulation techniques can serve as a powerful tool to accelerate the recovery of degraded areas and reduce the impact of climate change.

4.2 Land recovery through conservationist practices in agriculture

Conservationist practices in agriculture is a sustainable farming system that aims to conserve, improve, and efficiently utilize natural resources through integrated management of soil, water, and biological resources combined with external inputs. It involves minimizing soil disturbance, improving organic matter and soil cover, and using crop rotations and associations to reduce the impact of pests and diseases. Practices such as no-tillage systems, cover crops, terracing, and agroforestry aim to maintain soil, water, and biodiversity, and reduce soil erosion and compaction. These practices also positively impact the soil microbiome, influencing the composition and functioning of bacterial and fungal communities, and improving soil health (Banerjee et al., 2018; Tao et al., 2020). Furthermore, conservationist practices can promote disease resistance-inducing and growth-promoting beneficial microbes in the soil, potentially enhancing plant survival and crop production (Berendsen et al., 2018; Raaijmakers & Mazzola, 2016; Tao et al., 2020). However, the precise functional role of the soil microbiome in agricultural systems is still not fully understood, highlighting the need for further research to optimize conservationist practices for improved crops (Zhao et al., 2020). Overall, the integration of conservationist practices in agriculture is crucial for sustainable soil management and crop production, emphasizing the importance of understanding and managing the soil microbiome in agricultural systems (Zhao et al., 2020; Yuan et al., 2021).

No-till practices in agriculture have been shown to have a significant impact on soil microbiomes. These practices help maintain nutrients within the soil, reducing the need for supplements (Smith et al., 2016). Furthermore, no-till methods are critical for restoring and protecting soil health, contributing to the sustainability of global agriculture (Anderson, 2015). Additionally, agricultural practices such as tillage and cropping systems can lead to structural and functional changes in soil microbiomes, affecting soil health and plant disease resistance (Foo et al., 2017). The initial soil microbiome composition and functioning can predetermine future plant health, highlighting the long-term impact of agricultural practices on soil microbiomes (Wei et al., 2019). Overall, the adoption of no-till practices not only provides economic and environmental benefits but also contributes to the development of disease-suppressive soil microbiomes, supporting sustainable agriculture (Chen et al., 2022).

Another example of conservationist practices is crop rotation and cover cropping. These are agricultural practices that have demonstrated the ability to induce changes in soil microbiomes. These practices increase soil fertility, microbial diversity, and disease-suppressive capacity, leading to improved soil health and crop productivity (Castellano-Hinojosa & Strauss, 2020; Yuan et al., 2021). The intentional management of soil microbiomes through cover crop selection can be a strategy to enhance soil health and promote beneficial changes in soil chemical and biological attributes, such as increasing nitrogen and carbon availability and microbial diversity. Furthermore, crop rotation has been found to be effective in maintaining microbiome complexity and functioning in salinity-affected soils and in reducing the negative effects of salinity on soil microbiomes (Dasgupta, 2023). These practices have also been shown to regulate nutrition, reduce soilborne diseases, and improve soil fertility, making them important strategies for sustainable agriculture (Yuan et al., 2021).

Agroforestry practices, another conservation method, have been shown to alter soil microbial diversity and composition. This primarily occurs through the creation of a distinct tree row-associated microbiome, which differs from the crop row microbiome, thereby increasing overall beta diversity (Fahad et al., 2022). Additionally, agroforestry systems can promote diversity, alter the composition, and improve the resilience of root-associated fungal communities, which can have a positive feedback effect on the surrounding soil after the long-term establishment of tree enrichment (Ballauff et al., 2020). Furthermore, management practices in agro-systems generate spatial and temporal changes in soil physical and chemical properties, creating rapidly fluctuating environments that provide a wide range of niches for microbial growth (Dube et al., 2019). The impact of agroforestry on soil microbiomes is also influenced by factors such as nutrient input, which appears to be a major factor fostering the differentiation of soil microbiome between tree rows and arable land in agroforestry systems (Beule & Karlovsky, 2021). Moreover, the manipulation of the microbiome through agroforestry practices may reduce the need for fertilizer and pesticide use, leading to more sustainable agricultural practices (Liu et al., 2023).

Notably, conservationist practices in agriculture have immense potential for enhancing soil health, promoting biodiversity, and increasing crop productivity. These practices not only minimize soil erosion and compaction but also exert a significant influence on the soil microbiome. The microbiome plays a crucial role in nutrient cycling, disease suppression, and overall soil fertility, highlighting its significance in sustainable agriculture. In the pursuit of sustainable agriculture, integrating conservationist practices and managing soil microbiomes carefully are essential components. By harnessing the power of beneficial microbial communities and fostering harmonious relationships between crops and their microbial allies, we can pave the way toward a more resilient and environmentally friendly agricultural future.

5 Conclusion

Land-use management has a significant impact on soil microbiomes, influencing their composition, diversity, and functions. The changes in land-use types and agriculture practices significantly affect the variability of microbiomes in soil profiles. These changes can lead to distinct community assemblies and functions of microbiomes throughout soil profiles, highlighting the need for strategies and policies that consider the responses of vertical soil microbiomes to land-use changes. Furthermore, the role of land use in structuring the microbiomes of connected ecosystems has been emphasized, indicating that land-use types and water chemical properties play a crucial role in shaping the microbiomes of interconnected environments (Marmen et al., 2020).

Advanced molecular techniques, such as Next-Generation Sequencing, have provided a deeper understanding of the soil microbiome and its responses to land-use management. However, it is important to note that inherent biases in soil metagenomics present challenges in accurately defining the soil microbiome and its ecosystem function (Lemos et al., 2021a; Sanchez-Cid et al., 2022). Additionally, specialized metabolic functions of keystone taxa have been found to sustain soil microbiome stability, indicating the potential of specific microbial groups to maintain stability in the face of environmental perturbations caused by land-use management and other factors (Xun et al., 2021). Microbiome-based predictive models have been developed to understand how soil microbial biodiversity might be affected by future climate change scenarios, providing valuable insights into the potential responses of soil microbiomes to changing environmental conditions (Cowan et al., 2022). Moreover, the functional stability of the soil microbiome has been linked to the maintenance of ecosystem services such as primary production, soil carbon accumulation, and nutrient cycling, highlighting the importance of understanding and predicting the responses of soil microbiomes to land use management for ecosystem sustainability (Jiao et al., 2021).

Considering that climate change is pivotal in assessing the future perspectives of how land-use management affects soil microbiomes, it is essential to recognize that long-term shifts in average weather patterns can profoundly impact the soil microbiome and biodiversity. This emphasizes the necessity of acknowledging climate change as a significant factor influencing how soil microbiomes respond to land-use management practices (Mishra et al., 2022). Additionally, the responses of soil microbiomes to environmental changes display notable geographic patterns, indicating the intricate interplay among land use, climate, and the dynamics of soil microbiomes (Jiao et al., 2022).

In conclusion, future perspectives on how land-use management affects soil microbiomes are multifaceted, encompassing advanced molecular techniques, microbiome-based predictive models, climate change considerations, and ecosystem services assessment. Understanding the intricate relationships between land-use management and soil microbiomes is essential for sustainable land management practices and the preservation of ecosystem functions.

6 Where to look for further information

- Danielson RE, Rodrigues JL (2022) Impacts of land-use change on soil microbial communities and their function in the Amazon Rainforest. *Advances in Agronomy* 175:179–258. https://doi.org/10.1016/bs.agron.2022.04.001.
- Horwath W (Ed.) (2022) *Improving Soil Health*. Sawston: Burleigh Dodds Science Publishing.
- Lambin EF, Geist H (Eds.) (2010) *Land-Use and Land-Cover Change: Local Processes and Global Impacts*. Berlin and Heidelberg: Springer. https://doi.org/10.1007/3-540-32202-7.
- Otten W (Ed.) *Advances in Measuring Soil Health*. Burleigh Dodds Science Publishing. https://doi.org/10.1201/9781003048046.
- Pedrinho A, Mendes LW, Pereira APA, Araujo ASF, Vaishnav A, Karpouzas DG, Singh BK (2024) Soil microbial diversity plays an important role in resisting and restoring degraded ecosystems. *Plant and Soil*. https://doi.org/10.1007/s11104-024-06489-x.
- Pylro V, Roesch L (Eds.) (2017) *The Brazilian Microbiome: Current Status and Perspectives*. Cham: Springer. https://doi.org/10.1007/978-3-319-59997-7.
- Reicosky D (Ed.) (2018a) *Managing Soil Health for Sustainable Agriculture. Volume 1: Fundamentals*. Burleigh Dodds Science Publishing. https://doi.org/10.1201/9781351114530.
- Reicosky D (Ed.) (2018b) *Managing Soil Health for Sustainable Agriculture. Volume 2: Monitoring and Management*. Burleigh Dodds Science Publishing. https://doi.org/10.1201/9781351114585.
- Singh JS, Tiwari S, Singh C, Singh AK (Eds.) (2021) *Microbes in Land Use Change Management*. Elsevier. https://doi.org/10.1016/C2020-0-00650-X.
- van Elsas JD, Trevors JT, Soares Rosado A, Nannipieri P (Eds.) (2019) *Modern Soil Microbiology, Third Edition* (3rd ed.). CRC Press. https://doi.org/10.1201/9780429059186.
- Yadav AN (Ed.) (2021) *Soil Microbiomes for Sustainable Agriculture: Functional Annotation*. Cham: Springer. https://doi.org/10.1007/978-3-030-73507-4.

7 References

Alori ET, et al. (2017) Microbial phosphorus solubilization and its potential for use in sustainable agriculture. *Frontiers in Microbiology* 8, 971.

Anderson, R (2015) Integrating a complex rotation with no-till improves weed management in organic farming. A review. *Agronomy for Sustainable Development* 35(3), 967–974. https://doi.org/10.1007/s13593-015-0292-3

Arenberg MR, Arai Y (2019) Uncertainties in soil physicochemical factors controlling phosphorus mineralization and immobilization processes. *Advances in Agronomy* 154, 153–200. https://doi.org/10.1016/bs.agron.2018.11.005

Averill C, et al. (2022) Defending Earth's terrestrial microbiome. *Nature Microbiology* 7, 1717–1725.

Baccini A, et al. (2017) Tropical forests are a net carbon source based on aboveground measurements of gain and loss. *Science* 358(6360), 230–234. https://doi.org/10.1126/science.aam5962

Bach EM, et al. (2020) Soil biodiversity integrates solutions for a sustainable future. *Sustainability* 12, 2662.

Ballauff J, et al. (2020) Legacy effects overshadow tree diversity effects on soil fungal communities in oil palm-enrichment plantations. *Microorganisms* 8(10), 1577. https://doi.org/10.3390/microorganisms8101577

Banerjee S, et al. (2018) Keystone taxa as drivers of microbiome structure and functioning. *Nature Reviews Microbiology* 16(9), 567–576. https://doi.org/10.1038/s41579-018-0024-1

Berendsen R, et al. (2018) Disease-induced assemblage of a plant-beneficial bacterial consortium. *The Isme Journal*, 12(6), 1496–1507. https://doi.org/10.1038/s41396-018-0093-1

Beule L, Karlovsky, P (2021) Tree rows in temperate agroforestry croplands alter the composition of soil bacterial communities. *PLoS One*, 16(2), e0246919. https://doi.org/10.1371/journal.pone.0246919

Borneman J, Triplett EW (1997) Molecular microbial diversity in soils from Eastern Amazonia: evidence for unusual microorganisms and microbial population shifts associated with deforestation. *Applied and Environmental Microbiology* 63, 2647–2653.

Brossi MJL, et al. (2014) Assessment of bacterial bph gene in Amazon Dark Earth and their adjacent soils. *PLoS One* 9, e99597.

Camenzind T, et al. (2018) Nutrient limitation of soil microbial processes in tropical forests. *Ecological Monographs* 88(1), 4–21. https://doi.org/10.1002/ecm.1279

Cardinale BJ, et al (2012) Biodiversity loss and its impact on humanity. Nature, 486(7401), 59–67. doi: 10.1038/nature11148

Castellano-Hinojosa, A. and Strauss, S (2020) Impact of cover crops on the soil microbiome of tree crops. *Microorganisms* 8(3), 328. https://doi.org/10.3390/microorganisms8030328

Cenciani K, et al. (2009) Bacteria diversity and microbial biomass in forest, pasture and fallow soils in the southwestern Amazon basin. *Revista Brasileira de Ciência do Solo* 33, 907–916. https://doi.org/10.1590/S0100-06832009000400015

Ceola G, et al. (2021) Biogeographic patterns of arbuscular mycorrhizal fungal communities along a land-use intensification gradient in the subtropical atlantic forest biome. *Microbial Ecology*. https://doi.org/10.1007/s00248-021-01721-y

Chavarro-Bermeo JP, et al. (2022) Responses of soil phosphorus fractions to land-use change in Colombian Amazon. *Sustainability* 14(4), 2285. https://doi.org/10.3390/su14042285

Chen L, et al. (2022) The impact of no-till on agricultural land values in the united states midwest. *American Journal of Agricultural Economics* 105(3), 760–783. https://doi.org/10.1111/ajae.12338

Cheng G, et al. (2019) Selection and dissemination of antimicrobial resistance in Agri-food production. *Antimicrobial Resistance and Infection Control* 8, 158. https://doi.org/10.1186/s13756-019-0623-2

Conrad R (2007) Microbial ecology of methanogens and methanotrophs. *Advances in Agronomy* 96, 1–63. https://doi.org/10.1016/S0065-2113(07)96005-8

Cowan D, et al. (2022) Biogeographical survey of soil microbiomes across sub-saharan africa: structure, drivers, and predicted climate-driven changes. *Microbiome* 10(1).https://doi.org/10.1186/s40168-022-01297-w

Danielson RE, Rodrigues JL (2022) Impacts of land-use change on soil microbial communities and their function in the Amazon Rainforest. *Advances in Agronomy* 175, 179–258. https://doi.org/10.1016/bs.agron.2022.04.001

Dasgupta D (2023) Cover cropping reduces the negative effect of salinity on soil microbiomes. *Journal of Sustainable Agriculture and Environment* 2(2), 140–152. https://doi.org/10.1002/sae2.12054

Delory BM, et al. 2019. When history matters: the overlooked role of priority effects in grassland overyielding. *Functional Ecology* 33, 2369–2380. https://doi.org/10.1111/1365-2435.13455.

Dube J, et al. (2019) Differences in bacterial diversity, composition and function due to long-term agriculture in soils in the eastern free state of south africa. *Diversity* 11(4), 61. https://doi.org/10.3390/d11040061

Durigan M, et al. (2017) Soil organic matter responses to anthropogenic forest disturbance and land use change in the Eastern Brazilian Amazon. *Sustainability* 9, 379. https://doi.org/10.3390/su9030379

Durrer A, et al. (2021) Beyond total carbon: conversion of amazon forest to pasture alters indicators of soil C cycling. *Biogeochemistry* 152, 179–194. https://doi.org/10.1007/s10533-020-00743

Elrys AS, et al. (2023) Global soil nitrogen cycle pattern and nitrogen enrichment effects: tropical versus subtropical forests. *Global Change Biology* 29(7), 1905-1921. https://doi.org/10.1111/gcb.16603

Fahad S, et al. (2022) Agroforestry systems for soil health improvement and maintenance. *Sustainability*, 14(22), 14877. https://doi.org/10.3390/su142214877

Faoro H, et al. (2010) Influence of soil characteristics on the diversity of bacteria in the southern brazilian atlantic forest. *Applied and Environmental Microbiology* 76, 4744–4749. https://doi.org/10.1128/AEM.03025-09

Foo J, et al. (2017) Microbiome engineering: current applications and its future. *Biotechnology Journal* 12(3).https://doi.org/10.1002/biot.201600099

Forster P, et al. (2007) Changes in atmospheric constituents and in radiative forcing. In: Solomon, S., et al. (ed.) *Climate Change: The Physical Science Basis. International Nuclear Information System*. Cambridge University Press, Cambridge, United Kingdom and New York, NY, USA, 129–234.

Fujii K, et al. (2018) Plant–soil interactions maintain biodiversity and functions of tropical forest ecosystems. *Ecological Research* 33, 149–160. https://doi.org/10.1007/s11284-017-1511-y

Gama-Rodrigues AC, et al. (2014) An exploratory analysis of phosphorus transformations in tropical soils using structural equation modeling. *Biogeochemistry* 118, 453–469. https://doi.org/10.1007/s10533-013-9946-x

Germano MG, et al. (2012) Functional diversity of bacterial genes associated with aromatic hydrocarbon degradation in anthropogenic dark earth of Amazonia. *Pesquisa Agropecuaria Brasileira* 47, 654–664.

Goss-Souza D, et al. (2017) Soil microbial community dynamics and assembly under long-term land use change. *FEMS Microbiology Ecology* 93(10), fix109. https://doi.org/10.1093/femsec/fix109

Goss-Souza D, et al. (2019) Amazon forest-to-agriculture conversion alters rhizosphere microbiome composition while functions are kept. *FEMS Microbiology Ecology*. https://doi.org/10.1093/femsec/fiz009

Goss-Souza D, et al. (2020) Ecological processes shaping bulk soil and rhizosphere microbiome assembly in a long-term amazon forest-to- agriculture conversion. *Microbial Ecology* 79, 110–122. https:// doi.org/10.1007/s00248-019-01401-y

Gougoulias C, et al. (2014) The role of soil microbes in the global carbon cycle: tracking the below-ground microbial processing of plant-derived carbon for manipulating carbon dynamics in agricultural systems. *Journal of the Science of Food and Agriculture* 94(12), 2362–2371. https://doi.org/10.1002/jsfa.6577

Grossman JM, et al. (2010) An assessment of nodulation and nitrogen fixation in inoculated Inga oestediana, a nitrogen-fixing tree shading organically grown coffee in Chiapas, Mexico. *Soil Biology and Biochemistry* 38, 769–784.

Hamaoui Jr GS, et al. (2016) Land-use change drives abundance and community structure alterations of thaumarchaeal ammonia oxidizers in tropical rainforest soils in Rondônia, Brazil. *Applied Soil Ecology* 107, 48–56. https://doi.org/10.1016/j.apsoil.2016.05.012

He Y, et al. (2020) Antibiotic resistance genes from livestock waste: occurrence, dissemination, and treatment. *NPJ Clean Water* 3, 1–11. https://doi.org/10.1038/s41545-020-0051-0

Intergovernmental Panel on Climate Change – IPCC (2014) Climate change 2014: synthesis report. Pachauri RK, Allen MR, Barros VR, Broome J, Cramer W, Christ R, Church JA, Clarke L, Dahe Q, Dasgupta P, Dubash NK (eds) Contribution of Working Groups I, II and III to the fifth assessment report of the Intergovernmental Panel on Climate Change. Geneva, p. 151.

Jantz SM, et al. (2015) Future habitat loss and extinctions driven by land-use change in biodiversity hotspots under four scenarios of climate-change mitigation. *Conservation Biology* 29, 1122–1131. https://doi.org/10.1111/cobi.12549

Jesus EC, et al. (2009) Changes in land use alter the structure of bacterial communities in Western Amazon soils. *The ISME Journal* 3, 1004–1011. https://doi.org/10.1038/ismej.2009.47

Jiao S, et al. (2021) Core microbiota drive functional stability of soil microbiome in reforestation ecosystems. *Global Change Biology*, 28(3), 1038–1047. https://doi.org/10.1111/gcb.16024

Jiao S, et al. (2022) Core phylotypes enhance the resistance of soil microbiome to environmental changes to maintain multifunctionality in agricultural ecosystems. *Global Change Biology*, 28(22), 6653–6664. https://doi.org/10.1111/gcb.16387

Kaiser K, et al. (2016) Driving forces of soil bacterial community structure, diversity, and function in temperate grasslands and forests. *Scientific Reports* 6, 1–12. https://doi.org/10.1038/srep33696

Kroeger M. E., et al. (2021). Rainforest-to-pasture conversion stimulates soil methanogenesis across the Brazilian Amazon. *ISME Journal* 15:658–672. https://doi.org/10.1038/s41396-020-00804-x

Lal R (2020) Soil degradation as a reason for inadequate human nutrition: significance, causes, and remedies. *Sustainability* 12(6), 2336. doi: 10.3390/su12062336

Lauber CL, et al. (2013) Temporal variability in soil microbial communities across land-use types. *The ISME Journal* 7, 1641–1650. https://doi.org/10.1038/ ismej.2013.50

Lemos LN, et al. (2021a) Genome-resolved metagenomics is essential for unlocking the microbial black box of the soil. *Trends in Microbiology* 29, 279–282.

Lemos LN, et al. (2021b) Amazon deforestation enriches antibiotic resistance genes. *Soil Biology and Biochemistry* 153, 108110. https://doi.org/10.1016/j.soilbio.2020.108110

Liu J, et al. (2023) Integrated microbiome and metabolomics analysis reveal a closer relationship between soil metabolites and bacterial community than fungal community in pecan plantations. *Land Degradation and Development* 34(10), 2812–2824. https://doi.org/10.1002/ldr.4649

Lupatini M, et al. 2019. Moisture is more important than temperature for assembly of both potentially active and whole prokaryotic communities in subtropical grassland. *Microbial Ecology* 77, 460–470. https://doi.org/10.1007/s00248-018-1310-1.

Maia SM, et al. (2010) Soil organic carbon stock change due to land use activity along the agricultural frontier of the southwestern Amazon, Brazil, between 1970 and 2002. *Global Change Biology* 16(10), 2775–2788. https://doi.org/10.1111/j.1365-2486.2009.02105.x

Marmen S, et al. (2020) The role of land use types and water chemical properties in structuring the microbiomes of a connected lake system. *Frontiers in Microbiology* 11. https://doi.org/10.3389/fmicb.2020.00089

Martínez JL, et al. (2015) What is a resistance gene? Ranking risk in resistomes. *Nature Reviews Microbiology* 13(2), 116–123. https://doi.org/10.1038/nrmicro3399

Melo VF, et al. (2017) Land use and changes in soil morphology and physical-chemical properties in Southern Amazon. *Revista Brasileira de Ciência do Solo* 41, e0170034. https://doi.org/10.1590/18069657rbcs20170034

Mendes LW, et al. (2017) Using metagenomics to connect microbial Community biodiversity and functions. *Current Issues in Molecular Biology* 24, 103–118.

Mendes LW, et al. (2014) Taxonomical and functional microbial community selection in soybean rhizosphere. *The ISME Journal* 8, 1577–1587.

Mendes LW, et al. (2015a) Land-use system shapes soil bacterial communities in Southeastern Amazon region. *Applied Soil Ecology* 95, 151–160. https://doi.org/10.1016/j.apsoil.2015.06.005

Mendes LW, et al. (2015b) Soil-borne microbiome: linking diversity to function. *Microbial Ecology* 70, 255–265. https://doi.org/10.1007/s00248-014-0559-2

Merloti LF, et al. (2019) Forest-to-agriculture conversion in Amazon drives soil microbial communities and N-cycle. *Soil Biology and Biochemistry* 137:107567. https://doi.org/10.1016/j.soilbio.2019.107567

Meurer KH, et al. (2016) Direct nitrous oxide (N_2O) fluxes from soils under different land use in Brazil - a critical review. *Environmental Research Letters* 11(2), 023001. https://doi.org/10.1088/1748-9326/11/2/023001

Meyer KM, et al. (2017) Conversion of Amazon rainforest to agriculture alters community traits of methane-cycling organisms. *Molecular Ecology* 26, 1547–1556. https://doi.org/10.1111/mec.14011

Mishra A, et al. (2022) Unboxing the black box—one step forward to understand the soil microbiome: a systematic review. *Microbial Ecology* 85(2), 669–683. https://doi.org/10.1007/s00248-022-01962-5

Mirza B. S., et al. (2014). Response of Free-Living Nitrogen-Fixing Microorganisms to Land Use Change in the Amazon Rainforest. *Appl Environ Microbiol* 80. https://doi.org/10.1128/AEM.02362-13

Mitchard ETA (2018) The tropical forest carbon cycle and climate change. *Nature* 559, 527–534. https://doi.org/10.1038/s41586-018-0300-2

Myers N, et al. (2000) Bio-diversity hotspots for conservation priorities. *Nature* 403, 853–558. https://doi.org/10.1038/35002501.

Nascimento N, et al. (2019) What drives intensification of land use at agricultural frontiers in the Brazilian Amazon? Evidence from a decision game. *Forests* 10(6), 464. https://doi.org/10.3390/f10060464

Navarrete AA, et al. (2011) Land-use systems affect archaeal community structure and functional diversity in Western Amazon soils. *Revista Brasileira de Ciência do Solo* 35, 1527–1540.

Navarrete AA, et al. (2015) Soil microbiome responses to the short-term effects of Amazonian deforestation. *Molecular Ecology* 24(10), 2433–2448. https://doi.org/10.1111/mec.13172

Neill C, et al. (1997) Soil carbon and nitrogen stocks following forest clearing for pasture in the southwestern Brazilian Amazon. *Ecological Applications* 7(4), 1216-1225. https://doi.org/10.1890/1051-0761(1997)007[1216:SCANSF]2.0.CO;2

Nunes CA, et al. (2022) Linking land-use and land-cover transitions to their ecological impact in the Amazon. *Proceedings of the National Academy of Sciences* 119(27), e2202310119. https://doi.org/10.1073/pnas.2202310119

Obregon Alvarez D, et al. (2023) Shifts in functional traits and interactions patterns of soil methane-cycling communities following forest-to-pasture conversion in the Amazon Basin. *Molecular Ecology* 32(12), 3257–3275. https://doi.org/10.1111/mec.16912

Pajares S, Bohannan BJ (2016) Ecology of nitrogen fixing, nitrifying, and denitrifying microorganisms in tropical forest soils. *Frontiers in Microbiology* 7, 1045. https://doi.org/10.3389/fmicb.2016.01045

Paula FS, et al. (2014) Land use change alters functional gene diversity, composition and abundance in Amazon forest soil microbial communities. *Molecular Ecology* 23(12), 2988–2999. https://doi.org/10.1111/mec.12786

Pazinato JM, et al. (2010) Molecular characterization of the archaeal community in an Amazonian Wetland Soil and culture-dependent isolation of methanogenic Archaea. *Diversity* 2, 1026–1047.

Pedrinho A, et al. (2019) Forest-to-pasture conversion and recovery based on assessment of microbial communities in Eastern Amazon rainforest. *FEMS Microbiology Ecology* 95(3), fiy236. https://doi.org/10.1093/femsec/fiy236

Pedrinho A, et al. (2020) The natural recovery of soil microbial community and nitrogen functions after pasture abandonment in the Amazon region. *FEMS Microbiology Ecology* 96(9), fiaa149. https://doi.org/10.1093/femsec/fiaa149

Pedrinho A, et al. (2023) Impacts of deforestation and forest regeneration on soil bacterial communities associated with phosphorus transformation processes in the Brazilian Amazon region. *Ecological Indicators* 146, 109779. https://doi.org/10.1016/j.ecolind.2022.109779

Piccolo MC, et al. (1996) ^{15}N natural abundance in forest and pasture soils of the Brazilian Amazon Basin. *Plant Soil* 182, 249–258. https://doi.org/10.1007/BF00029056

Raaijmakers J, Mazzola M (2016) Soil immune responses. *Science* 352(6292), 1392–1393. https://doi.org/10.1126/science.aaf3252

Ribeiro MC, et al. (2009) The Brazilian Atlantic Forest: how much is left, and how is the remaining forest distributed? Implications for conservation. *Biological Conservation* 142, 1141–1153. https://doi.org/10. 1016/j.biocon.2009.02.021

Richardson AE, Simpson RJ (2011) Soil microorganisms mediating phosphorus availability update on microbial phosphorus. *Plant Physiology* 156(3), 989–996. https://doi.org/10.1104/pp.111.175448

Rodrigues JL, et al. (2013) Conversion of the Amazon rainforest to agriculture results in biotic homogenization of soil bacterial communities. *Proceedings of the National Academy of Sciences* 110(3), 988–993. https://doi.org/10.1073/pnas.1220608110

Sanchez-Cid C, et al. (2022) Sequencing depth has a stronger effect than dna extraction on soil bacterial richness discovery. *Biomolecules* 12(3), 364. https://doi.org/10.3390/biom12030364

Singh A, et al. (2020) Role of microorganisms in regulating carbon cycle in tropical and subtropical soils. In: Ghosh P, et al. (eds) *Carbon Management in Tropical and Sub-Tropical Terrestrial Systems*. Singapore: Springer. https://doi.org/10.1007/978–981-13-9628-1_15

Smith C, et al. (2016) Microbial community responses to soil tillage and crop rotation in a corn/soybean agroecosystem. *Ecology and Evolution* 6(22), 8075–8084. https://doi.org/10.1002/ece3.2553

Solomon D, et al. (2002) Phosphorus forms and dynamics as influenced by land use changes in the sub-humid Ethiopian highlands. *Geoderma* 105(1–2), 21–48. https://doi.org/10.1016/S0016-7061(01)00090-8

Soltangheisi A, et al. (2019) Forest conversion to pasture affects soil phosphorus dynamics and nutritional status in Brazilian Amazon. *Soil and Tillage Research* 194, 104330. https://doi.org/10.1016/j.still.2019.104330

Souza LF, et al. (2022) Maintaining grass coverage increases methane uptake in Amazonian pastures, with a reduction of methanogenic archaea in the rhizosphere. *Science of the Total Environment* 838, 156225.

Souza Braz AD, et al. (2013) Soil attributes after the conversion from forest to pasture in Amazon. *Land Degradation & Development* 24(1), 33-38. https://doi.org/10.1002/ldr.1100

Stein LY (2020) The long-term relationship between microbial metabolism and greenhouse gases. *Trends in Microbiology* 28(6), 500–511. https://doi.org/10.1016/j.tim.2020.01.006

Steudler PA, et al. (1996) Consequence of forest-to-pasture conversion on CH_4 fluxes in the Brazilian Amazon Basin. *Journal of Geophysical Research* 101(D13), 18547–18554. https://doi.org/10.1029/96JD01551

Sühs, RB, et al. 2020. Preventing traditional management can cause grassland loss within 30 years in southern Brazil. *Scientific Reports* 10, 1–9. https://doi.org/ 10.1038/s41598-020-57564-z

Taketani RG, et al. (2013) Bacterial community composition of anthropogenic biochar and Amazonian anthrosols assessed by 16S rRNA gene 454 pyrosequencing. *Antonie van Leeuwenhoek* 104, 233–242.

Tao C, et al. (2020) Bio-organic fertilizers stimulate indigenous soil pseudomonas populations to enhance plant disease suppression. *Microbiome* 8(1).https://doi.org/10.1186/s40168-020-00892-z

Tomazelli D, et al. (2023a) Pasture management intensification shifts the soil microbiome composition and ecosystem functions. *Agriculture, Ecosystems & Environment* 346, 108355. https://doi.org/10.1016/j.agee.2023.108355.

Tomazelli D, et al. (2023b) Goss-Souza, D. Natural grassland conversion to cultivated pastures increases soil microbial niche specialization with consequences for ecological processes. *Applied Soil Ecology* 188, 104913. https://doi.org/10.1016/j.apsoil.2023.104913.

United Nations (2017) Global Land Outlook. United Nations Convention to Combat Desertification, Bonn, Germany. https://www.unccd.int/resources/global-land-outlook

United Nations Environment Programme (UNEP) (2013) *Drawing Down N_2O to Protect Climate and Ozone Layer*. Nairobi, Kenya: UNEP.

Vieira ICG (2019) Land use drives change in Amazonian tree species. *Anais da Academia Brasileira de Ciências* 91, e20190186. https://doi.org/10.1590/0001-3765201920190186

Venturini A. M., et al. (2022) Increased soil moisture intensifies the impacts of forest-to-pasture conversion on methane emissions and methane-cycling communities in the Eastern Amazon. Environmental Research 212 (Part A): 113139. https://doi.org/10.1016/j.envres.2022.113139

Wang P, et al. (2021) Disturbances consistently restrain the role of random migration in grassland soil microbial community assembly. *Global Ecology and Conservation* 26, e01452 https://doi.org/10.1016/j. Gecco.2021.e01452.

Wei Z, et al. (2019) Initial soil microbiome composition and functioning predetermine future plant health. *Science Advances* 5(9) https://doi.org/10.1126/sciadv.aaw0759

Weinzettel J, et al. (2018) Human footprint in biodiversity hotspots. *Frontiers in Ecology and the Environment* 16, 447–452. https://doi.org/10.1002/fee.1825

Wright GD (2007) The antibiotic resistome: the nexus of chemical and genetic diversity. *Nature Reviews Microbiology* 5, 175–186. https://doi.org/10.1038/nrmicro1614

Wu SH, et al. (2019). The effects of afforestation on soil bacterial communities in temperate grassland are modulated by soil chemical properties. *PeerJ* 2019. https://doi.org/10.7717/peerj.6147.

Xie WY, et al. (2018) Antibiotics and antibiotic resistance from animal manures to soil: a review. *European Journal of Soil Science* 69(1), 181-195. https://doi.org/10.1111/ejss.12494

Xun W, et al. (2021) Specialized metabolic functions of keystone taxa sustain soil microbiome stability. *Microbiome* 9(1) https://doi.org/10.1186/s40168-020-00985-9

Yin Y, et al. (2019) Soil microbial character response to plant community variation after grazing prohibition for 10 years in a Qinghai-Tibetan alpine meadow. In: *Plant Soil*, pp. 175–189. https://doi.org/10.1007/ s11104-019-04044-7.

Yuan X, et al. (2021) Development of fungal-mediated soil suppressiveness against fusarium wilt disease via plant residue manipulation. *Microbiome* 9(1) https://doi.org/10.1186/s40168-021-01133-7

Zanella PG, et al. (2021) Grazing intensity drives plant diversity but does not affect forage production in a natural grassland dominated by the tussock-forming grass

Andropogon lateralis Nees. *Scientific Reports* 11, 1–11. https://doi. org/10.1038/s41598-021-96208-8.

Zhang FQ, et al. (2016) Dominance of ammonia-oxidizing archaea community induced by land use change from Masson pine to eucalypt plantation in subtropical China. *Applied Microbiology and Biotechnology* 100, 6859-6869. https://doi.org/10.1007/s00253-016-7506-8

Zhao Z, et al. (2020) Fertilization changes soil microbiome functioning, especially phagotrophic protists. *Soil Biology and Biochemistry* 148, 107863. https://doi.org/10.1016/j.soilbio.2020.107863

Index

www.ingramcontent.com/pod-product-compliance
Ingram Content Group UK Ltd.
Pitfield, Milton Keynes, MK11 3LW, UK
UKHW021008120926
8556IPUK00028B/157

9 781801 464741